Exploring AutoCAD Map 3D 2017

(7th Edition)

CADCIM Technologies

525 St. Andrews Drive
Schererville, IN 46375, USA
(www.cadcim.com)

Contributing Author
Sham Tickoo

Professor
Purdue University Northwest
Hammond, Indiana, USA

CADCIM Technologies

Exploring AutoCAD Map 3D 2017, 7th Edition
Sham Tickoo

CADCIM Technologies
525 St Andrews Drive
Schererville, Indiana 46375, USA
www.cadcim.com

ISBN 978-1-942689-45-4

DEDICATION

*To teachers, who make it possible to disseminate knowledge
to enlighten the young and curious minds
of our future generations*

*To students, who are dedicated to learning new technologies
and making the world a better place to live in*

SPECIAL RECOGNITION

*A special thanks to Mr. Denis Cadu and the ADN team of Autodesk Inc.
for their valuable support and professional guidance to
procure the software for writing this textbook*

THANKS

*To employees of CADCIM Technologies and
Tickoo Institute of Emerging Technologies for their valuable help*

Online Training Program Offered by CADCIM Technologies

CADCIM Technologies provides effective and affordable virtual online training on various software packages including Computer Aided Design, Manufacturing, and Engineering (CAD/CAM/CAE), computer programming languages, animation, architecture, and GIS. The training is delivered 'live' via Internet at any time, any place, and at any pace to individuals as well as the students of colleges, universities, and CAD/CAM/CAE training centers. The main features of this program are:

Training for Students and Companies in a Classroom Setting

Highly experienced instructors and qualified engineers at CADCIM Technologies conduct the classes under the guidance of Prof. Sham Tickoo of Purdue University Northwest, USA. This team has authored several textbooks that are rated "one of the best" in their categories and are used in various colleges, universities, and training centers in North America, Europe, and in other parts of the world.

Training for Individuals

CADCIM Technologies with its cost effective and time saving initiative strives to deliver the training in the comfort of your home or work place, thereby relieving you from the hassles of traveling to training centers.

Training Offered on Software Packages

CADCIM Technologies provides basic and advanced training on the following software packages:

CAD/CAM/CAE: CATIA, Pro/ENGINEER Wildfire (Creo), SolidWorks, Autodesk Inventor, Solid Edge, NX, AutoCAD, AutoCAD LT, Customizing AutoCAD, AutoCAD Electrical, EdgeCAM, Alias and ANSYS

Architecture and GIS: AutoCAD Map 3D, AutoCAD Civil 3D, AutoCAD Raster Design, Autodesk Revit (Structure/Architecture/MEP), Autodesk Navisworks, STAAD.PRO, ArcGIS, MS Project and Oracle Primavera P6.

Animation and Styling: Autodesk 3ds Max, Maya, Pixologic ZBrush, and The Foundry NukeX

Computer Programming: C++, VB.NET, Oracle, AJAX, and Java

*For more information, please visit the following link: **http://www.cadcim.com**.*

Note
If you are a faculty member, you can register by clicking on the following link to access the teaching resources: ***www.cadcim.com/Registration.aspx***. The student resources are available at ***www.cadcim.com***. We also provide **Live Virtual Online Training** on various software packages. For more information, write us at ***sales@cadcim.com***.

Table of Contents

Chapter 3: Working with Basic Tools and Coordinate Systems

Chapter 4: Working with Feature Data

Chapter 5: Styling and Querying Feature Data

Chapter 6: Creating Object Data and Attaching External Database

Chapter 7: Classifying Objects and Working with Classified Objects

Chapter 8: Removing Digitization Errors and Working with Topologies

Chapter 9: Data Analysis

Chapter 10: Working with Different Types of Data

This page is intentionally left blank

Preface

AutoCAD Map 3D 2017

AutoCAD Map 3D 2017, developed by Autodesk Inc., is a powerful tool used for creating, maintaining, and analyzing geospatial data. Built on the latest release of AutoCAD software, it has various tools that help in creating, editing, analyzing, and interpreting various kinds of spatial datasets effectively. AutoCAD Map 3D has a wide range of applications in infrastructure design, layout planning, and spatial analysis.

AutoCAD Map 3D has interoperability with major design and mapping software. This feature allows the Map 3D users to access CAD and GIS data to perform spatial analysis. In AutoCAD Map 3D, you can use the FDO Data Access to connect to various databases, such as Web Mapping Server and Spatial Database, with ease. In addition, you can use various object creation, and raster and vector data analysis tools to create customized datasets based on your project requirements. AutoCAD Map 3D also allows you to import survey data into your project, and perform query operations, thereby saving your time and effort considerably. Moreover, you can import point cloud data, such as LiDAR data, and then use it to generate 3D surfaces for 3D analysis. AutoCAD Map 3D provides you with the powerful tools for displaying and publishing spatial data through electronic and paper media.

Exploring AutoCAD Map 3D 2017 is a comprehensive textbook that has been written to cater to the needs of the students and the professionals. The chapters in this textbook are structured in a pedagogical sequence, which makes the learning process very simple and effective for both the novice as well as the advanced users of AutoCAD Map 3D. In this textbook, complex geospatial processes have been illustrated through easy-to-understand flow diagrams. Also, various processes such as creating feature and drawing objects, managing object data, and displaying spatial data have been covered in this textbook. This edition also introduces users to the concepts of industry model database for managing spatial data. The simple and lucid language used in this textbook makes it a ready reference for both the beginners and the intermediate users.

The salient features of the textbook are as follows:

- **Tutorial Approach**
  The author has adopted the tutorial point-of-view and learn-by-doing approach throughout the textbook. This approach guides the users through various processes involved in creating and analyzing spatial data. At the end of each chapter, tutorials are provided to practice the concepts learned in the chapter.

- **Real-World Projects as Tutorials**
  The author has used about 30 real-world GIS projects as tutorials in this book. This will enable the readers to relate the tutorials to the real-world projects in GIS industry. In addition, there are about 20 exercises based on the real-world GIS projects.

- **Tips and Notes**
  The additional information related to various topics is provided to the users in the form of tips and notes.

- **Learning Objectives**
  The first page of every chapter summarizes the topics that are covered in that chapter.

- **Self-Evaluation Test, Review Questions, and Exercises**
  The chapters end with Self-Evaluation Test so that the users can assess their knowledge of the chapter. The answers to Self-Evaluation Test are given at the end of the chapters. Also, the Review Questions and Exercises are given at the end of the chapters and they can be used by Instructors as test questions and exercises.

- **Heavily Illustrated Text**
  The text in this book is heavily illustrated with about 300 line diagrams and screen capture images.

Symbols Used in the Textbook

Note
The author has provided additional information in the form of notes.

Tip
The author has provided a lot of information to the users about the topic being discussed in the form of tips.

New
This symbol indicates that the command or tool being discussed is new in the current release of AutoCAD Map 3D 2017.

Formatting Conventions Used in the Textbook

Please refer to the following list for the formatting conventions used in this textbook.

- Names of tools, buttons, options, browser, palette, panels, and tabs are written in boldface

  Example: The **Create Coordinate System** tool, the **Remove** button, the **Map** panel, the **Home** tab, **Properties** palette, and so on.

- Names of dialog boxes, drop-downs, drop-down lists, list boxes, areas, edit boxes, check boxes, and radio buttons are written in boldface.

  Example: The **COGO Input** dialog box, the **Tables** drop-down list, the **Field Name** edit box of the **Define New Object Data Table** dialog box, and so on.

- Values entered in edit boxes are written in boldface.

  Example: Enter **Buildings** in the **Name** edit box.

- Names of the files saved are italicized.

  Example: *c03_tut1a.dwg*

- The methods of invoking a tool/option from the Ribbon, Application Menu, or the shortcut keys are given in a shaded box.

Ribbon:	Feature Edit > Split/Merge > Merge Feature
Command:	MAPFEATUREMERGE

Naming Conventions Used in the Textbook

Tool
If you click on an item in a panel of the Ribbon and a command is invoked to create/edit an object or perform some action, then that item is termed as **tool**.
For example: **Rotate** tool and **Connect** tool

If you click on an item in a panel of the Ribbon and a dialog box is invoked wherein you can set the properties to create/edit an object, then that item is also termed as **tool**, refer to Figure 1.
For example: **Assign** tool and **Attach/Detach Object Data** tool

Figure 1 *Tools in the Ribbon*

Button
The item in a dialog box that has a 3d shape like a button is termed as **Button**. For example, **OK** button, **Cancel** button, **Apply** button, **Open** button, and so on, refer to Figure 2. If the item in a Ribbon is used to exit a tool or a mode, it is also termed as button. For example, **Finish Edit Mode** button, **Cancel Edit Mode** button, and so on.

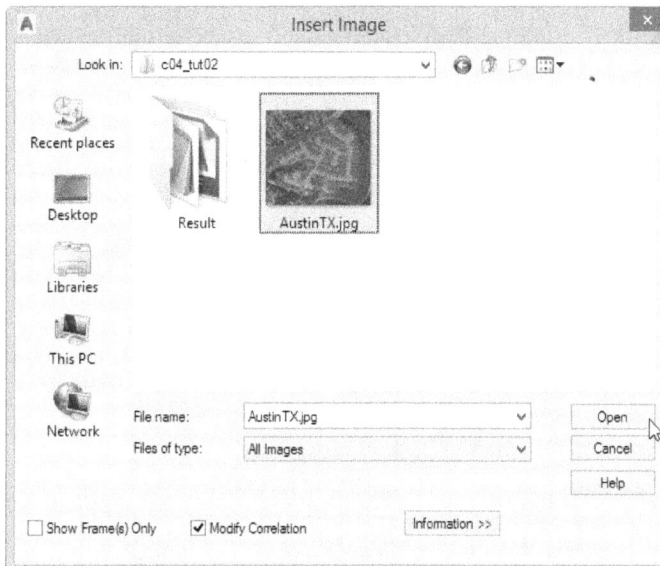

Figure 2 *Choosing the **Open** button*

Drop-down

A drop-down is the one in which a set of common tools are grouped together. You can identify a drop-down with a down arrow on it, refer to Figure 3. These drop-downs are given a name based on the tools grouped in them. For example, **Property** drop-down, **COGO** drop-down, and so on.

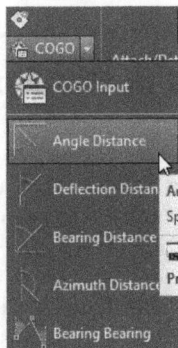

Figure 3   Choosing a tool from a drop-down

Drop-down List

A drop-down list is the one in which a set of options are grouped together. You can set various parameters using these options, refer to Figure 4. You can identify a drop-down list with a down arrow on it. For example, **Layers** drop-down list, **Units** drop-down list, and so on.

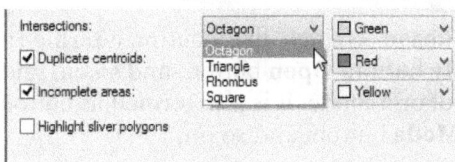

*Figure 4   Selecting an option from the **Intersection** drop-down list*

Options

Options are the items that are available in shortcut menus, dialog boxes, drop-down lists, and so on, refer to Figure 5. For example, choose the **Zoom Extents** option from the shortcut menu displayed on right-clicking in the drawing area.

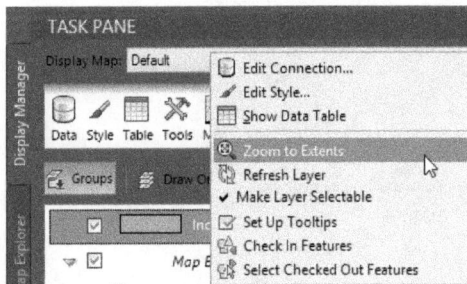

Figure 5   Choosing an option from the shortcut menu

Free Companion Website

It has been our constant endeavor to provide you the best textbooks and services at affordable price. In this endeavor, we have come out with a Free Companion website that will facilitate the process of teaching and learning of AutoCAD Map 3D 2017. If you purchase this textbook, you will get access to the files on the Companion website.

Faculty Resources

- **Technical Support**
  You can get online technical support by contacting *techsupport@cadcim.com*.

- **Instructor Guide**
  Solutions to all review questions and exercises in the textbook are provided in the instructor guide to help the faculty members test the skills of the students.

- **PowerPoint Presentations**
  The contents of the book are arranged in PowerPoint slides that can be used by the faculty for their lectures.

- **Map 3D Files**
  The Map3D files used in illustration, tutorials, and exercises are available for free download.

Student Resources

- **Technical Support**
  You can get online technical support by contacting ***techsupport@cadcim.com***.

- **Map 3D Files**
  The Map3D files used in illustrations and tutorials are available for free download.

If you face any problem in accessing these files, please contact the publisher at ***sales@cadcim.com*** or the author at ***stickoo@pnw.edu*** or ***tickoo525@gmail.com***.

Stay Connected

You can now stay connected with us through Facebook and Twitter to get the latest information about our textbooks, videos, and teaching/learning resources. To stay informed of such updates, follow us on Facebook (***www.facebook.com/cadcim***) and Twitter (***@cadcimtech***). You can also subscribe to our YouTube channel (***www.youtube.com/cadcimtech***) to get the information about our latest video tutorials.

This page is intentionally left blank

Chapter *1*

Introduction to AutoCAD Map 3D 2017

Learning Objectives

After completing this chapter, you will be able to:
- *Understand various terms associated with Map 3D*
- *Start AutoCAD Map 3D 2017*
- *Use workspaces*
- *Work with various components of AutoCAD Map 3D*
- *Use AutoCAD Map 3D Help*

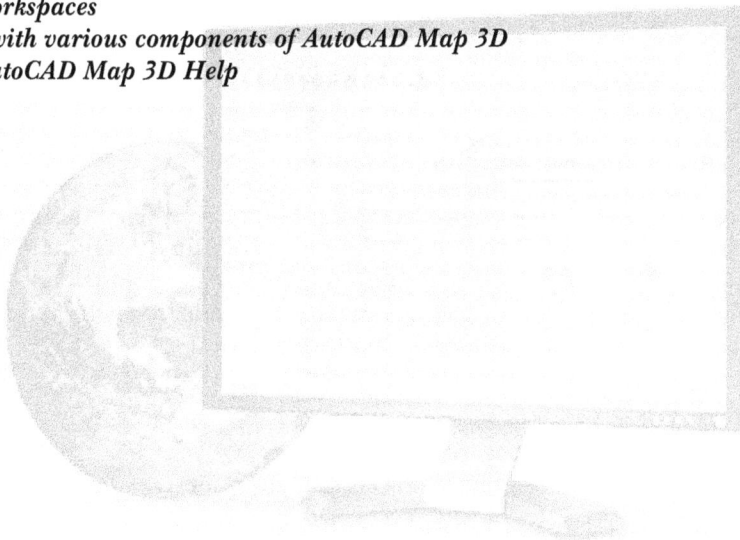

INTRODUCTION

AutoCAD Map 3D is a Geographic Information System (GIS) application developed by Autodesk. This application along with the standard AutoCAD drafting tools is equipped with tools for capturing, editing, updating, and analyzing geospatial data. Moreover, the interoperability feature of this application enables it to read, write, and convert data from one spatial data format to another. AutoCAD Map 3D supports various raster and vector data formats such as .tiff, .dem, .jpeg, .dwg, .shp, .dgn, and .tab. It also supports survey data in file formats such as .xyz, .gps, and .csv. The features in AutoCAD Map 3D enable you to connect to the spatial data using the relational database management system (RDBMS), thereby enhancing the productivity and helping in managing and analyzing a large spatial dataset efficiently.

The tools and options available in AutoCAD Map 3D help you to perform complex geospatial data analysis and obtain accurate results, thus making it ideal for infrastructure planning, management, and decision making.

In this chapter, you will be introduced to the concept of geospatial analysis. Next, you will learn about the user interface of AutoCAD Map 3D and some of the data types that are used in the field of Geographical Information System.

GEOSPATIAL ANALYSIS

Application of statistical analysis and other analytical techniques (such as network analysis, buffer analysis, and overlay analysis) to interpret the data related to a geographical area is known as geospatial analysis. Geospatial analysis includes various methods of data interpretation such as analyzing, interpreting, and presenting the GIS dataset using GIS software. This type of data analysis is widely used in urban planning, landscape designing, geographical mapping, utility management, navigation, and disaster management.

Various types of GIS data such as vector data, raster data, survey data, and point data are used in geospatial analysis. Different types of data are suitable for different types of geospatial analysis. As a result, the selection of the spatial data type depends on the scope and requirement of the project. In case the spatial data is not available in the required data type, the process of conversion of data from one data type into another is usually practiced. Some of the commonly used GIS data are discussed next.

Vector Data

Vector data is a GIS data structure that represents geographical features in point, line, and polygon geometry. It can also store the non-spatial information of the geographic feature in a data table. The vector data format is used to create, edit, analyze, and store large amount of spatial data. Vector data stores geographic features using three basic geometry types, namely point, line, and polygon. These feature data types are discussed next.

Point Feature

A point feature represents a spatial point for a specific object. A post box, street lamp, fire hydrant, and tree are some of the geographical objects that are represented as point objects.

Line or Polyline Feature

A line or polyline feature is used to represent a linear feature or a streamlined feature data such as roads, transmission lines, streams, rivers, pipe networks, and boundaries.

Polygon Feature

A polygon feature (parcel) is a closed polyline object with the attribute or property data attached to it. The polygon feature is used to represent an area feature such as council boundary, plots, farms, zones, wards, and water bodies such as lakes and ponds.

Raster Data

Raster data consists of a matrix of cells, also known as pixels, organized into rows and columns. Each cell in a raster contains a value that represents information, such as elevation and temperature. You can graphically display the data in the raster by using various rendering techniques. These techniques help you to render data in various color schemes.

You can insert a raster file into a drawing and then use it to collect information. In AutoCAD Map 3D, you can also import point files (elevation data) and LiDAR data into the workspace and then generate 3D raster surfaces. You can also analyze and display the raster data using various tools.

Drawing Object and Object Data

In GIS, there are three primary types of geometries: point, line, and polygon. These geometries are used to represent different geographic features in the drawing; for example, a road feature represented using a line segment. These geometries in the drawing are known as the drawing object. The drawing object may also have data associated with it. For example, a road feature may have data such as speed limit, name, and length associated with it. The associated data is known as object data.

Some of the terms used in this book, related to drawing objects, are explained next.

Property

The property of a drawing object refers to the display parameters such as color, thickness, and pattern of drawing object.

Attribute

Attribute refers to the non spatial data that is attached to a drawing object and does not provide any information about the display parameters of the drawing object it is attached to. For example, the area and population of a state attached to a closed polyline object (polygon), records of births and deaths pertaining to a geographical place attached to a point object, number of accidents attached to a line or polyline object (roads, streets, highways, or motor ways).

Data Table

Data table is a way of presenting property and attribute values in the form of a table. It is attached to an object, a layer, or a feature in a Workspace. Figure 1-1 shows various parts of a data table.

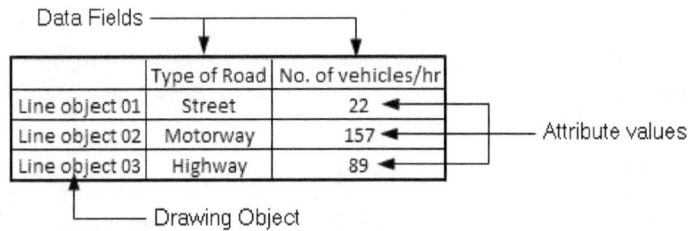

Figure 1-1 *Various parts of a data table*

Data Field
In a data table, the data field displays a group of attribute values related to a specific data object. A data field is also referred to as an attribute or a property.

Attribute Value
Attribute value of a drawing object is the value corresponding to the drawing object in various data fields. For example: attribute value of the **Line object02** drawing object in the **No. of vehicles/hr** data field is **157**, refer to Figure 1-1.

Survey Data
Survey data is the point, line, or polygon data obtained by locating specific survey on the earth surface. Instead of storing survey data in isolated, individual files: GIS allows you to store all survey data in one database. It basically stores 3D data of a particular geographic location. The survey data can be obtained by using survey instruments such as total station, GPS, and so on.

Industry Model Data
Industry Model Data is a new name given to the topobase database. Topobase is the database that includes features, parcels, and attribute data of a geographical location. Industry model data includes project settings, privileges, attribute data, and coordinate and projection systems related to a project. The projects such as infrastructure project management, utility designing, and facility management extensively use the industry model data for project management.

> **Note**
> *To create an industry model, you need to have the **Autodesk Infrastructure Administrator** application installed on your system. To install this application, select the check box corresponding to **Autodesk Infrastructure Administrator 2017** while installing the **AutoCAD Map 3D 2017** software.*

LAYERS
Layers are the overlays containing specific geometry, property, and attributes of a particular feature. You can transform data from one layer to another based on the file formats of data. In AutoCAD Map 3D, you can use layers in the form of the AutoCAD drawing layer *(.dwg)* and the vector or the feature layer *(.shp/.sdf)*. These two layer types are discussed next.

AutoCAD Layer
AutoCAD layer is a drawing layer that contains drawing objects (text, point, line, and polygon) and their properties. You can create and modify a layer within a drawing file. To use AutoCAD

layer in any other software, you will need to export or save the AutoCAD drawing layer in the file format that is recognized by the other software.

Vector or Feature Layer
Vector or feature layer is an independent layer that contains information about all feature data spatially related to a geographical location. This layer can be in the SHP or SDF file format. It is easy to transfer feature data from one software to another using these types of layers.

STARTING AutoCAD Map 3D 2017
From the **Start Menu**, choose **All Programs > Autodesk > AutoCAD Map 3D 2017** (Windows 7), as shown in Figure 1-2; the **AutoCAD Map 3D 2017** application will start. Alternatively, double click on the shortcut icon of **AutoCAD Map 3D 2017** on the desktop.

WORKSPACE
Workspace is a combination of menus, toolbars, Ribbon, palettes, and control panels. It is used to represent a customized drawing environment based on user requirements. You can also customize a workspace to suit the working environment of a task.

When you start AutoCAD Map 3D 2017 for the first time, the **AutoCAD Map 3D - Select Your Default Workspace** window will be displayed, as shown in Figure 1-3. In this window, you can read information about the three predefined workspaces that are available in AutoCAD Map 3D. These workspaces are **Planning and Analysis Workspace**, **Maintenance Workspace**, and **2D Drafting Workspace**.

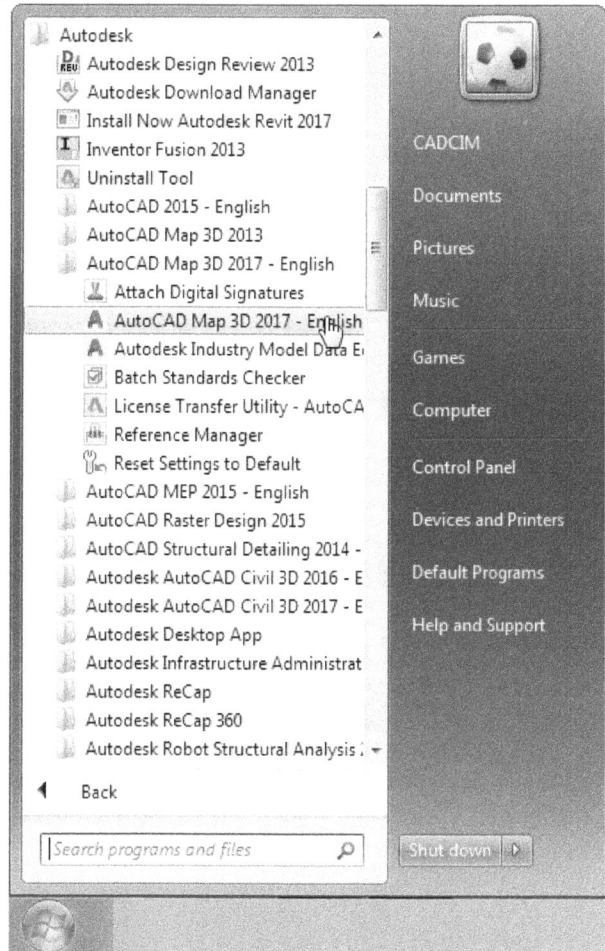

Figure 1-2 Starting AutoCAD Map 3D 2017

Tip
*You can choose a different workspace at any time after starting AutoCAD Map 3D application. To do so, select the required workspace option from the **Workspace** drop-down list in the Quick Access Toolbar.*

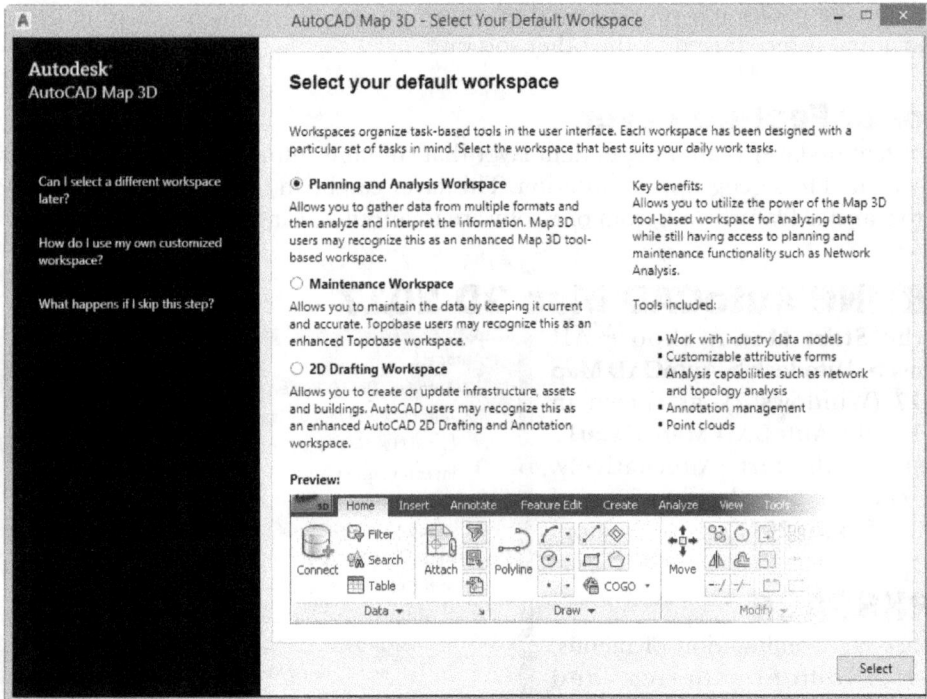

Figure 1-3 The AutoCAD Map 3D - Select Your Default Workspace window

To choose a workspace, select the radio button corresponding to the required workspace and then choose the **Select** button in the **AutoCAD Map 3D - Select Your Default Workspace** window; the window will be closed and the **AutoCAD Map 3D 2017** screen will be displayed with the selected workspace. The predefined workspaces in AutoCAD Map 3D are discussed next.

Planning and Analysis Workspace

By default, the **Planning and Analysis Workspace** radio button is selected in the **AutoCAD Map 3D - Select Your Default Workspace** window. As a result, when you choose the **Select** button in the window, AutoCAD Map 3D interface opens in the **Planning and Analysis Workspace**. In this workspace, the tools are grouped in different categories based on their uses in layout planning and GIS analysis. Figure 1-4 shows the Ribbon displayed in the **Planning and Analysis Workspace**.

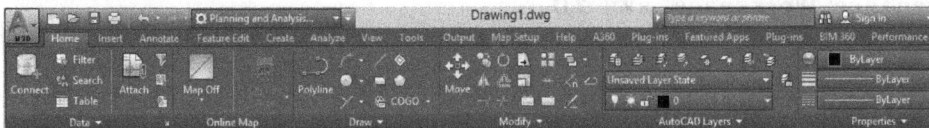

Figure 1-4 Partial view of the Ribbon in the Planning and Analysis Workspace

Note
The theory and tutorial sections of all the chapters in this textbook are discussed based on the Planning and Analysis Workspace Ribbon interface. Hence, it is recommended to retain default settings unless instructed otherwise.

Maintenance Workspace

The **Maintenance Workspace** is a Ribbon interface in which tools are grouped in different panels based on their usage in an industry model or project. The tools in this interface are very useful when you are working with an industry model data, an infrastructural project, or a utility project. Also, it helps to administer an entire project by specifying the required privileges to each member of the team. To invoke this workspace, select the **Maintenance Workspace** radio button in the **AutoCAD Map 3D - Select Your Default Workspace** window and then choose the **Select** button; the **Maintenance Workspace** will be displayed in the AutoCAD Map 3D interface. Figure 1-5 shows the Ribbon in the **Maintenance Workspace**. You can use various tools in this workspace to maintain an industry model data.

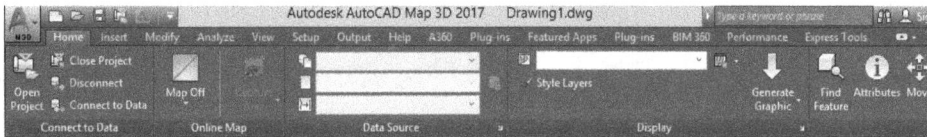

*Figure 1-5 Partial view of the Ribbon in the **Maintenance Workspace***

2D Drafting Workspace

The **2D Drafting Workspace** is a customized interface for drafters and designers. This workspace is very helpful for those who are familiar with AutoCAD Ribbon and mostly work with the drawing data. In this workspace, the panels are customized based on the use of each tool in drafting. To invoke this workspace, select the **2D Drafting Workspace** radio button in the **AutoCAD Map 3D - Select Your Default Workspace** window and then choose the **Select** button; the **2D Drafting Workspace** will be displayed in the AutoCAD Map 3D 2017 interface. You can use various tools in this workspace to maintain an industry model data.

SWITCHING WORKSPACES

After starting the AutoCAD Map 3D application, you can switch from one workspace to another workspace. To switch from one workspace to another, select the corresponding option from the **Workspace** drop-down list located in the Quick Access Toolbar, refer to Figure 1-6; the current workspace will be replaced by the selected workspace.

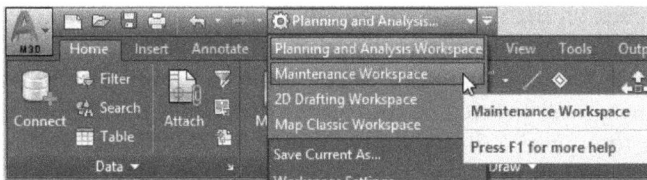

*Figure 1-6 Choosing the **Maintenance Workspace** option from the **Workspace** drop-down list*

Note
*The procedure of accessing tools and dialog boxes in the **Map Classic Workspace** are different from the procedure used in the other workspaces as mentioned previously in this chapter.*

AutoCAD Map 3D 2017 INTERFACE

AutoCAD Map 3D 2017 interface consists of drawing area, Ribbon, command line, Quick Access Toolbar, **TASK PANE**, model and layout tabs, and status bar, refer to Figure 1-7. The header section of the application interface displays the name of the current drawing file. The different components of the AutoCAD Map 3D 2017 interface are discussed next.

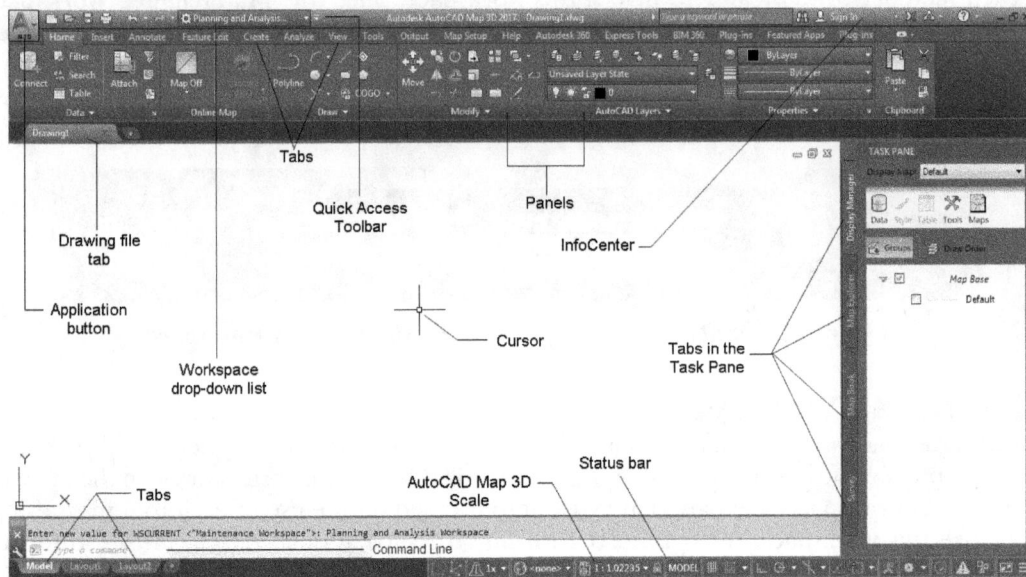

Figure 1-7 *AutoCAD Map3D 2017 interface*

Application Menu

The **Application** button is located at the top-left of the AutoCAD Map 3D 2017 screen. Choose the **Application** button; the Application Menu will be displayed, as shown in Figure 1-8. The menu contains some of the tools that are available in the **Standard** toolbar. Alternatively, press ALT+F to display the tools in the Application Menu. You can search a command using the search field on the top of the Application Menu. To search a command, enter the complete or partial name of the command in the search field; a list showing all possible commands will be displayed. You can click on the desired command from the list to activate it.

By default, the **Recent Documents** button is chosen in the Application Menu. As a result, the recently opened drawings will be listed on the right in the Application Menu. Click on the required file name in the list to open the file. To open a file that is not listed in this menu, choose the **Open** button in the Application Menu; the **Select File** dialog box will be displayed. Browse to the location of the required file. Click on the file name and then choose the **Open** button; the selected file will be opened in the drawing area. In AutoCAD Map 3D 2017, you can specify the settings of the **Display**, **User Preferences**, **Files**, and **Drafting** parameters in the **Options** dialog box. To invoke this dialog box, choose the **Options** button displayed at the bottom-right of the Application Menu. Next, use the options in this dialog box to specify the required parameters.

To exit AutoCAD Map 3D, choose the **Exit AutoCAD Map 3D 2017** button from the Application Menu.

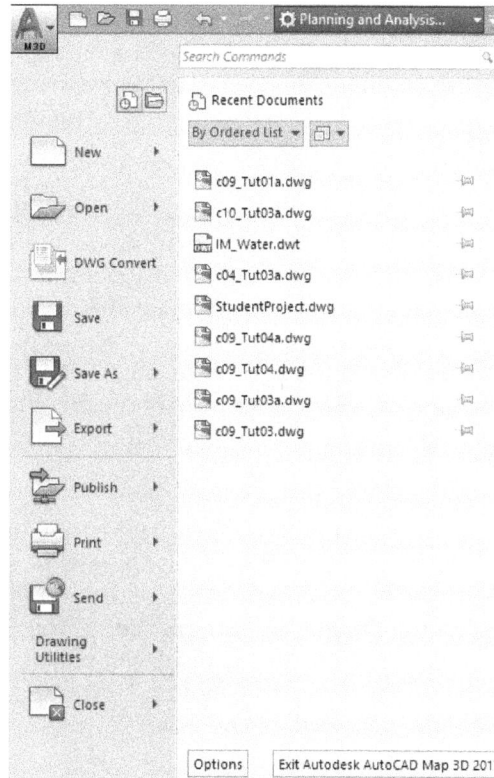

Figure 1-8 *The Application Menu*

Ribbon

Ribbon interface in AutoCAD Map 3D contains tools organized into various tabs and panels based on their functionality, refer to Figure 1-7.

When you start the AutoCAD session for the first time, by default the Ribbon is displayed horizontally below the Quick Access Toolbar. The Ribbon consists of various tabs. These tabs have different panels which in turn have tools arranged in rows. Some of the tools have a small black down arrow. This arrow indicates that the tools having similar functions are grouped together. To choose a tool, click on the down arrow next to them; a drop-down will be displayed. Choose the required tool from the drop-down displayed. Note that if you choose a tool from the drop-down, the corresponding command will be invoked and the tool that you have chosen will be displayed in the panel. For example, to draw a circle using the **2-Point** option, click on the down arrow next to the **Center, Radius** tool in the **Draw** panel of the **Home** tab; a drop-down will be displayed. Choose the **2-Point** tool from the drop-down and then draw the circle. You will notice that the **2-Point** tool is displayed in place of the **Center, Radius** tool. In this textbook, the tool selection sequence will be written as choose **2-Point** tool from **Home > Draw > Circle** drop-down.

The tools which are not displayed within the available area of the panel are placed in the expandable area of the panel. Panels with an expandable area have a down arrow displayed to the right of its name. To view the tools in the expandable area of the panel, you can choose the down arrow. You can click on the push-pin in the expanded panel to keep it in the expanded state. Also, some of the panels have an inclined arrow at the lower-right corner. When you click on the inclined arrow, a dialog box is displayed. You can define the settings of the corresponding panel in this dialog box.

AutoCAD Map 3D allows you to change the default location of the Ribbon interface. To do so, right-click on the blank space in the Ribbon; a shortcut menu is displayed. Next, choose the **Undock** option from this menu; the Ribbon is undocked. After undocking the Ribbon, you can move, resize, anchor, and turn on the auto-hide option for the display of the Ribbon. To do so, right-click on the heading strip in the Ribbon; a shortcut menu will be displayed. Choose the required option from this menu. For example, to vertically anchor the floating Ribbon to the left of the drawing area, right-click on the heading strip of the floating Ribbon; a shortcut menu is displayed. Next, choose the **Anchor Left <** option from the shortcut menu; the Ribbon will be anchored to the left.

You can also customize the display of tabs and panels in the Ribbon. To do so, right-click on any of the tools in it; a shortcut menu will be displayed. On moving the cursor over one of the options, a flyout will be displayed with a tick mark before all the options. Also, the corresponding tab or panel will be displayed in the Ribbon. Select/clear the appropriate option to display/hide a particular tab or panel. You can also reorder the display of panels in the tab. To do so, press and hold the left mouse button on the panel to be moved. Next, drag it to the required position and release the mouse button; the panel will be moved.

Drawing File Tabs

The Drawing File tabs displayed above the drawing area, refer to Figure 1-7, show the drawings that are currently opened. Using these tabs, you can quickly switch between drawings. The order in which these tabs are displayed is based on the sequence in which the files were opened.

Drawing Area

The drawing area covers the major portion of the screen. In this area, you can draw objects by using various tools/commands. To draw an object, you need to define coordinate points. You can do so by using the pointing device. The cursor represents the position of the pointing device on the screen. There is a coordinate system icon at the lower-left corner of the drawing area.

Drawing Status Bar

The **Drawing Status Bar** is displayed at the bottom of the drawing area and below the command window, refer to Figure 1-9. To customize the status bar options, choose the **Customization** button on the status bar, a flyout will be displayed. Choose options according to your needs. The **Drawing Status Bar** displays the **Coordinate System**, **Vertical Exaggeration**, **Isolate Object**, and **AutoCAD Map 3D Scale** buttons. Various options in the **Drawing Status Bar** are discussed next.

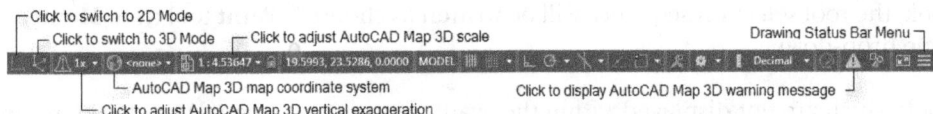

Figure 1-9 *The options in the **Drawing Status Bar***

2D Mode Button

The **2D Mode** button is used to make platform for 2D working environment. When you apply the *map2d.dwt* template settings to the current drawing, you will enter the 2D drafting and designing environment and the **2D Mode** button will be activated. The *map2d.dwt* is a template file that contains the settings of the 2D drawing environment.

3D Mode Button

The **3D Mode** button is used to switch from 2D to 3D working environment. When you apply the settings from the *map3d.dwt* template file, you will enter the 3D designing environment and the **3D Mode** button will be activated.

Vertical Exaggeration

The Vertical Exaggeration is used to change the vertical scale of the drawing. It is used to raise or lower the vertical features which might appear too small or too big relative to the horizontal scale in the drawing. To specify the exaggeration factor for visual enhancement, click on the down-arrow next to the **Vertical Exaggeration** button in the **Drawing Status Bar**; a flyout will be displayed. In this flyout, choose the required option; the display of the raster image will be enhanced based on the factor selected. Figure 1-10 shows the view of a raster image at the default vertical exaggeration (**1x**). Figure 1-11 shows an enhanced view of the raster image at vertical exaggeration of **2x**.

*Figure 1-10 Model vertically exaggerated to **1x***   *Figure 1-11 Model vertically exaggerated to **2x***

You can also specify a custom value for vertical exaggeration. To do so, click on the down-arrow corresponding to the **Vertical Exaggeration** option; a flyout will be displayed. In this flyout, choose the **Custom** option, as shown in Figure 1-12; a window will be displayed. In this window, enter the desired value in the **Enter Exaggeration Value** edit box and then choose the **OK** button; the window will be closed and the raster image will be vertically exaggerated to the specified value.

Warning Button

The **Warning** button is used to display the details of the errors that have occurred while performing an action. By default, this button is inactive. When an error occurs, this button becomes active (it will be highlighted in yellow color). To view the description of the error occurred, click on the **Warning** button; the **AutoCAD Map Messages** dialog box will be displayed. This dialog box has two areas: **Messages** and **Message details**. Next, select a message in the **Messages** area; the details pertaining to the selected

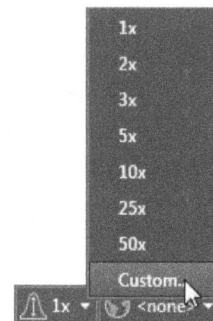

*Figure 1-12 Choosing the **Custom** option for vertical exaggeration*

message will be displayed in the **Message details** area. You can use the **Show Log File** button to view additional information about the selected option. On choosing this button, the log file of the error will be displayed in a text editor such as Notepad.

Click to link/unlink AutoCAD Map 3D stylization while zooming

After a style has been applied to a vector data, you can use this button to link or unlink the AutoCAD Map 3D stylization while zooming. If the lock in this button is closed, the style applied to the vector layer will be displayed at different zooming scales (stylization linked). If the lock in this button is open, the style applied to a vector layer will not be displayed (stylization is not linked).

Annotation Scale

The **Annotation Scale** option has a drop-down list that displays all the annotation scales available for the current drawing. The annotation scale is used to control the size and display of the annotative objects in the model space.

Annotation Objects

If this option is selected, then the annotated objects in the drawing are scaled automatically based on the scale selected in the **Annotation Scale** drop-down list.

Drawing Coordinates

The information about the coordinates is displayed on the left side of the **Status Bar**. You can set the display of coordinates in the **Status Bar** to static or dynamic. To change the coordinate display settings in the **Status Bar**, enter **COORDS** in the command line; you will be prompted to specify a new value. Specify a new **COORDS** value in the command line. If the value is set to 0, the coordinate display will be static, which means that the coordinate values displayed in the **Status Bar** will change only when a point is specified. Setting the value of the **COORDS** variable to 1, 2, or 3 will display the coordinate in dynamic mode. If the variable is set to 1, AutoCAD Map 3D constantly displays the absolute coordinates of the graphics cursor with respect to the UCS origin.

Note
*When you assign a global coordinate system to a workspace, the coordinates of a spatial point with reference to the assigned coordinate system will be displayed in the **Drawing Coordinates** area.*

Infer Constraints

You can choose the **Infer Constraints** button to automatically apply the geometric constraints while you create or edit a geometric object.

Snap Mode

The **Snap Mode** button is chosen to activate the grid snap mode. If this mode is activated, the cursor will move in fixed increments using the current settings of the snap grid. The F9 key acts as a toggle key to turn the snap mode off or on. To change the snap spacing, right-click on the **Snap Mode** button; a shortcut menu will be displayed. Choose the **Settings** option from the menu; the **Drafting Settings** dialog box will be displayed with the **Snap and Grid** tab chosen by default. In the **Snap spacing** area of this tab, specify the required snap spacing along the X and Y axes in the corresponding edit boxes. Next, choose the **OK** button to apply the settings.

Grid

The grid lines are used as reference lines to draw objects in AutoCAD Map 3D. Choose the **Grid** button to toggle the display of the grid in the drawing area. Alternatively, you can use the F7 key to toggle the grid display. To change the spacing of the grid lines, right-click on the **Snap Mode** button; a shortcut menu will be displayed. Choose the **Settings** option from the menu; the **Drafting Settings** dialog box will be displayed with the **Snap and Grid** tab chosen. In the **Grid spacing** area of this tab, specify the required spacing along the X and Y axes in the corresponding edit boxes. Next, choose the **OK** button to apply the settings.

Ortho Mode

This button is used to activate or deactivate the ortho mode. When this mode is active, you can draw lines at right angles only. Alternatively, choose the F8 key to turn this mode on or off.

Polar Tracking

If you turn the polar tracking on, the movement of the cursor is restricted along a path determined by the angle set as the polar angle. Choose the **Polar Tracking** button to turn the polar tracking on or off. You can also use the F10 key to turn this option on or off. Note that turning the polar tracking on, automatically turns off the ortho mode.

Object Snap

You can use the **Object Snap** button to turn the object snap mode on or off. Alternatively, you can use the F3 key to turn the object snap mode on or off. Note that the status of **OSNAP** (off or on) does not prevent you from using the immediate mode object snaps.

3D Object Snap

When this button is chosen, you can snap the key point on a solid or a surface. You can also use the F4 key to turn the 3D object snap on or off.

Object Snap Tracking

When you choose this button, the inferencing lines will be displayed. Inferencing lines are dashed lines that are displayed automatically when you select a sketching tool and track a particular key point on the screen. On choosing this button, the object snap tracking turns on or off.

Allow/Disallow Dynamic UCS

On choosing this button, you are allowed or disallowed the use of dynamic UCS. Allowing the dynamic UCS ensures that the XY plane of the UCS is dynamically aligned with the selected face of the model. You can also use the F6 key to turn the **DUCS** button on or off.

Dynamic Input

The **Dynamic Input** button is used to turn the **Dynamic Input** on or off. Turning it on facilitates the heads-up design approach because all the commands, prompts, and dimensional inputs will now be displayed in the drawing area and you do not need to look at the Command prompt all the time. This saves the design time and also increases the efficiency of the user. If the **Dynamic Input** mode is turned on, you will be allowed to enter the commands through the **Pointer Input** boxes and the numerical values through the **Dimensional Input** boxes. You will also be allowed to select the command options through the **Dynamic Prompt** options in the graphics window. To turn the **Dynamic Input** on or off, use the CTRL+D keys.

Show/Hide Lineweight

Choose this button to turn on or off the display of lineweights in the drawing. If this button is not chosen, the display of lineweight will be turned off.

Show/Hide Transparency

This button is used to turn on or off the transparency set for a drawing. You can set the transparency in the **Properties** panel or in the layer in which the sketch is drawn.

Quick Properties

If you select a sketched entity when this button is chosen in the **Status Bar**, the properties of the selected entity will be displayed in a panel.

Model or Paper source

The **Model or Paper source** button is used to switch between the model space and paper space while in the layout space environment. Switching to model space while in the layout environment will enable you to work on the drawing objects in the model. To switch to model space in the layout environment, choose the **Model or Paper source** button; the drawing objects in the model will become editable and the text on the button will change from **PAPER** to **MODEL**. To switch back to paper space, choose the **Model or Paper source** button; the paper space will be invoked and now you can work with the map elements in the layout.

Note
*You can invoke the Layout space environment or the Model space environment in the drawing window by choosing the **Model** or **Layout** tab displayed at the bottom left of the drawing window, refer to Figure 1-7.*

Toolbar/Window Positions Unlocked

The **Toolbar/Window Positions Unlocked** button is used to lock and unlock the positions of the toolbars and the windows. When you choose this button, a shortcut menu is displayed. Choosing the **Floating Toolbars/Panels** option allows you to lock the current position of the floating toolbars. Also, a check mark is displayed against those toolbars in the shortcut menu that are currently locked. Choosing the **Docked Toolbars/Panels** option from the shortcut menu allows you to lock the current position of all the docked toolbars. Similarly, you can lock or unlock the position of floating and docked windows such as the **Properties** window or the **Tool Palettes**. If you move the cursor on the **All** option, a cascading menu is displayed that provides the option to lock and unlock all the toolbars and windows.

Selection Cycling

You can use this button to select an object from the group of two or more overlapping or closely placed entities. To select an object from the overlapping entities, choose the **Selection Cycling** button in the **Status Bar**; the selection cycling mode will be activated. Next, click at the required location to select an object; a list box will be displayed. This list box contains a list of objects at the selected location. Choose the required object from the list to proceed.

Command Line

The **Command Line** is located at the bottom of the drawing area where you can enter commands to execute an action. It also displays the subsequent prompt sequences and messages. You can change the size of the **Command Line**. To do so, place the cursor on the top edge of the command line; the cursor will change into a double line bar known as the grab bar. Click and drag the command line to change the size. This way you can increase its size to see all the previously used commands. Alternatively, you can press CTRL+9 to show or hide the **Command Line**. You can also press the F2 key to display the **AutoCAD Text window** which displays the previous commands and messages, as shown in Figure 1-13.

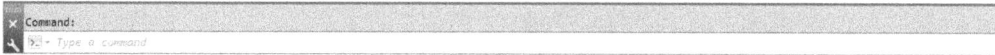

*Figure 1-13 The **Command Line***

TASK PANE

The **TASK PANE** is displayed on the right in the AutoCAD Map 3D 2017 screen, as shown in Figure 1-14. To hide/show the **TASK PANE** in the interface, choose the **Map Task Pane** button in the **Palettes** panel of the **View** tab.

You will find alternative options to the Ribbon interface in the **TASK PANE**. In the **TASK PANE**, there are four tabs: **Display Manager**, **Map Explorer**, **Map Book**, and **Survey**. These tabs are explained next.

Display Manager Tab

The options in the **Display Manager** tab are used to connect and display feature layers and then perform various types of analysis on these feature layers.

Map Explorer Tab

The options in the **Map Explorer** tab are used to assign the rights of a drawing to a user, attach new drawings to the current drawing, define queries, and so on.

Map Book Tab

The options in the **Map Book** tab are used to divide a large map into small tiles. You can then render each tile on a separate page. You can publish map book online or can have paper print.

*Figure 1-14 The **TASK PANE***

Survey Tab

The options in the **Survey** tab are used to import and export the survey data, and create a data store.

Navigation Bar

In AutoCAD Map 3D 2017, the commonly used tools for navigation are grouped together in a toolbar known as the **Navigation Bar**. By default, this toolbar is placed on the top right corner in the drawing area. Figure 1-15 shows the **Navigation Bar** available in the AutoCAD Map 3D 2017. The tools in the **Navigation Bar** are discussed next.

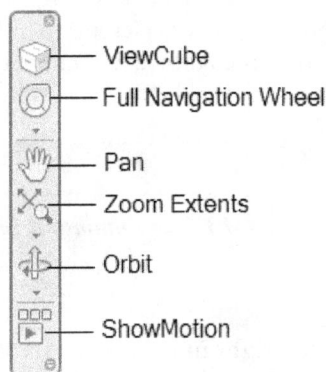

Figure 1-15 *The tools in the* ***Navigation Bar***

ViewCube

The **ViewCube** tool is used to switch between the standard and isometric views or to roll the current view.

Full Navigation Wheel

The **Navigation Wheel** has a set of navigation tools such as pan, zoom, and orbit. You can use any of these options to set the view of a drawing in the drawing window.

Pan

This tool allows you to view the portion of the drawing that is outside the current display area. To view the outside portion, choose this tool, press and hold the left mouse button, and then drag the drawing area. Press ESC to exit this tool.

Zoom

A list of various zoom tools is displayed in the **Zoom** drop-down. The options in the **Zoom** drop-down are used to enlarge or reduce a view in the drawing window without affecting the actual shape and size of the objects in this view and generally by default, zoom extents tab is chosen.

Orbit

The tools in the **Orbit** drop-down are used to rotate a view in the drawing window in the 3D space.

ShowMotion

Choose this button to capture different views in a sequence and animate them when required.

InfoCenter Bar

By default, the **InfoCenter** bar is located on the right side in the AutoCAD Map 3D application title bar. This bar contains various options for accessing the help and online resources for AutoCAD Map 3D. Figure 1-16 shows various tools and options in the **InfoCenter** bar. Some of the options in this bar are discussed next.

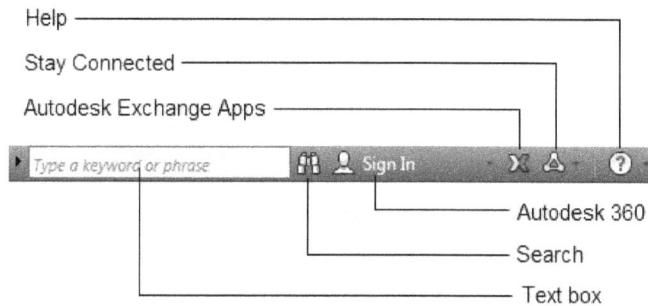

*Figure 1-16 The **InfoCenter** bar*

Autodesk 360

Autodesk 360 is a cloud computing platform introduced by Autodesk. This platform provides a set of cloud services and products that can help you share, simulate, visualize and design your work. You can access the Autodesk 360 services by using your Autodesk ID.

To login to your Autodesk 360 account, choose the **Sign In** button in the **InfoCenter** bar; a drop-down list will be displayed. Next, select the **Sign In to Autodesk 360** option from this list; the **Autodesk-Sign In** dialog box will be displayed, as shown in Figure 1-17. Enter your credentials in this dialog box and choose the **Sign In** button; Autodesk will validate your credentials and will provide access to your account.

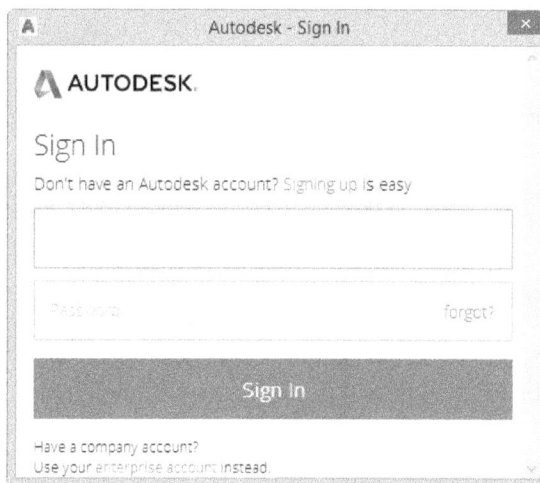

*Figure 1-17 The **Autodesk - Sign In** dialog box*

After you login to your account, you can use various Autodesk 360 services such as the Live maps, data sharing, and so on. The usage of these services is discussed in detail in further chapters.

> **Tip**
> *You can sign up for a free Autodesk ID by choosing the* **Need an Autodesk ID?** *link in the* **Autodesk - Sign In** *dialog box.*

Autodesk Exchange Apps

On choosing the **Autodesk Exchange Apps** button from the **InfoCenter** bar, the Autodesk Exchange Apps website will be opened. This website is an online resource from where you can browse and download e-books, models, training materials, and add-ons for your AutoCAD product. Some of the products available are free to download.

Stay Connected

The option in this drop-down list is used to view and manage your Autodesk profile. When you choose the **Autodesk Account** option from this drop down list, the Autodesk-Sign In page will be displayed wherein you can manage your profile.

InfoCenter Text Box and Search Button

If you need to search the AutoCAD Map 3D help based on a keyword, you can enter the required keyword in the text box of the **InfoCenter** bar and then choose the **Search** button. The application will search for the available topics related to the keyword entered, and will display the results in the **Autodesk AutoCAD Map 3D Help** window.

> **Note**
> *You can also invoke the* **Autodesk AutoCAD Map 3D Help** *window by pressing the* **F1** *key.*

Help

This drop-down list contains the options that provide links to various useful resources such as download offline help and language pack for Map 3D. Figure 1-18 shows various options available in the **Help** drop-down list.

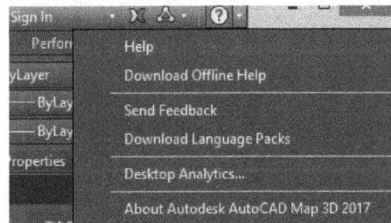

*Figure 1-18 The options in the **Help** drop-down list*

You can also invoke the **Autodesk AutoCAD Map 3D 2017 - Help** window by selecting the **Help** option from this drop-down list. Figure 1-19 shows the default view of the **Autodesk AutoCAD Map 3D Help** window.

The default view of the **Autodesk AutoCAD Map 3D 2017 - Help** window provides you with the links to various Autodesk resources. These resources are organized into two basic categories: **Learn** and **Resources**. You can explore the required resource by following the link displayed below the resource name.

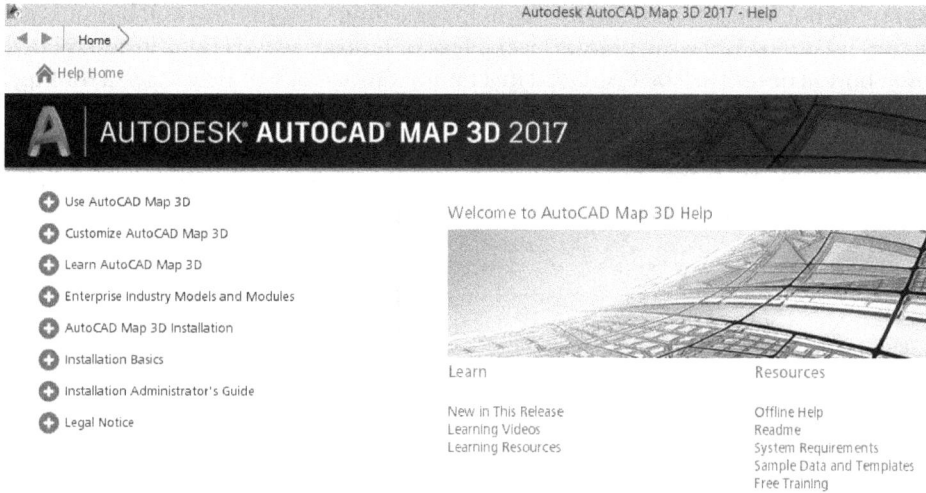

Figure 1-19 *Parial View of the* ***Autodesk AutoCAD Map 3D 2017 - Help*** *window*

You can also use the options in the various tabs displayed in the left pane of the **Autodesk AutoCAD Map 3D 2017 - Help** window to search and view the help topics in AutoCAD Map 3D. You can choose the **Contents** tab to browse through the help topics that are arranged in a hierarchical order. You can also search the available help for a topic of your interest. To do so, specify the required keyword in the search text box of the **Search** tab and then press ENTER; the topics related to the specified keyword will be listed along with a short description in the left pane. Next, choose the required topic from this list; the content will be displayed in the right pane.

More Help

In AutoCAD Map 3D, you can access help using various methods as per your requirement. Some of these methods are listed on the below:

1. **Access help from the command bar:** While specifying a command through the command line, AutoCAD Map 3D displays a list of matching commands. This list also displays the **Search in Help** and **Search on Internet** buttons for the option highlighted in the list. You can choose the **Search in Help** button to search for relevant information about the selected option in the AutoCAD Map 3D help. Alternatively, you can choose the **Search on Internet** button to search for information online. Figure 1-20 shows the **Search on Internet** button being chosen from the displayed list.

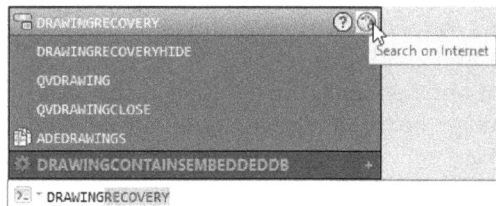

Figure 1-20 *Choosing the* ***Search on Internet*** *button*

2. **Displaying the Tooltip:** You can display help regarding various tools and buttons in form of tool tips. To do so, hover the pointer over a tool or button; a short description related to the chosen button or tool will be displayed in a tooltip. To activate or deactivate a tooltip, invoke the **Options** dialog box by entering the **OP** command in the command line. Next, choose the **Display** tab in the **Options** dialog box. In the **Window Elements** area of the **Display** tab, select or clear the **Show ToolTips** check box to activate or deactivate the tooltip, refer to Figure 1-21.

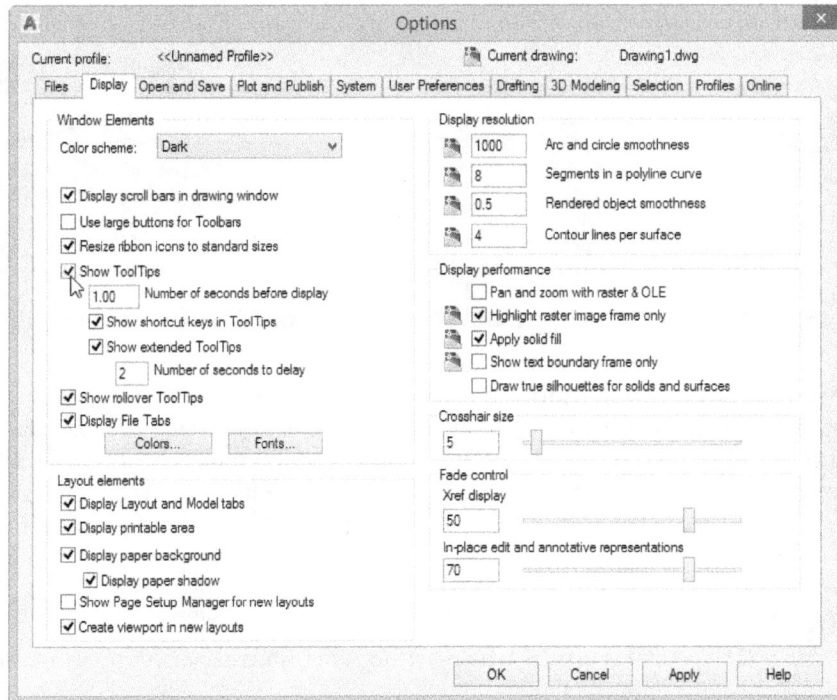

*Figure 1-21   Selecting the **Show ToolTips** check box in the **Options** dialog box*

3. Almost all dialog boxes in AutoCAD Map 3D have a **Help** button. You can access help on various topics by choosing this button. In case, the **Help** button is not displayed in a dialog box, press the F1 key on the keyboard; the help window for the dialog box will be displayed.

AutoCAD Map 3D DIALOG BOXES

There are certain commands which when invoked display a dialog box. A dialog box is a secondary window that allows the users to specify attributes and parameters to perform different tasks. It is also used to provide the users with information or progress feedback. A typical dialog box has a name assigned to it. It also contains controls such as the radio buttons, text or edit boxes, check boxes, slider bars, image boxes, and command buttons. Figure 1-22 shows some of the components in a dialog box.

The Title Bar displays the name of a dialog box. Tabs contain various sections with a group of related options under them. The check boxes are toggle options for making the corresponding option available or unavailable. The drop-down list displays a list of items to choose from. The text box is an area where you can enter any text such as file name. It is also called an edit box because you can make changes to the text entered in it. In some dialog boxes, there is the [**...**]

button which when chosen displays another related dialog box. There are certain command buttons (**OK**, **Cancel**, and **Help**) at the bottom of the dialog box.

The names of these buttons imply their functions. The button with a dark border is the default selected button. A dialog box may also have a **Help** button which when clicked will show related help on various features of the dialog box.

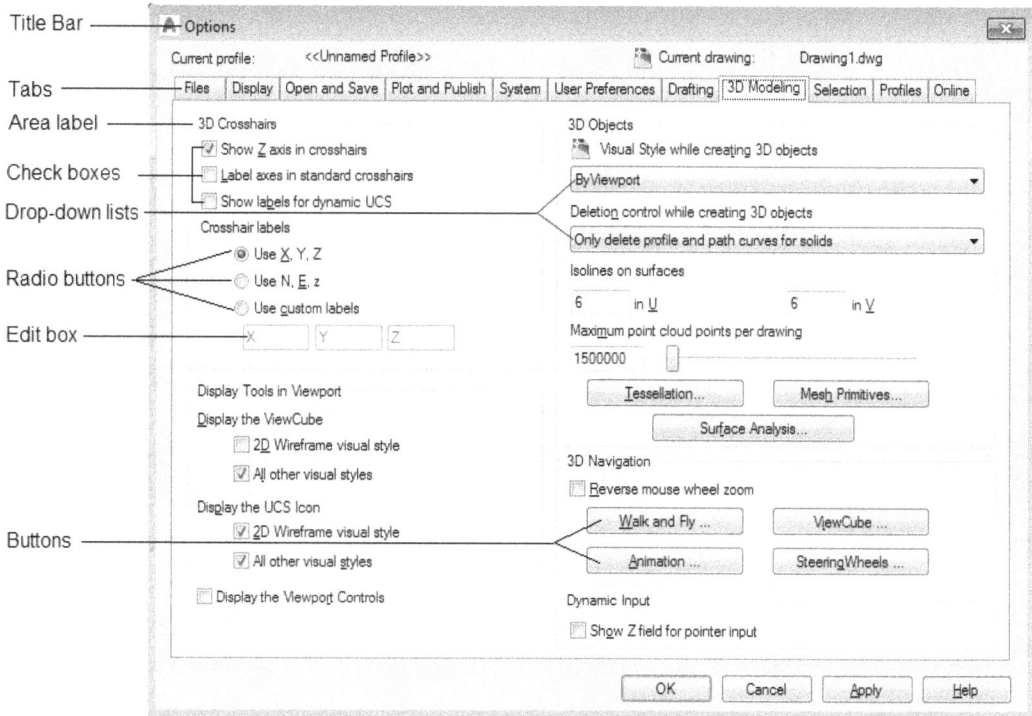

Figure 1-22 *Components of a dialog box*

COMMONLY USED ABBREVIATIONS IN GIS

Some of the commonly used file formats and abbreviations in the field of GIS are listed in the table below.

Abbreviation	Description
GIS	Geographical Information System
DWG	Drawing file (AutoCAD file format)
DWT	Drawing template file (AutoCAD file format)
DWF	Design Web Format (AutoCAD file format)
SHP	Shape file (ESRI file format)
SDF	Spatial Data File (Autodesk file format)
CSV	Comma Separated Values (MS Excel file format)
TXT	Text file
JPEG	Joint Photographic Experts Group (Image file)

TiFF	Tagged Image File Format (Image file)
ISD	Index file generated for point cloud data
LAS and LSD	LiDAR data file format
GPS	Global Positioning System
CAD	Computer Aided Design
TIN	Triangular Irregular Networks
DBMS	Database Management System
AM/FM	Automated Mapping/Facilities Management
PRJ	Projection
UTM	Universal Transverse Mercator
WGS	World Geodetic System
IDW	Inverse Distance Weighted
RMSE	Root Mean Square Error
GCP	Ground Control Points
COGO	Coordinate Geometry
NAD	North American Datum
DEM	Digital Elevation Model
DIME	Dual Independent Map Encoding
DPI	Dots Per Inch
CRS	Coordinate Reference System

Self-Evaluation Test

Answer the following questions:

1. The _____ **Workspace** is used for layout planning and GIS analysis.

2. You can specify the settings of the **Display**, **User Preferences**, **Files**, and **Drafting** parameters in the _____ dialog box.

3. The _____ button helps to snap the key point on a solid or a surface.

4. The **COORDS** command is used to change the coordinate display settings in the **Status Bar**. (T/F)

5. The options in the **Survey** tab are used to divide a large map into small tiles. (T/F)

Answers to Self-Evaluation Test
1. Planning and Analysis, 2. Options, 3. 3D Object Snap, 4. T, 5. F

Chapter 2

Getting Started with AutoCAD Map 3D 2017

Learning Objectives

After completing this chapter, you will be able to:
- *Start a new drawing*
- *Open existing drawing*
- *Add data to the project*
- *Generate output in different ways*
- *Create sheet sets*

INTRODUCTION

This chapter aims to introduce you to the interoperability feature of AutoCAD Map 3D. In this chapter, you will learn to create a new drawing file using a drawing template file. You will also learn to open an existing drawing file and load various types of GIS data into your drawing using the **Connect** tool.

Moreover, you will learn different ways of producing data output such as printing, plotting to a file, and publishing a map file. This chapter also discusses the procedure to create a standard sheet set using various methods in AutoCAD Map 3D.

STARTING A NEW DRAWING

To start a new drawing file, choose **New > Drawing** from the Application Menu, as shown in Figure 2-1; the **Select template** dialog box will be displayed, as shown in Figure 2-2. In this dialog box, browse to the **Template** folder. Next, select the **map2d** template file from the list of drawing templates in this folder, refer to Figure 2-2. Choose **Open** to apply template settings to your drawing; the new drawing will be opened in a 2D mapping environment. Similarly, you can select other drawing templates for mapping based on your project requirement.

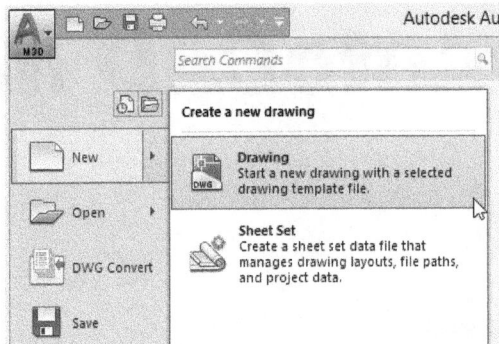

Figure 2-1 *Choosing the* ***Drawing*** *option from the Application Menu*

> **Tip**
> *You can directly invoke the* ***Select template*** *dialog box by choosing the* ***New*** *option from the Application Menu.*

OPENING AN EXISTING DRAWING FILE

You can also start a project by using the data in an existing drawing file. To open a drawing file, choose the **Open** button from the Application Menu; the **Select File** dialog box will be displayed, as shown in Figure 2-3. In this dialog box, select one of the following file extensions *.dwg*, *.dws*, *.dxf*, or *.dwt* from the **Files of type** drop-down list. Next, browse to the appropriate folder containing the required file by using the **Look in** drop-down list. Once you open the required folder, select the data file and then choose the **Open** button; the contents of the selected file will be displayed in the drawing window.

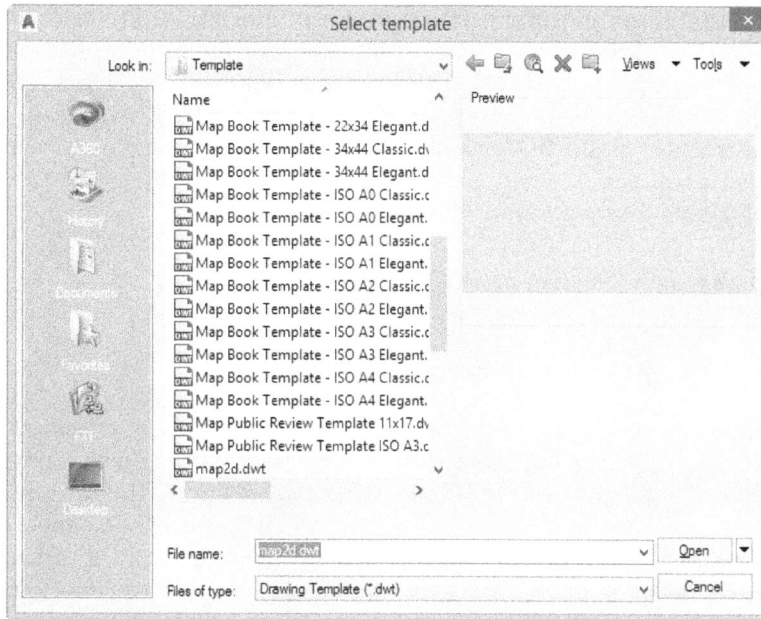

*Figure 2-2 The **Select template** dialog box with the **map2d** drawing template selected*

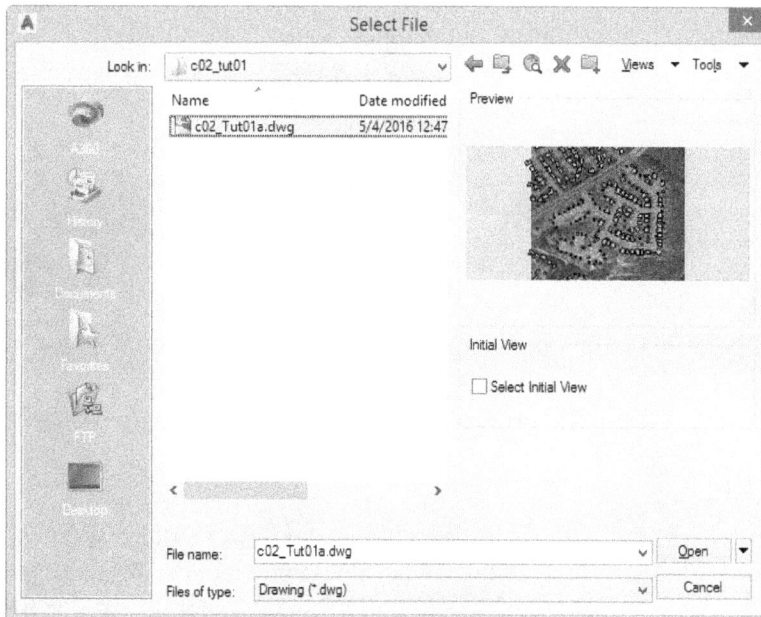

*Figure 2-3 The **Select File** dialog box*

ADDING DATA TO THE PROJECT

AutoCAD Map 3D supports a wide range of GIS and CAD data formats. This interoperability feature of AutoCAD Map 3D enables you to use various GIS data types and file formats.

In AutoCAD Map 3D, you can use the data available in other CAD files by attaching them as external reference files. You can also import the data available in other file formats into the AutoCAD's .dwg file format. AutoCAD Map 3D has incorporated the FDO data access technology into Autodesk geospatial products. This technology provides an effective means of data-access and data-management. Using FDO, you can connect directly to a data source and then work in its native format without converting into the .dwg format. Some of the methods that are commonly used to connect data in your drawing are discussed next.

Loading Data by Using the Connect Tool

Ribbon:	Home > Data > Connect
Task Pane:	Display Manager > Data > Connect to Data

You can load various types of data into the Workspace without affecting their structure and accuracy by using the **Connect** tool. This tool allows you to connect raster files, vector data files, databases, DTMs, and other type of data to the Workspace. Figure 2-4 shows the flow diagram of loading data into your Workspace by using the **Connect** tool. Different datasets such as vector data, raster data, open geospatial data source, and WMS can be combined into one group and loaded into the Workspace by using this tool. You can also adjust the settings of the datasets such as coordinate system, connection name, source folder, and so on as per your requirement. The method of loading various data types using the **Connect** tool is discussed next.

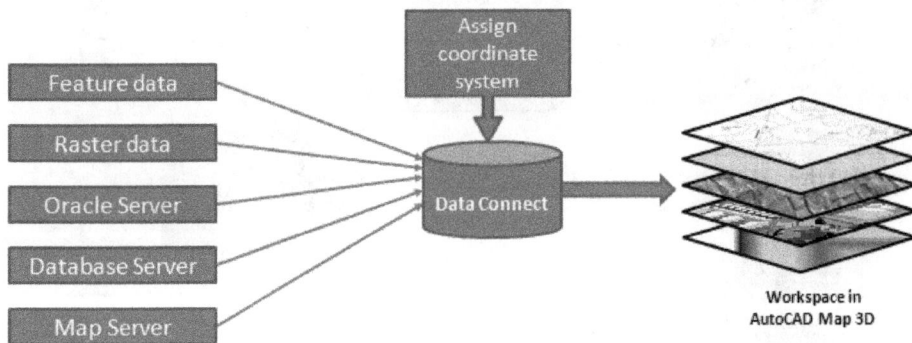

Figure 2-4 Flow diagram displaying the loading of data into the Workspace

Loading SHP Data

You can load SHP data to the current drawing by using the **DATA CONNECT** wizard. To do so, choose the **Connect** tool from the **Data** panel of the **Home** tab; the **DATA CONNECT** wizard will be displayed, as shown in Figure 2-5. Alternatively, you can invoke this wizard from the **TASK PANE**. To do so, choose the **Data** button from the **Display Manager** tab; a flyout will be displayed. Choose the **Connect to Data** option from the flyout; the **DATA CONNECT** wizard will be displayed.

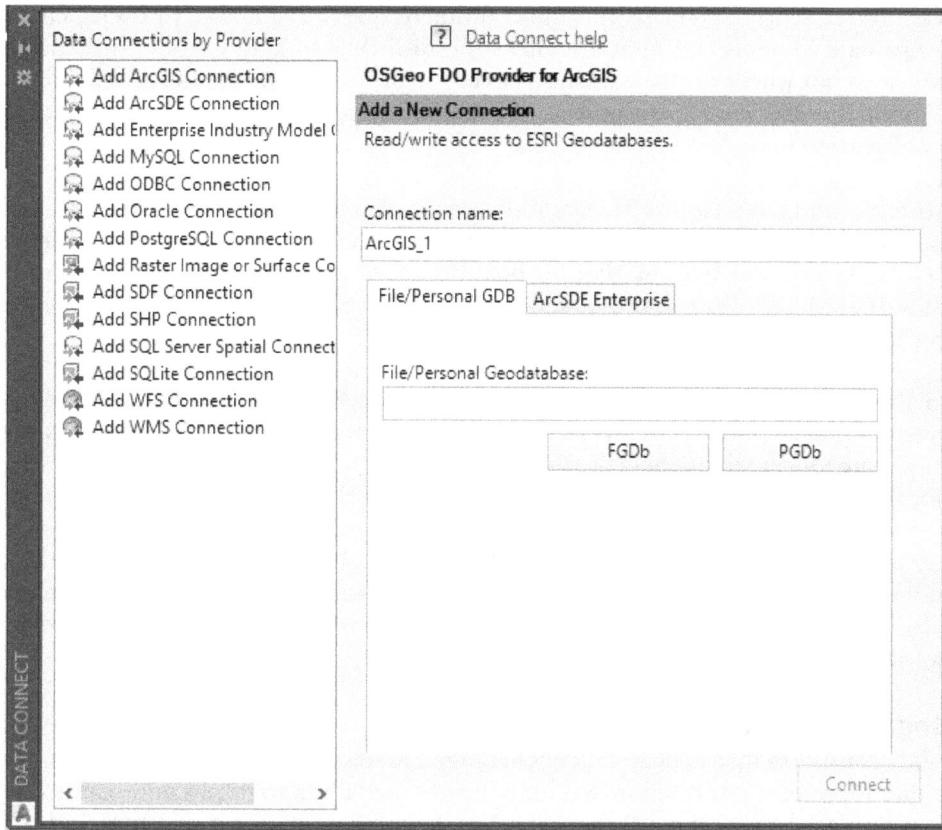

*Figure 2-5 The **DATA CONNECT** wizard*

In this wizard, to connect and add SHP data to the existing drawing, select the **Add SHP Connection** option from the **Data Connections by Provider** list box; the right pane of the wizard will display the page containing the options for adding data. For example, to load the Land cover map data (SHP file) into the Workspace, select the **Add SHP Connection** option from the **Data Connections by Provider** list box; the **OSGeo FDO Provider for SHP** page will be displayed in the right pane. Enter the name of the connection in the **Connection name** edit box.

Next, to connect a single SHP file, choose the button to the right of the **Source file or folder** edit box with the text SHP; the **Open** dialog box is displayed. In this dialog box, select the required SHP file and choose the **Open** button; the path of the selected SHP file will be displayed in the **Source file or folder** edit box and the **Connect** button will become active.

Choose the **Connect** button; the selected SHP file will be displayed in the **Add Data to Map** list box. In case you want to connect multiple SHP files that are located within the same folder, choose the Browse button; the **Browse For Folder** dialog box is displayed. In this dialog box, select the folder containing the required SHP files. Next, choose the **OK** button; the path of the selected folder will be displayed in the **Source file or folder** edit box and the **Connect** button will become active. Choose the **Connect** button; the list of all SHP files contained within the selected folder is displayed in the **Add Data to Map** list box. The list box displays the SHP files with their corresponding coordinate system.

If data is not referenced, you can add the coordinate system manually. To do so, choose the **Edit Coordinate Systems** button at the top of the **Add Data to Map** list box; the **Edit Spatial Context** dialog box will be displayed. Specify a coordinate system to georeference the data using the options in this dialog box. The process of georeferencing the data to a particular coordinate system is discussed in the next chapter.

After georeferencing the data, you will add the data to the map. To do so, select the check box corresponding to the required SHP file. Next, choose the **Add to Map** option from the drop-down displayed below the **Add Data to Map** list box; the contents of the SHP file will be displayed in the drawing window. Note that the name of the added SHP file is displayed in the **Display Manager** tab of the **TASK PANE**.

You can also apply data filters while adding data from the SHP file. To do so, choose the **Add to Map with Query** option from drop-down that is displayed below the **Add Data to Map** list box; the **Create Query** window will be displayed. Use this window to frame a spatial or aspatial query for filtering the data. On framing the query, choose the **OK** button; the filtered data will be added to the drawing window and the layers will be displayed in the list box of the **Display Manager** in the **TASK PANE**. Then, close the **DATA CONNECT** wizard. Next, select the added layer in the **Display Manager** list box, and right-click on the feature selected; a shortcut menu will be displayed. Now, choose the **Zoom to Extents** option from the shortcut menu; the drawing window will display the data in the added layers.

Loading Raster Data

Raster data consists of matrix of raster cells arranged in rows and columns. Every raster cell has a value that represents information. Satellite images, aerial photographs and scanned images are few examples of raster data. Raster data has the most simple data structure and is widely used in GIS as background layer.

You can load raster data into your drawing by using the **DATA CONNECT** wizard. To load the raster data, invoke the **DATA CONNECT** wizard by choosing the **Connect** tool from the **Data** panel. In the wizard, choose the **Add Raster Image or Surface Connection** option from the left pane; the **Autodesk FDO Provider for Raster** page will be displayed. In this page, specify the name for the connection in the **Connection name** edit box. Next, to load a single raster dataset, choose the button to the right of the **Source file or folder** edit box; the **Open** dialog box will be displayed. In this dialog box, browse to the location and select the required raster dataset. Next, choose the **Open** button in the dialog box; the **Open** dialog box closes and the name of the selected raster will be displayed in the **Source file or folder** edit box of the **DATA CONNECT** wizard. Next, choose the **Connect** button displayed below the **Source file or folder** edit box; the **Raster Image or Surface** page of the **DATA CONNECT** wizard is displayed.

To connect to multiple raster data from a folder using one connection, choose the button to the far right of the **Source file or folder** edit box; the **Browse to the folder** dialog box will be displayed. Use this dialog box to browse and select the folder containing the required raster datasets. Next, choose the **OK** button; the **Browse to the folder** dialog box will be closed and the name of the selected folder will be displayed in the **Source file or folder** edit box of the **DATA CONNECT** wizard. Next, choose the **Connect** button displayed below the **Source file or folder** edit box; the **Raster Image or Surface** page of the **DATA CONNECT** wizard will be displayed. The **Schema** list in this page will now display a list of all the rasters in the selected folder.

Select the required raster data from the **Schema** list by selecting the check box. Figure 2-6 shows multiple raster data selected in the **Schema** list of the **DATA CONNECT** wizard. If required, you can assign or modify the coordinate system of the raster data. The method of assigning the coordinate system to a data is discussed in detail later in this textbook.

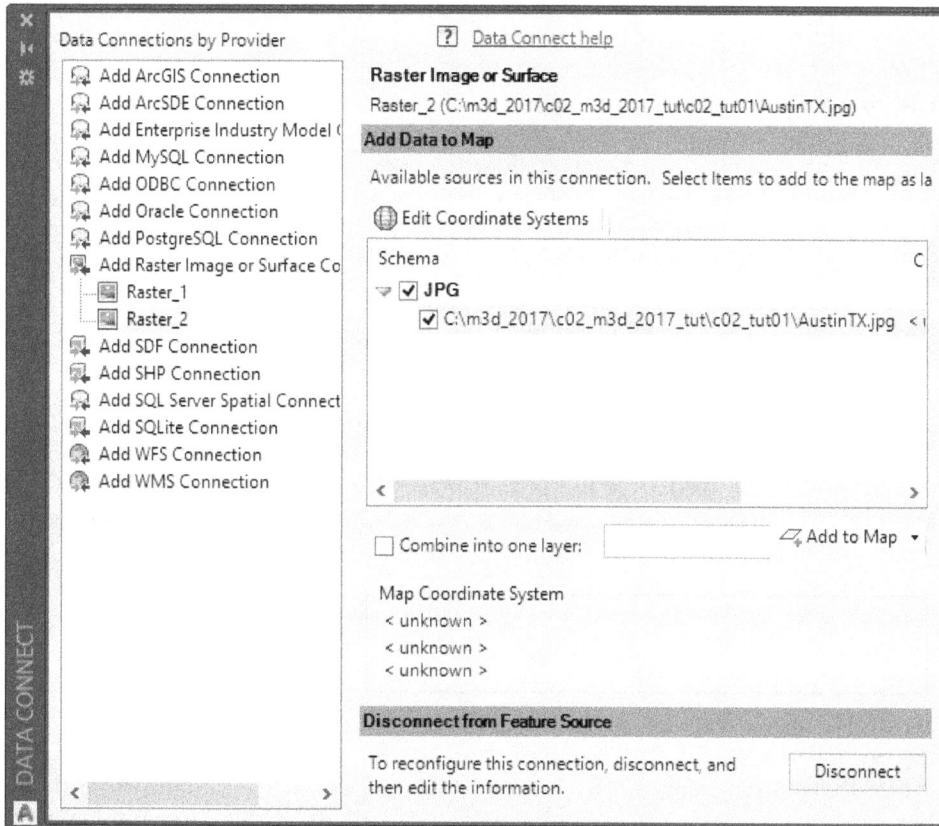

*Figure 2-6 Multiple raster data selected in the **DATA CONNECT** wizard*

On selecting the required raster data, choose the **Add to Map** button displayed below the **Schema** list; the selected raster will be added to the drawing. You can also use the **Image** tool from the **Image** panel of the **Insert** tab to insert raster image into your drawing. The method of inserting the raster using this tool is discussed later in this book.

Loading SDF Data

SDF is the spatial data file format developed by Autodesk to store different geospatial data. You can load SDF drawing file into AutoCAD Map 3D. To do so, choose the **Connect** tool from the **Data** panel; the **DATA CONNECT** wizard will be displayed, refer to Figure 2-5. In this window, choose the **Add SDF Connection** option from the **Data Connections by Provider** list box; the **OSGeo FDO Provider for SDF** page will be displayed in the right pane of the wizard. Specify the name of the connection in the **Connection name** edit box. Then, choose the button next to the **Source file** edit box; the **Open** dialog box will be displayed. In this dialog box, select the required file and then choose the **Open** button; the dialog box will be closed and the file location

will be displayed in the **Source file** edit box in the **OSGeo FDO Provider for SDF** page of
the wizard. Next, choose the **Connect** button. On doing so, the list of data files in the
source connection will be displayed in the **Add Data to Map** list box. Select the check box
corresponding to the data that you want to add to the map. Next, choose the **Add to Map** button
from the drop-down below the **Add Data to Map** list box. The added data will be displayed in
the drawing window.

Loading WFS Data

WFS data (Web Feature Service) provides geospatial feature data. AutoCAD Map 3D allows you
to use WFS data using the **Connect** tool. To do so, choose the **Connect** tool from the **Data** panel;
the **DATA CONNECT** wizard will be displayed. In this window, choose the **Add WFS Connection**
option from the **Data Connections by Provider** list box; the **OSGeo FDO Provider for WFS**
area will be displayed. Specify the name of the connection in the **Connection name** edit box.
Enter the server name or the URL in the **Server name** edit box. Alternatively, you can select
the previously used URL from the drop-down list. Select the relevant version from the **Version**
drop-down list. You can also select the **Default version** option from the drop-down list. Next,
choose the **Connect** button; the **User Name & Password** dialog box will be displayed. Enter
the credentials in the dialog box and choose the **Login** button; the list of data files in the source
connection will be displayed in the **Add Data to Map** list box. Select the data that you want to
add to the map from the **Add Data to Map** list box; the drop-down list below the **Combined
Layer Info** list box will be activated. Choose the required option from this drop-down list; the
selected data will be added to the drawing window.

Note
*When you select the **Default version** option from the **Version** drop-down list, AutoCAD
Map 3D will load the latest version of the WFS data.*

Loading WMS Data

Using the WMS data (Web Map Service) connection, you can incorporate the raster data that is
published on a public web server into your map. To bring the WMS data into AutoCAD Map 3D,
choose the **Connect** tool as explained in the previous section; the **Data Connect** window will be
displayed. In this window, choose the **Add WMS Connection** option from the **Data Connections
by Provider** list box; the **OSGeo FDO Provider for WMS** page will be displayed. Specify the
connection name and server name or URL in the **Connection name** and **Server name or URL**
text box, respectively. Next, choose the **Connect** button; the **User Name and Password** dialog
box will be displayed. Enter the credentials and choose the **OK** button. On connecting to the
WMS server, the list of available schema will be displayed in the list box of the **OSGeo FDO
Provider for WMS** page. Select the required schema from the list box and choose the **Add Data
to Map** option to load data into the drawing window.

Note
*OSGeo FDO Provider for WFS or WMS data of AutoCAD Map 3D provides read only access to
these data types.*

Loading Data from Databases

AutoCAD Map 3D provides inbuilt data connections to connect to various databases such as
Microsoft Access, Oracle, MySQL, and ArcSDE. To access data from Microsoft Access, Excel, or

dbase, select the **Add ODBC Connection** option from the **Data Connections by Provider** list box; the **OSGeo Provider for ODBC** page is displayed in the right pane of the **DATA CONNECT** wizard. In this page, specify the connection name in the **Connection name** edit box. Select the relevant source type from the **Source type** drop-down list. If you select the **Data Source Name (DSN)** option from the **Source type** drop-down list, then you need to enter the name of the data source. To do so, choose the Browse button next to the **Source** edit box; the **Select Data Source Name** dialog box will be displayed. Select the required data source name from the **Data Source Names** list box and then choose the **Select** button; the source name will added to the **Source** edit box. You can also select the **Connection String** option and add the connection string to the **Source** edit box. Next, choose the **Test Connection** button; the **User Name & Password** dialog box will be displayed. Enter the user name and password to login. Now, connect to the database and add the data to the Map as explained previously. Similarly, you can connect to the data store in MySQL, Oracle and PostgreSQL. To connect to different data store in these databases, specify the connection name and service name and then select the data store that you want to add to the map in the drawing window. To load the data from ArcSDE, you need to specify the connection name, server name, and the instance name. On doing so, you will be able to connect to the database and add the data store to the map.

Loading Enterprise Industry Model

Enterprise Industry Models are different types of schemas that are stored in Oracle database. These models can be for storm water system, wastewater sewage system, or gas pipeline network. To load the data from the Enterprise Industry Model, choose the **Connect** tool from the **Data** panel; the **Data Connect** window will be displayed. In this window, choose the **Add Enterprise Industry Model Connection** option from the **Data Connections by Provider** list box; the **OSGeo FDO Provider for Enterprise Industry Models** page will be displayed. Specify the connection name and service name in the **Connection name** and **Service name** edit boxes, respectively. Next, specify the Map 3D or the computer user name and password in the **Map 3D Main or System User name** and **Map 3D Main or System User password** edit boxes, respectively. On doing so, you can select the relevant industry model from the **Industry model** drop-down and add data to the map.

Tip
Schema defines the structure of the different feature classes. It defines the rules, feature type, data type, feature properties, and so on. Also, schema of a particular enterprise industry model can be edited and updated as per the user requirement in AutoCAD Map 3D.

Loading Data by Using the Map Explorer Tab

You can load drawing data by using the options in the **Map Explorer** tab of the **TASK PANE**. In this tab, you can add the drawing data file by using the select, drag, and drop methods. To do so, choose the **Map Explorer** tab from the **TASK PANE**; the options in the **Map Explorer** tab will be displayed, as shown in Figure 2-7. Next, open the folder containing the source data file. Select the source data file, and then drag and drop the file into the **Map Explorer** tab; the data source file will become visible at the top of the list in the **Map Explorer** tab.

You can attach drawings to the Workspace by using the options in the **Map Explorer** tab. To do so, right-click on the **Drawings** folder in the tab; a shortcut menu will be displayed. Choose the **Attach** option from the shortcut menu; the **Select drawings to attach** dialog box will be displayed. In this dialog box, select the source drawing file and then choose the **Add** button; the added drawing files will be displayed in the **Selected drawings** list box. If the selected file is not needed, choose the **Remove** button to remove the drawing file from the list. Next, choose the **OK** button; the list of selected drawings will be displayed under the **Drawings** node in the **Map Explorer** tab of the **TASK PANE**. Select a file name in **Drawings** node and right-click on it; a shortcut menu will be displayed. Next, choose the **Zoom Extents** option from the shortcut menu; the selected drawing will be displayed in the drawing window.

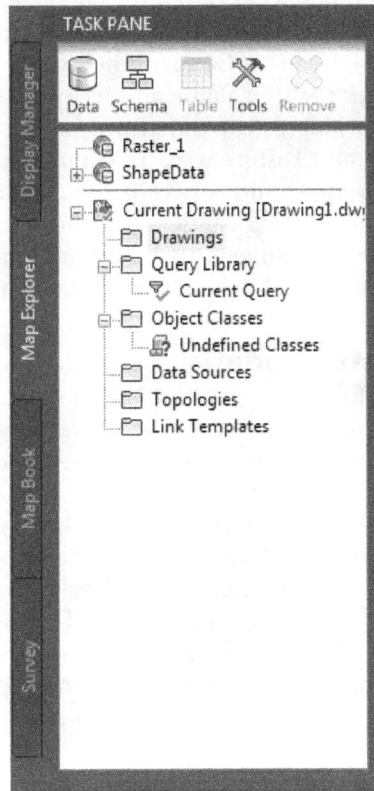

Figure 2-7 The TASK PANE with the Map Explorer tab chosen

Note
*You can also attach drawings to the Workspace by using the **Attach** tool from the **Data** panel in the **Home** tab.*

Importing GCP Data by Using the Survey Tab

You can import the Ground Control Points (GCP) data or the survey data into the Workspace by using the **Survey** tab in the **TASK PANE**. Before you import a survey data into your project, you must connect to an existing survey data store that contains data. You can also create a new survey data store to reposit survey data. To do so, choose the **Data** button in the **Survey** tab; a flyout will be displayed. From this flyout, choose the **New Survey Data Store** button; the **New Data Store** dialog box will be displayed, as shown in Figure 2-8.

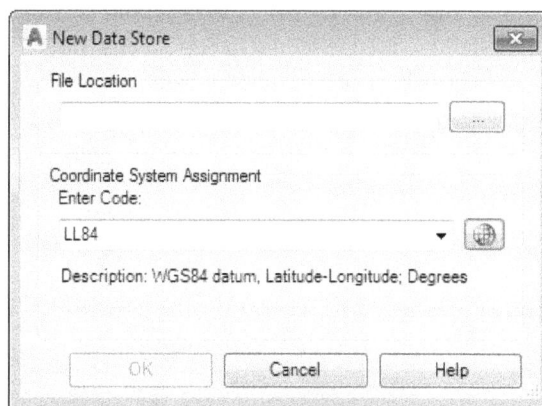

Figure 2-8 The New Data Store dialog box

In this dialog box, choose the Browse button in the **File Location** area; the **Create new Survey Data Store** dialog box will be displayed. In this dialog box, browse to the required location and specify a name for the new data store in the **File name** edit box and then choose the **OK** button; the **Create new Survey Data Store** dialog box will be closed and the path for the data store will be displayed in the **File Location** text box of the **New Data Store** dialog box. In the **Coordinate System Assignment** area of the **New Data Store** dialog box, select a suitable Coordinate Reference System (CRS), either by entering a value in the **Enter Code** edit box or by using the Browse button located on the right of the **Enter Code** edit box. Next, choose the **OK** button in the **New Data Store** dialog box; a survey data store will be created. The created data store is displayed in the **Display Manager** tab and the **Map Explorer** tab of the **TASK PANE**. The created data store is also selected in the **Current Data Store** drop-down in the **Survey** tab and the **Survey Data Store** node is added to the list box of this tab. To import points to this data store, right-click on the **Survey Data Store** node; a shortcut menu is displayed. Choose the **Import LandXML** or the **Import ASCII Points** option from the shortcut menu to import point data into the project. Next, right-click on the **Survey Data Store** node; a shortcut menu is displayed. Choose the **Zoom to Extents** option from this menu; the survey data will be displayed in the drawing window.

Importing an External File Format Data by Using the Map Import Tool

Ribbon:	Insert > Import > Map Import
Command:	MAPIMPORT

Using the data import feature of AutoCAD Map 3D, a Map 3D user can import spatial data into the Map 3D environment. Using this feature, you can import spatial data available in various file formats such as ASCII, ESRI shape file, MapInfo TAB, TIFF, GML, and DGN into the Map 3D environment. To import the datasets into the AutoCAD Map 3D, choose the **Map Import** tool from the **Import** panel in the **Insert** tab; the **Import Location** dialog box will be displayed. In this dialog box, select the required file extension from the **Files of type** drop-down list. Next, browse and select the required dataset and then choose the **OK** button; the **Import Location** dialog box will be closed and the **Import** dialog box will be displayed. Figure 2-9 shows the **Import** dialog box for a .shp file.

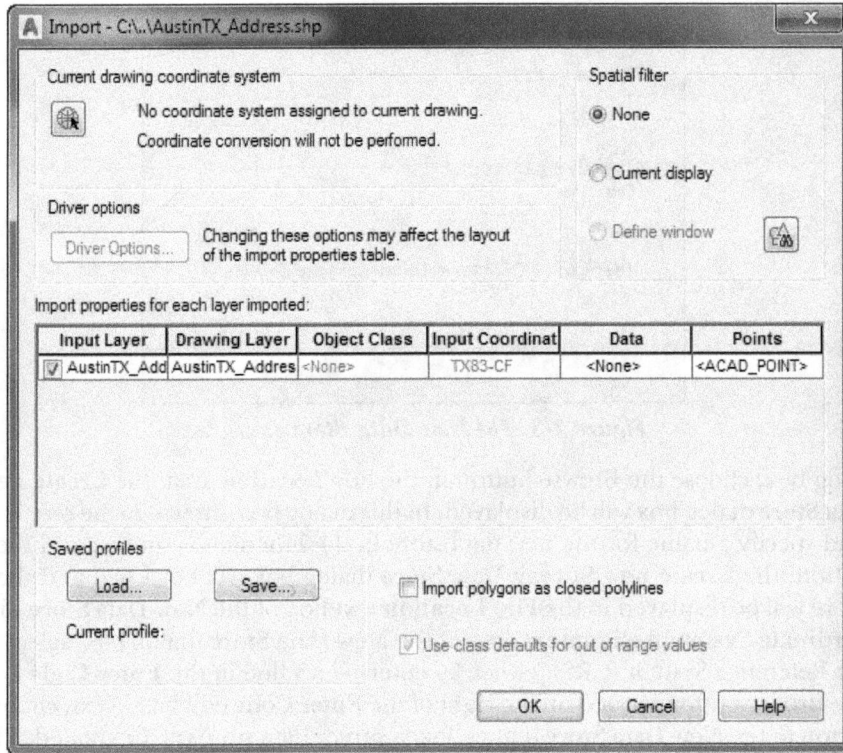

*Figure 2-9 The **Import** dialog box*

Note
*The options displayed in the **Import** dialog box will depend on the type of data file selected in the*
***Import Location** dialog box.*

In this dialog box, for importing the SHP file, the **Current drawing coordinate system** area
displays the coordinate system of the drawing. You can change the coordinate system of the
drawing using the **Assign Global Coordinate System** dialog box. To invoke this dialog box,
choose the button in the **Current drawing coordinate system** area of the **Import** dialog
box. While importing the data, you can also set the spatial filter, which can be used to
import data within a required region. To set a spatial filter, select a radio button corresponding
to the required option in the **Spatial filter** area of the **Import** dialog box. The table in the **Import
properties for each layer imported** area allows you to manage the imported data. You can
specify properties such as the layer drawing for the imported data. You can also choose whether
to include or exclude the object data in the drawing. While including the object data, AutoCAD
Map 3D allows you to assign the table name and manage its field content.

After specifying the required parameters in the **Import** dialog box, choose the **OK** button to
add the imported layer with default settings. The options used in the **Import** dialog box are
discussed later in this book.

WORKING WITH GRIPS

New

AutoCAD Map 3D 2017 allows you to edit the vector features by using grips. Grips are convenient means of editing objects. These are small geometrical shape entities like sqaures and rectangles that are displayed on the key points of an object on selecting the object, as shown in the Figure 2-10. Using grips, you can stretch, move, rotate, scale, and mirror objects, and change properties.

Figure 2-10 *Grips displayed on a polyline feature*

On hovering the cursor over the square grip at the end of the polyline feature, three options will be displayed: **Stretch Vertex**, **Add Vertex**, and **Remove Vertex**. On selecting the **Stretch Vertex** option, the entity will be enabled for editing. You can stretch it in polar direction with respect to the entity or you can specify angular direction for stretching. You can also add new vertex to the existing feature by using the **Add Vertex** option. For omitting the unwanted feature you can use the **Remove Vertex** option. On selecting this option, the feature created previously will be deleted.

On hovering the cursor over the rectangle grip at the middle of the polyline feature, three options will be displayed: **Stretch**, **Add Vertex**, and **Convert to Arc**. You can stretch the existing feature by using the **Stretch** option. On selecting the **Add Vertex** option you will be prompted to specify the new vertex point. Click on the desired point, the polyline feature will be changed with an additonal vertex. On selecting the **Convert to Arc** option, you will be prompted to specify midpoint of the arc segment, click at the desired point; the entity will be changed to an arc.

DATA OUTPUT METHODS

AutoCAD Map 3D allows you to generate the output of your project in several ways, such as saving the file in various formats, sending mail through web services, publishing the map using various mediums, and so on. In this section, you will learn about the data output methods such as saving files, exporting data, printing data, plotting and publishing, and sending files through web services. These methods are discussed next.

Generating Data by Using the Save As Option

You can save a drawing file in AutoCAD Map 3D's native file format .DWG or in various other formats by using the Application Menu. To save a drawing file in the AutoCAD file format, choose the **Save As** option from the Application Menu; the **Save Drawing As** dialog box will be displayed. In this dialog box, browse to the location where you want to save the file. Next, enter a name in the **File name** edit box and select the file format from the **Files of type** drop-down list. Choose the **Save** button to save the file with assigned name and file type.

Generating Data by Using the Export Option

You can use the **Export** option in the Application Menu for exporting a drawing file into different file formats. Different file formats in the **Export** option are discussed next.

AutoCAD Map 3D uses the DWF (Design Web Format) to distribute data efficiently. Drawings can be exported into the DWF format by using the **Export** option. To do so, choose **Export > DWF** from the Application Menu; the **Save As DWF** dialog box will be displayed, as shown in Figure 2-11. In the **Save As DWF** dialog box, the export file properties such as **Type**, **Override Precision**, **Layer Information**, and so on will be displayed in the **Current Settings** area. To change these user settings for the file to be exported, choose the **Options** button in this area; the **Export to DWF/PDF Options** dialog box will be displayed. The user settings are categorized as **General DWF/PDF Options** and **DWF data options**. Specify the required settings and then choose the **OK** button to apply the settings to the file being exported.

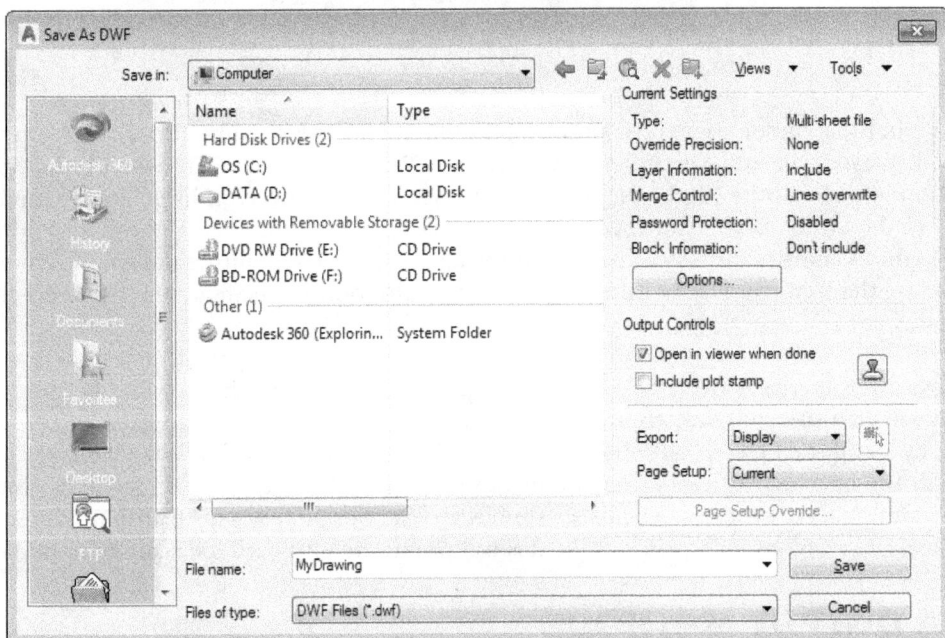

*Figure 2-11 The **Save As DWF** dialog box*

The options in the **Output Controls** area of the **Save As DWF** dialog box are used to set the display properties of an export file. To display a file in a viewer after exporting it, select the **Open in viewer when done** check box. To apply plot stamp to the export file, select the **Include plot stamp** check box. To modify the plot stamp settings, choose the **Plot Stamp Settings** button located on the right of the **Include plot stamp** option; the **Plot Stamp** dialog box will be displayed. Modify the settings as per your requirements in this dialog box, and then choose the **OK** button to apply the settings and close the **Plot Stamp** dialog box. To choose the extents for the file to be exported, select the desired option from the **Export** drop-down list. To export the entire content of the drawing, select the **Extents** option from the drop-down list. On selecting the **Display** option, all the objects that are visible in the current drawing window will be exported. To export the objects in a particular region, select the **Window** option from the **Export** drop-down list; the **Select Window** button on the right of this option will be activated. Next, choose the **Select Window** button; the **Save As DWF** dialog box is closed. Select the area in the drawing window to be exported. On selecting the area, the **Save As DWF** dialog box will be displayed again. To change the page settings of the export file, select the **Override** option from the **Page Setup** drop-down list; the **Page Setup Override** button will become active. To modify the page settings, choose the **Page Setup Override** button; the **Page Setup Override** dialog box will be displayed. In this dialog box, apply the page settings as required. Choose the **OK** button to apply the settings and close the **Page Setup Override** dialog box.

In the **Save As DWF** dialog box, enter the name of the exported file in the **File name** edit box, and then choose the **Save** button; the file will be saved as a DWF file at the specified location. Using the similar exporting process, you can create files with the DWFx and PDF file extensions. More file extensions are available in the dialog box when you choose the **Other GIS Format** option to export the drawing from the Application Menu.

Generating Data by Using the DWG Convert Option

The **DWG Convert** option in the Application Menu is used to change drawings from one AutoCAD drawing type to another. To convert a drawing, choose **DWG Convert** from the Application Menu; the **DWG Convert** dialog box will be displayed, as shown in Figure 2-12. Different options in this dialog box are discussed next.

Files Tree and Files Table Tabs

In the **DWG Convert** dialog box, the **Files Tree** tab is chosen by default. This tab displays the list of drawings in a hierarchical order. The selected check box against the file name indicates that the file will be converted. Clear the check box to exclude the file from conversion. To add files to the list, drag and drop the required files into the list or use the **Add file** button displayed below the list. The **Files Table** tab displays the list of drawing files to be converted into a table format. Five buttons are located at the bottom of the **Files Tree** and **Files Table** tabs. The use of these buttons is discussed next.

Add file

The **Add file** button is used to add drawing files for conversion process. To do so, choose this button; the **Select File** dialog box will be displayed. In the dialog box, browse to the folder containing the required drawing file. Next, select the file/s and then choose the **Open** button; the dialog box will be closed and the selected file/s will be added to the list in the **DWG Convert** dialog box.

New list

The **New list** button clears the existing list in the tab and starts a fresh list. On choosing the **New List** button, the **DWG Convert** message box will be displayed. To save the existing list, choose the **Yes** button in the message box; the message box will be closed and the **Save Conversion List** dialog box will be displayed. In the dialog box, enter a name in the **File name** edit box and then choose the **Save** button; the list will be saved with an extension.

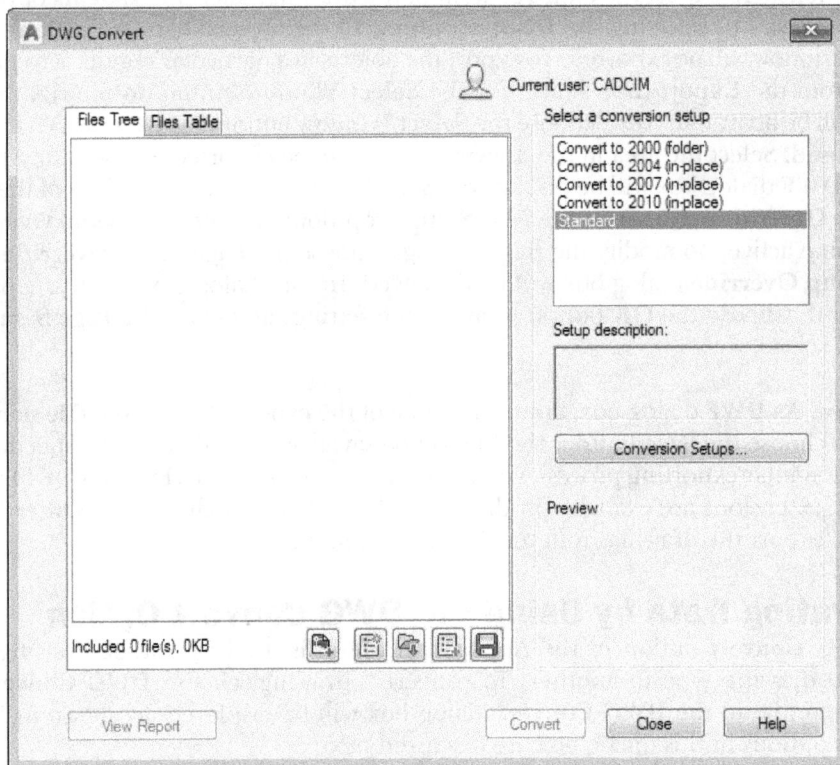

*Figure 2-12 The **DWG Convert** dialog box*

Open list

The **Open list** button is used to open an existing conversion list of drawing files. To do so, choose the **Open list** button; the **Open Conversion List** dialog box will be displayed. In the dialog box, select the desired list/s from the folder and then choose the **Open** button; the selected list will be added to the active tab.

Append list

The **Append list** button is used to add an existing list to the current list of drawings. To do so, choose the **Append list** button; the **Append Batch Control List** dialog box will be displayed. In the dialog box, select an option from the list box and then choose the **Open** button; the dialog box will be closed and the drawings in the selected list will be added to the current list of drawings in the active tab.

Save list

The **Save list** button is used to save the current list in the form of a batch conversion list (**.bcl*).

Select a Conversion Setup

The **Select a conversion setup** area displays a list of conversion setup. These setups contain the settings that will be used while converting the drawing files. Choose the required option from the list box and then choose the **Convert** button to begin the file conversion process. To modify or create the conversion settings, choose the **Conversion Setups** button in this area; the **Conversion Setups** dialog box will be displayed, as shown in Figure 2-13.

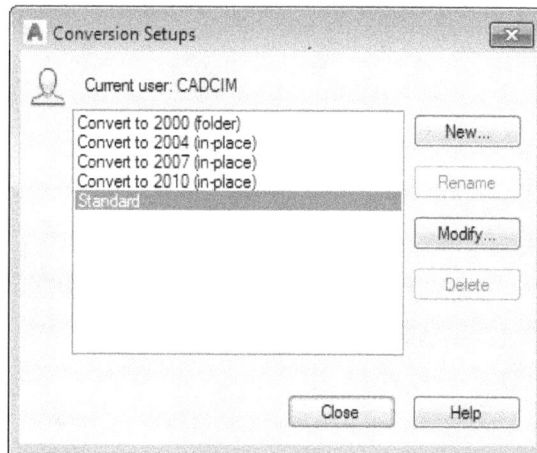

*Figure 2-13 The **Conversion Setups** dialog box*

Using the options in this dialog box, you can create, rename, modify, or delete a conversion setup. To create a new conversion setup, choose the **New** button from the dialog box; the **New Conversion Setup** window will be displayed. In the window, enter a name in the **New Conversion setup name** edit box and select an option from the **Based on** drop-down list. Next, choose the **Continue** button; the window will be closed and the **Modify Conversion Setup** dialog box will be displayed. Customize the settings based on your requirement in this dialog box and then choose the **OK** button to close it. Then, choose the **Close** button to close the **Conversion Setups** dialog box.

After specifying all options in the **Files Tree** and **Files Table** tabs, and the **Select a conversion setup** area, choose the **Convert** button; the selected drawing file or entire list of drawing files will be converted into the specified file format.

Generating Data by Using the Printing Tools

Printing paper copies of the drawing is another way of providing the project output. The Application Menu provides various useful options such as **Page Setup**, **Manage Plotters**, **Manage Plot Styles** and **Plot** for managing printing processes. Some of the printing tools available in the Application Menu are discussed next.

Plot Tool

You can use the options in the **Plot** dialog box to specify the settings for plotting. To invoke this dialog box, choose the **Print** option from the Application menu; a cascading menu will be displayed. Next, choose the **Plot** tool from the displayed menu; the **Plot** dialog box will be displayed. Figure 2-14 shows the **Plot-Model** dialog box. The options in the **Plot** dialog box are discussed next.

Page setup Area

In the **Page setup** area, the **Name** drop-down list displays the name of the current page setup. To select a previously saved page setup, choose the required option from the **Name** drop-down list. You can also import the page setup. To do so, choose the **Import** option in the **Name** drop-down list; the **Select Page Setup From File** dialog box will be displayed. You can use the options in this dialog box to browse and open the required page setup.

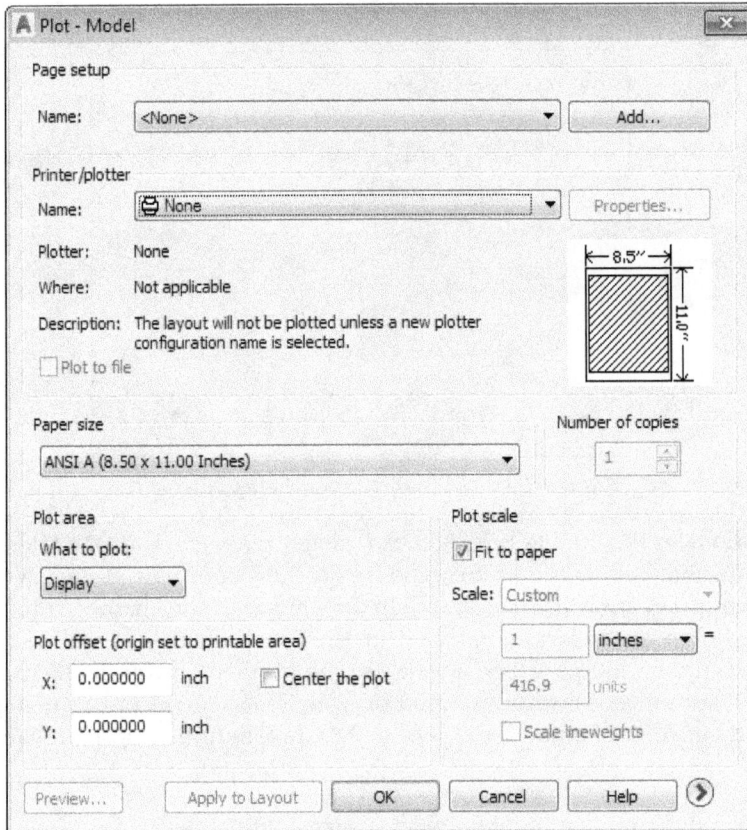

*Figure 2-14 The **Plot-Model** dialog box*

To save the current page setup for future reference, choose the **Add** button in the **Page setup** area; the **Add Page Setup** dialog box will be displayed. Specify a name for the page setup in the text box and choose the **OK** button. The page setup will be saved and the name of the saved page setup will be added to the **Name** drop-down list of the **Plot** dialog box.

Printer/plotter Area

The options in the **Printer/plotter** area are used to modify the properties of the selected printer or plotter template. When you select a page setup from the **Name** drop-down list, the **Properties** button will get activated. You can use the **Properties** button to modify the settings of the printer/plotter template as per your requirement. If you need to apply settings from an existing template file, select a template from the **Name** drop-down list; the **Name**, **Where**, and **Description** properties of the **page setup** will be displayed in the **Printer/plotter** area. Also, the preview of the modified settings will appear in the display box below the **Properties** button. You can select the **Plot to file** check box to plot the drawing to a file, instead of printing on a paper.

Paper size Area

The drop-down list in the **Paper size** area is used to specify standard paper dimensions. You can modify the existing paper dimension of a standard template by selecting the required option from the drop-down list in this area. You can specify the number of copies to be printed using the **Number of copies** spinner.

Plot area

The option in the **Plot area** is used to specify the area in the drawing for plotting. To do so, select an option from the **What to plot** drop-down list. If you select the **Window** option from this drop-down list, the cursor will change to a crosshair. You can use the crosshair to select the required plot region in the drawing. Selecting the **Limits** option from the **What to plot** drop-down list will plot all the drawing objects in the selected paper area of the current layout. You can select the **Display** option to plot all the drawing objects that are currently displayed in the drawing window.

Plot offset Area

The options in the **Plot offset (origin set to printable area)** are used to set the position of the drawing with respect to the printing paper. To specify the distance along x and y directions from the lower left corner of the paper, enter the values in the **X** and **Y** edit boxes, respectively. To keep the drawing file at the center of the paper source, select the **Center the plot** check box.

Plot scale Area

The options in the **Plot scale** area are used to adjust the scale of the drawing up to the extents of the paper. Select the **Fit to paper** check box to fit the drawing proportionally to the paper source. Alternatively, you can apply custom settings for the plot scale by clearing the **Fit to paper** check box. On doing so, the **Scale** drop-down list and other options will become active. Now, you can specify the required scale by selecting an option from the **Scale** drop-down list and entering the required values in the corresponding edit boxes.

After specifying the properties in the **Plot** dialog box, choose the **OK** button; the plotter will start plotting the specified file. But, in case you have selected the **Plot to file** check box in the **Plot** dialog box, the **Browse for the Plot File** dialog box will be displayed on choosing the **OK** button. Specify the plot file location and name in the **Browse for Plot File** dialog box and choose the **Save** button.

Note
*You will notice a message box on the lower right corner in the **Application Status bar** displaying the information about the plotting process. To view the details of the printing process, click on the link in the message box.*

Tip
*You can also preview the result of plotting by choosing the **Preview** button in the **Plot** dialog box.*

Batch Plot Tool

The **Batch Plot** tool in the **Print** panel of the Application Menu is used to publish multiple sheets or drawings. To invoke this tool, choose **Print > Batch Plot** from the Application Menu; the **Publish** dialog box will be displayed, as shown in Figure 2-15.

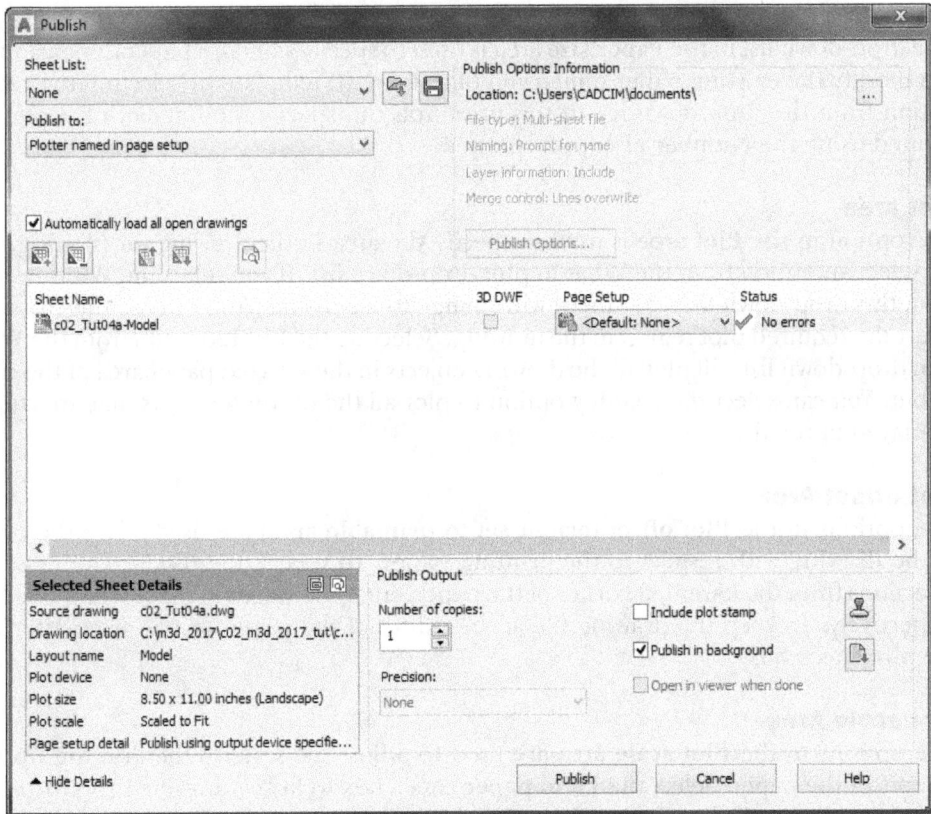

*Figure 2-15 The **Publish** dialog box*

In this dialog box, select the required sheet set data file from the **Sheet List** drop-down list. Alternatively, you can select the sheet set data by choosing the **Load Sheet List** button on the right of this drop-down list. Next, to apply a file format for publishing the sheet set data file, select an option from the **Publish to** drop-down list. Additionally, select the **Automatically load all open drawings** check box to add all the currently open drawings to the publishing list.

On applying the required settings, the information regarding the file to be published will appear in the **Publish Options Information** display area. To edit the user publishing details, choose the **Publish Options** button; the **Publish Options** dialog box will be displayed. Change the user settings by modifying the settings in this dialog box and then choose the **OK** button.

Tip
*You can also invoke the **Publish** dialog box by choosing the **Batch Plot** tool from the **Plot** panel of the **Output** tab in the Ribbon.*

To add a sheet set to the publishing list, choose the **Add Sheets** button above the sheet list box. On doing so, the **Select Drawings** dialog box will be displayed. Next, browse to the location of the required file/s and choose the **Select** button; the selected file/s will be displayed in the list box showing the added sheet set for publishing. The options in the **Publish Output** area are used to set the property for publishing the sheet set data. To change the number of copies of a published set, specify the value in the **Number of copies** spinner. To add plot stamp to the sheet set data, select the **Include plot stamp** check box in the **Publish Output** area. To modify the plot stamp settings, choose the **Plot Stamp Settings** button. Select the **Publish in background** check box to hide the **Publish Job Progress** dialog box while publishing selected sheets. If you select the **Open in viewer when done** check box, the sheet set will be displayed in a viewer when it is published. The details of the properties of the sheet set file selected will be displayed in the **Selected Sheet Details** display box.

After specifying the required settings in the **Publish** dialog box, choose the **Publish** button; the **Publish Job Progress** dialog box will be displayed. This dialog box will display the progress of publishing. In case the **Publish in background** check box is selected in the **Publish** dialog box, the **Plot - Processing Background Job** dialog box will be displayed. Choose the **Close** button to close this dialog box. Note that the plot processing will continue in the background.

Page Setup Tool

You can specify the settings for the page setup by using the **Page Setup** tool. To do so, choose **Print > Page Setup** from the Application Menu; the **Page Setup Manager** dialog box will be displayed, as shown in Figure 2-16. The options in this dialog box are discussed next.

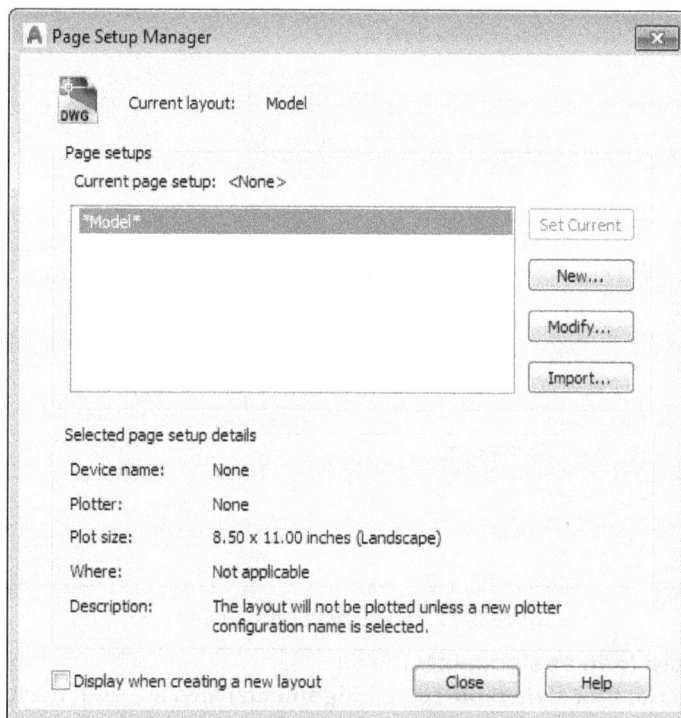

Figure 2-16 The Page Setup Manager dialog box

Page setups Area

The options in the **Page setups** area are used to create a new page setup template, modify an existing template, and remove an unused template. To create a new page setup, choose the **New** button on the right of the **Current page setup** list box; the **New Page Setup** dialog box will be displayed. In this dialog box, enter the desired name of the page setup in the **New page setup name** edit box and then choose the **OK** button; the **Page Setup - Model** dialog box will be displayed. In this dialog box, apply all required printer settings and then choose the **OK** button; the **Page Setup - Model** dialog box will be closed and the newly created page setup will be added to the **Current page setup** list box of the **Page Setup Manager** dialog box.

Selected page setup details area

The **Selected page setup details** area displays the settings of a selected page setup such as device name, plotter, plot size, and description.

To modify the settings in an existing page setup file, select the name of the page setup in the **Current page setup** list box and then choose the **Modify** button; the **Page Setup** dialog box will be displayed, as shown in Figure 2-17. Some of the options in this dialog box are same as those discussed in the **Plot** dialog box. The remaining options in this dialog box are discussed next.

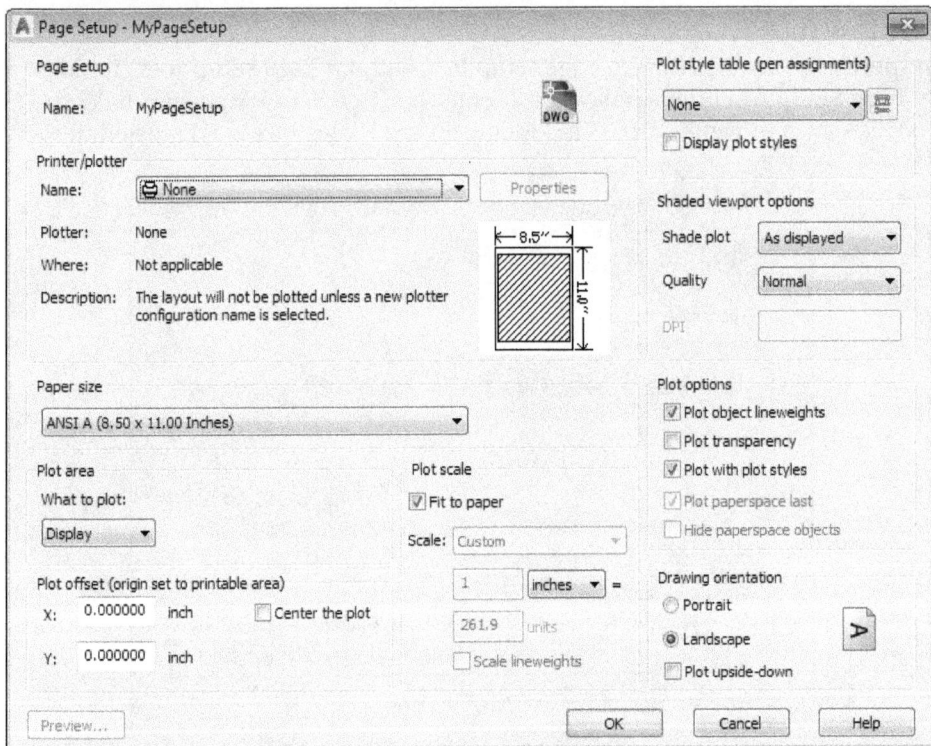

*Figure 2-17 The **Page Setup** dialog box*

Plot style table (pen assignments) Area

The options in the **Plot style table (pen assignments)** area are used to change the plotting style of drawings and layers in layouts. To apply the standard plot style table to the current layout, select the required option from the **Plot style table** drop-down list. You can modify the plot style table settings as per your requirements. To do so, choose the **Edit** button next to the **Plot style table** drop-down list; the **Plot Style Table Editor** dialog box will be displayed. Specify the required values in this dialog box and then choose the **Save & Close** button to apply the settings and to close the dialog box. To display the plot style settings in the drawing, select the **Display plot styles** check box in this area.

Shaded viewport options Area

The options in the **Shaded viewport options** area are used to specify the values for the shading of the layout. To specify how the plot should be viewed, select the required option from the **Shade plot** drop-down list. To adjust the quality of the view in the layout, select an option from the **Quality** drop-down list.

Plot options Area

The **Plot options** area is used to set plotting styles in a layout. To ignore line weights while plotting objects, clear the **Plot object lineweights** check box. To plot the layout by using the chosen plot style, select the **Plot with plot styles** check box. To plot objects in paper source before styling objects in model space, clear the **Plot paper source last** check box. To hide objects in paper space while plotting, select the **Hide paperspace objects** check box. You can also apply transparency to the plot style. To do so, select the **Plot transparency** check box in this area.

Drawing orientation Area

The **Drawing orientation** area is used to orient the drawing layout to the paper layout. To apply the portrait or landscape orientation to the drawing layout, select the **Portrait** or **Landscape** radio button. To plot the drawing layout upside down with respect to the plotting paper, select the **Plot upside-down** check box.

After specifying the properties in the **Page Setup** dialog box, choose the **OK** button to apply the settings and close the dialog box. The newly created page setup will be added to the **Current page setup** list box of the **Page Setup Manager** dialog box. Next, choose the **Close** button to close the **Page Setup Manager** dialog box.

Manage Plotters Tool

The **Manage Plotters** tool is used to edit or modify the settings of the added plotters. To invoke this tool, choose **Print > Manage Plotters** from the Application Menu; the **Plotters Manager** dialog box will be displayed. In this dialog box, edit and adjust the plotter settings as per your requirement.

Manage Plot Styles Tool

The **Manage Plot Styles** tool is used to create and edit the plot styles. To invoke this tool, choose **Print > Manage Plot Styles** from the Application Menu; the **Plot Styles Manager** dialog box will be displayed. Set the required options in this dialog box to create a new plot style.

> **Note**
> *The **Plotter Manager** and **Plot Style Manager** dialog boxes will be available for editing only after you install a plotter.*

View Plot and Publish Details Tool

The **View Plot and Publish Details** tool is used to interpret the details of plotting and publishing. To know the details of plotting and publishing, choose **Print > View Plot and Publish Details** from the Application Menu; the **Plot and Publish Details** dialog box will be displayed. The options in this dialog box are used to check the details of plotting or plotting errors, as shown in Figure 2-18.

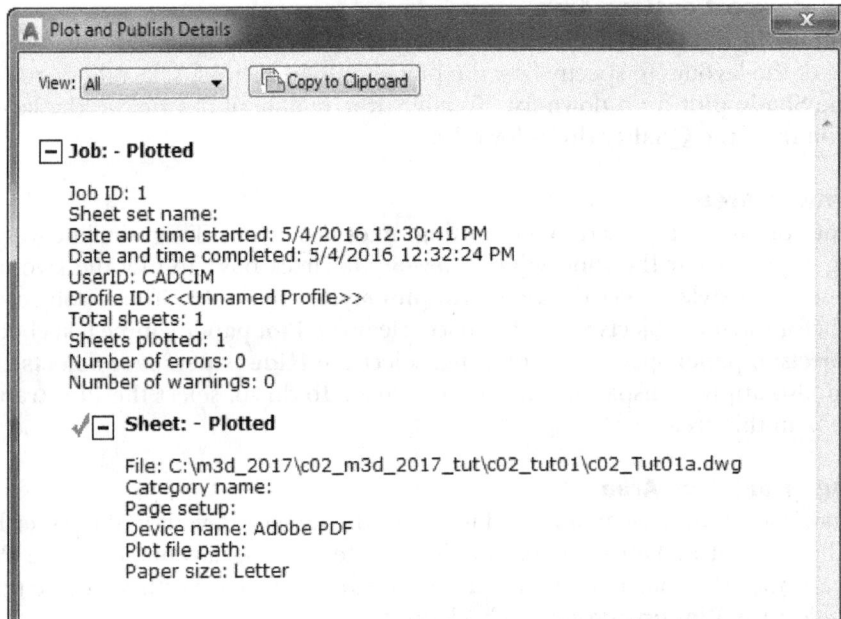

*Figure 2-18 The **Plot and Publish Details** dialog box with plot and publish details*

In the **Plot and Publish Details** dialog box, you can display the details of all printing jobs or the one that encountered errors while executing printing by selecting the **All** or **Errors Only** option from the **View** drop-down list. To copy the displayed information from the **Plot and Publish Details** dialog box, choose the **Copy to Clipboard** button. You can now paste the text in any text editing application such as **Notepad** or **Wordpad**.

Tip
*You can also invoke the **Plot and Publish Details** dialog box by choosing the **View Details** tool from the **Plot** panel of the **Output** tab.*

Generating Data Using the Publish Option

The **Publish** option in the Application Menu allows the user to generate AutoCAD Map 3D output using various tools and techniques. These data sharing techniques in the **Publish** option are discussed next.

Using the Publish to Map Server Tool

AutoCAD Map 3D allows user to publish the native DWG file directly to the Autodesk Infrastructure Map Server, without the loss of attribute data or visual styling of the DWG drawing file. This enables the user to share the data quickly and cost effectively through a web browser. To publish a drawing to the AutoCAD Infrastructure Map Server, choose **Publish > Publish to Map Server** tool from the Application Menu; the **Publish to Infrastructure Map Server** dialog box will be displayed. In this dialog box, you can select and configure the schema, classes, and attributes of the drawing to be published. To do so, select the required drawing features in the **Configure DWG Element For** list box and then, choose the browse button next to the **Select Destination Folder** edit box; the **Select Destination Folder** dialog box will be displayed. In this dialog box, enter the address of the map server in the **Connect to site** text box and choose the **Connect** button; the **Connect to Infrastructure Map Server Site** dialog box is displayed. Enter the login credentials in this dialog box and choose the **OK** button. The contents of the map server repository will be displayed in the **Select Destination folder** list box of the **Select Destination Folder** dialog box. Select the folder from the list box and choose the **OK** button; the **Select Destination Folder** dialog box will be closed. In the **Publish to Infrastructure Map Server** dialog box, select the **Show map in web browser after publishing** check box to display the map after publishing. Next, choose the **Publish** button; the map will be published and the default web browser will display the published map.

Using the Send to 3D Print Service Tool

The **Send to 3D Print Service** tool is used to send 3D solid objects or 3D water tight meshes to a 3D printer service. To invoke this tool, choose **Publish > Send to 3D Print Service** from the Application Menu; the **3D Printing - Prepare Model for Printing** window will be displayed. In this window, choose the **Continue** option; the cursor will change into a selection box. Now, you can select 3D objects or water tight meshes by using the selection box. For example, if you select a 3D box object from a drawing by using the selection box and then press ENTER, the **Send to 3D Print Service** dialog box will be displayed, as shown in Figure 2-19.

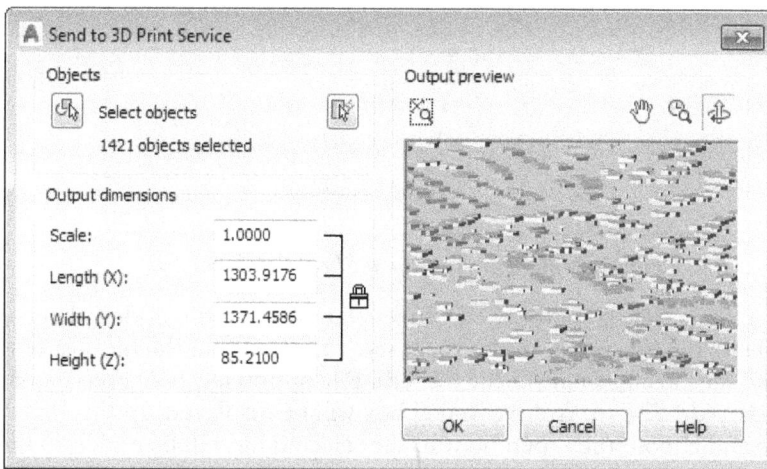

*Figure 2-19 The **Send to 3D Print Service** dialog box displaying a 3D object*

The 3D objects in the drawing can be selected by using the options in the **Objects** area. To select more objects, choose the **Select objects** button in this area; the **Send to 3D Print Service** dialog box will be closed. Select the required object in the drawing area by clicking on individual objects or by drawing a selection box. Next, press ENTER; the **Send to 3D Print Service** dialog box will be displayed again. The selected 3D objects will be displayed in the **Output preview** area of this dialog box. Use the zoom controls at the top of the display box to adjust the display extent of the selected 3D objects. The true dimensions of the selected 3D objects will be displayed in the **Output dimensions** area. The **Scale** edit box displays the scale of the selected object. The **Length**, **Width**, and **Height** edit boxes display the length, width, and height of the selected 3D object, respectively. You can change the dimensions in the **Output dimensions** area. Next, choose the **OK** button; the **Create SLT File** dialog box will be displayed. Specify the SLT file name and location in the dialog box and choose the **Save** button to save the file. You need to send this SLT file to the 3D printing service for printing.

Using the Archive Tool

The **Archive** tool is used to store the entire sheet or a part of it. To invoke this tool, choose **Publish > Archive** from the Application Menu; the **Archive a Sheet Set** dialog box will be displayed, as shown in Figure 2-20.

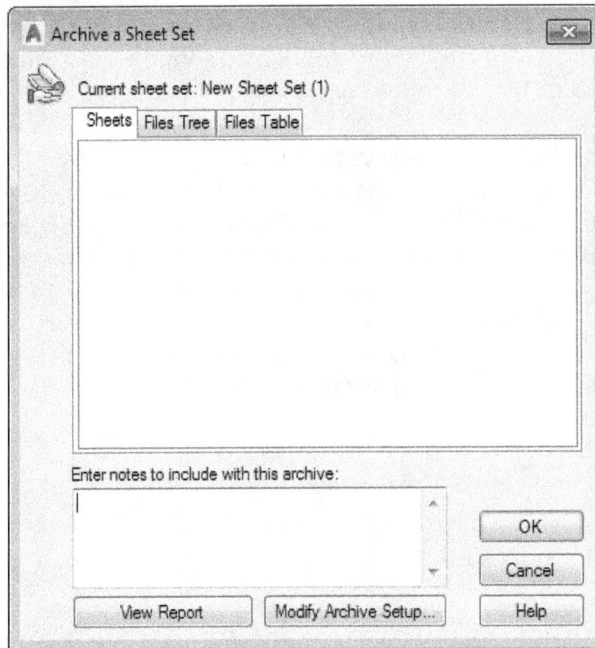

Figure 2-20 The Archive a Sheet Set dialog box

In the **Archive a Sheet Set** dialog box, the **Sheets** tab displays the added sheet set data. The **Files Tree** tab shows the list of files in the sheet set. To add files in the files list of the sheet set, choose the **Files Tree** tab and then choose the **Add File** button located at the lower right corner of the list box; the **Add File to Archive** dialog box will be displayed. In this dialog box, select the required file and choose the **Open** button; the selected file will be added to the list of files. To see details of the files in the sheet set data, choose the **Files Table** tab. Files can be added in this table as well. To do so, choose the **Add a File** button from the lower right corner of the list box; the **Add File to Archive** dialog box will be displayed. In this dialog box, select the required file and choose the **Open** button; the selected file will be added to the list of files.

You can also add notes related to sheet set data in the **Enter notes to include with this archive** text box below the **Sheets** tab in the **Archive a Sheet Set** dialog box. To view details of a sheet set file, choose the **View Report** button; the **View Archive Report** dialog box will be displayed. This dialog box displays the properties of the sheet set such as **Archive Report**, **Sheet Set**, **Files**, and **Sheet Set Data File**. To change the setup of the archive, choose the **Modify Archive Setup** button; the **Modify Archive Setup** dialog box will be displayed. In this dialog box, you can modify the archive package type, file format, file location/path, file name, and include options as per your requirement. Choose the **OK** button to apply the settings and close the **Modify Archive Setup** dialog box. Next, choose the **OK** button to close the **Archive a Sheet Set** dialog box; the **Specify Zip File** dialog box will be displayed. Specify the archive file name and location for saving the file in this dialog box. Next, choose the **Save** button; the archive file will be created and saved at the specified location.

Sending Files as Packages

This is an additional feature available in AutoCAD Map 3D. This option enables you to transfer data files in form of different packages. The different methods for sending the file package through web services are discussed next.

Sending Files by Using the eTransmit Tool

The **eTransmit** tool is used to create a package by adding data and dependent files to it. This package can be transmitted through internet. To create an eTransmit data package, choose **Send > eTransmit** from the Application Menu; the **Create Transmittal** dialog box will be displayed, as shown in Figure 2-21.

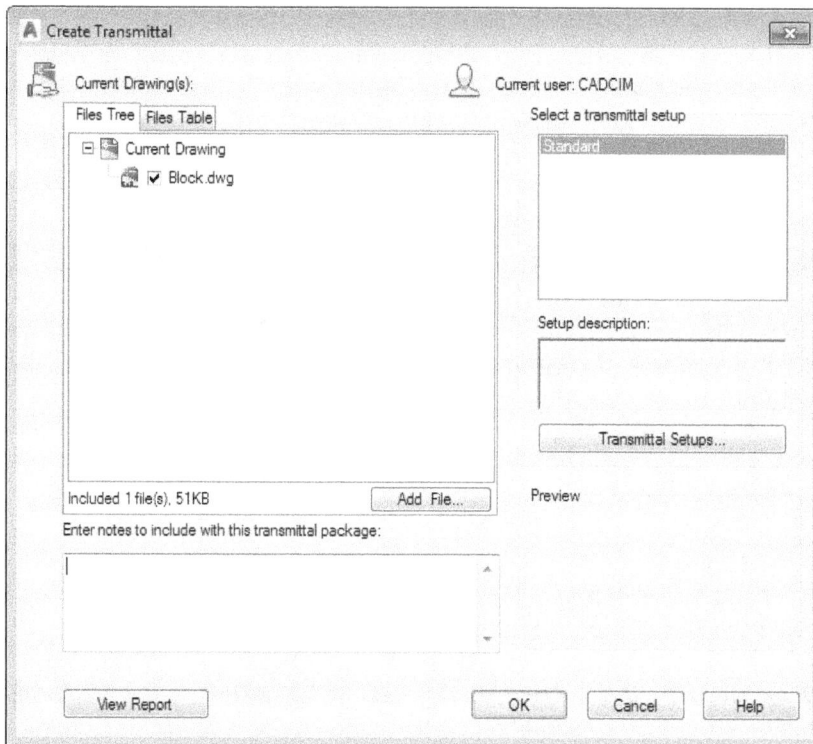

*Figure 2-21 The **Create Transmittal** dialog box displaying an added data file*

By default, the **Files Tree** tab is chosen in the **Create Transmittal** dialog box. To add data files to a package, choose the **Add File** button located at the lower right corner of the **Files Tree** tab; the **Add File to Transmittal** dialog box will be displayed. In this dialog box, select the required file and then choose the **Open** button; the selected file will be displayed in the **Files Tree** list box. To check details of each file added in the package, choose the **Files Table** list box. You can also add data file by choosing the **Add File** button located at the lower right corner of the **Files Table** list box.

A list of transmittal setup is available in the **Select a transmittal setup** list box. To apply a transmittal setup to the current drawing, select the required option from the **Select a transmittal setup** list box; the information about the transmittal setup will be displayed in the **Setup description** display box. To create a new setup or modify an existing setup, choose the **Transmittal**

Setups button located at the bottom of the **Select a transmittal setup** area; the **Transmittal Setups** dialog box will be displayed. The options in this dialog box are used to create a new transmittal setup or modify an existing transmittal setup. On modifying or creating a new setup, choose the **Close** button in the **Transmittal Setups** dialog box to close it. You can enter the description of the package in the **Enter notes to include with this transmittal package** text box below the **Tree Files** list box. The **Preview** display box shows the preview of the drawing selected in the **Tree Files** or **Files Table** tabs. To read the details of the transmittal package, choose the **View Report** button at the bottom left of the **Create Transmittal** dialog box. After specifying all the required parameters of the transmittal package, choose the **OK** button to create the transmittal package (zip file).

Sending E-mail with Compressed File Attachment

The **Email** tool is used to attach the transmittal package as an attachment with a mail. To use this tool, you must set up the **User profile** using the **Mail** tool in the **Control Panel** of the Windows operating system. After creating the **User profile**, choose **Send > Email** from the Application Menu; the MS Outlook window with transmittal package zip file attached to the mail will be displayed. You can send this mail as a normal e-mail with an attachment.

UNDERSTANDING THE CONCEPT OF SHEET SETS

A sheet is a layout formed from a drawing file. A sheet set is an organized collection of sheets from several drawing files. A sheet set organizes maps systematically and efficiently by providing easy access to various drawing files. It is very easy to plot and publish all drawings in a sheet set. You can manage and create sheet sets by using the **SHEET SET MANAGER** window. To invoke the **SHEET SET MANAGER** window, choose the **Sheet Set Manager** button from the **Palettes** panel in the **View** tab. Alternatively, press CTRL+4 keys; the **SHEET SET MANAGER** window will be displayed. In this window, you can set the properties of the sheet set data.

In AutoCAD Map, you can create a sheet set in two different ways by using either an example sheet set or an existing drawing. Both these methods are discussed next.

Creating a Sheet Set by Using an Example Sheet Set

To create a sheet set by using an example sheet set, choose **New > Sheet Set** from the Application Menu; the **Create Sheet Set** wizard will be displayed. By default, the **Begin** page will be displayed in the wizard, as shown in Figure 2-22. Alternatively, to invoke the **Create Sheet Set** wizard, choose the **New Sheet Sets** tool from the **SheetSet** drop-down in the **SHEET SET MANAGER** window.

> **Tip**
> *You can also use the NEWSHEETSET command to invoke the Create Sheet Set wizard.*

Select the **An example sheet set** radio button from the **Begin** page of the **Create Sheet Set** wizard, and then choose the **Next** button; the **Sheet Set Example** page of the wizard will be displayed, as shown in Figure 2-23.

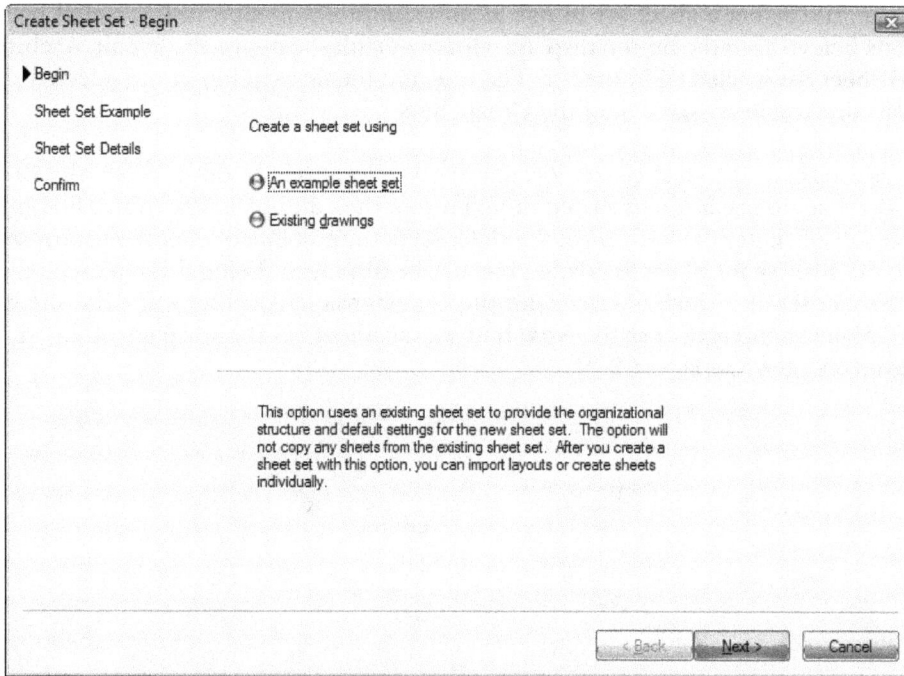

*Figure 2-22 The **Begin** page of the **Create Sheet Set** wizard*

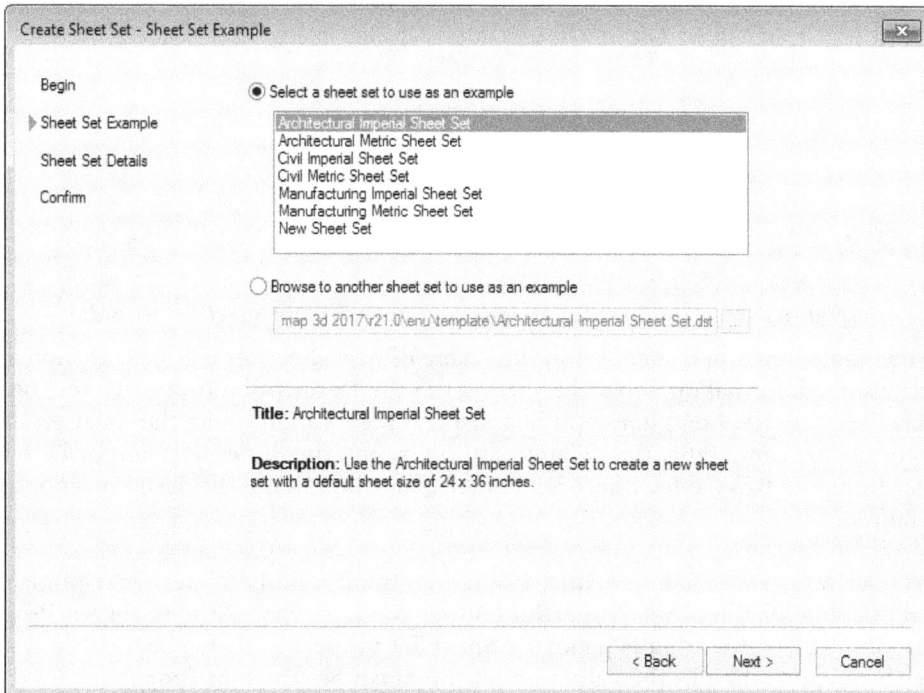

*Figure 2-23 The **Sheet Set Example** page of the **Create Sheet Set** wizard*

In this page, the **Select a sheet set to use as an example** radio button is selected by default. The list box below this radio button displays a list of default sheet sets. By default, **Architectural Imperial Sheet Set** is selected in the list. The title and the description of the selected sheet set are displayed at the lower portion of the **Create Sheet Set** wizard.

You can also select a custom sheet set located in other location. To do so, select the **Browse to another sheet set to use as an example** radio button; the edit box below this radio button will be activated. Next, enter the location of the sheet set in the edit box or choose the browse button next to it; the **Browse for Sheet Set** dialog box will be displayed. Using this dialog box, you can locate the sheet set file, which is saved with the *.dst* extension. After selecting the sheet set file to be used as an example, choose the **Next** button; the **Sheet Set Details** page of the wizard will be displayed, as shown in Figure 2-24.

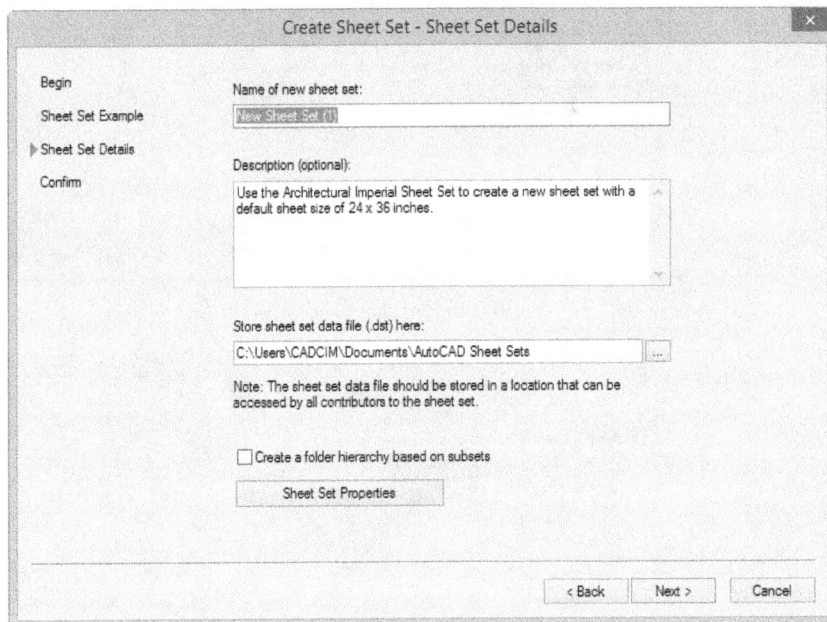

*Figure 2-24 The **Sheet Set Details** page of the **Create Sheet Set** wizard*

Specify the name of the new sheet set in the **Name of new sheet set** edit box. Optionally, you can specify some information about the sheet set in the **Description (optional)** text box. The **Store sheet set data file (.dst) here** edit box displays the location where the sheet set data file will be stored. You can modify the default location by entering a new location or by selecting the folder using the **Browse for Sheet Set Folder** dialog box that is displayed on choosing the Browse button.

The sheet set properties such as name, storage location, template, and description can be modified by using the **Sheet Set Properties** button. To do so, choose the **Sheet Set Properties** button from the **Sheet Set Details** page; the **Sheet Set Properties** dialog box will be displayed with the sheet name. To edit the values under the **Sheet Set Custom Properties** head of the dialog box, choose the **Edit Custom Properties** button; the **Custom Properties** dialog box will be displayed. Now, add or delete the required properties and choose the **OK** button to exit this dialog box. Next, choose the **OK** button to exit the **Sheet Set Properties** dialog box. After specifying the parameters in the **Sheet Set Details** page of the **Create Sheet Set** wizard, choose the **Next** button; the **Confirm** page of the wizard will be displayed, as shown in Figure 2-25.

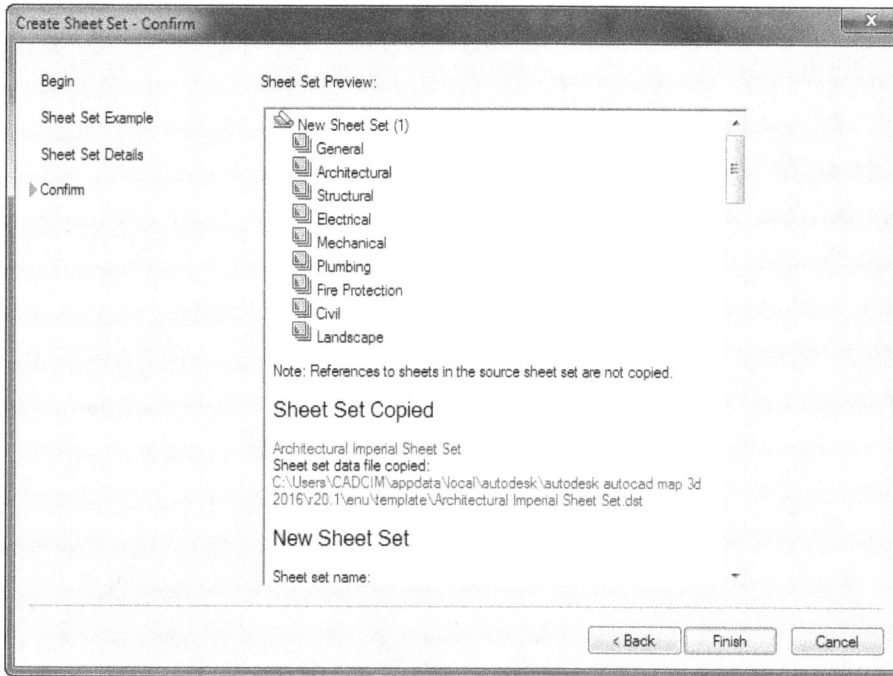

*Figure 2-25 The **Confirm** page of the **Create Sheet Set** wizard*

This page shows the detailed structure of the sheet set and also lists its parameters and properties. Choose the **Finish** button; the **Create Sheet Set** wizard will be closed and the **Sheet Set Manager** window will be displayed with the new sheet added to the drop-down list. The **Sheet List** tab of the **Sheet Set Manager** window will display the new sheet structure. To view the sheet, choose the **Sheet List** tab. Next, right-click in the blank space; a shortcut menu will be displayed. Choose the **Preview/Details Pane** option from the shortcut menu; the **Details** pane will be displayed with the details of the new sheet set. To view the model, choose the **Model Views** tab; the details of the model will be displayed in the **Details** pane. Also, you can add new drawing or folder location in this tab. To do so, choose the **Add New Location** node and right-click; a shortcut menu will be displayed. Choose the **Add New Location** option from the shortcut menu; the **Browse for Folder** dialog box will be displayed. Select the required drawing or folder and choose the **Open** button; the location will be added to the tab.

Note
The path of the file shown in Figures 2-22, 2-23, and 2-24 may vary from user to user, depending upon the location of the sheet set data saved.

Creating a Sheet Set by Using Existing Drawings
As mentioned earlier, this type of sheet set is used to organize and archive an existing set of drawings. To create this sheet set, select the **Existing drawings** radio button from the **Begin** page of the **Create Sheet Set** wizard and then choose the **Next** button; the **Sheet Set Details** page of the

wizard will be displayed. On the **Sheet Set Details** page, enter the name and description of the sheet set. After setting the required parameters on this page, choose the **Next** button; the **Choose Layouts** page will be displayed. Choose the **Browse** button from this page and browse to the folder in which the files to be included in the sheet set are saved. All drawing files along with their initialized layouts will be displayed in the list box below the **Browse** button.

TUTORIALS

General instructions for downloading tutorial files:
Before starting the tutorials, you need to download the tutorial data to your computer. To do so, follow the steps given below:

1. Log on to *www.cadcim.com* and browse to *Textbooks > Civil/GIS > Map 3D > Exploring AutoCAD Map 3D 2017*. Next, select *c02_m3d_2017_tut.zip* file from the **Tutorial Files** drop-down list. Next, choose the corresponding **Download** button to download the data file.

2. Extract the contents of the zip file to the following location:

    C:\m3d_2017

    Notice that the *c02_m3d_2017_tut* folder is created in the *m3d_2017* folder.

Tutorial 1	Using the Connect Tool - I

In this tutorial, you will connect various datasets to your drawing by using the **Connect** tool.
(Expected time: 30 min)

The following steps are required to complete this tutorial:

a. Create a new drawing file.
b. Load the image and shape files by using the **Connect** tool.
c. Save the drawing file.

Creating a New Drawing File

1. Choose the **New** option from the Application Menu; the **Select template** dialog box is displayed.

2. In this dialog box, ensure that the **map2d.dwt** template file is selected. Choose the **Open** button; the **map2d.dwt** template settings are applied to the modeling space.

Loading the Data Using the Data Connect Wizard

In this part of the tutorial, you will load dataset to the drawing by using the **Connect** tool.

1. Choose the **Connect** tool from the **Data** panel in the **Home** tab; the **DATA CONNECT** wizard is displayed.

2. In the **DATA CONNECT** wizard, select the **Add Raster Image or Surface Connection** option from the **Data Connection by Provider** list box; the **Autodesk FDO Provider for Raster** page is displayed in the right pane of the wizard.

3. In this page, enter **Base Map** in the **Connection name** edit box, and then choose the button next to the **Source file or folder** edit box; the **Open** dialog box is displayed.

4. In the **Open** dialog box, browse to the following location:

    C:\m3d_2017\c02_m3d_2017_tut\c02_tut01

5. Select the **AustinTX.jpg** file and then choose the **Open** button in this dialog box; the **Open** dialog box is closed. Also, in the **DATA CONNECT** wizard, the **Connect** button is activated and the path of the selected file is displayed in the **Source file or folder** edit box.

6. Next, choose the **Connect** button; the **Raster Image or Surface** page is displayed in the right pane of the wizard, as shown in Figure 2-26.

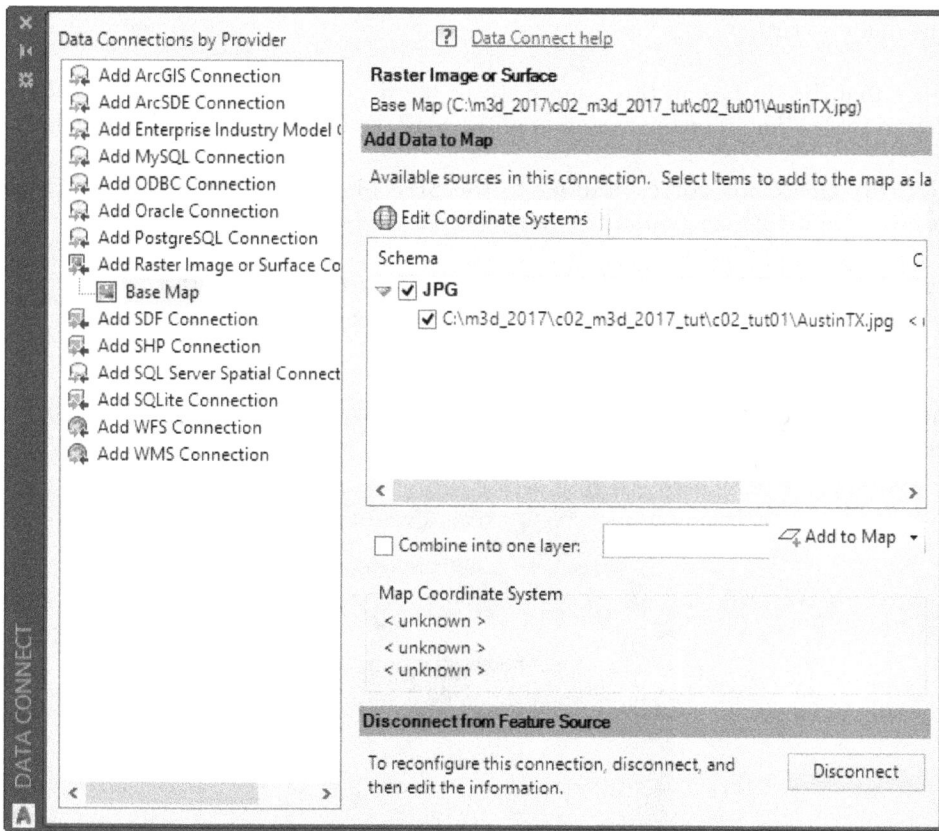

*Figure 2-26 The **Raster Image or Surface** page of the **DATA CONNECT** wizard*

7. Choose the **Add to Map** button; the image file **AustinTX.jpg** is added to the drawing and is displayed in the drawing window. Note that the **AustinTX** layer is added to the list box in the **Display Manager** tab of the **TASK PANE**.

    Next, you will add SHP data to the drawing using the **DATA CONNECT** wizard. To do so, open the **DATA CONNECT** wizard in case you have closed it.

8. In the **DATA CONNECT** wizard, select the **Add SHP Connection** option from the **Data Connections by Provider** list box; the **OSGeo FDO Provider for SHP** page is displayed in the right pane of the wizard.

9. Enter **ShapeData** in the **Connection name** edit box. Next, choose the button displayed next to the SHP button from the **OSGeo FDO Provider for SHP** area; the **Browse For Folder** dialog box is displayed.

10. Browse to *C:\m3d_2017\c02_m3d_2017_tut* and then select the *c02_tut01* folder. Next, choose the **OK** button in the **Browse For Folder** dialog box; the dialog box is closed. Also, the path of the selected folder is displayed in the **Source file or folder** edit box and the **Connect** button is also activated in the wizard.

11. Next, choose the **Connect** button; the **SHP** page is displayed in the right pane of the wizard.

    Notice that the list box in this page displays the list of all the SHP files available in the selected folder. The coordinate system of the SHP data is also displayed in the list box.

12. Select the check box corresponding to the SHP files **AustinTX_Address**, **AustinTX_BuildingFootprints**, and **AustinTX_street** in the list box.

13. Next, choose the **Add to Map** button; the shape files are added into the drawing. Close the **DATA CONNECT** wizard to see the data in the drawing window.

14. The **Display Manager** tab of the **TASK PANE** displays the name of the shape data added to the drawing, as shown in Figure 2-27.

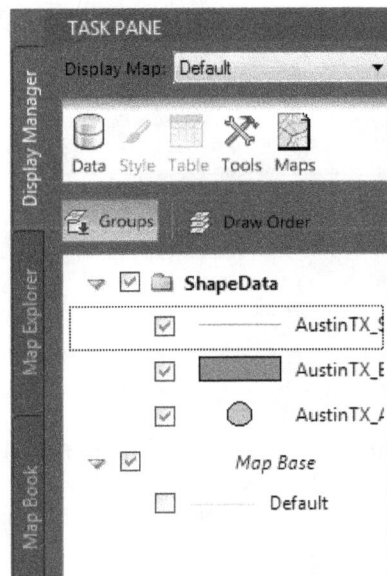

*Figure 2-27 The **Display Manager** tab showing the added SHP file data*

Saving the Drawing File

1. Choose the **Save As** tool from the Application Menu; the **Save Drawing As** dialog box is displayed.

2. In the **Save Drawing As** dialog box, browse to the location *C:\m3d_2017\c02_m3d_2017_tut\ c02_tut01* and enter **c02_Tut01a.dwg** in the **File name** edit box.

3. Choose the **Save** button; the drawing file is saved.

Tutorial 2   Using the Connect Tool - II

In this tutorial, you will connect and load the data from a WMS. You will also query and add vector data to the drawing using the **Connect** tool.   **(Expected time: 25 min)**

The following steps are required to complete this tutorial:

a. Create a new drawing file.
b. Connect to the link given below using the **Connect** tool.
    http://bhuvan-noeda.nrsc.gov.in/cgi-bin/hazard.exe
c. Add data from WMS and SHP files using the **Connect** tool.
d. Save the drawing file.

Creating a New Drawing File

1. Choose the **New** option from the Application Menu; the **Select template** dialog box is displayed.

2. In the **Select template** dialog box, select the **map2d** template file from the list box below the **Look in** drop-down list, and then choose the **Open** button; the **map2d** template is applied to the modeling space.

Loading Data from Web Map Service (WMS) Data and SHP File

In this section, you will connect and load data from a web map service using the WMS client in AutoCAD Map 3D.

1. Choose the **Connect** tool from the **Data** panel in the **Home** tab; the **DATA CONNECT** wizard is displayed.

2. In the **DATA CONNECT** wizard, select the **Add WMS Connection** option from the **Data Connections by Provider** list box; the **OSGeo FDO Provider for WMS** page is displayed in the right pane of the wizard.

3. In this page, enter **WMS_AssamHazardPortal** in the **Connection name** edit box. Next, in the **Server name or URL** edit box, enter the link given below:

    http://bhuvan-noeda.nrsc.gov.in/cgi-bin/hazard.exe

4. Ensure that the **Default version** option is selected in the **Version** drop-down list and then choose the **Connect** button; the **User Name & Password** dialog box is displayed. The selected WMS is a free resource and no login credentials are required to access the data.

5. Choose the **Login** button; the **User Name & Password** dialog box is closed and the **Connecting** message box is displayed. On establishing a connection with the web map server, the **WMS** page is displayed in the **DATA CONNECT** wizard. The list box in this page shows various schemas available in the selected WMS, refer to Figure 2-28.

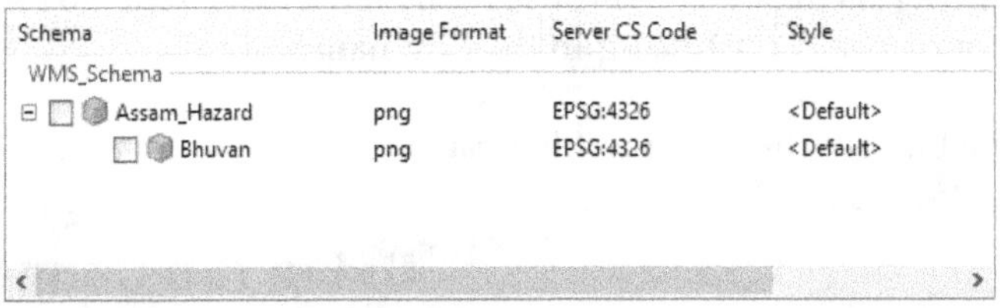

Schema	Image Format	Server CS Code	Style
WMS_Schema			
☐ ⚪ Assam_Hazard	png	EPSG:4326	<Default>
☐ ⚪ Bhuvan	png	EPSG:4326	<Default>

Figure 2-28 The list of schemas available on the web map server

6. Select the check boxes corresponding to **Assam_Hazard** and **Bhuvan**   ⊿ Add to Map ▾
    and then choose the **Add to Map** button; Map 3D will communicate with
    the WMS and add the selected data into the map.

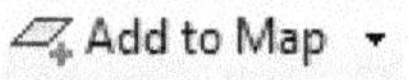

Note
Loading data from WMS may require some time depending on your internet speed and the volume of data traffic on the web map server.

When the loading process is completed, the map is displayed in the drawing window, as shown in Figure 2-29, and the name of the added layers is displayed in the **Display Manager** tab of the **TASK PANE**. You can toggle the display of layers in the drawing by selecting the check box corresponding to the layer name that you want to display.

Figure 2-29 The WMS data displayed in the drawing window

Next, you will add the vector map of Europe to your map.

7. In the **DATA CONNECT** wizard, select the **Add SHP Connection** option from the **Data Connections by Provider** list box; the **OSGeo FDO Provider for SHP** page is displayed in the right pane of the wizard.

8. In this page, enter **Assam** in the **Connection name** edit box. Next, choose the **SHP** button from the **OSGeo FDO Provider for SHP** area; the **Open** dialog box is displayed.

9. Browse to the *C:\m3d_2017\c02_m3d_2017_tut\ c02_tut02* and choose the **India_Adm.shp** file.

10. Next, choose the **Open** button in the **Open** dialog box; the dialog box is closed and the path of the selected SHP file is displayed in the **Source file or folder** edit box. Notice that the **Connect** button has also been activated.

11. Next, choose the **Connect** button; the **SHP** page is displayed in the right pane of the wizard.

12. Choose the **Add to Map with Query** tool from the drop-down displayed below the list box; the **Create Query** window is displayed.

13. Enter **NAME_1 = 'Assam'** in the edit window, as shown in the Figure 2-30, and choose the **OK** button in the **Create Query** window; the query is executed and the filtered map objects are added to the drawing window. Next, close the **DATA CONNECT** wizard.

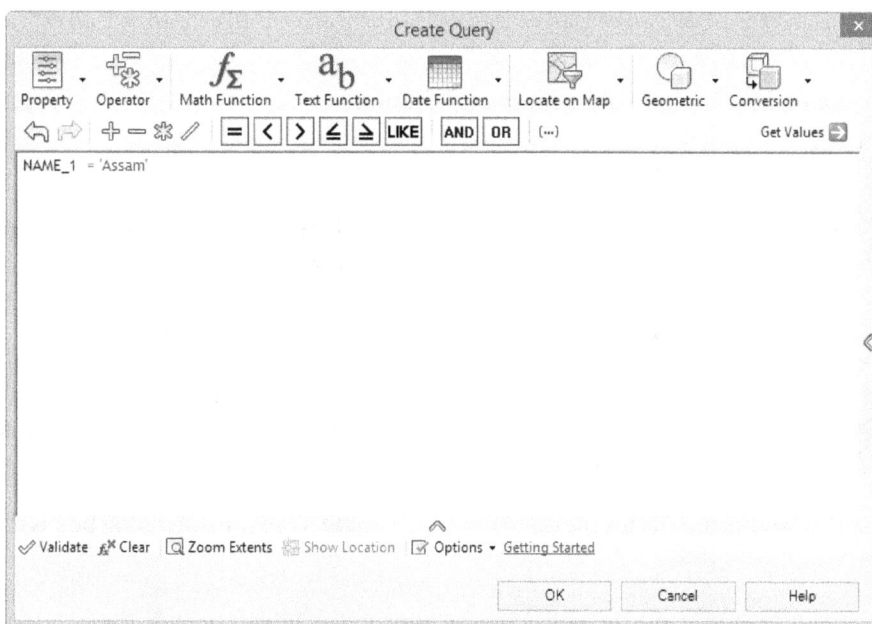

Figure 2-30 The **Create Query** *window displaying the query*

Saving and Closing the Drawing File

1. Choose the **Save** tool from the Application Menu; the **Save Drawing As** dialog box is displayed.

2. In this dialog box, browse to the location *C:\m3d_2017\c02_m3d_2017_tut\c02_tut02* and enter **c02_Tut02a.dwg** in the **File name** edit box.

3. Select the **AutoCAD 2013 Drawing (*.dwg)** option in the **Files of type** drop-down list located at the bottom of the **Save Drawing As** dialog box, if not selected by default.

4. Choose the **Save** button; the drawing file is saved.

Tutorial 3	Using the eTransmit Tool

In this tutorial, you will transfer drawing files through web services by using the **eTransmit** tool.   **(Expected time: 20 min)**

The following steps are required to complete this tutorial:

a. Open the drawing file.
b. Create a transmittal package by using the **eTransmit** tool.
c. Send an e-mail with transmittal package attached to it.

Opening the Drawing File

1. Choose the **Open** button from the Quick Access Toolbar; the **Select File** dialog box is displayed.

2. In this dialog box, browse to the following location:

    C:\m3d_2017\c02_m3d_2017_tut\c02_tut03

3. Select the **c02_Tut03.dwg** file from the **c02_tut03** folder and then choose the **Open** button in this dialog box; the drawing is displayed in the drawing window.

Creating the Transmittal Package by Using the eTransmit Tool

In this section of the tutorial, you will create a zip file (transmittal package) for sharing data.

1. Choose **Send > eTransmit** from the Application Menu; the **eTransmit - Save Changes** message box is displayed prompting you to save the drawing before continuing.

2. Choose the **Yes** button in this message box; the **Create Transmittal** dialog box is displayed, as shown in Figure 2-31.

3. In this dialog box, expand the **JPEG Image** node in the **Files Tree** tab; the files in this node are displayed.

4. Clear the check box corresponding to the **AustinTX.jpg** subnode.

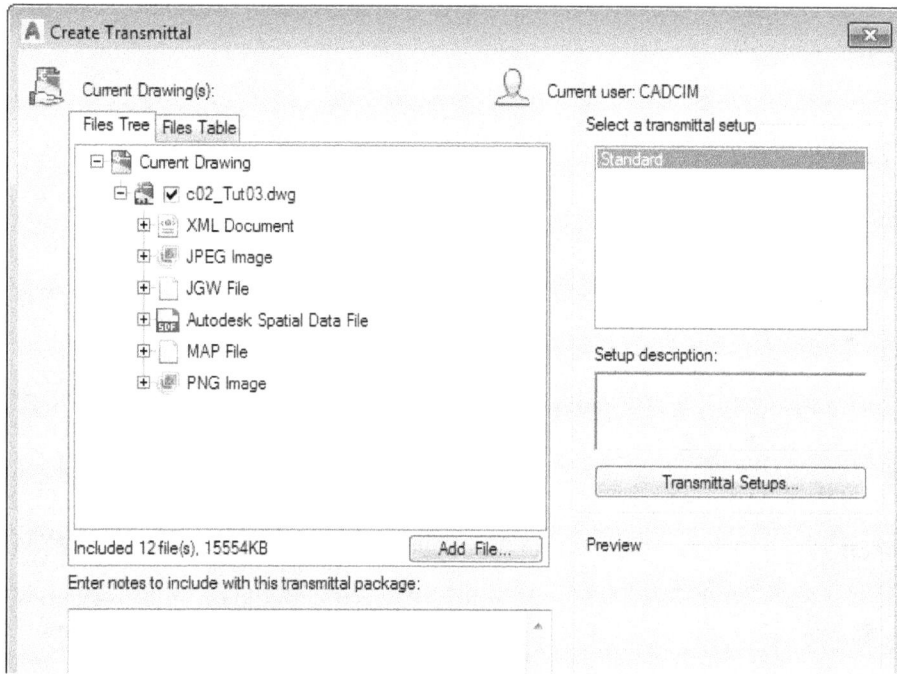

*Figure 2-31 Partial view of the **Create Transmittal** dialog box with files available for transmittal*

5. Choose the **OK** button in the **Create Transmittal** dialog box; the **Specify Zip File** dialog box is displayed.

6. In this dialog box, browse to the following location:

    C:\m3d_2017\c02_m3d_2017_tut\c02_tut03

7. Enter **c02_Tut03a** in the **File name** edit box and then choose the **Save** button; the **Archive Package Creation is in Progress** message box is displayed showing the progress of the file creation process. After completion of the process, the message box automatically closes and the compressed file is saved in the specified folder.

Sending the E-mail with the Current Drawing File Attached to It

1. Choose **Send > Email** option from the Application Menu, the **Microsoft Outlook** window will open, refer to Figure 2-32.

Note
*To send a file using the **Email** tool in the Application Menu of **AutoCAD Map 3D**, you need to install and configure Microsoft Outlook.*

2. Enter the email address of the recipients and then send the mail. Close the drawing by choosing **Close** from the Application Menu.

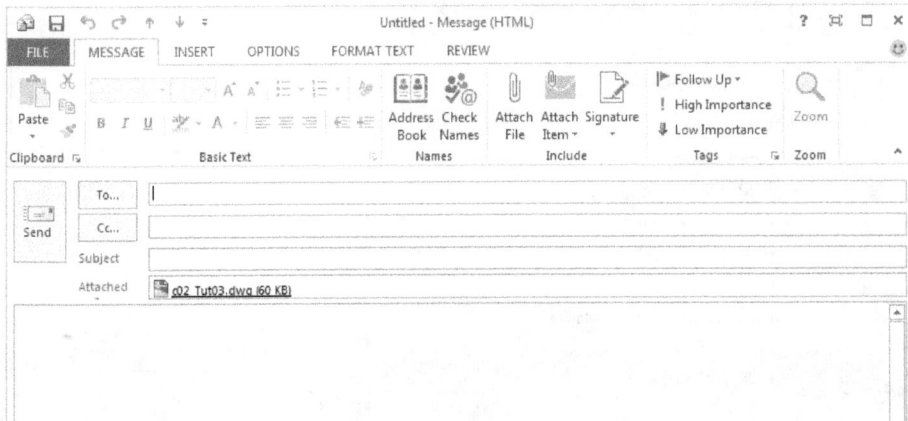

Figure 2-32 *Partial view of the* ***Microsoft Outlook*** *window*

Tutorial 4   Loading Survey Data in the Data Store

In this tutorial, you will load the Ground Control Points (GCP) file or the survey point data by using the **Survey** tab in the **TASK PANE**.   **(Expected time: 30 min)**

The following steps are required to complete this tutorial:

a.   Start a new drawing file.
b.   Create a Data Store with the name **Map-3D-2017_Data-Store** and assign a coordinate system.
c.   Import the survey point file.
d.   Save the file.

Starting a New Drawing File

1.   Choose the **New** option from the Application Menu; the **Select template** dialog box is displayed.

2.   In this dialog box, select the **map2d** template file in the list box below the **Look in** drop-down list, and then choose the **Open** button; the **map2d** template is applied to the modeling space.

Creating a New Data Store

In this section of the tutorial, you will create a survey data store for repositing survey data.

1.   Choose the **Survey** tab in the **TASK PANE**, if not chosen by default.

2.   Choose the **Data** button in the **Survey** tab of the **TASK PANE**; a pop-up menu is displayed.

3.   From this pop-up menu, choose the **New Survey Data Store** option; the **New Data Store** dialog box is displayed, as shown in Figure 2-33.

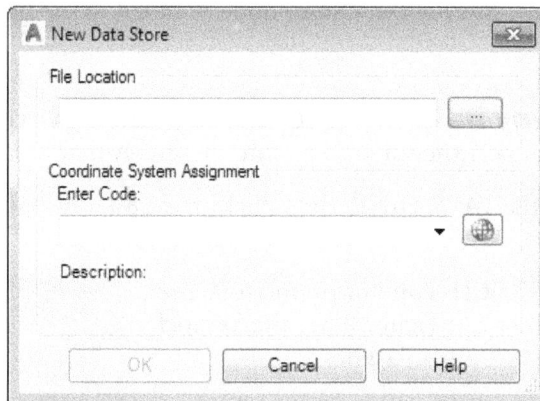

Figure 2-33 The New Data Store dialog box

4. In this dialog box, choose the browse button in the **File location** area; the **Create New Survey Data Store** dialog box is displayed.

5. In this dialog box, browse to the following location:

    C:\m3d_2017\c02_m3d_2017_tut\c02_tut04

6. Enter **Map-3D-2017_Data-Store** in the **File name** edit box.

7. Select the **SDF files (*.sdf)** option from the **Save as type** drop-down, list if not selected by default.

8. Choose the **OK** button; the **Create new Survey Data Store** dialog box is closed and the path of the file to be saved is displayed in the **File location** edit box of the **New Data Store** dialog box.

9. Next, choose the **Select Global Coordinate** button in the **Coordinate System Assignment** area in the dialog box; the **Coordinate System Library** dialog box is displayed.

10. In this dialog box, type **43 N** in the **Search** edit box; a list of available coordinate systems for the specified search parameter will be displayed in the list box below the **Search** edit box.

11. In the list box, select the **WGS72.UTM-43N** code with the description **WGS 72/UTM zone 43N**.

12. Choose the **Select** button in the dialog box; the **Coordinate System Library** dialog box is closed and the **WGS72.UTM-43N** code is displayed in the edit box of the **Coordinate System Assignment** area in the **New Data Store** dialog box. Also, the coordinate description is displayed below the edit box in the **Coordinate System Assignment** area.

13. Next, choose the **OK** button in the **New Data Store** dialog box; the **Map-3D-2017_Data-Store** is created and added to the **Current Data Store** drop-down list in the **Survey** tab of the

TASK PANE. Also, the **Survey Data Store** node is added to the list box in the **Survey** tab of this pane.

Importing the Survey Data into the Survey Data Store
In this section, you will import the survey data into the survey data store.

1. In the **Survey** tab of the **TASK PANE**, right-click on the **Survey Data Store** node; a shortcut menu is displayed.

2. Choose the **Import ASCII Points** option from the shortcut menu, as shown in Figure 2-34; the **Import ASCII Points** dialog box is displayed.

3. In this dialog box, choose the Browse button in the **File location** area; the **Import ASCII File** dialog box is displayed.

4. In the **Import ASCII File** dialog box, select the *CHIK_TotalStation.asc* file from the following location:

    C:\m3d_2017\c02_m3d_2017_tut\c02_tut04

5. Next, choose the **OK** button; the **Import ASCII File** dialog box is closed. Notice that the path of the selected file is displayed in the **File Location** edit box of the **Import ASCII Points** dialog box. Also, preview of the selected data is displayed in the **Preview** area, as shown in Figure 2-35.

*Figure 2-34 Choosing the **Import ASCII Points** option from the shortcut menu*

6. Choose the **OK** button; the **Import ASCII Points** dialog box is closed and the survey points or the Ground Control Points are displayed in the drawing window, as shown in Figure 2-36.

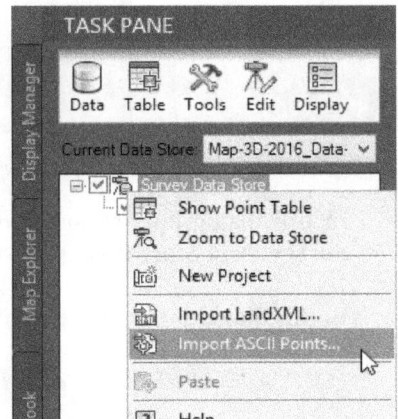

Note
*If the survey data is not displayed in the drawing window, then right-click on the Map-3D-2017_Data-Store vector layer in the **Display Manager** tab of the **TASK PANE**; a shortcut menu will be displayed. From this shortcut menu, choose the **Zoom to Extents** option; the survey data will be displayed in the drawing window.*

Saving the Drawing File
1. Choose the **Save As** tool from the Application Menu; the **Save Drawing As** dialog box is displayed.

2. In this dialog box, browse to the location *C:\m3d_2017\c02_m3d_2017_tut\c02_tut04* and enter **c02_Tut04a.dwg** in the **File name** edit box.

3. Select the **AutoCAD 2013 Drawing (*.dwg)** option from the **Files of type** drop-down list located at the bottom of the **Save Drawing As** dialog box if not selected by default.

4. Choose the **Save** button; the drawing file is saved.

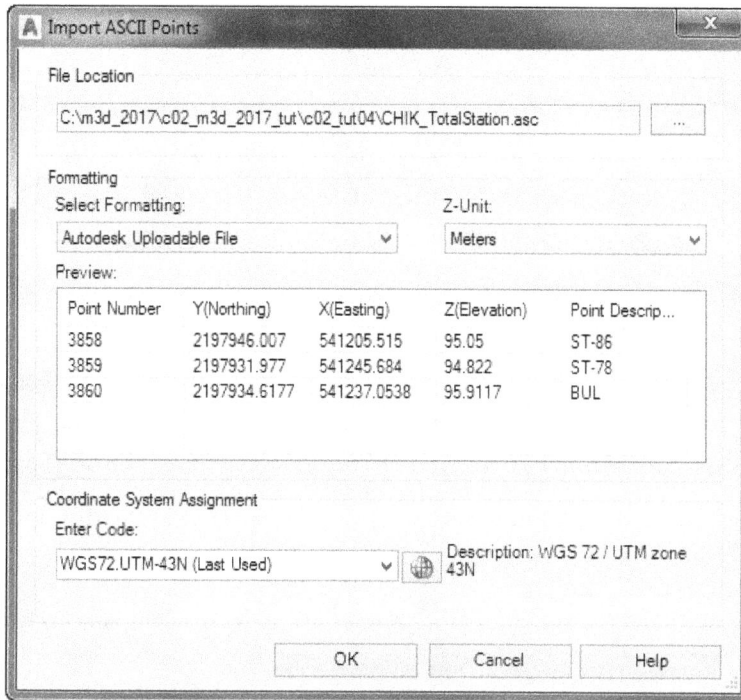

*Figure 2-35 The **Import ASCII Points** dialog box displaying the preview of the selected data*

Figure 2-36 The Ground Control Points displayed in the drawing window

Tutorial 5   Exporting DWG data as SDF data

In this tutorial, you will export the DWG data from the AutoCAD drawing file to SDF data
format.   **(Expected time: 45 min)**

The following steps are required to complete this tutorial:

a. Open a drawing file with the name *m3d_c02_Tut05.dwg*.
b. Create and export the drawing objects as SDF feature classes using the **MADDWGTOSDF** command.
c. Start a new drawing file.
d. Explore the exported SDF file.
e. Save the file as *c02-m3d-2017-Tut05a.dwg*.

Opening the Drawing File
1. Choose **Open > Drawing** from the Application Menu; the **Select file** dialog box is displayed.

2. In the **Select file** dialog box, select the **m3d_c02_Tut05.dwg** file from the following location:

    C:\m3d_2017\c02_m3d_2017_tut\c02_tut05

3. Next, choose the **Open** button; the drawing file is displayed in the drawing window.

Creating Feature Classes
In this section, you will create a set of feature classes using the drawing objects from the drawing file.

1. Enter the command **MAPDWGTOSDF** in the command line and press ENTER; the **Export Location** dialog box is displayed.

2. In this dialog box, enter **New SDF File** in the **File name** edit box. Next, browse to the following location:

    C:\m3d_2017\c02_m3d_2017_tut\c02_tut05

3. Now, choose the **OK** button; the dialog box is closed and the **Export** dialog box is displayed with the **Selection** tab chosen, as shown in Figure 2-37.

4. In the **Select objects to export** area of the **Selection** tab, choose the **Select manually** radio button.

5. Next, choose the button next to the **Select manually** radio button; the **Export** dialog box closes and you are prompted to select objects to be exported.

6. Draw a selection box in the drawing window to select all the drawing objects. Press ENTER to end selection; the **Export** dialog box is displayed again.
The objects for creating SDF data have now been selected. Next, you need to specify how these objects will be mapped in the SDF data.

7. Choose the **Feature Class** tab in the **Export** dialog box. In the **Object to Feature Class Mapping** area of this tab, select the **Create multiple classes based on a drawing object** radio button.

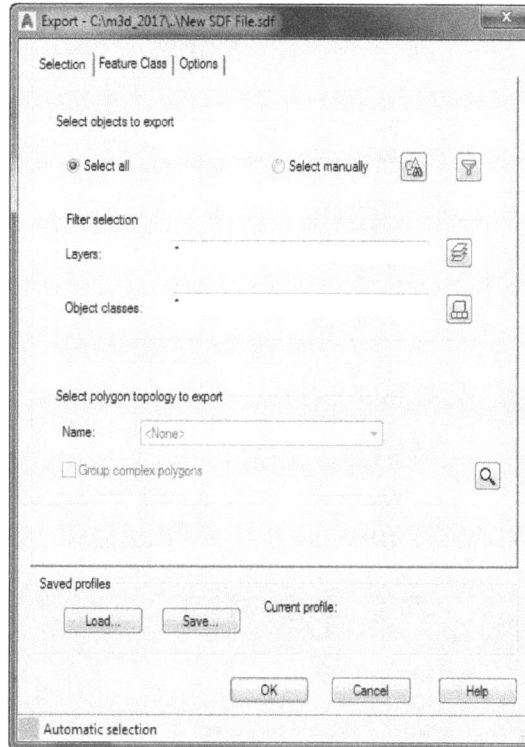

Figure 2-37 *The* *Export* *dialog box with the* *Selection* *tab chosen*

Note that all the layers in the drawing are displayed in the list box.

8. Next, clear the check box corresponding to **0** drawing object in the **Export** dialog box.

9. Click in the **Geometry** cell corresponding to the **Internal Road** drawing object; a drop-down list will be displayed.

10. Next, select the **Line** option from this drop-down list.

11. Repeat the procedure given in steps 9 and 10 and select the **Polygon** and **Line** geometry options for the **Parcels** and **Utility Lines** drawing objects respectively, refer to Figure 2-38.

12. Choose the **Select Attributes** button in the **Export** dialog box; the **Select Attributes** dialog box is displayed. Expand the node **Object Properties > AcDbEntity > General**. Select the check box corresponding to the **Color**, **Linetype**, and **Lineweight** options, refer to Figure 2-39.

13. Next, choose the **OK** button; the dialog box is closed.

14. Next, click on the **Internal Roads** cell of the **Feature Class** column in the table; a browse button is displayed in the cell, as shown in Figure 2-40.

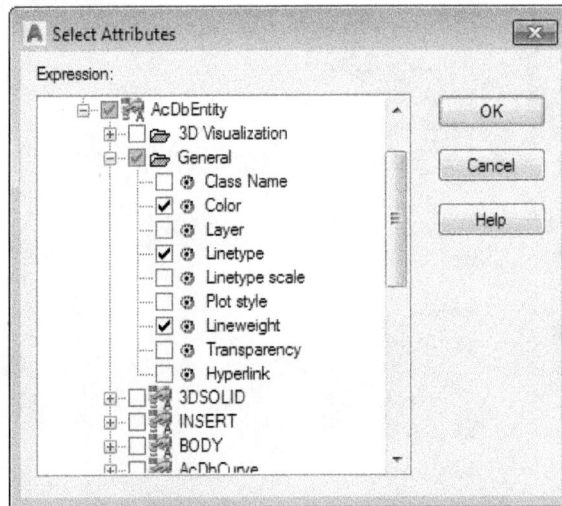

Figure 2-38 *The **Select Attributes** dialog box showing various options selected*

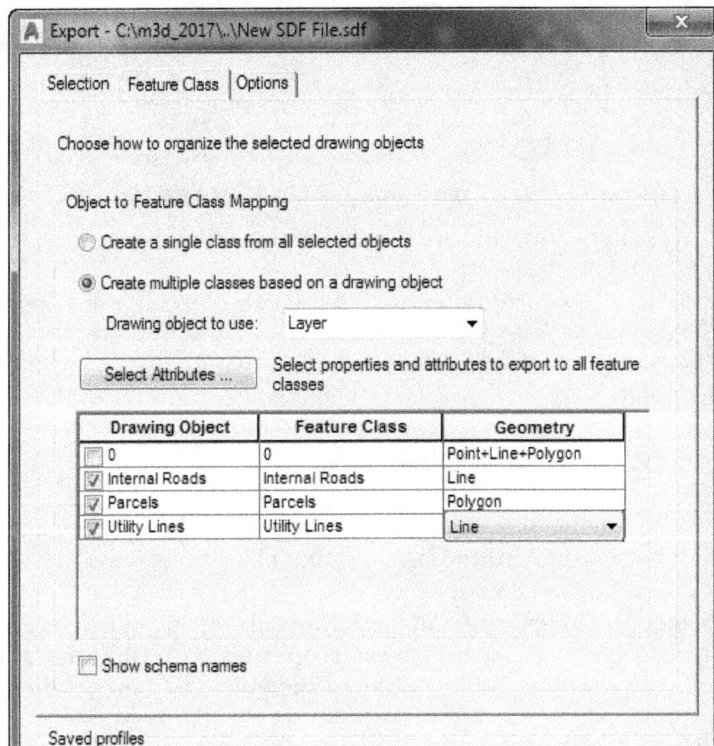

Figure 2-39 *Partial view of the **Export** dialog box showing the **Geometry** option selected for the drawing object*

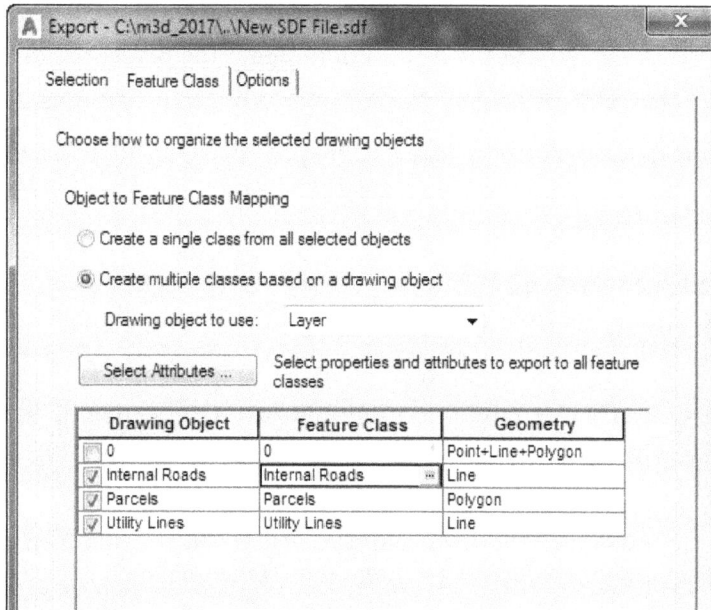

Figure 2-40 *The partial view of the* ***Export*** *dialog box showing the browse button displayed in the* ***Internal Roads*** *cell*

15. Choose the displayed Browse button; the **Feature Class Property Mapping - Internal Roads** dialog box is displayed.

16. In this dialog box, choose the **Select Attributes** button; the **Select Attributes** dialog box is displayed. In this dialog box, expand the **Object Data** node and then select the **Roads** check box.

17. Now, choose the **OK** button; the **Select Attributes** dialog box is closed. Note that the data columns in the Roads object data table are displayed in the **Feature Class Property Mapping - Internal Roads** dialog box.

18. Next, choose the **OK** button from the **Feature Class Property Mapping - Internal Roads** dialog box; the dialog box is closed.

19. Repeat the procedure given in steps 14 to 18 and create feature classes for the **Parcels** and **Utility Lines**. In the **Select Attributes** dialog box, select the **Parcels** check box for the **Parcels** feature class and the **Waterlines** check box for the **Utility Lines** feature class.

    The drawing objects are now mapped to the feature class. Now, you need to specify the options for exporting.

20. Choose the **Options** tab in the **Export** dialog box. In the **Other** area of this tab, select the **Treat closed polylines as polygons** check box.

21. Next, choose the **OK** button; the **Export** dialog box is closed and the **Export Process** message box is displayed showing the progress of the data being exported.

Starting a New Drawing File

1. Choose **New > Drawing** from the Application Menu; the **Select template** dialog box is displayed.

2. In the **Select template** dialog box, select the **map2d.dwt** template file in the list box below the **Look in** drop-down list and then choose the **Open** button; the **map2d.dwt** template is applied to the modeling space.

Exploring the Exported SDF file

1. Choose the **Connect** tool from the **Data** panel in the **Home** tab; the **DATA CONNECT** wizard is displayed.

2. In the **DATA CONNECT** wizard, select the **Add SDF Connection** option from the **Data Connections by Provider** list box; the **OSGeo FDO Provider for SDF** page is displayed in the right pane of the wizard.

3. In this page, enter **New SDF** in the **Connection name** edit box and then choose the browse button next to the **Source file** edit box; the **Open** dialog box is displayed.

4. In the dialog box, browse to the location *C:\m3d_2017\c02_m3d_2017_tut\c02_tut05* and select **New SDF File**. Next, choose the **Open** button from this dialog box; the path of the selected file is displayed in the **Source file** edit box.

5. Choose the **Connect** button; the **SDF** page is displayed in the right pane of the wizard.

6. Next, select the **Schema1** check box in this page and then choose the **Add to Map** button; the **New SDF file** is displayed in the drawing window. Figure 2-41 shows the **Display Manager** tab of the **TASK PANE** with the SDF data file. Next, close the **DATA CONNECT** wizard.

Figure 2-41 Partial view of the Display Manager tab in the TASK PANE

Saving and Closing the Drawing File

1   Choose the **Save As** tool from the Application Menu; the **Save Drawing As** dialog box is displayed.

2.   In the **Save Drawing As** dialog box, browse to the location *C:\m3d_2017\c02_m3d_2017_tut\ c02_tut05* and enter **c02_Tut05a.dwg** in the **File name** edit box.

3.   Choose the **Save** button; the drawing file is saved.

Self-Evaluation Test

Answer the following questions and then compare them to those given at the end of this chapter:

1.   Which of the following data connections is used to access Autodesk's spatial data format?

   (a) **Add SDF Connection**   (b) **Add SHP Connection**
   (c) **Add WFS Connection**   (d) **Add WMS Connection**

2.   Which of the following tabs in the **TASK PANE** displays the layer of the feature data in the map?

   (a) **Survey**   (b) **Map Book**
   (c) **Display Manager**   (d) **Map Explorer**

3.   Which of the following tools in the Application Menu is used to edit or modify the settings of the added plotters?

   (a) **Plot Preview**   (b) **Plot**
   (c) **Page Setup**   (d) **Manage Plotters**

4.   Which of the following tools in the **Map Setup** tab is used to assign coordinate system to the drawing?

   (a) **Define**   (b) **Assign**
   (c) **Attach**   (d) **Connect**

5.   Which of the following data connections is used to access the shape file format?

   (a) **Add SDF Connection**   (b) **Add SHP Connection**
   (c) **Add WFS Connection**   (d) **Add WMS Connection**

6.   You can load the survey Ground Control Points by using the _____ tab from the **TASK PANE**.

7.   You can use the tools in the _____ panel of the **Insert** tab to bring the data into your map.

8. The _____ tool in the Application Menu creates a zip file for the drawing and its dependencies.

9. The _____ tool in the Application Menu is used to publish drawings in the Autodesk Infrastructure Map Server.

10. Using the _____ option in the **DATA CONNECT** wizard, you can filter data that is to be added to the map.

11. You can create a sheet set by using an example sheet set or by using the existing drawings. (T/F)

12. You can load various datasets such as raster, vectors and database file into the Workspace by using the **Connect** tool. (T/F)

13. You can use the **Display Manager**, **Map Explorer**, **Map Book** or **Survey** tab to create a new survey data store. (T/F)

14. The **Page Setup** tool is used to set the properties of the plotting device, paper size, and page layout for plotting. (T/F)

15. You can publish data to the online server using the tools in the **Publish** option of the Application Menu. (T/F)

Review Questions

Answer the following questions:

1. Which of the following tools in the **Output** tab of the Ribbon is used to publish drawings using plotters?

    (a) **Preview**   (b) **Plot**
    (c) **Page Setup Manager**   (d) **Plotter Manager**

2. Which of the following tools in the **Home** tab is used to attach a drawing file to the current drawing?

    (a) **Export**   (b) **Assign**
    (c) **Attach**   (d) **Connect**

3. To insert a georeferenced image into the drawing space, use the _____ tool from the **Image** panel of the **Insert** tab.

4. To export data to several external file formats, use the _____ option from the Application Menu.

5. You can set the printing and plotting settings as per the user requirement. (T/F)

6. Using the **Connect** tool, you can combine raster data and vector data. (T/F)

7. You can use the **Connect** button in the **Map Explorer** tab of the **TASK PANE** as an alternative to the **Connect** tool in the Ribbon. (T/F)

8. The coordinate system used for conducting survey can be different from the one used in the drawing window. (T/F)

9. A drawing file can be saved with only *.dwg extension. (T/F)

10. You can create a new sheet style using the options in the **Sheet Set Manager** window. (T/F)

EXERCISES

Download *c02_m3d_2017_exe.zip* from *www.cadcim.com* and extract it for the following exercises.

Exercise 1   Loading Shape File

Extract the **c02_exr01** folder from *c02_m3d_2017_exe.zip* and then load the *Municipal.shp* shape file from the extracted folder by using the **Connect** tool. Next, save the file as a drawing file. Create a transmittal file using the **eTransmit** tool.   **(Expected time: 45 min)**

Exercise 2   Loading Survey Data

Extract the **c02_exr02** folder from *c02_m3d_2017_exe.zip*. Next, create a survey data store and then load the *c02-m3d-2017-exr02.txt* file from the extracted folder into the model space by using the options in the **Survey** tab.   **(Expected time: 30 min)**

Exercise 3   Exporting drawing As SDF

Extract the **c02_exr03** folder from *c02_m3d_2017_exe.zip*. Next, open the *c02-m3d-2017-exr03.dwg* file from the extracted folder. Use the **MAPDWGTOSDF** command to export the file as an SDF file. Save the file with the name **c02-m3d-2017-exr03a**.   **(Expected time: 30 min)**

Answers to Self-Evaluation Test

1. a, **2.** c, **3.** d, **4.** b, **5.** b, **6.** Survey, **7.** Import, **8.** eTrasmit, **9.** Publish to Map Server, **10.** Add to Map with Query, **11.** T, **12.** T, **13.** F, **14.** T, **15.** T

Chapter 3

Working with Basic Tools and Coordinate Systems

Learning Objectives

After completing this chapter, you will be able to:

• *Use basic tools for mapping*
• *Use coordinate systems*
• *Modify the settings and alignment of the UCS triad*
• *Create and modify coordinate systems*
• *Assign coordinate system to drawings and datasets*

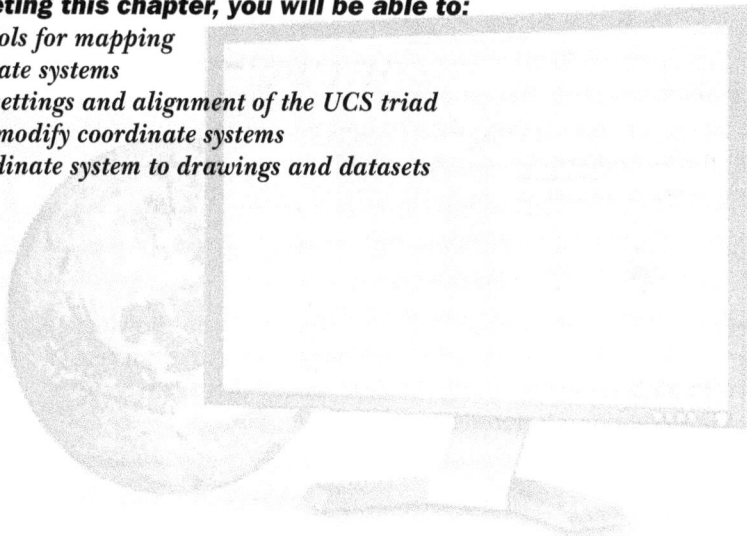

INTRODUCTION

In this chapter, you will learn some of the basic navigation and productivity enhancement tools, such as **Zoom**, **Pan**, and **Snap**, which will be frequently used while creating and analyzing the drawing data. These tools provide a flexible viewing and editing environment thereby enhancing the productivity and accuracy of the spatial database.

In this chapter, you will learn to perform various operations such as zooming, panning, and scaling the drawings. In Map 3D, while creating drawing objects, you will also learn to use the Ortho mode and the snap options.

Further in this chapter, you will learn about the User Coordinate System (UCS) and the World Coordinate System (WCS) in AutoCAD Map 3D. Also, the coordinate system for referencing drawing objects to the real world will be discussed in this chapter.

USING THE BASIC NAVIGATION TOOLS

The mapping procedures, such as editing and reviewing data, frequently require the drawing data to be scaled, zoomed, and panned. These operations require the use of the **Pan** and **Zoom** tools which makes them the most extensively used tools in AutoCAD Map 3D. Some of these frequently used navigation tools are discussed next.

Zoom Tools

Command: ZOOM or Z

In GIS, sometimes you may need to check the minute spatial details of the data displayed, and alter it, if required. For carrying out such tasks, using an enlarged view of the objects in the drawing area would be advantageous. The zoom tools are used to enlarge or reduce the view of a drawing without affecting the actual size of the drawing objects. You can use various zoom tools such as **Extents**, **Realtime**, **Window**, **Previous**, **Object**, and **All** to view a particular area of the drawing in an enlarged view/window.

To use different zoom tools in the drawing process, click on the down-arrow next to the **Extents** tool in the **Navigate** panel of the **View** tab; all the zoom tools will be displayed in a drop-down. Figure 3-1 shows the zoom tools available in the drop-down. Choose the required zoom tool from the drop-down to proceed. Some of the important zoom tools are discussed next.

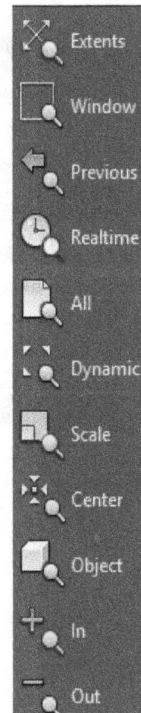

Extents

The **Extents** tool is used to display all the objects in the drawing within the extent of the current drawing window. To display all the drawing objects within the drawing, choose the **Extents** tool in the **Navigate** panel from the **View** tab; the selected data will get enlarged.

Figure 3-1 *Various zooming tools*

> **Tip**
> *You can use the **Zoom to Extents** option to display all the data in the required feature layer. To do so, right-click on the required layer in the **TASK PANE**; a shortcut menu will be displayed. Choose the **Zoom to Extents** option from the displayed menu; the drawing will zoom to display all the data within the extent of the selected layer.*

Window

You can zoom in a region of a drawing by drawing a window around the area to be zoomed. To do so, choose the **Window** tool from **View > Navigate > Extents** drop-down; you will be prompted to specify the first corner of the area to be zoomed. Click in the drawing; you will be prompted to specify the opposite corner. Specify the other corner. You can also specify the two corner points of the rectangular area by entering the coordinates in the Command prompt. On specifying the area, the center of the specified area (window) will become the center of the new display screen. When you use this zoom tool, the area inside the window gets magnified and fills the drawing window.

Previous

While working on a complex drawing, you may need to zoom in on a portion of the drawing to edit some minute details of the drawing object. After you have completed the editing, you may want to return to the previous view. This can be done by using the **Previous** tool. AutoCAD remembers the last ten views, and these views can be restored by using the **Previous** tool.

Realtime

You can use the **Realtime** tool to zoom in and zoom out a drawing interactively by using the left mouse button. To do so, choose the **Realtime** tool from **View > Navigate > Extents** drop-down; the **Realtime** tool will be invoked. Next, press and hold the left mouse button and then drag the cursor up or down to zoom in or zoom out the drawing. To exit the **Realtime** zooming mode, right-click in the drawing window; a shortcut menu will be displayed. Choose the **Exit** option from this menu. You can also press ESC or ENTER to exit the mode.

> **Tip**
> *You can also use the scroll wheel to zoom in or out of the drawing.*

All

This tool is used to zoom to the drawing limits or the extents of the objects in the drawing window, whichever is greater. Whenever you increase the limits, the current display is not affected and hence it is not displayed. In this case, you need to use the **All** tool to display the limits of the drawing. Sometimes it is possible that the objects are drawn beyond the limits. In such a case, the **All** tool zooms to fill the drawn objects in the drawing area, irrespective of their limits.

Scale

This tool is commonly used while plotting a drawing. This tool allows you to zoom into the drawing using the specified scale. To zoom a drawing using this option, choose the

Scale tool from **View > Navigate > Extents** drop-down; you will be prompted to specify the scale factor. Enter the scale factor and then press ENTER; the drawing will be zoomed to the specified scale.

Center

This tool is used to increase or decrease the magnification of the view in the current viewport. To zoom in or zoom out from the view using this option, choose the **Center** tool from **View > Navigate > Extents** drop-down; you will be prompted to specify the center. Click in the drawing to specify the center point; you will be prompted to specify the magnification or height. Specify the magnification and press ENTER; the drawing will zoom to the area with the specified point at its center and at the given magnification.

Object

You can use the **Object** tool from the **Extents** drop-down to zoom the selected object/s. When you choose the **Object** tool, you will be prompted to select the object/s. You can select the required objects in the current drawing by using the selection box or by clicking on individual object/s. After selecting the object, press ENTER; the drawing window will zoom the selected objects.

In

You can use the **In** tool from the **Extents** drop-down to increase the apparent size of the object. When you choose this tool, you will be prompted to specify corner of the window or the scale factor. Note that here the absolute units of the objects will remain unchanged and onl their apparent size will increase.

Out

You can use the **Out** tool from the **Extents** drop-down to decrease the apparent size of the object. When you choose this tool, you will be prompted to specify corner of the window or the scale factor. Note that here the absolute units of the objects will remain unchanged and only their apparent size will decrease.

Pan Tool

Ribbon:	View > Navigate > Pan
Command:	PAN

You can use the **Pan** tool to pan a drawing interactively. This means you can shift a drawing by sliding it, and then placing it at the required position. To slide a drawing, choose the **Pan** tool from the **Navigate** panel; the cursor will change into a hand symbol, indicating that you are in the pan mode. Place the cursor in the drawing area and then press and hold the left mouse button. Now, you can drag the cursor to the required place on the screen to move the drawing. To exit the pan mode, right-click in the drawing; a shortcut menu will be displayed. Next, choose the **Exit** option from displayed menu. You can also press the ESC or ENTER key to exit the pan mode.

AutoCAD Map 3D Scale

Status Bar: AutoCAD Map 3D Scale

The **AutoCAD Map 3D Scale** tool in the Status Bar is used to set the scale of a layer or a view. If you change this scale, the zoom level of the drawing will also get changed. To zoom using the **AutoCAD Map 3D Scale** tool, click on the down arrow in the Status Bar displayed on the right of the **AutoCAD Map 3D Scale** tool; a pop-up menu will be displayed, as shown in

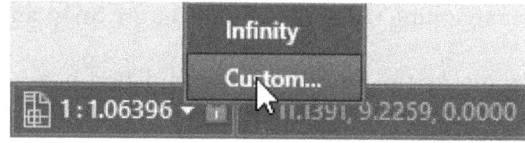

*Figure 3-2 The **AutoCAD Map 3D Scale** tool with the pop-up menu displayed and the **Custom** option chosen*

Figure 3-2. Choose the **Custom** option from the pop-up menu; the scale editing window will be displayed. In this window, you can specify the scale of the drawing viewport. To do so, enter the required value in the **Enter scale value** edit box. Next, choose the **OK** button in this window; the drawing will be displayed at the modified scale value.

If you increase the scale value by using the **AutoCAD Map 3D Scale** tool, the size of the drawing view will be reduced in the drawing window, as shown in Figure 3-3. If you decrease the scale value by using this tool, the size of the drawing view in the drawing window will be magnified.

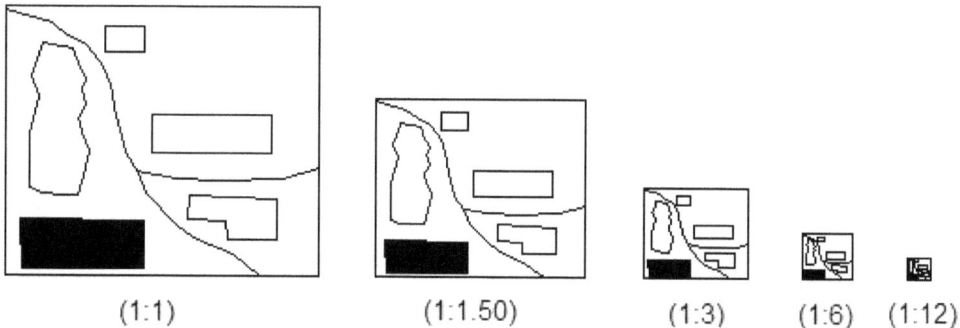

(1:1)   (1:1.50)   (1:3)   (1:6)   (1:12)

Figure 3-3 The drawing views at different scale values

Note
*1. If a drawing view is altered by using a **Zoom** tool, the scale value in the **AutoCAD Map 3D Scale** tool will also be altered accordingly.*

*2. Any modification made in a drawing view by using the **AutoCAD Map 3D Scale** or **Zoom** tool will not affect the physical measurements of the drawing.*

USING THE SNAP FUNCTIONS IN MAP 3D

In AutoCAD Map 3D, you can snap the cursor or snap to a desired point in an object using the snap functions. If you activate the snap function while in the drawing mode, you can track some of the known object points, vertices, or nodes in the drawing. AutoCAD Map 3D provides two basic types of snap functionalities, grid snap and object snap. These snap functions are discussed next.

Snapping Cursor (Using Snap Spacing)

The **Snap Mode** button from the Status Bar is used to activate the function that allows the
user to snap to the points using specified snap spacing. You can specify the parameters
for snapping using the options in the **Snap and Grid** tab of the **Drafting Settings** dialog box.

To invoke the **Drafting Settings** dialog box, click on the down-arrow next to the **Snap Mode**
button; a menu will be displayed. Choose the **Snap Settings** option; the **Drafting Settings** dialog
box will be displayed, as shown in Figure 3-4. The method of setting the properties in the **Snap
and Grid** tab of this dialog box is discussed next.

Note
*Choosing the **Grid Mode** button from the Status Bar will result in the display of the grid
along the XY plane in the drawing window.*

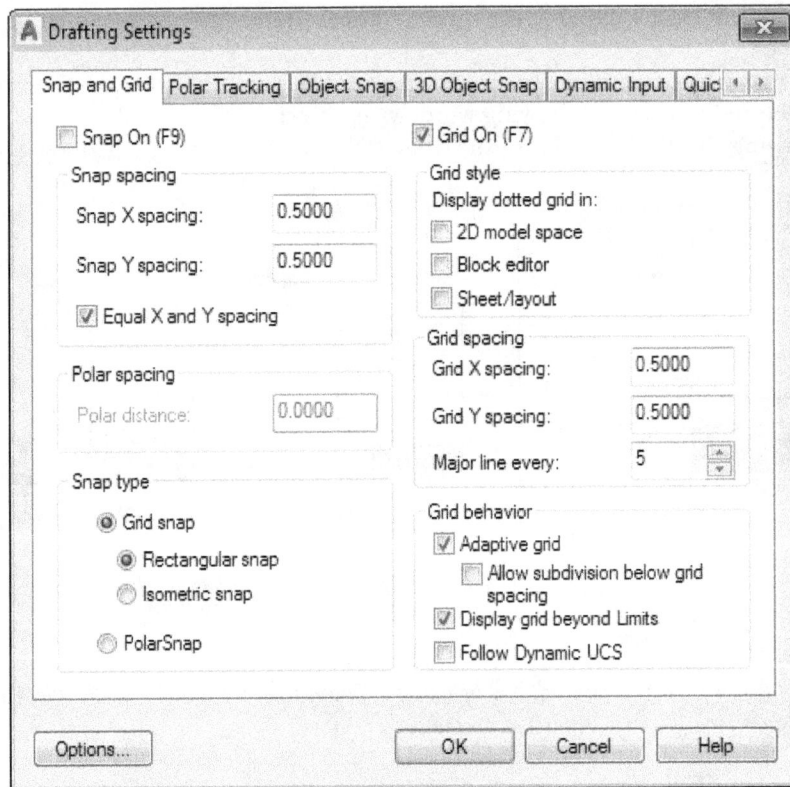

*Figure 3-4 The **Drafting Settings** dialog box with the **Snap and Grid** tab chosen*

Snap and Grid Tab

Choose the **Snap and Grid** tab in the **Drafting Settings** dialog box to display various options.
The **Snap On (F9)** check box in this tab is clear by default, which allows the cursor to move
freely in the drawing window. Selecting the **Snap On (F9)** check box will activate the Snap Mode.
As a result, the cursor will snap to the points at the specified snap spacing. The **Grid On (F7)**
check box is also clear by default in this tab. Therefore, the grid pattern will not be displayed.

You can select the **Grid On (F7)** check box to display the grid pattern in the drawing window. The different areas in this tab are discussed next.

> **Tip**
> *You can display the grid lines in any 2D drawing space by choosing the **Grid Mode** button in the Status Bar. Similarly, you can invoke the snapping mode by choosing the **Snap Mode** button in the Status Bar.*

Grid style Area

The options in the **Grid style** area are used to specify the display style of the dotted grid pattern in the 2D model space, block editor, or the sheet or layout modeling space. You can select the **2D model space** check box to display the dotted grid pattern in the 2D drawing environment. Similarly, you can select the **Block editor** check box to display the dotted grid pattern while editing blocks.

Grid spacing Area

The options in the **Grid spacing** area are used to adjust the spacing between grid points along the X and Y axes. To modify the spacing between grid points along the X axis, enter the required value in the **Grid X spacing** edit box. Similarly, to modify the spacing between grid points along the Y axis, enter the required value in the **Grid Y spacing** edit box. To modify the number of minor lines between two major lines while working in the 3D sketching mode, enter a value in the **Major line every** edit box or set the value using the spinner located next to it.

Grid behavior Area

The options in the **Grid behavior** area are used to modify the settings for the grid display. The **Adaptive grid** check box in this area is selected by default. As a result, the number of grid lines or grid line density between the major grid lines will be limited during zoom out. Select the **Allow subdivision below grid spacing** check box in the **Adaptive grid** option in the **Grid Behaviour Area** if you want to display subdivision lines in the minor grid lines, while zooming in the drawing. To display grids beyond the limits of window, select the **Display grid beyond Limits** check box. To attach the grid plane to the current UCS, select the **Follow Dynamic UCS** check box.

Snap spacing Area

The options in the **Snap spacing** area are used to set the snap spacing in the grid mode. To modify the snapping distance between two snap points along the X-axis, enter the required value in the **Snap X spacing** edit box. Similarly, to set the snapping distance between two snap points along the Y-axis, enter the required value in the **Snap Y spacing** edit box. To set different snapping spaces between snap points along the X and Y axes, clear the **Equal X and Y spacing** check box. Various methods to customize the snap spacing are discussed next.

Snap to Grid: In the snap to grid mode, if you move the cursor in the XY-plane, the cursor will snap to the adjacent grid point, refer to **(1)** in Figure 3-5.

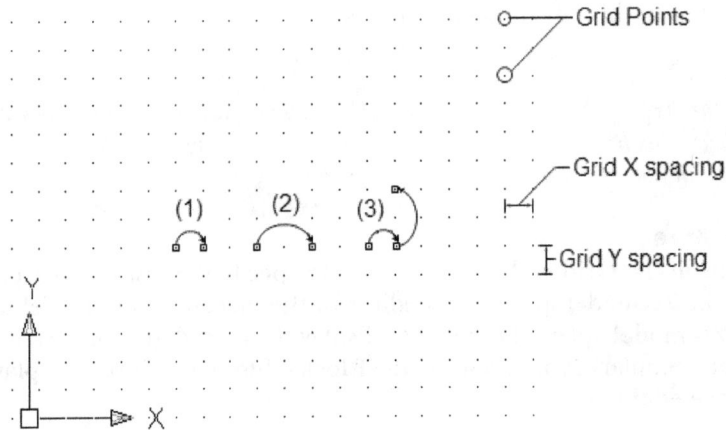

Figure 3-5 Different grid snapping options with grid details

Snap to Spacing: In the snap to spacing mode, if you modify the snap spacing without changing the grid spacing, the cursor will snap to the point located at the modified snap spacing, refer to **(2)** in Figure 3-5.

Snap to Unequal X Spacing and Y Spacing: In the snap to unequal X spacing and Y spacing, if the snap spacing in the X axis and the Y axis are different, then the cursor will trace the next grid point with the corresponding snap spacing given in the X and Y axes, refer to **(3)** in Figure 3-5.

Snap type Area

The options in the **Snap type** area are used to specify the snap type. To snap the cursor to grid points in the X and Y axes, select the **Grid snap** radio button. In the grid snap mode, you can choose an option depending on your drawing requirement. To draw a geometric model in the rectangular snap mode, select the **Rectangular snap** radio button. To draw a geometric model in the isometric snap mode, select the **Isometric snap** radio button. While working in the **Polar Tracking** mode, you can set the cursor to snap along the polar alignment angles by selecting the **PolarSnap** radio button.

Object Snap

The object snapping helps the users to snap to the precise locations on the objects. You can activate object snapping by pressing the F3 function key. AutoCAD Map 3D identifies various locations such as end point, midpoint, center, node and intersection as snap locations. You can choose the required snap modes from the **Object Snap** tab of the **Drafting Settings** dialog box. The various options in the **Object Snap** tab are discussed next.

Object Snap Tab

The options in the **Object Snap** tab are used to snap to geometric points on a drawing object. In this tab, the **Object Snap On (F3)** and **Object Snap Tracking On (F11)** check boxes are selected by default. Additionally, you can toggle the object snap and object snap tracking on or off. The options in the **Object Snap modes** area are used to control the object snap tracking types. Some of the frequently used object snap tracking options are discussed next.

Tip
*You can also turn on an object snap option by right-clicking on the **Object Snap** button in the Status Bar, and then choosing the required snapping option from the shortcut menu displayed.*

Endpoint

The **Endpoint** object snap tracking option is used to draw an object with reference to an endpoint of another drawing object. If you select the **Endpoint** check box, the crosshair will trace the endpoint of the drawing object in the drawing mode. The endpoint object snapping is illustrated in Figure 3-6.

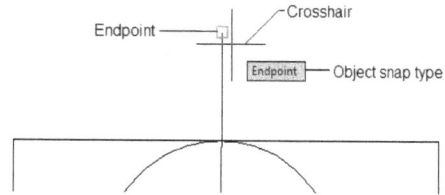

Figure 3-6 The Endpoint object snapping

Midpoint

The **Midpoint** object snap option is used to draw an object with reference to the middle point of a line drawing. If you select the **Midpoint** check box, the crosshair will snap to the middle point of a nearby line drawing. The midpoint object snapping is illustrated in Figure 3-7.

Center

The **Center** object snap option is used to track the center of a circle, arc, ellipse, or elliptical arc and then draw objects with reference to the center. To use this option in the drawing, select the **Center** check box. Figure 3-8 shows an example of the center object snapping.

Tangent

The **Tangent** object snap option is used to track a tangent point along an arc and a circular geometry. To invoke this option, select the **Tangent** check box. The tangent object snapping is illustrated in Figure 3-9.

Quadrant

The **Quadrant** object snap option is used to create a drawing with respect to one of the four quadrant points. To invoke this option, select the **Quadrant** check box. Figure 3-10 illustrates an example of using the **Quadrant** object snapping option.

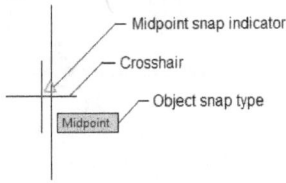

Figure 3-7 *The Midpoint object snapping*

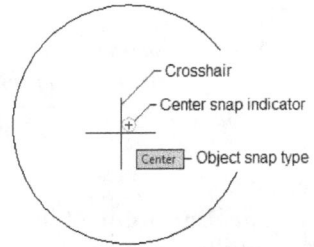

Figure 3-8 *The Center object snapping*

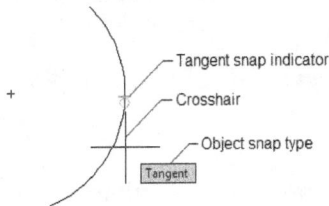

Figure 3-9 *The Tangent object snapping*

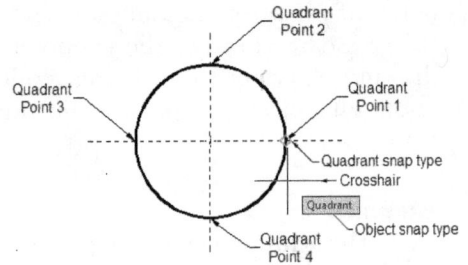

Figure 3-10 *The Quadrant object snap located at the first quadrant*

Node

The **Node** option is used to snap to a point object in the drawing mode. To apply the node object snapping to the drawing mode, select the **Node** check box. You can draw an object from the point object by using this option. An example of drawn line using three point (nodes) is shown in Figure 3-11.

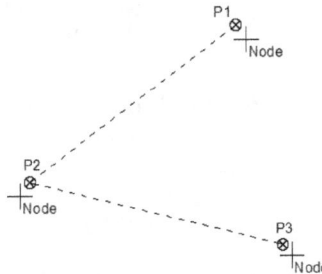

Figure 3-11 *The node object snapping*

Note
*The other object snap options such as **Intersection**, **Extension**, **Insertion**, **Perpendicular**, **Nearest**, **Apparent intersection**, and **Parallel** are used to snap crosshair at different points along the drawing objects in the drawing mode.*

WORKING IN THE ORTHO MODE

You can turn the Ortho mode on or off by choosing the **Ortho Mode** button in the Status Bar or by using the F8 key. The Ortho mode allows you to draw lines at right angles in 2D drawings only. Whenever you use the pointing device to specify the next point, the movement of the rubber-band line connected to the cursor will be either in the horizontal (parallel to the X axis) or vertical (parallel to the Y axis) direction. To draw a line in the Ortho mode, specify the starting point at the **Specify first point** prompt. To specify the second point, move the cursor with the pointing device and click at the desired point. The line thus drawn will be either vertical or horizontal depending on the direction of movement of the cursor, refer to Figures 3-12 and 3-13.

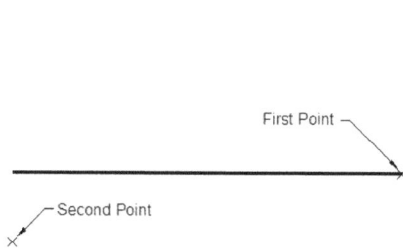

Figure 3-12 Drawing a horizontal line using the Ortho mode

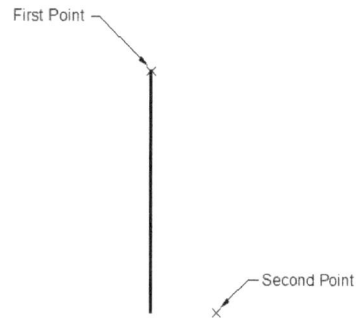

Figure 3-13 Drawing a vertical line using the Ortho mode

Note
You can turn the Ortho mode on and off at any time during drawing and editing. But in 3D views, the Ortho mode restricts the cursor movement in up and down directions.

COORDINATE REFERENCE SYSTEM (CRS)

A coordinate reference system (CRS) defines how an object relates spatially to the locations on the Earth's surface. CRS is a mathematical model used to locate geographical entities. Using a coordinate reference system, you can integrate multiple datasets with different CRS into your project. In AutoCAD Map 3D, there are two coordinate system: a fixed reference system called world coordinate system (WCS) and a user defined coordinate system called user coordinate system (UCS). CRS is referred to as coordinate system or global coordinate system. Assigning appropriate coordinate system is the most essential part of the data preparatory work.

A CRS associates a coordinate system with an object by means of a datum. As a result, the definition of a CRS must encompass the definition of a coordinate system and a datum. Everest 1830, NAD 83 and WGS 72 are some of the commonly used datum.

A geographic data can be represented using the Geographic Coordinate System and the Projected Coordinate System. A Geographic Coordinate System is defined as a 3D surface model and measured in latitudes and longitudes. A Projected Coordinate System is a model that is defined by a flat 2D surface and can be measured in meters and feet. Coordinate systems provide a framework for defining real world locations.

The Prime Meridian located at the Royal Observatory, Greenwich is used as a reference point for measuring Longitudes or East/West angles. Longitude (λ) of a point is defined as the angle from the prime meridian to the meridian plane of a given point while latitude (∅) is the angle between the equatorial plane and the perpendicular to the ellipsoid through a given point. Figure 3-14 shows the latitude and longitude of point P.

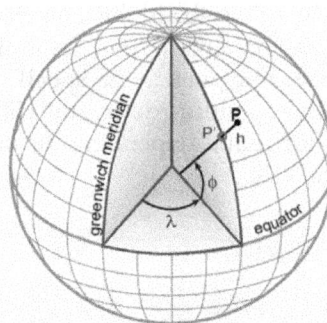

Next, you will learn to define a new coordinate reference system using the **Create** tool in AutoCAD Map 3D.

Figure 3-14 Latitude and Longitude of point P

Creating a Coordinate Reference System

Ribbon:	Map Setup > Coordinate System > Create drop-down > Create Coordinate System
Command:	MAPCSCREATE

You can define a new datum, ellipsoid, or a coordinate system using the **Create Coordinate System** tool in the **Coordinate System** panel of the **Map Setup** tab. The new coordinate system can be created by modifying an existing coordinate system or by defining the parameters for the coordinate system.

Note

It is recommended that a user must thoroughly know the concepts of the terrestrial and celestial reference systems or map projections. If you do not have sufficient knowledge and accurate defining parameters, avoid defining a new global coordinate system.

AutoCAD Map 3D saves all the defined coordinate reference systems in a single folder. You will require a written permission to save the created CRS in this folder. Follow the steps given next to change the permission for this folder.

1. Close the **AutoCAD Map 3D 2017** software application if it is running.

2. In the **Windows Explorer,** browse to the following folder location:

    C:\ProgramData\Autodesk

3. In the **Autodesk** folder, right-click on the **Geospatial Coordinate Systems <version>** folder; a menu will be displayed. In the menu, choose the **Properties** option; the **Geospatial Coordinate Systems <version> Properties** dialog box will be displayed.

4. In this dialog box, choose the **Security** tab; the options in this tab will be displayed. Choose the **Edit** button in this tab; the **Permissions for Geospatial Coordinate Systems <version>** dialog box will be displayed.

5. In this dialog box, select a user name from the **Group or user names** list box; the permissions assigned to the selected user will be displayed in the **Permissions for <Users>** list box.

6. In the **Permissions for <Users>** list box, select the check box corresponding to the **Full Control** option in the **Allow** column.

7. Next, choose the **Apply** button and then choose the **OK** button from this dialog box; the **Permissions for Geospatial Coordinate Systems <version>** dialog box will be closed and settings will be saved.

8. Again, choose the **OK** button in the **Geospatial Coordinate Systems <version> Properties** dialog box to close it.

After setting the user access, you can define a coordinate system based on the mapping, modeling, or project requirements. You can define a new coordinate system by creating a coordinate system definition, datum, ellipsoid, geodetic transformation, and geodetic transformation path. In addition to creating a coordinate system, you can create a new coordinate system category. The method of creating a coordinate system by using the **Create a coordinate system definition** option is discussed next.

Creating a Coordinate System by Using the Coordinate System Definition Option

The **Create a coordinate system definition** option is used to create a coordinate system by defining the parameters of the required coordinate system. While creating a coordinate system, you need to specify various options in different pages of the **Create Coordinate System** wizard. To create a coordinate system, choose the **Create Coordinate System** tool from the **Map Setup > Coordinate System > Create** drop-down; the **Create Coordinate System** wizard will be displayed, as shown in Figure 3-15. The different pages of this wizard are discussed next.

Create coordinate system Page

By default, the **Create coordinate system** page is displayed in the wizard with the **Create a coordinate system definition** radio button selected.

To create the coordinate system definition, choose the **Next** button on this page; the **Specify starting point** page of the wizard will be displayed.

Specify starting point Page

In this page, by default, the **Start with an ellipsoid** radio button is selected. As a result, you are prompted to create a coordinate system based on ellipsoid. Also, you can select the **Start with a coordinate system** or **Start with a datum** radio button to redefine a coordinate system or define a coordinate system based on the existing datum. To continue with the method used for creating a coordinate system, retain the default setting in this page and then choose the **Next** button; the **Specify ellipsoid** page will be displayed.

Specify ellipsoid Page

In this page, you can specify the method of creating an ellipsoid. You can choose to create an ellipsoid either by defining a new ellipsoid or by modifying the parameters for an existing ellipsoid.

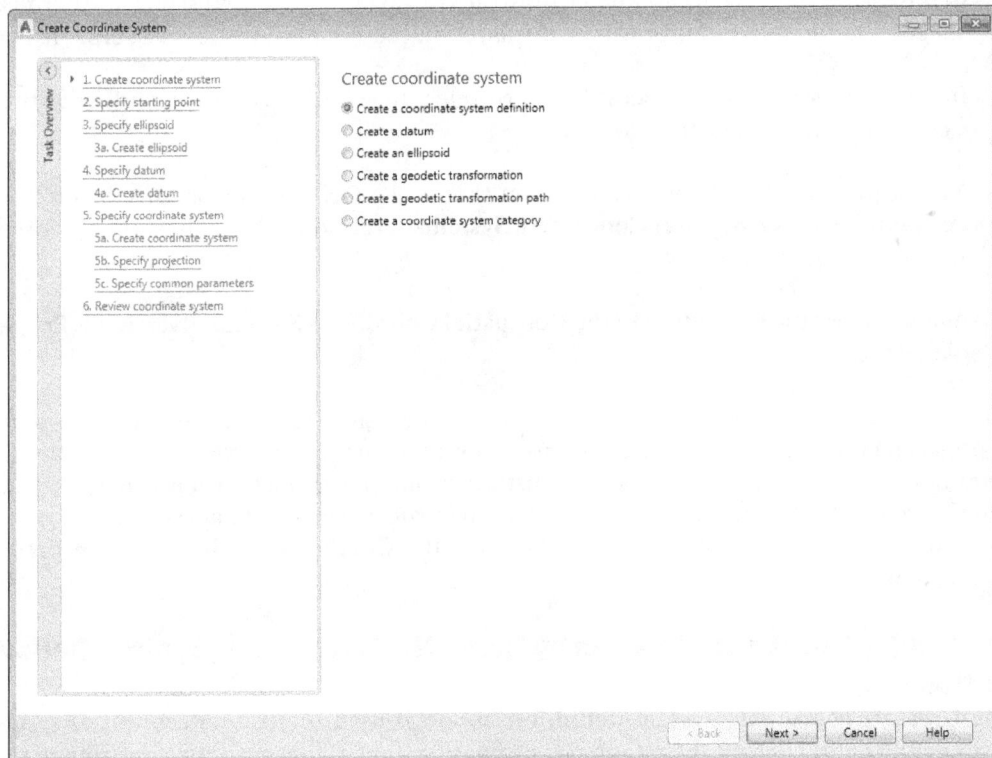

*Figure 3-15 The **Create coordinate system** wizard*

To create an ellipsoid by modifying the parameters for an existing ellipsoid, select the **Create a new ellipsoid from an existing ellipsoid** radio button; the **Select** button in the **Specify ellipsoid** page will be activated.

Choose the **Select** button; the **Coordinate System Library** dialog box will be displayed, as shown in Figure 3-16. Choose the required ellipsoid from the list box in this dialog box and then choose the **Select** button; the name of the selected ellipsoid will be displayed in the **Ellipsoid** text box. Choose the **Next** button; the **Modify ellipsoid** page will be displayed. This page will display parameters of the selected ellipsoid. You can modify the required parameters in this page.

To create an ellipsoid by defining all the ellipsoidal parameters, select the **Create a new ellipsoid** radio button from the **Specify ellipsoid** page of the wizard. Next, choose the **Next** button; the **Create ellipsoid** page will be displayed. The various options in this page are discussed next.

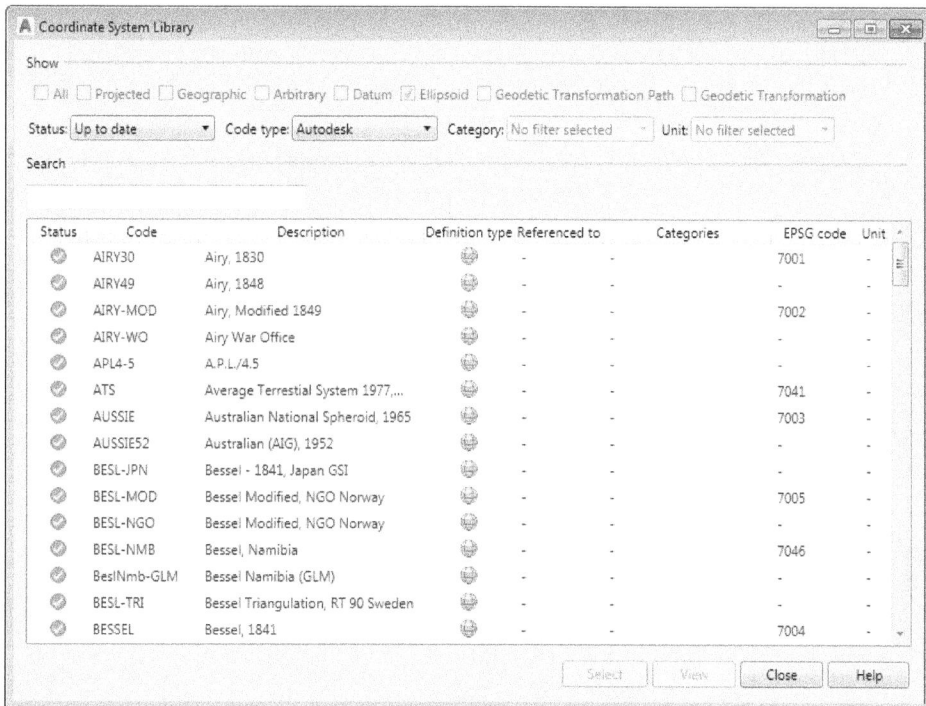

Figure 3-16 The Coordinate System Library dialog box

Create ellipsoid Page

In this page, assign a desired code name to the ellipsoid in the **Code** edit box. You can also enter a brief description about the ellipsoid in the **Description** text box.

The **Reference this ellipsoid (rather than a datum) in the coordinate system** check box is cleared by default. As a result, the current ellipsoid will be used for reference in this coordinate system definition. If you select this check box, the links to the page for specifying datum will be removed. In the **Ellipsoid Dimensions** area, you can specify whether you want to create an ellipsoid or a sphere by selecting the radio button corresponding to the required option. Depending on the selection of the radio button, the edit boxes for defining the ellipsoid parameters will be enabled.

Next, you need to specify values in the **Equatorial radius (meters)**, **Inverse flattening**, **Polar radius (meters)**, and **Eccentricity squared** edit boxes. Figure 3-17 shows the **Create ellipsoid** page with the defined parameters.

After specifying all values in this page, choose the **Next** button; the **Specify datum** page of the **Create Coordinate System** wizard will be displayed.

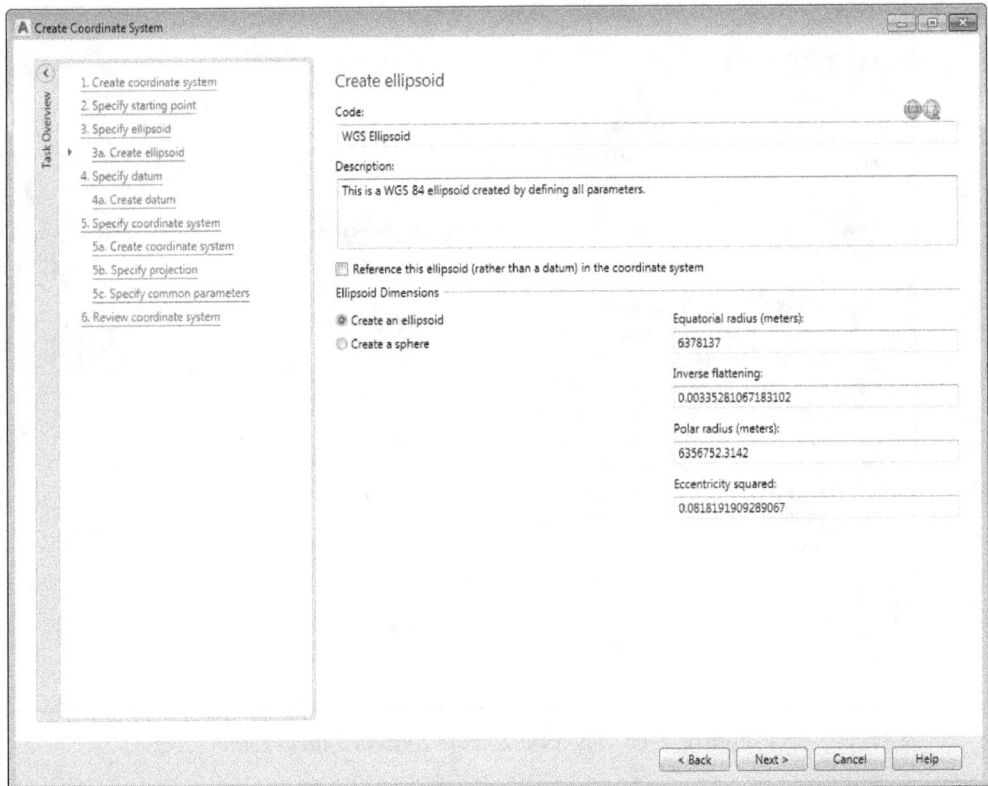

*Figure 3-17 The **Create ellipsoid** page displaying the parameters for creating an ellipsoid*

Specify datum Page

Similar to the **Create ellipsoid** page, you can select the method of creating datum in the **Specify datum** page. To create a new datum by defining its parameters, select the **Create a new datum** radio button and then choose the **Next** button; the **Create datum** page of the wizard will be displayed.

Note

*To create a datum by modifying parameters of an existing datum, select the **Create a new datum from an existing datum** radio button in the **Specify datum** page; the **Datum** edit box and the **Select** button will be activated. Next, choose the **Select** button; the **Coordinate System Library** dialog box will be displayed. In this dialog box, choose the required datum and then choose the **Select** button; the **Coordinate System Library** dialog box will be closed and the selected datum name will be displayed in the **Datum** edit box of the **Specify datum** page. Choose the **Next** button in the **Specify datum** page to proceed.*

Create datum Page

In the **Create datum** page, you can create a new datum with reference to an existing ellipsoid. To create a new datum, enter a suitable code in the **Code** edit box. You can also add a brief description about the datum in the **Description** text box. By default, the name of the ellipsoid specified earlier is displayed in the **Ellipsoid** edit box. If you want to select another ellipsoid, choose the **Select** button; the **Coordinate System Library** dialog box will be displayed. Select the required ellipsoid from the list box in the **Coordinate System Library** dialog box and then choose the **Select** button; the dialog box will be closed and the name of the selected ellipsoid will be displayed in the **Ellipsoid** edit box. Next, enter the name of the data source containing the information about selected ellipsoid in the **Source** edit box. After specifying all the settings in this page, choose the **Next** button; the **Specify coordinate system** page will be displayed.

Specify coordinate system Page

You can use the options in the **Specify coordinate system** page to create a new coordinate system by using your own settings or create a new coordinate system from an existing coordinate system.

By default, the **Create a new coordinate system** radio button is selected in the **Specify coordinate system** page. Select an option from the **Coordinate system type** drop-down list to specify the type of coordinate you wish to create. You can create a **Projected**, **Geographic**, or **Arbitrary** coordinate system. On specifying all the options in the **Specify coordinate system** page, choose the **Next** button; the **Create coordinate system** page will be displayed, as shown in Figure 3-18. Based on the type selected, an icon will be displayed on upper right corner of this page.

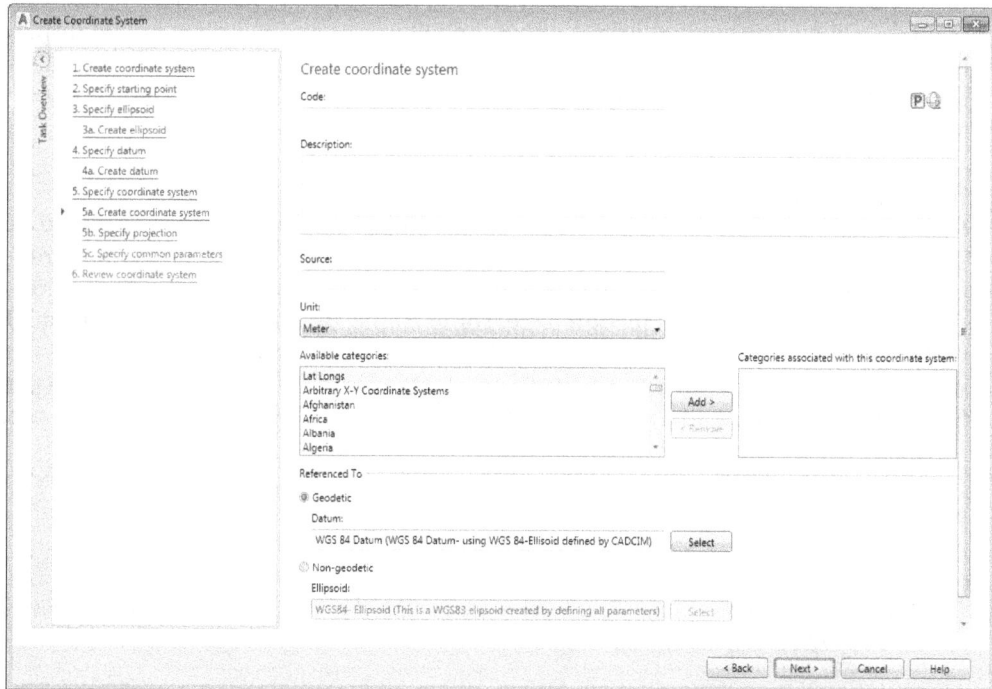

*Figure 3-18 The **Create Coordinate System** wizard with the **Create coordinate system** page*

Note
*If you select the **Create a new coordinate system from an existing coordinate system** radio button in the **Specify coordinate system** page, the **Select** button will be activated. You can use this button to select an existing coordinate system.*

Create coordinate system Page

In this page, you can specify the settings for a coordinate system based on your project requirements. To apply a code to the coordinate system to be created, enter a suitable code in the **Code** edit box and then enter the desired information in the **Description** text box. Next, enter the name of the data source that contains information about current coordinate system in the **Source** edit box. To specify a measuring unit for the current coordinate system, select an option from the **Unit** drop-down list. To associate the current coordinate system with an existing category, select the option from the **Available categories** list box and then choose the **Add** button; the selected category will be added to the **Categories associated with this coordinate system** list box. To remove a category from this list box, select the unwanted category and then choose the **Remove** button; the selected category will be removed.

You need to reference the current coordinate system with a datum or an ellipsoid. By default, the **Geodetic** radio button is selected in the **Referenced To** area of this page. As a result, the **Datum** edit box and the **Select** button on its right are activated. Also, the code of the datum created in earlier pages is displayed in the **Datum** edit box. To reference a different datum, use the **Select** button.

You can select the **Non-geodetic** radio button from the **Referenced To** area to reference an ellipsoid for the current coordinate system. On selecting this radio button, the **Ellipsoid** edit box will be activated and the code of the ellipsoid created earlier will be displayed in it. You can use the **Select** button to specify a different ellipsoid. After specifying all settings, choose the **Next** button; the **Specify projection** page of the **Create Coordinate System** wizard will be displayed.

Specify projection Page

In this page, you need to specify the settings for the projection parameters. To apply a projection to the current coordinate system, select an option from the **Projection** drop-down list. Depending on the option selected in this drop-down list, a list of parameters will be displayed in the **Specify projection** page. For example, on selecting the **Albert Equal Area Conic** option in the **Projection** drop-down list, you need to specify the northern and southern standard parallel, and the origin latitude and longitude in the **Northern standard parallel**, **Southern standard parallel**, **Origin longitude**, and **Origin latitude** edit boxes, respectively.

After specifying all the parameters in this page, choose the **Next** button; the **Specify common parameters** page will be displayed.

Note
*The options displayed in the **Parameters** area will be based on the projection type selected from the **Projection** drop-down list.*

Specify common parameters Page

In this page, you can specify the parameters used for projecting given map coordinates. To apply a scale to the current coordinate system, enter the scale value in the **Map (paper) scale** edit box. Next, select the relevant option from the **X increases to the** and **Y increases to the** drop-down lists in the **Quadrant** area. Next, you need to specify the geographic limits of the location in terms of latitudes and longitudes. The values entered must be in the degree format and measured with respect to Greenwich. To specify the range for the geographic location under study, enter the minimum and maximum values of the latitudes and longitudes in their respective edit boxes in the **Useful Range: Geographic** area.

In the **Minimum Non-Zero Coordinate Values** area, you can specify the geographic limits of the location under study in terms of Cartesian coordinates. To do so, enter the non-zero x and y values in the **Non-zero X** and **Non-zero Y** edit boxes. To specify the limits in the form of Cartesian coordinates, choose the button corresponding to the **Useful Range:Cartesian** option; four edit boxes will be displayed. Next, enter the limits of the geographic location in their respective edit boxes. After specifying all parameters in this page, choose the **Next** button; the **Review coordinate system** page will be displayed.

Coordinate system Page

In this page, you can review the values, parameters, and settings applied to the current coordinate system. To modify any setting, choose the **Edit** button in the area corresponding to the option to be modified. On doing so, the page corresponding to the options to be modified will be displayed. In the page displayed, modify the settings of parameters based on project requirement and then choose the **Review coordinate system** link from the **Task Overview**; the **Coordinate System** page will be displayed with the modified settings. After reviewing all the parameters in this page, choose the **Finish** button; the **Create Coordinate System** dialog box will be closed and a coordinate system will be created in the specified category.

Assigning a Coordinate Reference System to Data

Ribbon: Map Setup > Coordinate System > Assign
Command: MAPCSASSIGN

As mentioned earlier, the coordinate reference system defines how an object spatially relates to a location on the Earth's surface. A spatial data with a defined CRS can be easily integrated into your project and can be used for the purpose of spatial analysis.

You can start a new project by defining its coordinate reference system and then create datasets. The procedure to define coordinate reference system to a dataset while importing it into your drawing environment and the procedure to assign coordinate reference system to your drawing (project) is discussed next.

Assigning CRS to the Drawing

You can use the **Assign** tool to define a coordinate system to the current drawing. To assign a CRS to the drawing, choose the **Assign** tool from the **Coordinate System** panel; the **Coordinate System - Assign** dialog box will be displayed, as shown in Figure 3-19. The options in different areas of the dialog box are discussed next.

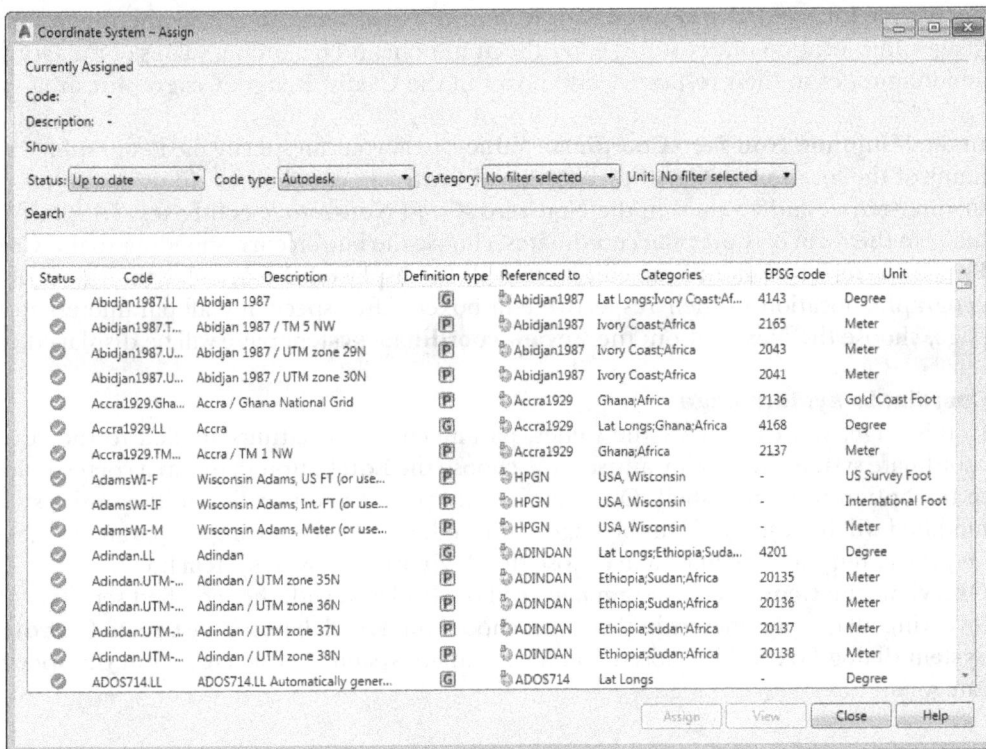

*Figure 3-19 The **Coordinate System - Assign** dialog box*

Currently Assigned Area

The **Code** and **Description** in this area display code and description of the coordinate system (CRS) that has been currently assigned to the drawing. In case no CRS has been defined, the **Code** and **Description** labels will display **N/A**.

Show Area

The **Show** area of the **Coordinate System - Assign** dialog box contains four drop-down lists, namely **Status**, **Code type**, **Category**, and **Unit**. You can use the options in these drop-down lists to filter the list of available coordinate system. These drop-down lists are discussed next.

Status: You can filter the coordinate system (CRS) based on the status by selecting the relevant option from the **Status** drop-down list in the **Show** area. To view the list of updated coordinate systems, select the **Up to date** option from the **Status** drop-down list. Similarly, to view the list of out of date coordinate systems, select the **Out of date** option from the drop-down list. You can also select the **User defined** option from the drop-down list to view the coordinate system created by the user.

Code: To filter the coordinate systems based on the code, select the relevant option from the **Code type** drop-down list. To filter the coordinate systems based on the EPSG (European Petroleum Survey Group) code, select the **EPSG** option from the drop-down list else select the **Autodesk** option from the drop-down list.

Category: This drop-down list contains the list of available categories of coordinate systems. To filter the coordinate systems based on the category, select the required option from this drop-down list. For example, to display the list of geographic coordinate system, select the **Lat Longs** option from the drop-down list. Similarly, you can select the country name to filter the coordinate systems based on the country. For example, to view the list of coordinate systems for Australia, select the **Australia** option from the **Category** drop-down list. The coordinate system can also be filtered based on the arbitrary coordinate system, obsolete coordinate system, UTM coordinate system, and so on.

Unit: To view the list of coordinate system based on a particular unit, select the required option from the **Unit** drop-down list. For example, to view the list of coordinate systems based on the US Survey foot, select the **US Survey foot** option from the drop-down list. Similarly, you can view the list of coordinate systems based on meters, international foot, degrees, and so on.

Search Area

You can use the edit box in the **Search** area to search the required coordinate system from the coordinate system library. To find a coordinate system, enter the code, description, details, or information of the coordinate system in this edit box. The list of coordinate systems displayed below the search edit box will dynamically update as you type in the search edit box.

After applying filters, select the required coordinate system from the list displayed below the search text box. To read the details about the selected coordinate system, choose the **View** button; the **Coordinate System** window will be displayed, refer to Figure 3-20. Choose the **Close** button in this window to exit. Next, to assign the selected coordinate system to the drawing, choose the **Assign** button from the **Coordinate System - Assign** dialog box; the dialog box will be closed and the coordinate system will be assigned to the current drawing.

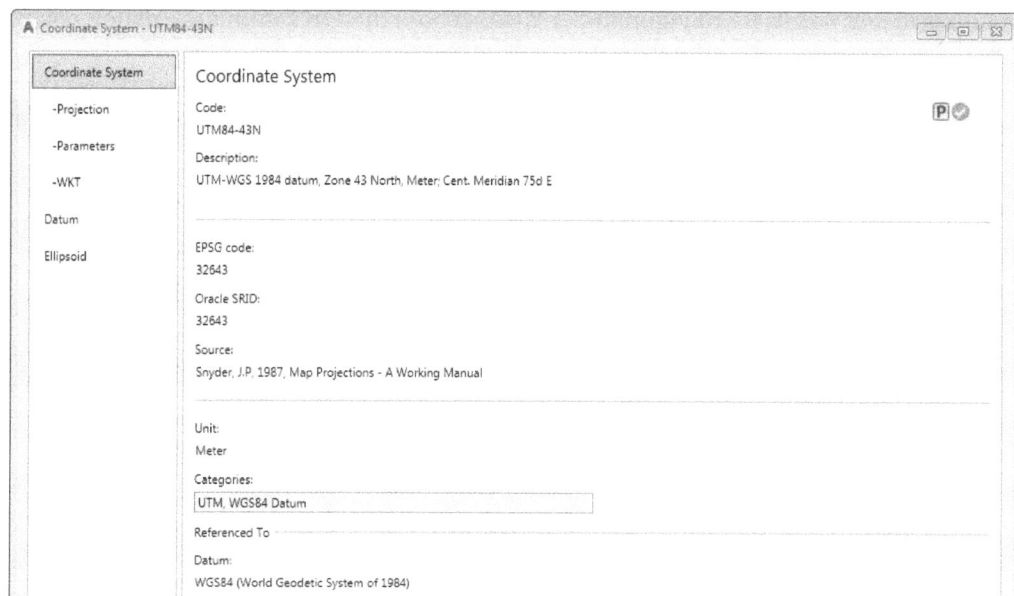

*Figure 3-20 Partial view of the **Coordinate System** window with the **Coordinate System** page*

Using the Connect Tool to Assign a Coordinate System to a Dataset

A coordinate system can be assigned to a dataset using the **Connect** tool and also to load the data into the Workspace. To do so, select data in an external data format by using the **Connect** tool as explained in the previous chapter, and then choose the **Edit Coordinate Systems** button located at the top of the list box; the **Edit Spatial Contexts** dialog box will be displayed, as shown in Figure 3-21.

🌐 Edit Coordinate Systems

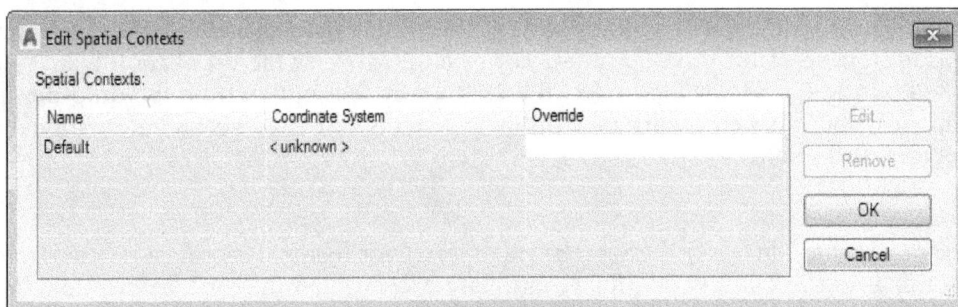

*Figure 3-21 The **Edit Spatial Contexts** dialog box*

To assign a new coordinate system or edit an existing coordinate system, double-click on the first record in the **Spatial Contexts** list box; the **Coordinate System Library** dialog box will be displayed. Alternatively, to invoke the **Coordinate System Library** dialog box, select the record in the **Spatial Contexts** list box and then choose the **Edit** button; the **Coordinate System Library** dialog box will be displayed. Select the required coordinate system from the **Coordinate System Library** dialog box and then choose the **Select** button; the dialog box will be closed and the code of the selected coordinate system will be displayed in the **Override** column of the **Edit Spatial Contexts** dialog box, refer to Figure 3-22. Next, choose the **OK** button; the dialog box will be closed and the selected coordinate system will be assigned to the dataset.

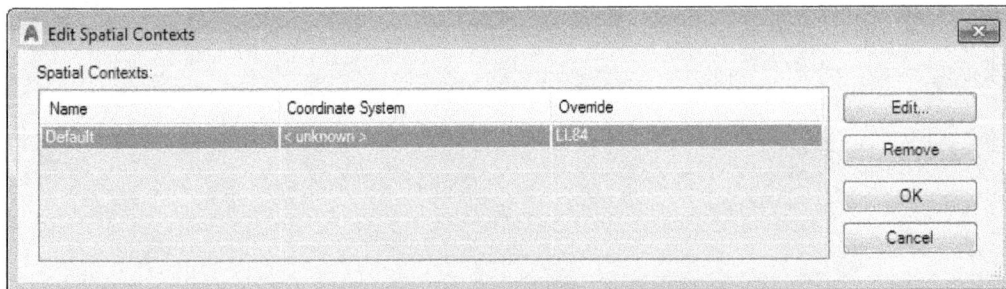

*Figure 3-22 The **Code** of the selected coordinate system displayed in the **Override** column*

Tip
*You can also invoke the **Edit Spatial Contexts** dialog box by choosing the **Edit Coordinate System** option from the shortcut menu. To display the shortcut menu, right-click on the dataset name in the DATA CONNECT wizard.*

USER COORDINATE SYSTEMS IN AutoCAD

AutoCAD Map 3D application is developed using an AutoCAD platform. In AutoCAD, there are two coordinate systems: **World Coordinate System** (**WCS**) and **User Coordinate System** (**UCS**). These coordinate systems are also available in AutoCAD Map 3D. Whenever you start the AutoCAD Map 3D application, the drawing will be set to WCS. WCS is a fixed coordinate system and the X axis in WCS is the horizontal axis, the Y axis is vertical and the Z axis is perpendicular to the XY plane. The origin is defined at the intersection of X and Y axis (0, 0) in the lower left corner of the drawing. Using WCS, you can create 2D drawings and surface models. However, creating 3D models with multiple features using WCS is a cumbersome task. In such a scenario, a custom defined coordinate system may help accelerate the drafting process. A custom defined coordinate system is called as the User Coordinate System (UCS). You can define a UCS in terms of the WCS. While defining the UCS, you can specify its origin and the orientation of its axis. This helps you to specify points in three-dimensions. A brief description of working with UCS is given next.

Moving the UCS Triad

| **Ribbon:** | View > Coordinates > UCS |
| **Command:** | UCS |

To move the UCS triad, choose the **UCS** tool from the **Coordinates** panel of the **View** tab; the cursor changes into crosshair in the drawing window. Place the crosshair at the required point and then click; the UCS triad moves to the specified point. Next, press ENTER; the position of the UCS triad will be fixed at the specified point.

Rotating the UCS Triad about the X Axis

| **Ribbon:** | View > Coordinates > X |
| **Command:** | UCS |

You can rotate the UCS about the X axis in the YZ plane. To rotate the UCS icon in the YZ plane while keeping the X axis fixed, choose the **X** tool from the **Coordinates** panel; the cursor will change to a crosshair in the drawing window and you will be prompted to specify the rotation angle. Enter the rotation angle and press ENTER; the UCS will rotate about the X axis. The angle of rotation can be applied to the UCS icon, as shown in Figure 3-23.

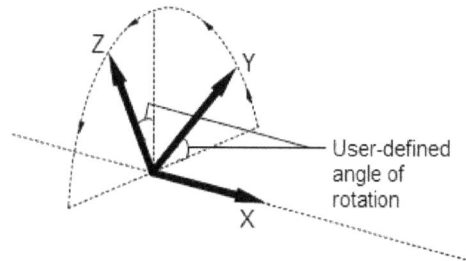

Figure 3-23 *Rotating the UCS icon keeping the X axis fixed*

Similarly, you can use the **Y** and **Z** tools from the **Coordinates** panel of the **View** tab to rotate the UCS icon in the XZ and XY plane, respectively.

Using a Named UCS for Drawings

Ribbon: View > Coordinates > UCS, UCS Settings Button

You can assign a name to the UCS and then use it anytime during the project. The UCS settings that need to be applied to many drawing objects in the project can be saved as a template so that you can use the new template for many drawing objects as per project requirement. In the previous section of this chapter, you learned how to realign a UCS as per your requirement. To specify a named UCS for the drawings in the project, choose the **UCS**, **UCS Settings** button from the **Coordinates** panel; the **UCS** dialog box will be displayed, as shown in Figure 3-24. The **Named UCSs**, **Orthographic UCSs**, and **Settings** tabs in the **UCS** dialog box are discussed next.

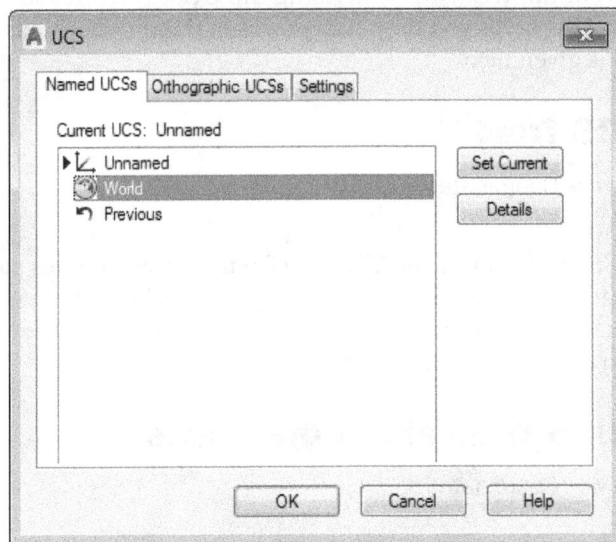

Figure 3-24 The UCS dialog box

Named UCSs Tab

The options in the **Named UCSs** tab are used to specify a name for the modified UCS settings and to set a named UCS as the current UCS. In the **UCS** dialog box, choose the **Named UCSs** tab, if it is not chosen by default. To rename a UCS, select it from the **Current UCS** list box and then right-click on the **Unnamed** option; a shortcut menu will be displayed. Choose the **Rename** option from this shortcut menu; the **Unnamed** option will change into an edit box. In this edit box, enter a name, and then press ENTER. To set a particular UCS as the current UCS, select it from the **Current UCS** list box and right-click; a shortcut menu will be displayed. Choose the **Set Current** option from the shortcut menu and the selected UCS will be set as the current UCS. Alternatively, to set the UCS as the current UCS, select the relevant UCS and then choose the **Set Current** button; the selected UCS will be set as the current UCS. To delete a particular UCS, choose the **Delete** option from the shortcut menu. You can also choose the **Details** option from the shortcut menu to view the details of the selected UCS.

Note
*The **Unnamed** option will appear only if you have modified the alignment of the default UCS icon.*

Orthographic UCSs Tab

The options in the **Orthographic UCSs** tab of the **UCS** dialog box are used to apply the orthographic coordinate system to the current drawing with reference to the available coordinate system. To apply a particular orthographic coordinate system to the current drawing, choose an option from the **Name** list box in the **UCS** dialog box. To apply the orthographic coordinate system with reference to an available coordinate system, choose an option from the **Relative to** drop-down list. To apply the modified settings to a coordinate system, choose the **Set Current** button. Next, choose the **OK** button; the UCS triad will change to the settings specified in the drawing window.

Settings Tab

The options in the **Settings** tab are used to specify the displaying and aligning properties of the UCS triad. These settings are discussed next.

UCS Icon settings Area

The options in the **UCS Icon settings** area are used to set the properties of the UCS icon. To hide the UCS icon from the drawing window, clear the **On** check box. To display the UCS icon at the lower left corner of the drawing window instead of displaying it at the origin of the coordinate system, clear the **Display at UCS origin point** check box. To apply the modified UCS settings to all viewports in the current drawing, select the **Apply to all active viewports** check box. To enable the selection of the UCS icon, select the **Allow Selecting UCS icon** check box. Note that, by default, this check box is selected.

UCS settings Area

The options in the **UCS settings** area are used to set viewports with UCS. The **Save UCS with viewport** check box is selected by default in this area. As a result, the modified UCS settings will be saved with the current viewport. If you clear the **Save UCS with viewport** check box, the modified settings will be applied to the entire drawing. To apply the current UCS settings to the viewport and ignore the UCS settings applied to a particular viewport, clear the **Save UCS with viewport** check box. To restore the plan view when the coordinate system in the viewport is changed, select the **Update view to Plan when UCS is changed** check box. Next, choose the **OK** button to close the dialog box.

TUTORIALS

General instructions for downloading tutorial files:
To complete the tutorials, you need to download the tutorial data to your computer. To download the tutorial data, follow the steps given below:

1. Log on to *www.cadcim.com* and browse to *Textbooks > Civil/GIS > Map 3D > Exploring AutoCAD Map 3D 2017*. In this page, select *c03_m3d_2017_tut.zip* file from the **Tutorial Files** drop-down list. Next, choose the corresponding **Download** button to download the data file.

2. Extract the contents of the zip file at the following location:

   C:\m3d_2017

Notice that the *c03_m3d_2017_tut* folder is created in the *m3d_2017* folder.

Tutorial 1   Assigning CRS

In this tutorial, you will start with a new drawing file. Next, you will assign a Coordinate Reference System (CRS) to the current drawing.   **(Expected time: 30 min)**

The following steps are required to complete this tutorial:

a.   Start a new drawing file.
b.   Assign the **NE83 - NAD83 Nebraska State Planes, Meter** coordinate system to the current Workspace.
c.   Save the file.

Starting a New Drawing

1.   Choose **New > Drawing** from the Application Menu; the **Select template** dialog box is displayed.

2.   In the **Select template** dialog box, select the *map2d.dwt* template file and then choose the **Open** button; the **map2d** template is applied to the Modelspace.

Assigning a Global Coordinate System to the Current Workspace

1.   Choose the **Assign** tool from the **Coordinate System** panel in the **Map Setup** tab; the **Coordinate System - Assign** dialog box is displayed.

Assign

2.   In the **Show** area of the dialog box, select the **USA, Nebraska** option from the **Category** drop-down list; a list of coordinate systems in this category is displayed in the list box of the dialog box.

3.   Next, select the **NE83** code with description **NAD83 Nebraska State Planes, Meter** from the list box, and then choose the **Assign** button; the selected coordinate system is assigned to the Workspace.

Saving the Drawing File

1.   Choose the **Save As** option in the Application Menu; the **Save Drawing As** dialog box is displayed.

2.   In the **Save Drawing As** dialog box, enter **c03_Tut01a** in the **File name** edit box and make sure that the **AutoCAD 2013 Drawing (*.dwg)** option is selected in the **Files of type** drop-down list.

3.   Next, choose the **Save** button; the drawing file is saved at the specified location.

Tutorial 2   Creating a Coordinate Reference System

In this tutorial, you will create a Coordinate Reference System using the parameters given below and then assign it to the current drawing.   **(Expected time: 45 min)**

Parameters of ellipsoid:
Code: CADCIM-GRS1980
Equatorial radius = 6378137 m
Polar radius = 6356752.31414035 m
Inverse flattening= 0.00335281068118332
Eccentricity squared= 0.0818191910428279

Parameters of datum:
Code: CADCIM- NAD83
Source: CADCIM

Parameters of new coordinate system:
Coordinate system type: Projected
Code: Austin TX
Source: CADCIM
Unit: US Survey Foot
Category: test only
Referenced to: (Geodetic) CADCIM- NAD83 datum

Projection: Lambert Conformal Conic, double standard parallel
Northern standard parallel: 31.8833333333333
Southern standard parallel: 30.1166666666667
Origin longitude: -100.333333333333
Origin latitude: 29.6666666666667
False easting: 2296583.33333333
False northing: 9842500

Common Parameters:
Map scale: 1
X Increases to the: East,
Y Increases to the: North
Minimum longitude: -108.3
Maximum longitude: -91.8666666666667
Minimum latitude: 29.5333333333333
Maximum latitude: 32.5
Non zero X:
Non zero Y: 0

The following steps are required to complete this tutorial:

a. Verify folder permissions.
b. Start a new drawing file.
c. Create a coordinate system definition using the given parameters.
d. Assign the coordinate system to the current drawing.
e. Save the file.

Verifying Folder Permissions

AutoCAD Map 3D saves the created CRS in the **Geospatial Coordinate Systems** folder. As a result, you need to have write permission for this folder. In this part of the tutorial, you will verify the permissions for this folder and change them if necessary.

1. Close the **AutoCAD Map 3D 2017** software application, if it is running.

2. In the **Windows Explorer**, browse to the following location:

    C:\ProgramData\Autodesk

    Note
    *The **ProgramData** folder is a hidden folder. To view this folder, you need to turn on the **Show hidden files, folders and drives** option for your Windows operating system.*

3. In the **Autodesk** folder, right-click on the **Geospatial Coordinate Systems <version>** folder; a shortcut menu is displayed. In this menu, choose the **Properties** option; the **Geospatial Coordinate Systems <version> Properties** dialog box is displayed.

4. In this dialog box, choose the **Security** tab from this dialog box; the options in this tab are displayed. Choose the **Edit** button in this tab; the **Permissions for Geospatial Coordinate Systems <version>** dialog box is displayed.

5. In this dialog box, select a user name from the **Group or user names** list box; the permissions assigned to the selected user are displayed in the **Permissions for Users** list box.

6. In the **Permissions for Users** list box, select the check box corresponding to the **Full Control** option in the **Allow** column; all the check boxes in the **Allow** button are selected by default.

7. Next, choose the **Apply** and then the **OK** button from this dialog box; the **Permissions for Geospatial Coordinate Systems <version>** dialog box is closed and the settings are saved.

8. Again, choose the **OK** button in the **Geospatial Coordinate Systems <version> Properties** dialog box to close it.

Starting the AutoCAD Map 3D Application

1. Start the AutoCAD Map 3D 2017 application.

Creating a Coordinate System Definition Using the Given Parameters

In this part of the tutorial, you will define the coordinate system using the **Create Coordinate System** tool. You will first define an ellipsoid and datum. Next, using the defined ellipsoid, datum, and the parameters given in the tutorial data, you will create the coordinate reference system.

1. Choose the **Create Coordinate System** tool from **Map Setup > Coordinate System > Create** drop-down; the **Create Coordinate System** wizard with the **Create coordinate system** page is displayed.

2. In this page, select the **Create a coordinate system definition** radio button if it is not selected by default. Next, choose the **Next** button; the **Specifying starting point** page of the wizard is displayed.

3. In this page, select the **Start with an ellipsoid** radio button, if it is not selected by default. Next, choose the **Next** button; the **Specify ellipsoid** page is displayed.

4. In this page, select the **Create a new ellipsoid** radio button, if it is not selected by default and then choose the **Next** button; the **Create ellipsoid** page is displayed.

5. In the **Create ellipsoid** page, enter **CADCIM - GRS1980** in the **Code** edit box.

6. Next, enter **Based on Geodetic Reference System of 1980** in the **Description** text box.

7. Make sure that the **Reference this ellipsoid (rather than a datum) in the coordinate system** check box is cleared and the **Create an ellipsoid** radio button is selected in the **Ellipsoid Dimensions** area.

8. Enter **6378137** as the equatorial radius, **0.00335281068118332** as the inverse flattening ratio and **6356752.31414035** as the polar radius in the **Equatorial radius (meters)**, **Inverse flattening**, and **Polar radius (meters)** edit boxes, respectively. Notice that the value for the eccentricity squared parameter is calculated and displayed automatically in the **Eccentricity squared** edit box, refer to Figure 3-25.

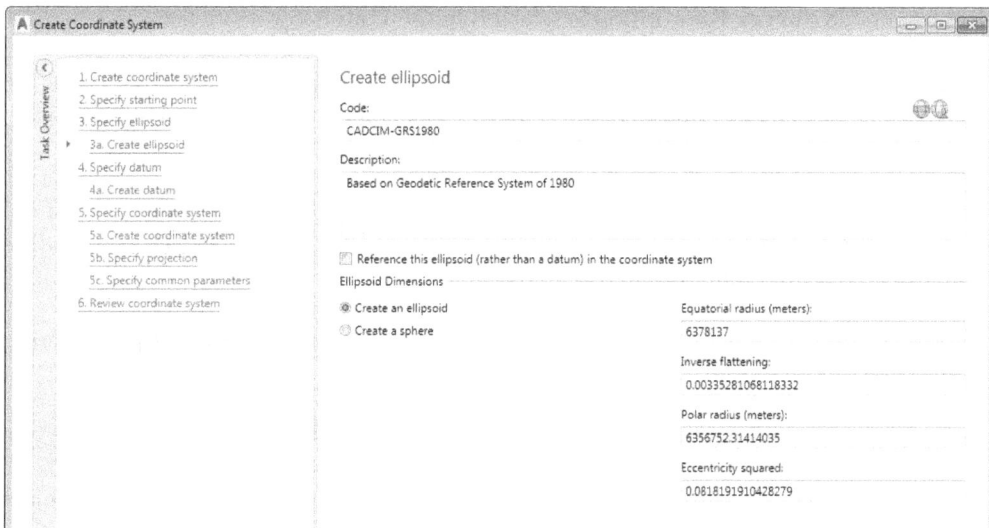

*Figure 3-25 Partial view of the **Create ellipsoid** page of the **Create Coordinate System** wizard*

9. Choose the **Next** button; the **Specify datum** page of the wizard is displayed.

10. Make sure that the **Create a new datum** radio button is selected in this page and then choose the **Next** button; the **Create datum** page is displayed.

Notice that the ellipsoid defined earlier as **CADCIM - GRS1980** is selected and displayed as **CADCIM - GRS1980 (Based on Geodetic Reference System of 1980)** in the **Ellipsoid** text box of the **Create datum** page.

11. Next, enter **CADCIM - NAD83** in the **Code** edit box and **NAD 1983 Datum - Ellipsoid GRS1980** in the **Description** text box.

12. Next, enter **CADCIM** in the **Source** edit box.

13. Choose the **Next** button in the **Create datum** page; the **Specify coordinate system** page of the wizard is displayed.

14. In this page, select the **Create a new coordinate system** radio button if it is not selected by default.

15. Select the **Projected** option from the **Coordinate system type** drop-down list, if it is not selected by default, and then choose the **Next** button; the **Create coordinate system** page is displayed in the wizard.

16. In this page, enter **AUSTIN-TX** in the **Code** edit box and **User defined CRS - Austin city** in the **Description** text box.

17. Enter **CADCIM** in the **Source** edit box.

18. Ensure the **US Survey Foot** option is selected in the **Unit** drop-down list.

19. Choose the **Test Only** option from the **Available categories** list box and then choose the **Add** button; the selected category is added to the **Categories associated with this coordinate system** list box.

20. Make sure that the **Geodetic** radio button is selected in the **Referenced To** area of this page and the **Datum** text box displays **CADCIM - NAD83 (NAD 1983 Datum - Ellipsoid GRS1980)**, refer to Figure 3-26.

21. Choose the **Next** button in the **Create coordinate system** page; the **Specify projection** page of the wizard is displayed.

    In this page, you will specify the projection parameters for the coordinate reference system.

22. Select the **Lambert Conformal Conic, double standard parallel** option from the **Projection** drop-down list; the **Create Coordinate System - Projection** message box is displayed. Choose the **Yes** button in this message box to close it.

23. In the **Parameters** area of the **Specify projection** page, specify **31.8833333333333** and **30.1166666666667** in **Northern standard parallel** and **Southern standard parallel** edit boxes, respectively.

24. Next, in the **Projection Origin** area, specify **-100.333333333333** and **29.6666666666667** in the **Origin longitude** and **Origin latitude** text boxes, respectively.

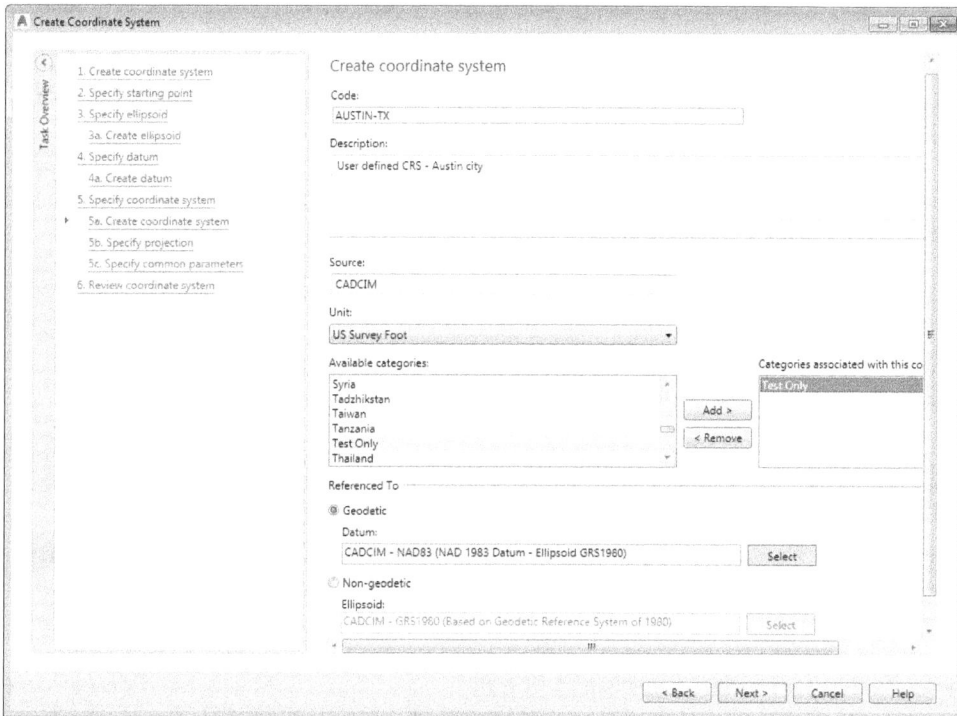

Figure 3-26 The Create coordinate system page displaying various parameters for creating CRS

25. Specify **2296583.33333333** and **9842500** in the **False easting** and **False northing** text boxes, respectively. Figure 3-27 shows the **Specify projection** page of the wizard with the projection parameters.

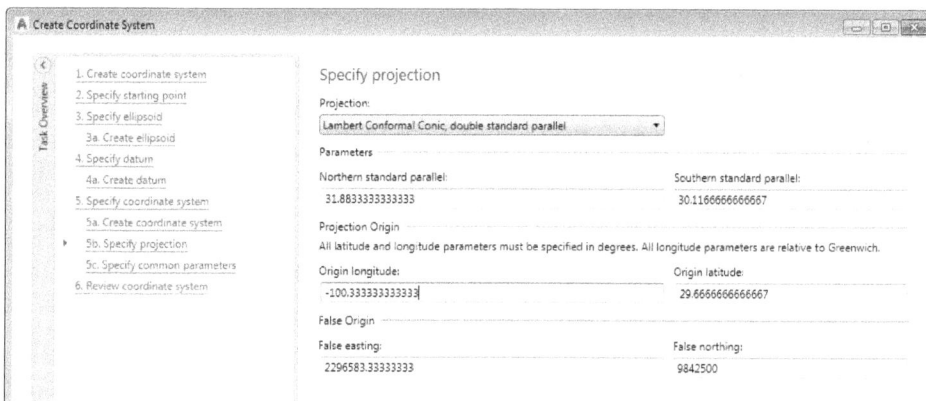

*Figure 3-27 Partial view of the **Specify projection** page*

26. Choose the **Next** button in the wizard; the **Specify common parameters** page is displayed.

27. In the **Scaling** area, specify **1** in the **Map (paper) scale** edit box.

28. Next, select the **East** and **North** options in the **X increases to the** and **Y increases to the** drop-down lists, respectively.

29. In the **Useful Range: Geographic** area, enter **-108.3** and **-89.8666666666667** in the **Minimum longitude** and **Maximum longitude** edit boxes, respectively.

30. Specify **29.5333333333333** and **32.5** as the minimum and maximum latitude values in the **Minimum latitude** and **Maximum latitude** edit boxes, respectively.

31. Retain **0** in both the **Non-zero X** and **Non-zero Y** edit boxes.

32. Choose the **Next** button in the **Specify common parameters** page; the **Coordinate system** page with a summary of the coordinate system is displayed.

33. Choose the **Finish** button; the **Create Coordinate System** wizard is closed and the coordinate system is saved in the **Geospatial Coordinate Systems** folder.

Assigning the Created Coordinate System to the Current Drawing

1. Choose the **Assign** tool from the **Coordinate System** panel in the **Map Setup** tab; the **Coordinate System-Assign** dialog box is displayed.

2. In the dialog box, select the **User defined** option from the **Status** drop-down list in the **Show** area; the list of user-defined coordinate systems is displayed in the list box below the **Search** area.

3. Select the **AUSTIN-TX** code from the displayed list and then choose the **Assign** button; the dialog box is closed and the selected coordinate system is assigned to the current drawing.

Saving the Drawing File

1. Choose the **Save As** option in the Application Menu; the **Save Drawing As** dialog box is displayed.

2. In the **Save Drawing As** dialog box, enter **c03_Tut02a** in the **File name** edit box and select the **AutoCAD 2013 Drawing (*.dwg)** option in the **Files of type** drop-down list, if it is not selected by default. Now, choose the **Save** button next to the **File name** edit box; the current drawing file is saved with the given name.

Tutorial 3   Assigning CRS while Loading Data

In this tutorial, you will assign a coordinate system (CRS) to a dataset while loading it using the **Connect** tool.   **(Expected time: 20 min)**

The following steps are required to complete this tutorial:

a. Start a new drawing file and then assign the **IslandsNET1993.LL ISN93** coordinate system to the dataset while loading data by using the **Connect** tool.

b. Save the file.

Starting a Drawing File

1. Choose **New > Drawing** from the Application Menu; the **Select template** dialog box is displayed.

2. In the **Select template** dialog box, select the *map2d.dwt* template file and then choose the **Open** button; the **map2d** template is applied to the Modelspace.

Assigning the Coordinate System to the Dataset Using the Data Connect Tool

1. Choose the **Connect** tool from the **Data** panel in the **Home** tab; the **DATA CONNECT** wizard is displayed.

2. In the **DATA CONNECT** wizard, select the **Add SHP Connection** option from the **Data Connections by Provider** list box; the **OSGeo FDO Provider for SHP** page is displayed in the right pane of the wizard.

3. In the **OSGeo FDO Provider for SHP** page, choose the browse button next to the **Source file or folder** edit box; the **Open** dialog box is displayed. In this dialog box, browse to the following location: *C:\m3d_2017\c03_m3d_2017\c03_tut03*.

4. Next, select the **c03-m3d-2017-tut03.shp** file from the list box and then choose the **Open** button; the **Open** dialog box is closed and the path of the shape file is added to the **Source file or folder** edit box in the **OSGeo FDO Provider for SHP** page.

5. Choose the **Connect** button; the selected shape file is displayed in the **Add Data to Map** list box.

6. Choose the **Edit Coordinate Systems** button in the **Add Data to Map** area; the **Edit Spatial Contexts** dialog box is displayed.

7. In the **Edit Spatial Contexts** dialog box, select the first row and then choose the **Edit** button; the **Coordinate System Library** dialog box is displayed.

8. In the **Coordinate System Library** dialog box, select the **Iceland** option from the **Category** drop-down list; a list of coordinate systems in the selected category is displayed in the list box below the **Search** edit box.

9. Next, select the **IslandsNet1993.LL** code with the **ISN93** description option from the list box.

10. Choose the **Select** button in the **Coordinate System Library** dialog box; this dialog box is closed and the **IslandsNet1993.LL** code is displayed in the **Override** column of the **Edit Spatial Contexts** dialog box.

11. Choose the **OK** button in the **Edit Spatial Contexts** dialog box; the dialog box closes and the selected coordinate system is assigned to the shape file.

12. In the **DATA CONNECT** wizard, choose the **Add to Map** button; the shape file gets connected to the Workspace. Note that the **Display Manager** tab of the **TASK PANE** displays the name of the added feature class.

13. Close the **DATA CONNECT** wizard by choosing the Close [**X**] button from the upper left corner; this wizard is closed and the geometry of the shape file is displayed in the drawing window, refer to Figure 3-28.

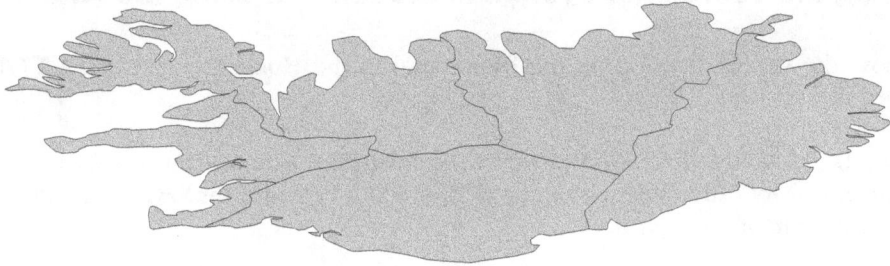

Figure 3-28 *Model created by combining various features*

Saving the Drawing File

1. Choose the **Save As** option in the Application Menu; the **Save Drawing As** dialog box is displayed.

2. In the **Save Drawing As** dialog box, enter **c03_Tut03a** in the **File name** edit box and select the **AutoCAD 2013 Drawing (*.dwg)** option in the **Files of type** drop-down list. Now, choose the **Save** button next to the **File name** edit box; the current drawing file is saved with the given name.

Self-Evaluation Test

Answer the following questions and then compare them to those given at the end of this chapter:

1. Which of the following zoom options displays the previous zoom state of the drawing?

    (a) **Window**   (b) **Realtime**
    (c) **Previous**   (d) **Object**

2. Which of the following tools is used to specify a coordinate system for the dataset?

    (a) **Connect**   (b) **Create**
    (c) **Assign**   (d) **Export**

3. Which of the following shortcut keys is used to toggle the display of the grid in the drawing window?

    (a) **F7**   (b) **F8**
    (c) **F10**   (d) **F6**

4. Which of the following tabs in the **Drafting Settings** dialog box is used to specify the settings for the snap grid?

    (a) **Polar Tracking**   (b) **Object Snap**
    (c) **Dynamic Input**   (d) **Snap and Grid**

5. Which of the following functional keys are used to turn on the **Object Snap** in the drawing window?

    (a) F11   (b) F9
    (c) F3   (d) F7

6. You can use the _____ object snap option to track the end point of a line segment.

7. The _____ option in the **Navigate** panel of the **View** tab is used to zoom to the limits of the selected layer in the drawing window.

8. The _____ object snap option is used to snap to the center of an arc, circle, ellipse, or elliptical arc.

9. To view the list of coordinate systems that are created by the user, select the _____ option from the **Status** drop-down list of the **Coordinate System** dialog box.

10. The world coordinate system is a _____ coordinate system that cannot be modified.

11. As you increase the scale value in the **AutoCAD Map 3D Scale** option, the view of the drawing gets enlarged. (T/F)

12. When you work in the Ortho mode, you can draw lines in any direction. (T/F)

13. You can define a new global coordinate system as per your requirement. (T/F)

14. You cannot align the User Coordinate System (UCS) in any direction. (T/F)

15. You cannot assign a coordinate system to a data from an external source. (T/F)

Review Questions

Answer the following questions:

1. The _____ tool in the **Coordinates** panel of the **View** tab is used to rotate the UCS icon in the YZ plane.

2. The **Midpoint** object snap mode is used to track the _____ point of a line segment.

3. The projected coordinate systems are derived from the _____ coordinate systems by using the map projections.

4. The angle between the prime meridian and the meridian plane of any point on earth's surface is known as _____.

5. You can invoke the **Edit Spatial Contexts** dialog box by choosing the _____ button in the **DATA CONNECT** wizard.

6. The Longitude is an angle between the ellipsoidal normal and the equatorial plane. (T/F)

7. The grid spacing and snapping cannot be adjusted as per the user requirement. (T/F)

8. You can create a user-defined coordinate system using an existing coordinate system. (T/F)

9. By specifying projection parameters, you can transform a coordinate system from one system to another. (T/F)

10. Map projections always results in Cartesian Coordinate System. (T/F)

EXERCISE

Exercise 1

Create a geographic coordinate system using the parameters given below and then assign it to the current drawing.   **(Expected time: 40 min)**

Units: degree
Minimum Longitude -180   Maximum longitude 180
Minimum Latitude -90   Minimum Latitude 90

Ellipsoid (WGS 84)
Code- UserDefinedWGS84   Equatorial radius: 6378137
Polar radius: 6356752.3142   Inverse Flattening: 0.00335281067183102
Eccentricity: 0.0818191909289067

Hint:
This coordinate reference system is based on LL84.

Chapter 4

Working with Feature Data

Learning Objectives

After completing this chapter, you will be able to:

• *Open the scanned maps and raster images*
• *Use the COGO functions while drawing*
• *Create and edit the properties in the Schema Editor*
• *Create, edit, and save feature layer*
• *Digitize point, line, and polygon features*
• *Add and edit the attribute data*

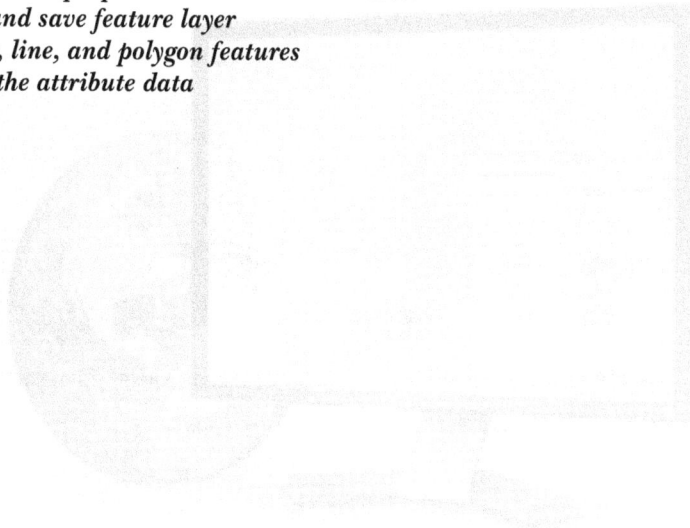

INTRODUCTION

In the previous chapters, you explored the user interface and various coordinate reference systems in AutoCAD Map 3D. You also learned to load various types of data into your project and use the basic navigation tools to explore the drawing data.

In this chapter, you will first learn to insert and correlate a raster image using the **Image** command and then create a feature data store by defining its schema. Next, you will learn the procedure for creating and editing feature data in the feature class. In addition, you will learn to digitize feature data by using the georeferenced raster images as the base map. The various COGO input routines such as the **Distance Distance**, **Angle Distance**, **Azimuth Distance**, and **Bearing Bearing** that help in attaining precision while digitizing, are also discussed in this chapter.

INSERTING AND CORRELATING RASTER IMAGE

Raster images, such as satellite and aerial photographs, are used in drawings as background for spatial data creation. They are also used for change detection, spatial analysis, and mapping. It is therefore required that the image coordinate should correlate to the real-world coordinates. This correlation information can be embedded in the image or stored as a file.

In AutoCAD Map 3D, you can use the **Image** tool to insert a raster image into the drawing. While inserting the image, you can use the correlation information of the image (stored within the image or as a separate correlation file) to insert it at its correct location. You can also insert the image in the drawing with respect to its real world coordinates by specifying the correlation parameters, such as image insertion point, angle of rotation, scale, and pixel size.

The process of inserting a raster image using the **Image** tool involves selecting the required image. Next, you will require to specify the image correlation (insertion) parameters. The insertion parameters can be specified by selecting the correlation source for the image. You can also do so by specifying them manually by defining insertion point, angle of rotation, scale, units, and pixel density. Figure 4-1 shows the illustration of specifying correlation parameters while inserting an image.

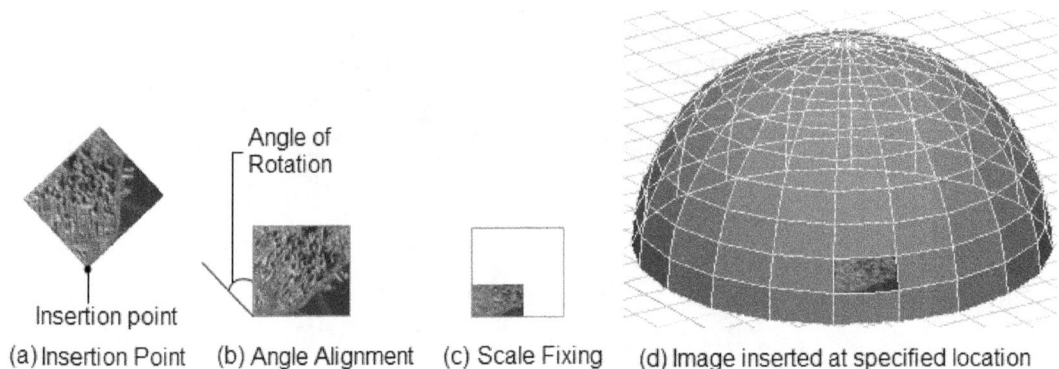

Figure 4-1 *Inserting a raster image*

Inserting a Raster Image Using the Image Tool

Ribbon: Insert > Image > Image
Command: MAPIINSERT

To insert a raster image, choose the **Image** tool from the **Image** panel; the **Insert Image** dialog box will be displayed. In this dialog box, select the desired image file format from the **Files of type** drop-down list. Next, browse to the required folder, select the image file, and choose the **Open** button; the **Insert Image** dialog box will be closed and the **Image Correlation** dialog box will be displayed, as shown in Figure 4-2. You can use the options in this dialog box to specify the correlation parameters or to choose the correlation source. The options and various tabs in this dialog box are discussed next.

Note
*To modify or view the insertion parameters of the image file before inserting it into the drawing, select the **Modify Correlation** check box located below the **Files of type** drop-down list in the **Insert Image** dialog box. If the **Modify Correlation** check box is cleared, the image will be directly inserted into the drawing without displaying the **Image Correlation** dialog box. Note that AutoCAD Map 3D inserts the image by using its default insertion parameters.*

Source Tab

In the **Image Correlation** dialog box, the **Source** tab is chosen by default. Note that, whenever an image is inserted using the **Image** tool, AutoCAD Map 3D searches for the correlation information associated with the image.

In the **Source** tab of the **Image Correlation** dialog box, you can select an option from the **Correlation Source** drop-down list to select the correlation source of the image.

On selecting the correlation source option from the drop-down list, the correlation information of the selected source will be displayed in the **Insertion Values**, **Density** and **Units for Insertion Point and Density** areas of the **Source** tab.

The **Insertion Values** area of the **Source** tab displays the coordinates of the insertion point, rotation angle, and scale of the insertion image. You can modify these parameters by editing the values in their corresponding edit boxes. To modify the insertion point of the image file,

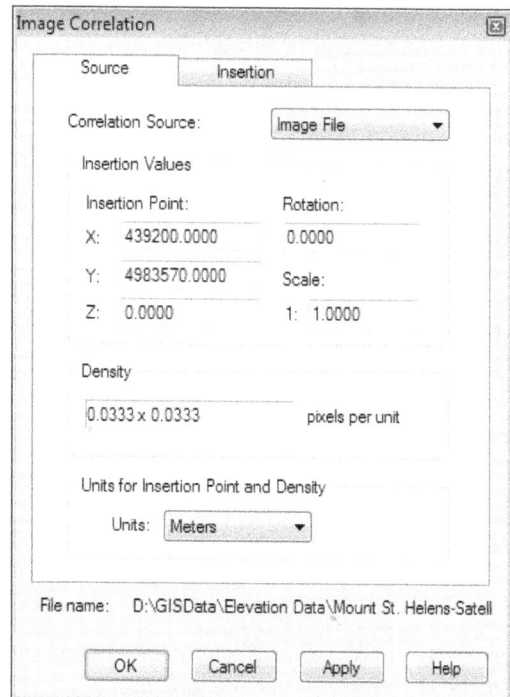

*Figure 4-2 The **Image Correlation** dialog box*

specify the X, Y, and Z coordinates of the point in the **X, Y**, and **Z** edit boxes, respectively. To rotate the image about the insertion point, specify the rotation angle in the **Rotation** edit box (use the positive or negative rotation angle value to specify angle in the clockwise or counterclockwise direction, respectively). To adjust the scale of the image, enter the desired scale value in the **Scale** edit box.

The edit box in the **Density** area displays the number of digits or pixels per square unit area (no. of rows × no. of columns per square unit area).

In the **Units for Insertion Point and Density** area, you can specify the measuring unit of the image. To do so, select an option from the **Units** drop-down list.

Insertion Tab

The **Insertion** tab shows how the image correlation settings specified in the **Source** tab apply to the current drawing. Figure 4-3 shows the **Image Correlation** dialog box with the **Insertion** tab chosen. You can modify the insertion parameters for the image by specifying the values in the corresponding edit boxes in this tab.

You can also graphically specify the insertion parameters of the image. To do so, choose the **Pick** button located below the **Z** edit box in this tab; the cursor will change to a crosshair with the image frame attached to it in the first quadrant. Now, click in the drawing window at the desired location; the image frame will be fixed at the lower left corner and you will be prompted to specify rotation. To specify the rotation, enter the required angle value in the command line. Alternatively, click in the drawing area to specify the rotation graphically; the image will be rotated by the specified angle and you will be prompted to specify the image scale. Next, specify the image scale by entering the value in the Command prompt. To graphically specify the scale, move the cursor in the drawing area. Notice that the frame scale changes along with the cursor movement. Click in the drawing area; the scale of the image will be specified. Also, the **Image Correlation** dialog box will be displayed again along with the insertion values specified graphically.

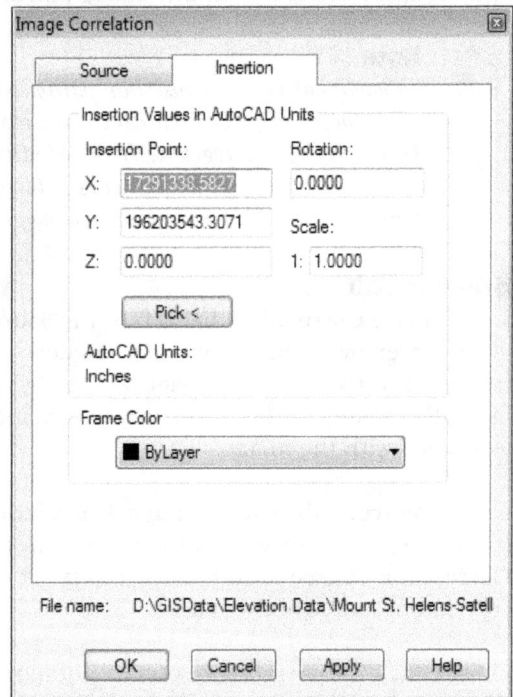

*Figure 4-3 The **Insertion** tab of the **Image Correlation** dialog box*

Note

*Modifying the values in the **Insertion** tab will also change the insertion (correlation) values in the **Source** tab of the **Image Correlation** dialog box.*

After specifying the correlation parameters, choose the **OK** button; the image will be inserted in the drawing. If the inserted image is not visible, invoke the **Extents** tool from the **Navigate** panel of the **View** tab; the drawing will zoom to its extent and the inserted image will be visible in the drawing window.

Tip

*The raster images and scanned maps can also be inserted by using the **Connect** tool from the **TASK PANE**, as explained in Chapter 2.*

FEATURE DATA

After inserting a raster image, you need to define or create feature objects that represent various real-world objects. These feature objects can be created using various geometries such as point, line, and polygon. The feature objects are created in a feature class. A feature class defines the type of real world objects such as lakes, administrative district, roads, street lights poles, and fire hydrants. For example, you can create polygon objects that represent water bodies (lakes, ponds) in a feature class named 'Lake'.

The feature classes are arranged in a feature data store. The feature data store (data store) in AutoCAD Map 3D is a collection of feature classes contained in a single storage location. A data store can be file based, such as SDF or it can be database based, such as oracle, MySQL etc. To add features from a feature class into your drawing, you first need to connect to the data store and then add the feature class to your drawing.

CREATING FEATURE DATA STORE

You can create a dataset in a format that complies with a specific Feature Data Object (FDO) provider, such as SDF, SHP, or Oracle. In order to create a dataset in these formats, you need to create a new data store. You can also create a new data store to migrate the existing data to a new FDO provider.

In AutoCAD Map 3D, you can create two types of data stores: database data store and file data store. To create a database data store, you will need user privileges for the database system.

After creating a data store, you need to define a schema for the store. A schema defines the available feature classes and their properties in the data store. You can define a new schema or import an existing schema definition in the store. The procedures to create an SHP and SDF data stores are discussed next.

Creating SHP Data Store

Ribbon: Create > Feature Data Store > SHP

Using AutoCAD Map 3D, you can create a file based data store in an SHP or SDF format. To create an SHP file based data store, choose the **SHP** tool from the **Feature Data Store** panel of the **Create** tab; the **Choose Shape File** dialog box will be displayed. In the **Choose Shape File** dialog box, browse to the required folder location and then specify the name for the shape file in the **File name** edit box. Next, choose the **Save** button; the **Choose Shape File** dialog box will be closed and the **Specify Coordinate System** dialog box will be displayed, as shown in Figure 4-4.

In this dialog box, you can assign a coordinate system to the new shape file. To do so, enter the code of the coordinate system in the **Coordinate System** edit box. Alternatively, you can select a coordinate system from the CRS library. To do so, choose the browse button to the right of the **Coordinate System** edit box; the **Coordinate System Library** dialog box will be displayed. Specify the coordinate system in this dialog box, as discussed in Chapter 3.

After the required coordinate system is selected, its code name will be displayed in the **Coordinate System** edit box in the **Specify Coordinate System** dialog box. Next, choose the **OK** button;

the dialog box will be closed and the **Schema Editor** dialog box will be displayed, as shown in the Figure 4-5.

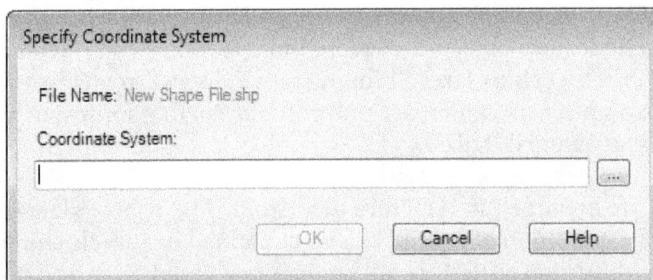

*Figure 4-4 The **Specify Coordinate System** dialog box*

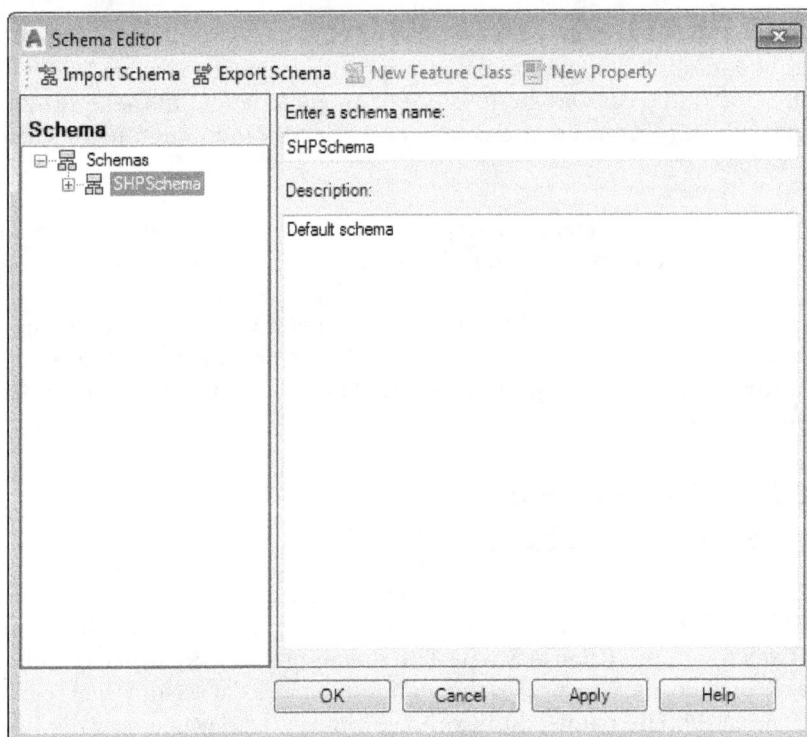

*Figure 4-5 The **Schema Editor** dialog box*

Creating a Schema for the SHP Data Store

You can create a new schema by using the options in the **Schema Editor** dialog box. A schema defines the structure of the data store. The data store schema stores information about the available feature class and its properties. You can create a new schema by defining its parameters or you can import an existing schema to create a new one.

Note
In Map 3D, you cannot edit or change the schema for a shape file after it has been created. It can be defined while creating a new file based SHP data store.

To name the default schema in the **Schema Editor** dialog box, specify the name of the new schema in the **Enter a schema name** edit box. Optionally, enter a description in the **Description** text box.

Importing an Existing Schema

You can import settings from an existing schema into the current schema. To do so, choose the **Import Schema** button from the **Schema Editor** dialog box; the **Open** dialog box will be displayed. In the **Open** dialog box, select the required schema file with *.XML extension, and then choose the **Open** button to include the schema in the drawing space; the imported schema will be added to the schemas parent node in the **Schema** list box.

Exporting a Schema

A schema can be exported to an external location and then can be used for another dataset. To export a schema in the *.xml* file format, choose the **Export Schema** button from the **Schema** dialog box; the **Save As** dialog box will be displayed. In this dialog box, enter the name in the **File name** edit box, and then choose the **Save** button to save the file.

Deleting an Existing Schema

To delete a schema, right-click on the schema name under the **Schemas** node in the left pane; a shortcut menu will be displayed. Choose the **Delete Schema** option from the shortcut menu to delete the schema.

Tip
*To edit a schema, right-click on the schema name and choose the **Edit Schema** option from the shortcut menu displayed.*

Adding a New Feature Class

To add a new feature class to an SHP data store, choose the **New Feature Class** button in the **Schema Editor** dialog box; a new feature class will be created and added to the schema node in the left pane. Also, the **Logical Feature Class** tab will be displayed in the right pane of the **Schema Editor** dialog box, as shown in Figure 4-6.

Note
*A SHP file can contain only one feature class. If a feature class exists in the SHP data store, the **New Feature Class** button in the **Schema Editor** dialog box will be disabled.*

To delete an existing feature class, right-click on the feature class name; a shortcut menu will be displayed. Choose the **Delete Feature Class** option from this menu; the selected feature class will be deleted.

In the **Logical Feature Class** tab, specify a name for the new feature class in the **Name** edit box. Optionally, you can provide details of the feature class in the **Description** text box. To specify the type, select an option from the **Type** drop-down list. By default, the **Feature class** option is selected in the **Type** drop-down list. To specify non spatial data, select the **Non-Feature class** option from the drop-down list. To specify whether the feature class will be abstracted or not, select the **Yes** or **No** radio button. Specify the identifier properties in the **Specify identifier**

propert(ies) and the order list box. You can also specify constraints by selecting the relevant option from the **Specify unique constraint(s) and the order** drop-down list.

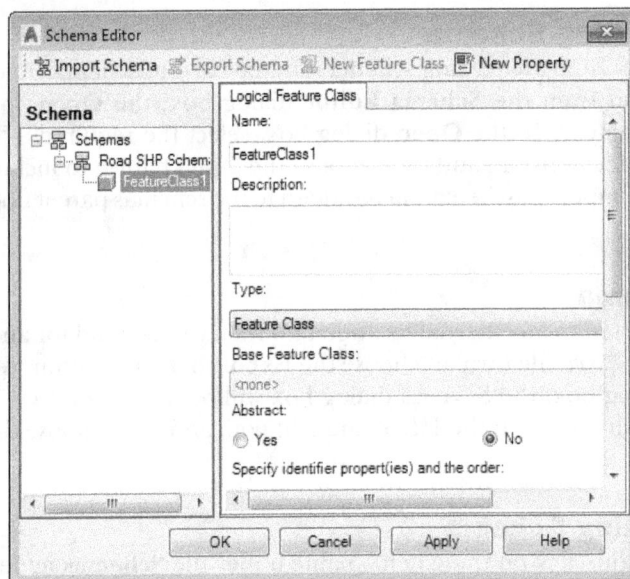

*Figure 4-6 The **Logical Feature Class** tab displayed in the **Schema Editor** dialog box*

Once the feature class is created, you can add properties to it. You can specify the geometry type of the feature class and add more properties to hold other attribute data. The procedure of defining the feature class property is discussed next.

Note
Once you made the changes to the properties of the feature class and saved it, you cannot edit further the properties of the feature class.

Adding Property to the Feature Class

Using the properties, you can define the attribute of the feature class. To add a new property, select the feature class node in the left pane of the **Schema Editor** dialog box; the **New Property** button will be activated. Choose this button; a new property with default property name will be added to the **Schema** list box. The **Logical Property** tab in the **Schema Editor** dialog box will display the parameters of this newly created property. Enter the name of the property in the **Name** edit box in the **Logical Property** tab. Select the relevant option from the **Type** drop-down list to specify the type of property (**Geometry** or **Data**); the **Data Attributes** area in the **Logical Property** tab will display the list of parameters corresponding to the option selected in the **Type** drop-down list.

Note
*A shape file does not support multiple geometry type for representing feature object. As a result, you can define only one property with **Type** set as **Geometry** in the **Schema Editor** dialog box.*

Tip

*You can also add a new property in the **Schema Editor** dialog box using the shortcut menu. To do so, select the feature class in the **Schema** list box and right-click; a shortcut menu will be displayed. Next, choose the **New Property** option from the displayed shortcut menu.*

If the **Geometry** option is selected in the **Type** drop-down list of the **Logical Property** tab, the **Data Attributes** area will display the list of attributes required to define the geometry of the feature class. To define the type of geometry for the feature class, click in the cell corresponding to **Geometric Types** in the **Data Attributes** list box; a drop-down list is displayed. Select the required geometry type from this drop-down list, as shown in Figure 4-7.

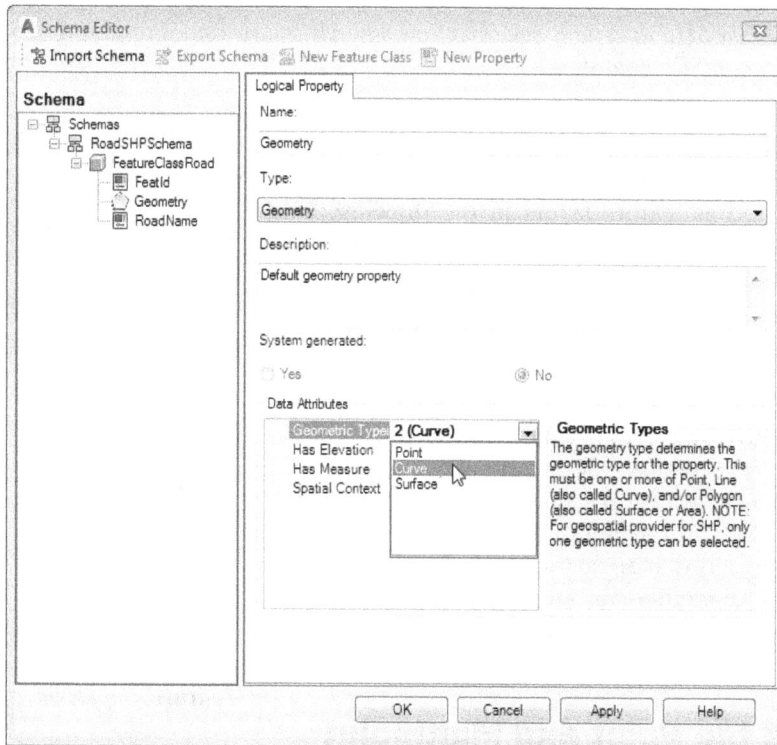

*Figure 4-7 Selecting the **Curve** option from the **Geometric Types** drop-down list*

You can add properties for defining the spatial data of the feature class. To do so, select the **Data** option from the **Type** drop-down list; the **Data Attributes** area will display parameters required to define this (**Data**) type of property. To specify the type of data stored in this property, click in the cell corresponding to the **Data Type** in the **Data Attributes** area; a drop-down list will be displayed. Choose the required data type from this drop-down list. Specify the other parameters corresponding to the selected data type.

After setting the required properties, choose the **OK** button from the **Schema Editor** dialog box; the **Submit Changes** message box will be displayed. Choose the **Yes** button from the message box; the **Schema Editor** dialog box will be closed, and the SHP file with properties will be added as a feature layer in the **Display Manager** tab of the **TASK PANE**.

Creating SDF Data Store

Ribbon: Create > Feature Data Store > SDF

You can create the SDF data store similar to the way as you created the SHP data store. To create an SDF data store, choose the **SDF** tool from the **Feature Data Store** panel of the **Create** tab; the **Choose Spatial Database File** dialog box will be displayed. In this dialog box, browse to the required location and assign a name to the data store to be created in the **File name** edit box. Next, choose the **Save** button; the **Specify Coordinate System** dialog box will be displayed. Now, follow the procedure explained in the previous sections to specify coordinate system, add schema, and create feature class.

Note
Unlike the SHP data store, you can create multiple feature classes consisting of multiple geometry types in an SDF data store.

CREATING A FEATURE DATA

After creating a data store, the feature classes in the data store are added as feature layers in the drawing. The **Display Manager** tab of the **TASK PANE** displays the name and style of the added feature layers. Now you can begin creating feature data in the feature layer.

To create a feature data, select the required feature layer in the **Display Manager** tab of the **TASK PANE**; the **Vector Layer** tab (contextual) will be displayed in the ribbon. Next, choose the required feature creation tool from the **New Feature** drop-down in the **Create** panel of the **Vector Layer** tab (contextual). Figure 4-8 shows various tools that are available in the **New Feature** drop-down. The method for creating a point, line, and polygon feature is discussed next.

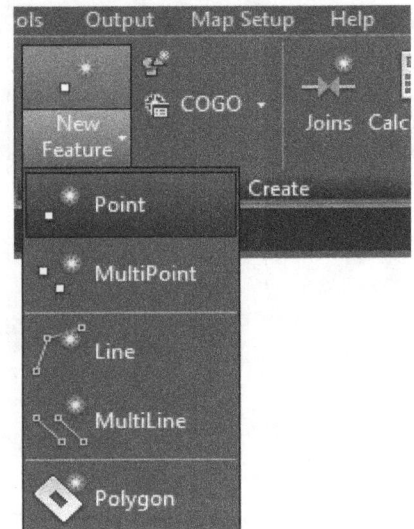

*Figure 4-8 The tools in the **New Feature** drop-down*

Creating the Point and Multipoint Feature Data

To create a point feature, choose the **Point** tool from the **Vector Layer > Create > New Feature** drop-down; you will be prompted to specify the point. Click in the drawing to specify the point; a point object will be created at the specified location and the **DATA TABLE** is displayed with a new row corresponding to the created feature. In this table, specify the attribute data of the created feature. Next, you will need to check in feature to save the modifications to the feature source. To check in created feature, right-click on the feature layer in the **Display Manager** tab of the **TASK PANE**; a shortcut menu will be displayed. Choose the **Check In Features** option from the displayed menu; the specified parameters for the new feature will be saved along with the updates in the DATA TABLE. Figure 4-9 shows an example of the point feature created.

Figure 4-9 *An example of the point feature created*

You can also create multipoint features in the drawing. Multipoints features are composed of more than one point. They are used to represent points with the same properties and attributes. To create multiple points, choose the **MultiPoint** tool from the **Vector Layer >** **Create > New Feature** drop-down; you will be prompted to specify points. Specify the required points by clicking in the drawing area. To end creating points, right-click in the drawing area; a shortcut menu will be displayed. Choose **Enter** from the displayed menu; the **DATA TABLE** will be displayed. Enter the attribute data in the **DATA TABLE**. Next, check in the created features to save them as explained earlier.

Note
*The **DATA TABLE** can be edited even after creating the feature class dataset. The methods of editing the **DATA TABLE** are discussed in the later section of this chapter.*

Creating the Line and MultiLine Feature Data

The line feature is used to denote a set of linear points connected by segments such as roads, railways, pipelines, canals, streams, and electric cables. The line feature is used to represent stream line geometry along with its properties in a tabular form. You can create a line or multiline feature using AutoCAD Map 3D.

To add a new line feature in the feature layer, select the feature layer in the **Display** **Manager** of the **TASK PANE**; the **Vector Layer** contextual tab will be displayed in the ribbon. Next, choose the **Line** tool from **Vector Layer > Create > New Feature** drop-down;

the cursor will change into a crosshair in the drawing window and you will be prompted to specify the start point of the line. Click in the drawing to specify the start point of the line; you will be prompted to specify the next point. Create the required linear geometry by specifying points. To end creating line geometry, right-click in the drawing area; a shortcut menu will be displayed. Choose **Enter** from the displayed menu or press ENTER; the **Data Table** is displayed. Enter the attribute data in the **DATA TABLE** and press ENTER; the line feature will be created. Figure 4-10 shows an example of the line feature created. Now, to save the changes to the feature source, right-click on the line feature in the **Display Manager** tab of the **TASK PANE**; a shortcut menu will be displayed. In this shortcut menu, choose the **Check In Features** option; the modified settings will be saved.

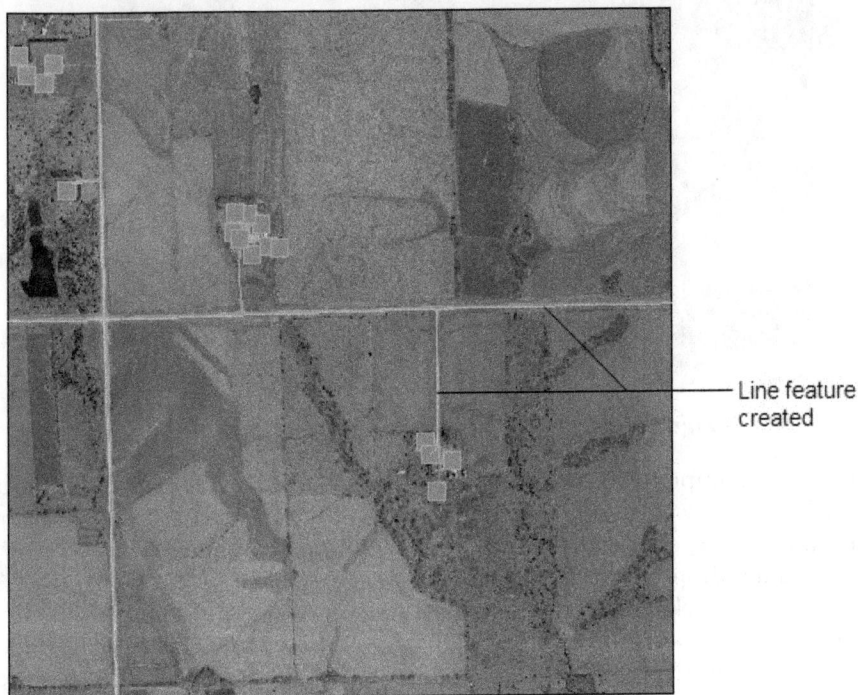

Figure 4-10 An example showing the line feature created

Note
*You can create a multi-line feature using the **MultiLine** tool. To do so, choose the*
***MultiLine** tool from **Vector Layer** (contextual) **> Create > New Feature**
drop-down.

Creating the Polygon and MultiPolygon Feature Data

Polygon features are drawing objects created by using the closed polyline geometry. They are used to represent area features such as plots, ponds, building perimeter, regions, and council boundary. Similar to the point and line features, you can create the polygon and multipolygon features using AutoCAD Map 3D.

To add new polygon features to a polygon feature class, choose the **Polygon** tool from **Vector Layer > Create > New Feature** drop-down; the cursor will change into a crosshair in the drawing window and you will be prompted to specify the first point. Click in the drawing to specify the start point; you will be prompted to specify the second point. Digitize the required polygon geometry in the drawing. To close the polygon automatically, enter **C** in the Command prompt; the polygon will be closed. Next, press ENTER; the **DATA TABLE** will be displayed. Specify the attribute for the polygon in the **DATA TABLE**.

Tip
*You can also create a polygon within a polygon to represent an island or a hole. To do so, select the **Ring** option from the dynamic input and then specify the geometry for the ring (hole). Next, press ENTER; the hole will be created within the polygon.*

Next, to save the changes in the feature layer, right-click on the line feature in the **Display Manager** tab of the **TASK PANE**; a shortcut menu will be displayed. In this shortcut menu, choose the **Check In Features** option; the modified settings will be saved. Figure 4-11 shows the polygon feature created.

Polygon feature created

Figure 4-11 The polygon feature created

Tip
*You can also save the created features by choosing the **Check In** tool from the **Edit Set** panel of the **Vector Layer** (contextual) tab.*

You can create a multi-polygon feature by using the **MultiPolygon** tool. To do so, choose the **MultiPolygon** tool from **Vector Layer** (contextual) **> Create > New Feature** drop-down.

ENHANCING PRECISION USING COGO INPUT TOOLS

Ribbon: Home > Draw > COGO drop-down > COGO Input
Command: MAPCOGOFUNCTION

You can use various routines in the **COGO** (Coordinate Geometry) **Input** tool to enhance drafting precision. These routines are used to locate coordinate points geometrically with reference to existing drawing entities such as point, line, and edge. The **COGO Input** tool will enhance the accuracy while drafting. They are also used to retrieve the geometric alignment information between two points. To invoke a routine, choose the **COGO Input** tool from the **Draw** panel of the **Home** tab; the **COGO Input** dialog box will be displayed, as shown in Figure 4-12. The left pane of the dialog box displays various routines available in this dialog box. The methods of using different routines from this dialog box are discussed next.

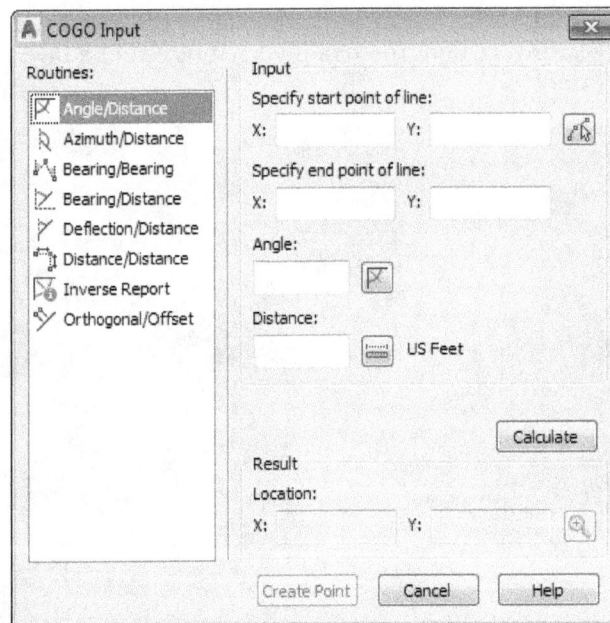

Figure 4-12 The COGO Input dialog box

Using the Angle/Distance Routine

The **Angle/Distance** routine is used to locate a new point based on the calculation of angle and distance measured from an existing line. The **Angle/Distance** routine is selected by default in the **Routines** list box of the **COGO Input** dialog box. As a result, the **Input** and **Result** areas corresponding to this routine are displayed in this dialog box. To specify a point based on the **Angle/Distance** routine, enter the start and end points of the reference line. To specify the coordinates (x, y) of the start point of the line, enter the coordinate value in the **X** and **Y** edit boxes below the **Specify start point of line** label and press TAB; a tripod will be displayed at the specified point. Next, to specify the endpoint of the line, enter the coordinates in the **X** and **Y** edit boxes below the **Specify end point of line** label.

Alternatively, you can graphically specify the coordinates of the start and end points of the reference line. To do so, choose the button displayed on the right of the **Y** edit box in the **COGO Input** dialog box; the dialog box will be closed, the cursor will change to a selection box in the drawing window and you will be prompted to select a line or points in the drawing window. Select the required line in the drawing window; AutoCAD Map 3D will read the start and end point coordinates of the selected line and specify them in the corresponding **X** and **Y** edit boxes in the **COGO Input** dialog box.

Note

AutoCAD Map 3D will display tripod at the start and end points of the reference line irrespective of the method used for specifying the coordinates of the points.

After specifying the coordinates for the start and end points, specify the value for the angle in the **Angle** edit box. This angle is the angle that the point will make with the reference line.

To apply the clockwise measuring angle from the selected reference line, enter the required angle value in the **Angle** edit box. To specify the distance of a new point from the start point, enter the distance in the **Distance** edit box. To locate the required coordinate point using these parameters, choose the **Calculate** button; the calculated x and y coordinates of the point will be displayed in the **X** and **Y** text boxes in the **Result** area. To align the new calculated point at the center of the drawing window, choose the zoom button next to the **Y** text box in the **Result** area; the calculated new point will be displayed in the drawing window. Figure 4-13 shows the use of the **Angle/Distance** routine to locate a new point with reference to the reference line.

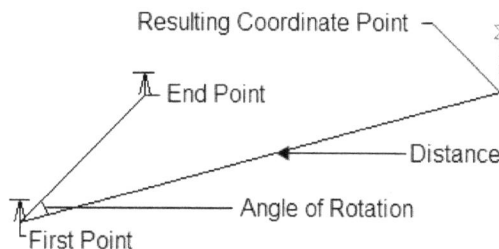

*Figure 4-13 Locating a new point by using the **Angle/Distance** routine*

Tip

*You can also invoke the **Angle/Distance** routine from **Home > Draw > COGO** drop-down. On choosing this routine, the cursor will change into a selection box in the drawing window. Now, select the desired line; the **Angle/Distance** sketching mode will be invoked and you will be prompted to specify angle. Next, align the line vector along the required angle and click to specify the angle; you will be prompted to specify distance. Click again to specify distance, as shown in Figure 4-14.*

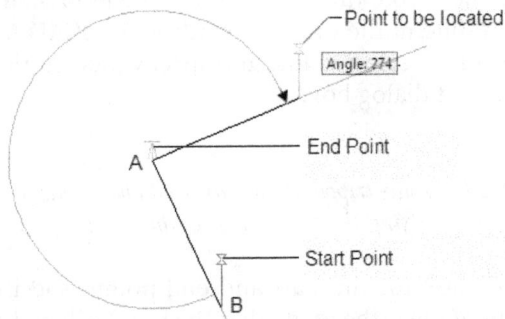

*Figure 4- 14 Specifying the parameters for the **Angle/Distance** routine*

Using the Azimuth/Distance Routine

The **Azimuth/Distance** routine is used to locate a new point based on the distance from a specific point and the angle measured clockwise from the North. To use the **Azimuth/Distance** routine, choose the **Azimuth/Distance** option in the **Routines** list of the **COGO Input** dialog box; the **Input** area will display the edit boxes for the parameters required to use the **Azimuth/Distance** routine.

To set the point of reference, enter the coordinates of the point in the **X** and **Y** edit boxes. Alternatively, to specify the point in the drawing, choose the button next to the **Y** edit box; the **COGO Input** dialog box will be closed and you will be prompted to specify the point. Click at the desired location in the drawing area to specify the point; the coordinates of the specified point will be displayed in the **X** and **Y** edit boxes in the **COGO Input** dialog box.

To specify the angle with respect to the North direction, enter the angle value in the **Azimuth** edit box. To apply the distance to the point, enter the value in the **Distance** edit box. Now, to locate a point based on the parameters given in the **Input** area, choose the **Calculate** button; the resulting coordinates of the point will be displayed in the **X** and **Y** text boxes. To align the located point to the center of the drawing window, choose the button next to the **Y** text box. Figure 4-15 shows a new point located by using the **Azimuth/Distance** routine.

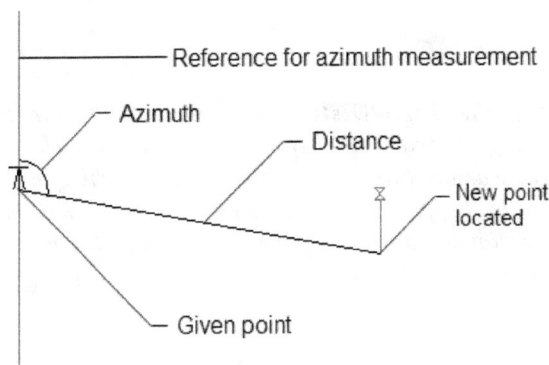

*Figure 4-15 Locating a new point by using the **Azimuth/Distance** routine*

Tip
*You can invoke the **Azimuth/Distance** routine also by choosing the **Azimuth Distance** tool from **Home > Draw > COGO** drop-down. Alternatively, you can enter the **ZD** command in the command line.*

Using the Bearing/Bearing Routine

The **Bearing/Bearing** routine is used to locate a new point at the intersection of bearings measured from two points. The angle of measurement or bearing is specified with respect to a quadrant. The directions of measurement of the bearing in four quadrants are shown in Figure 4-16. To locate a new point by using bearings from specified points or lines, choose the **Bearing/Bearing** routine in the **Routines** list of the **COGO Input** dialog box; the **Input** area in the dialog box will display the edit boxes for the parameters required to use the **Bearing/Bearing** routine.

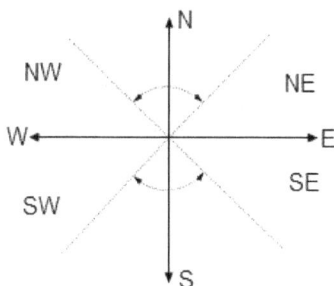

Figure 4-16 Four quadrants with the direction of angle measurement

Enter the coordinates of the two points in the corresponding **X** and **Y** edit boxes. Next, select the quadrant for the bearing by choosing the required option from the **Quadrant** drop-down list. Also, specify the bearing angle corresponding to each point in the **Bearing** edit box. You can also specify the bearing graphically. To do so, choose the button next to the **Bearing** edit box; the **COGO Input** dialog box will be closed and you will be prompted to specify the center of the quadrant. Click in the drawing to specify the center of the quadrant. The drawing window will display the quadrants and you will be prompted to enter angle. Click in the drawing to specify the required angle and quadrant; the **COGO Input** dialog box will be displayed again with the specified bearing and angle.

After specifying the required parameters, choose the **Calculate** button; the coordinates of the new point will be displayed in the **X** and **Y** text boxes in the **Result** area. To align the new point at the center of the drawing window, choose the button next to the **Y** text box in the **Result** area. Figure 4-17 shows a new point located by using the **Bearing/Bearing** routine.

New point created at the intersection of bearing

Bearing   Bearing

Point 1 (x,y)   Point 2 (x,y)

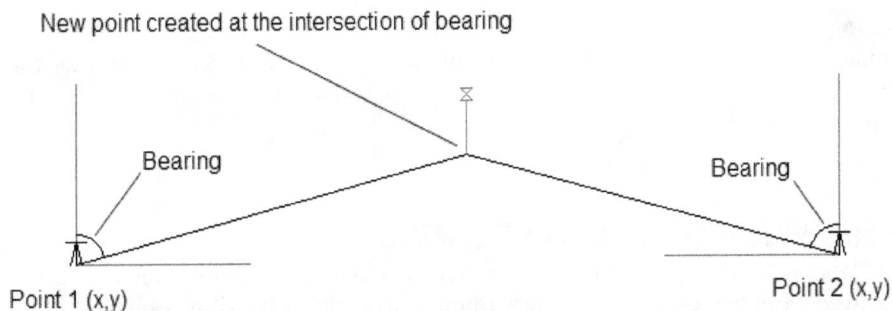

Figure 4-17 Locating a new point based on bearings from two points

Tip
*You can invoke the **Bearing/Bearing** routine by choosing the **Bearing Bearing** tool from **Home > Draw > COGO** drop-down. Alternatively, enter **BB** command in the command line.*

Using the Bearing/Distance Routine

The **Bearing/Distance** routine is used to locate a new point by using the bearing and distance from an existing point or line. To locate a new point from a point object, choose the **Bearing/Distance** routine from the **COGO Input** dialog box; the **Input** area in the dialog box will display the edit boxes for the parameters required to use the **Bearing/Distance** routine.

To select a reference point, enter the coordinates of the reference point in the **X** and **Y** edit boxes. Alternatively, you can select an existing point or locate a new point from the drawing window. To do so, choose the button next to the **Y** edit box in the **Input** area; the cursor will change into the crosshair. Now, you can use the crosshair to select an existing point or place a new point in the drawing window, as explained in previous section; the coordinates of the located point will be displayed in the **X** and **Y** edit boxes in the **COGO Input** dialog box.

To apply a suitable quadrant to bearing, select an option from the **Quadrant** drop-down list in the **Input** area. To use the required bearing for locating the point in the selected quadrant, enter the value in the **Bearing** edit box. To specify the distance, enter a value in the **Distance** edit box. To calculate the coordinates of the new point from the given parameters, choose the **Calculate** button; the coordinates will be displayed in the **X** and **Y** text boxes in the **Result** area. Figure 4-18 shows a new point located by using the **Bearing/Distance** routine.

Tip
*You can invoke the **Bearing Distance** routine by choosing the **Bearing Distance** tool from **Home > Draw > COGO** drop-down. Alternatively, enter the **BD** command in the Command prompt.*

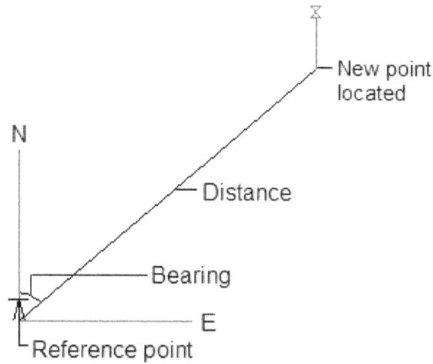

Figure 4-18 *Locating a new point by using the* ***Bearing/Distance*** *routine*

Using the Deflection/Distance Routine

The **Deflection/Distance** routine is used to locate a new point based on the angle of deflection from the selected line and distance. To do so, choose the **Deflection/Distance** option from the **Routines** list box of the **COGO Input** dialog box; the **Input** area in the dialog box will display the edit boxes for the parameters required to use the **Deflection/Distance** routine. To specify the coordinates of the start point and the endpoint of the reference line, enter the coordinate values in the **X** and **Y** edit boxes corresponding to the specific points. Alternatively, select the reference line by choosing the button next to the **Y** edit box; the cursor will change to the selection box, which can be used to select the reference line in the drawing window. To set the deflection angle with respect to the selected reference line, enter the angle in the **Deflection** edit box. Alternatively, you can select the deflection angle from the geometric alignment between two lines by choosing the button next to the **Deflection** edit box; the cursor will change to a crosshair. Now, you can specify the angle of deflection between two lines by aligning the line vector displayed to the required angle. Next, specify the distance in the **Distance** edit box or choose the button next to this edit box to specify the distance graphically in the drawing window.

After entering all the parameters, choose the **Calculate** button; the coordinates of the new location point will be displayed in the **X** and **Y** edit boxes in the **Result** area. Now, choose the button next to the **Y** text box; the new location point will align to the center of the drawing window. Figure 4-19 shows a new point located by using the deflection and distance parameters.

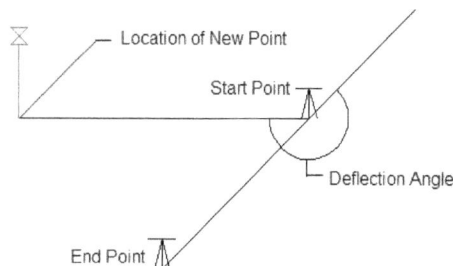

Figure 4-19 *A new point located by using the* ***Deflection/Distance*** *routine*

Using the Distance/Distance Routine

The **Distance/Distance** routine is used to locate a point based on the distances measured from two reference points. To select two points for measuring the distance, enter the coordinate values for the point1 and point2 in the corresponding **X** and **Y** edit boxes in the **COGO Input** dialog box. Next, specify the distance value from point 1 and point 2 in the **Distance** edit boxes corresponding to points. After entering the required parameters, choose the **Calculate** button; the drawing window will display two circles with point 1 and point 2 as their center. The radius of these circles will be equal to the value specified in their corresponding **Distance** edit boxes, refer to Figure 4-20. Also, the coordinates of the point of intersection corresponding to the selected radio button will be displayed in the **X** and **Y** edit boxes of the **Result** area.

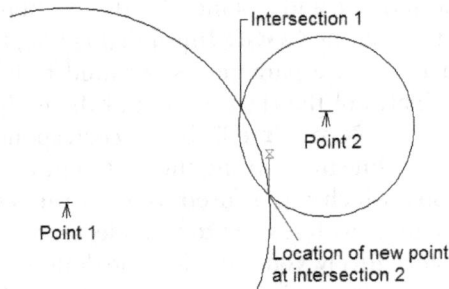

*Figure 4-20 Locating a new point by using the **Distance/Distance** routine*

Note
*The circles in the drawing area and the radio buttons in the **Result** area will be displayed only if the circles intersect each other. In case you do not see the circles, change the value in the **Distance** edit box/es in the **COGO Input** dialog box and then choose the **Calculate** button.*

Now, to locate a specific point from the two intersection points, select the **Intersection 1** or **Intersection 2** radio button; the coordinates of the new located point will be displayed in the **X** and **Y** display boxes in the **Result** area. Now, choose the button next to the **Y** display box; the new point located will align to the center of the drawing window.

The Inverse Report Routine

The **Inverse Report** routine is used to retrieve the geometric information between two point objects. To find the geometric details, choose the **Inverse Report** routine from the **Routines** list box in the **COGO Input** dialog box, and then enter the coordinates for the first and second points in the **X** and **Y** edit boxes, respectively. Alternatively, you can choose the button next to the **Y** edit box to select the required points from the drawing window. After you enter the coordinates for the two points in the **Input** area, choose the **Calculate** button; the geometric details between the two points will be displayed in the **Report** area. The geometric parameters displayed in this area are: **Bearing**, **Quadrant**, **Horizontal Distance**, **Vertical Distance**, **Slope Distance**, **Vertical Angle**, and **Percent Slope**. Figure 4-21 shows the geometric alignment information displayed in the **Report** area of the **COGO Input** dialog box.

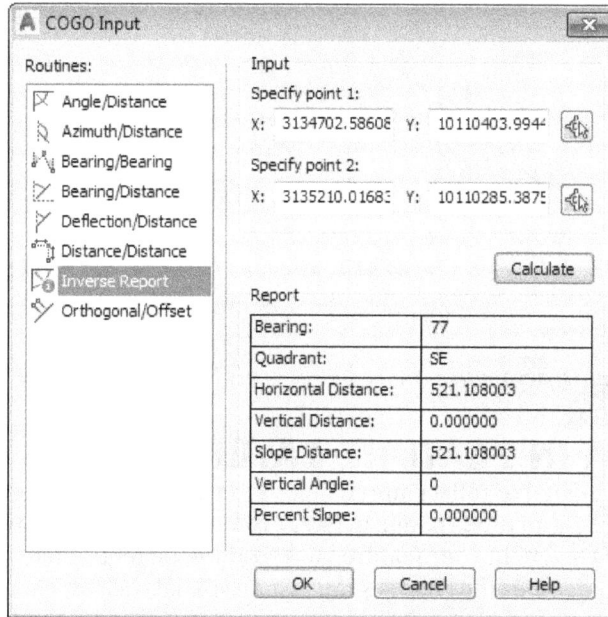

*Figure 4-21 The **COGO Input** dialog box displaying the geometric information*

Tip
*You can also invoke the **Inverse Report** routine by choosing the **Inverse Report** routine from Home > Draw > COGO drop-down.*

Using the Orthogonal/Offset Routine

The **Orthogonal/Offset** routine is used to locate a new point from the specified start point at the given offset. To locate a new point by using this routine, select the **Orthogonal/Offset** option from the **Routines** list box in the **COGO Input** dialog box; the **Input** area will display parameters required for calculating the new point using this routine. To specify the reference points, enter the coordinates in the **X** and **Y** edit boxes, respectively. Alternatively, you can select the reference line from the drawing window by choosing the button next to the **Y** edit box; the cursor will change to a selection box in the drawing window. Next, choose the required reference line by using this selection box; the coordinates of the start point and endpoint will be displayed in the respective edit boxes. To locate the position of offset from the start point along the line, enter the value in the **Distance** edit box. To specify perpendicular distance between the reference line and the new point, enter the value in the **Offset** edit box.

After entering all required parameters in the **COGO Input** dialog box, choose the **Calculate** button; the coordinates of the new point will be displayed in the **X** and **Y** text boxes. Now, to zoom the newly located point at the center of the drawing window, choose the button next to the **Y** text box. Figure 4-22 shows a new point located by using the **Orthogonal/Offset** routine.

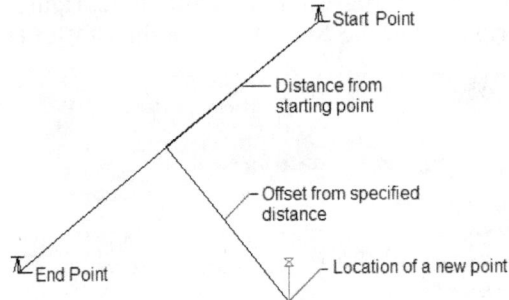

*Figure 4-22 Locating a new point by using the **Orthogonal/Offset** routine*

Note
*The **Orthogonal/Offset** routine can be invoked only by using the **COGO Input** dialog box.*

EDITING DATA IN FEATURE LAYER

GIS data needs to be updated to reflect any changes to a feature. The change can be spatial such as alteration to a building structure, change in river profile, and change in area of administration. To include this type of change in your feature layer, you will be required to edit the geometry of the existing feature. Sometimes the changes can be non spatial such as, change in the ownership of the building or change in the speed limit on a road. In this case, the building or the road geometry remains unchanged, but its non spatial property (ownership, speed limit) requires to be updated. This non spatial data attribute or property data can be edited using the **DATA TABLE**.

Figure 4-23 shows a flow diagram that describes the process for editing a feature (vector) layer in AutoCAD Map 3D. To edit a feature in a feature layer, select the feature layer corresponding to the feature in the **Display Manager** tab of the **TASK PANE**; the **Vector Layer** (contextual) tab will be displayed. Next, choose the **Check Out** tool from the **Edit Set** panel of the **Vector Layer** tab; the feature data in the layer becomes editable.

Note
By default, when you edit features, AutoCAD Map 3D checks out the features automatically. As a result, checking out features may not be necessary.

You can edit the geometry of the feature in the drawing window and its attribute in the **DATA TABLE** window. The various geometry editing and attribute updating procedures are discussed next.

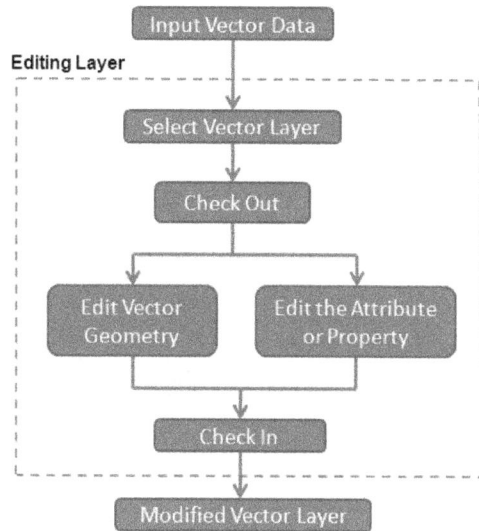

Figure 4-23 The flow diagram for editing the feature layer

Editing Feature Attribute

Ribbon: Home > Data > Table
Command: MAPDATATABLE

Table

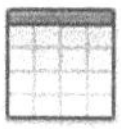

The **DATA TABLE** is an attribute table, which stores data of the feature objects in the feature layer. In this table, you can view and edit the attribute data of features. Using this geospatial data in the **DATA TABLE**, you can perform queries and spatial analysis, which are discussed in the later chapters. You can display this table by choosing the **Table** tool from the **Data** panel of the **Home** tab. Figure 4-24 shows an example of a **DATA TABLE** window for a feature layer.

FeatId	FIPS_ADMIN	GMI_ADMIN	ADMIN_NAME	FIPS_CNTRY	GMI_CNTRY	CNTRY_NAME	REGION	CONTINENT	POP_ADMIN	SC
1	IC	ISL-AST	Austurland	IC	ISL	Iceland	Northern Europe	Europe	14991	218
2	IC	ISL-NVS	Nordhurland V...	IC	ISL	Iceland	Northern Europe	Europe	12396	129
3	IC	ISL-SDH	Sudhurland	IC	ISL	Iceland	Northern Europe	Europe	22897	240
4	IC	ISL-VSL	Vesturland	IC	ISL	Iceland	Northern Europe	Europe	17355	939
5	IC	ISL-VST	Vestfirdhir	IC	ISL	Iceland	Northern Europe	Europe	12219	966
6	IC24	ISL-RYK	Reykjavik	IC	ISL	Iceland	Northern Europe	Europe	157415	181
7	IC	ISL-NYS	Nordhurland Ev...	IC	ISL	Iceland	Northern Europe	Europe	29967	220

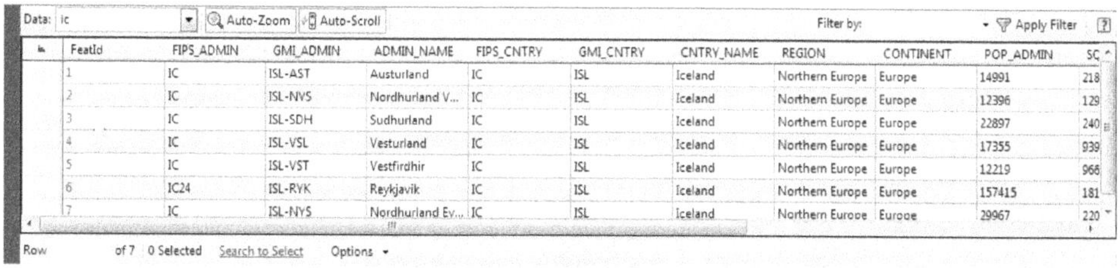

*Figure 4-24 The **DATA TABLE** displaying attribute data of a feature layer*

To edit the text or value in a particular cell of the **DATA TABLE**, select the required cell by clicking on it; the selected cell and the corresponding feature will be highlighted in the drawing area. Figure 4-25 shows a row selected in the **DATA TABLE** window and its corresponding feature highlighted in the drawing window. Notice the arrow symbol displayed on the left of the row. This symbol indicates that the current row or a cell in this row has been selected in the **DATA TABLE**. Again click in the cell; the cursor is displayed in the cell indicating that the cell is in the editable mode. Enter the required value in the selected cell. When you are done with editing, press ENTER; the row will be modified, but the feature is not checked in. Next, you need to save the changes back to the feature source. To do so, choose the **Check In** tool from the **Edit Set** panel of the **Vector Layer** tab.

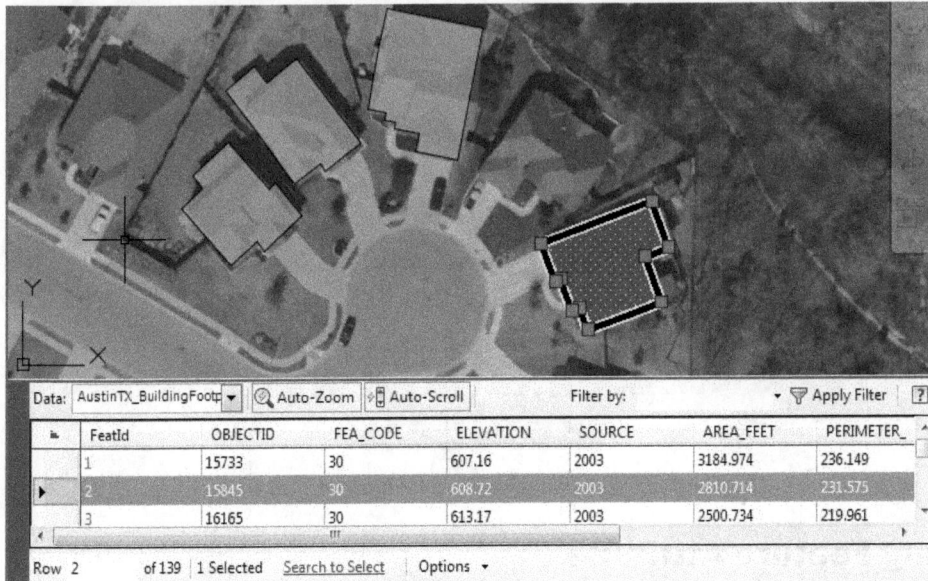

*Figure 4-25 Polygon feature highlighted corresponding to the selected row in the **DATA TABLE***

> **Tip**
> *The various symbols displayed on the left of the row in the **DATA TABLE** represents various operations being performed:*
>
> 1. *The symbol ▶   shows that a cell is selected in a row for editing.*
> 2. *The symbol ∂   shows that a cell is being edited in this row.*
> 3. *The symbol   ⬦ shows that a row has been recently modified, but the feature is not checked in.*

Editing the Feature Data Geometry

To edit the feature geometry, choose the **Check Out** tool from the **Edit Set** panel of the **Vector Layer** tab (contextual); you will be prompted to select the feature. Click on the required feature in the drawing window, the selected feature will now be editable and is highlighted. The selected feature will also display grips at its vertices. In Figure 4-25, the selected polygon is highlighted and the grips are displayed on the vertices of the polygon. You can select and move these grips to alter the shape of the selected feature. This type of editing is known as grip editing. In some cases you may require to split or merge the existing polygons. The method of splitting and merging polygons is discussed next.

Splitting the Geospatial Feature

| **Ribbon:** | Feature Edit > Split/Merge > Split Feature |
| **Command:** | MAPFEATURESPLIT |

To split a checked out feature into two or more parts, choose the **Split Feature** tool from the **Split/Merge** panel; you will be prompted to select the feature to be split. Click on the required feature in the drawing window to select it for splitting and then press ENTER. On doing so, you will be prompted to choose an option for creating a new feature or create a multipart feature after accomplishing the splitting of the selected feature. Choose the required option from the interactive input menu displayed in the drawing window; you will be prompted to specify whether you want to create a new feature ID or use the existing ID. Again, choose the required option from the interactive menu; you will be prompted to select the option to draw or select a line for splitting feature.

If you choose the **Select** option in the interactive menu, you will be prompted to specify the line that will be used to split the polygon feature. Select the required line and press ENTER to split the polygon along the selected line. If you choose the **Draw** option, you will be prompted to draw the line to split the feature. Draw the line feature in the drawing and press ENTER; the feature will be split along the line that you have drawn. Figure 4-26 shows the drawing window with the split feature and its corresponding rows in the **DATA TABLE** created as a result of splitting.

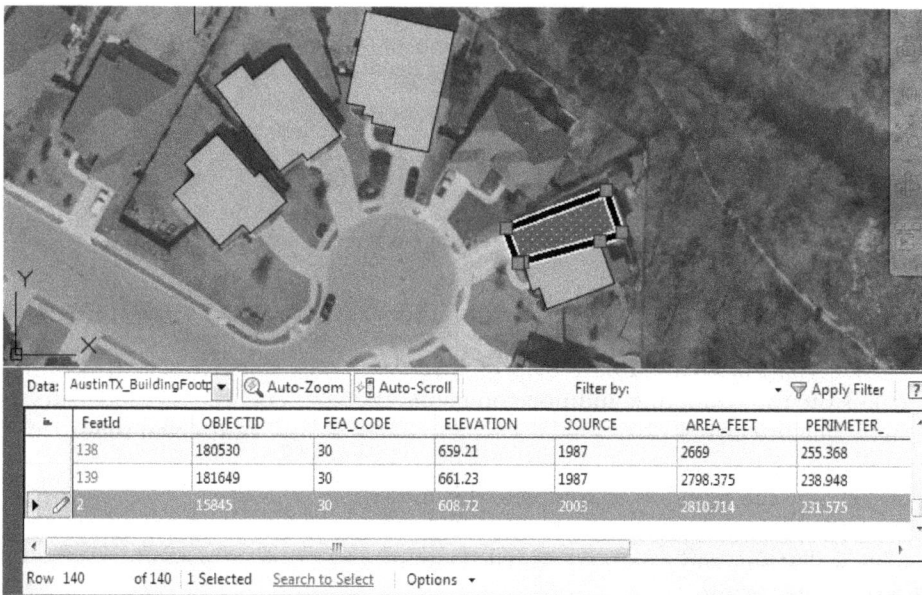

Figure 4-26 *The drawing window displaying the split features and its corresponding data in the* ***DATA TABLE***

In the **DATA TABLE**, the rows corresponding to the newly created feature data will be added below the existing rows. Next, you can edit the values in these new rows, as discussed in the earlier section. To complete the editing of the attribute data, choose the **Check In** tool from the **Edit Set** panel of the **Feature Edit** tab; the editing process will finish and the feature layer will be updated with new edits.

Merging the Geospatial Feature

Ribbon:	Feature Edit > Split/Merge > Merge Feature
Command:	MAPFEATUREMERGE

To merge feature, choose the **Merge Feature** tool from the **Split/Merge** panel; you will be prompted to select the features to be merged. Click on the required features in the drawing area to select them for merging and then press ENTER. On doing so, you will be prompted to specify whether you want to create a new feature ID or use the existing one. Specify it by choosing the required option from the interactive menu and press ENTER; the selected objects will be merged to form a single object. Next, you can edit the feature data in the **DATA TABLE**. After editing the attribute values in the **DATA TABLE**, choose the **Check In** tool from the **Edit Set** panel of the **Feature Edit** tab; the editing process will be completed and the modifications applied to the feature will be saved.

TUTORIALS

General instructions for downloading tutorial files:

Before starting the tutorials, you need to download the tutorial data to your computer. To do so, follow the steps given below:

1. Log on to *www.cadcim.com* and browse to *Textbooks > Civil/GIS > Map 3D > Exploring AutoCAD Map 3D 2017*. Next, select *c04_m3d_2017_tut.zip* file from the **Tutorial Files** drop-down list. Next, choose the corresponding **Download** button to download the data file.

2. Extract contents of the zip file to the following location:

    C:\m3d_2017

Notice that the *c04_m3d_2017_tut* folder is created within the *m3d_2017* folder.

Tutorial 1	**Creating Data Store and Feature Data**

In this tutorial, you will create a base map by inserting a raster image. Next, you will create an SHP data store for point, line and polygon geometry type. Finally, you will create and save features in the feature layers.   **(Expected time: 1hr 15min)**

The following steps are required to complete this tutorial:

a. Start a new drawing and assign coordinate system.
b. Insert the raster image.
c. Using the raster image, digitize the point, line, and polygon features, as shown in Figure 4-27.
d. Specify attribute data for the created features.
e. Save the drawing file.

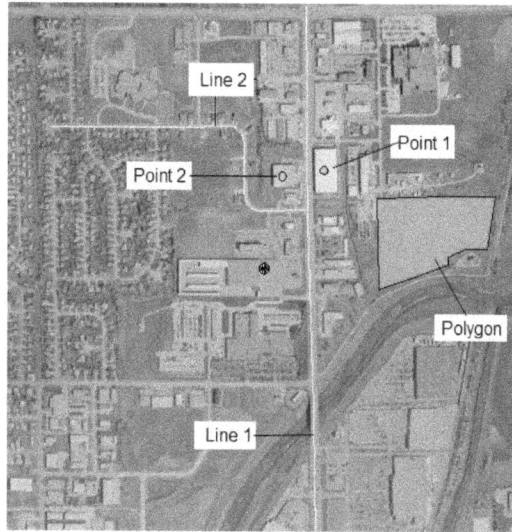

Figure 4-27 Point, line, and polygon features to be created

Starting a New Drawing

In this section of the tutorial, you will start with a new drawing and will assign a coordinate system to it.

1.   Choose the **New** button from the Quick Access Toolbar of AutoCAD Map 3D; a new drawing is started.

2.   Choose the **Assign** tool from the **Coordinate System** panel in the **Map Setup** tab; the **Coordinate System - Assign** dialog box is displayed.

3.   In this dialog box, select the **USA, Nebraska** option from the **Category** drop-down list; the coordinate systems in the selected category are displayed in the list box of this dialog box.

4.   Choose the **NE83** coordinate system with the description **NAD83 Nebraska State Planes, Meter** option from the list box.

5.   Choose the **Assign** button in the **Coordinate System - Assign** dialog box; the dialog box is closed and the selected coordinate system is assigned to the current Workspace.

Loading the Raster Image by Using the Connect Tool

1.   Choose the **Connect** tool from the **Data** panel in the **Home** tab; the **DATA CONNECT** wizard is displayed.

2.   In the **DATA CONNECT** wizard, choose the **Add Raster Image or Surface Connection** option from the **Data Connections by Provider** list box; the **Autodesk FDO Provider for Raster** page is displayed.

3. Enter **Raster** in the **Connection name** edit box, if not displayed by default. Next, choose the Browser button corresponding to the **Source file or folder** edit box; the **Open** dialog box is displayed.

4. In the **Open** dialog box, browse to the following location:

    C:\m3d_2017\c04_m3d_2017_tut\c04_tut01

5. Now, select the *14tpl945240.jpg* image file and choose the **Open** button; the dialog box is closed and the path of the selected image file is displayed in the **Source file or folder** edit box of the **Autodesk FDO Provider for Raster** page.

6. Choose the **Connect** button below the **Source file or folder** edit box; the image file is added to the list box in the **Raster Image or Surface** page.

7. Next, choose the **Edit Coordinate Systems** button in the wizard; the **Edit Spatial Contexts** dialog box is displayed.

8. Select the first row in this dialog box, if not selected by default and then choose the **Edit** button; the **Coordinate System Library** dialog box is displayed.

9. In the **Coordinate System Library** dialog box, select the **USA, Nebraska** option from the **Category** drop-down list; the coordinate systems in the selected category are displayed in the list box in this dialog box.

10. Choose the **NE83 NAD83 Nebraska State Planes, Meter** option from the list box and then choose the **Select** button; the **Coordinate System Library** dialog box is closed and the code **NE83** is displayed in the first row of the **Override** column in the **Edit Spatial Contexts** dialog box. Now, choose the **OK** button; the **Edit Spatial Contexts** dialog box is closed and the selected coordinate system is assigned to the source image file.

11. Next, choose the **Add to Map** button from the **DATA CONNECT** wizard displayed below the **Add Data to Map** list box; the image file is added to the **Display Manager** tab of the **TASK PANE** and is displayed in the drawing window.

12. Close the **DATA CONNECT** wizard.

Creating a Point Shape File for Storing Point Data

In this section of the tutorial, you will create a file based data store with geometry type point. You will also create additional property to store attribute data.

1. Choose the **SHP** tool from the **Feature Data Store** panel in the **Create** tab; the **Choose Shape File** dialog box is displayed.

2. In the **Choose Shape File** dialog box, browse to the following location:

    C:\m3d_2017\c04_m3d_2017_tut\c04_tut01

3. Enter **Point_Feature** in the **File name** edit box and then choose the **Save** button in the dialog box; the dialog box is closed and the **Specify Coordinate System** dialog box with the **NE83** code in the **Coordinate System** edit box is displayed. (Make a new folder if same feature files are showing in the window)

4. Choose the **OK** button in this dialog box; the dialog box is closed and the **Schema Editor** dialog box is displayed.

5. In the **Schema Editor** dialog box, expand the **SHP Schema** node in the **Schema** list box and select the **FeatureClass 1** feature class; the **Logical Feature Class** tab is displayed in the right pane of the dialog box. Next, expand the **FeatureClass 1** node in the **Schema** list box by clicking on the corresponding **[+]** symbol; the properties in this feature class are displayed.

6. Enter **Imp_places** in the **Name** edit box of the **Logical Feature Class** tab. Also, enter **Point feature class showing the location of places** in the **Description** text box. Next, select the **Feature Class** option from the **Type** drop-down list if it is not selected by default.

7. Select the **FeatId** property from the **Schema** list box; the **FeatureClass 1** name is changed to **Imp_places** and the **Logical Property** tab is displayed in the right pane of this dialog box, as shown in Figure 4-28.

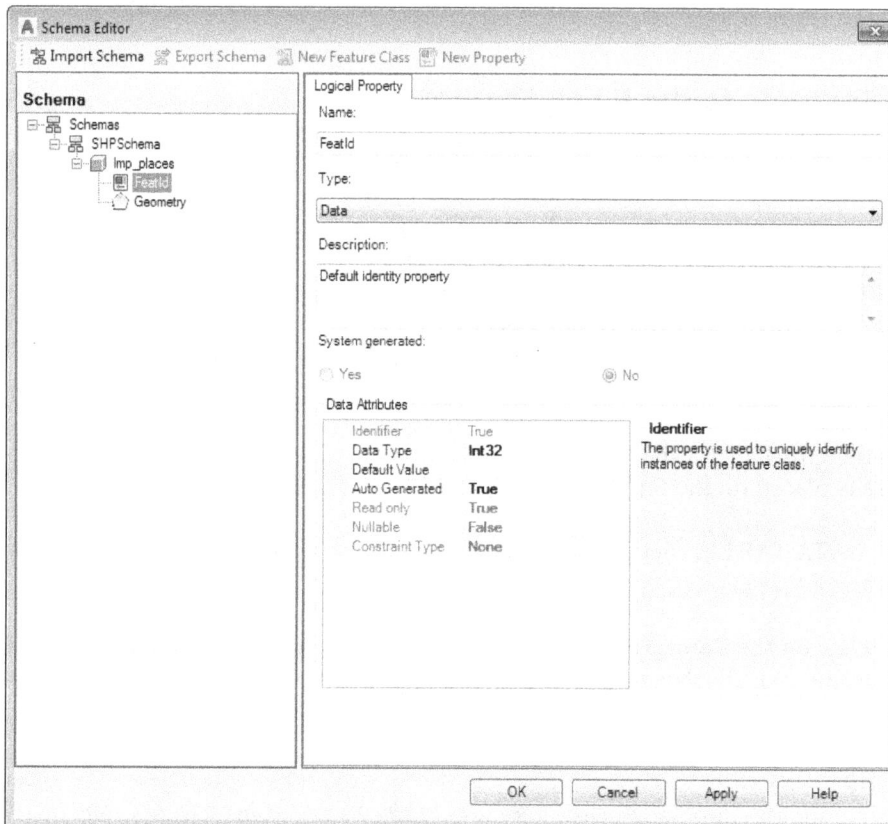

Figure 4-28 *The **Logical Property** tab displayed in the **Schema Editor** dialog box*

8. Next, select the **Geometry** property in the **Schema** list box; the **Logical Property** tab corresponding to the **Geometry** property is displayed in the right pane of the **Schema Editor** dialog box.

9. In the **Data Attributes** area, select the **Point** option from the **Geometric Types** drop-down list; **1(Point)** is displayed in the **Geometric Types** drop-down list.

10. Select the **Imp_places** feature class in the **SHP Schema** list box and then choose the **New Property** button located above the **Logical Feature Class** tab; a new property is created and **Property1** is added to the **Imp_places** node in the **Schema** list box of the **Schema Editor** dialog box.

    Note that the **Logical Property** tab in the **Schema Editor** dialog box displays options corresponding to **Property 1**.

11. Enter **Name** in the **Name** edit box and then select the **Imp_places** node in the **Schema** list box; the name **Property1** changes to **Name** in the **Schema** list box and the **Logical Feature Class** tab is displayed in the right pane of this dialog box.

12. Choose the **OK** button in the **Schema Editor** dialog box; the **Submit Changes** message box is displayed. Next, choose the **Yes** button in this message box; both the dialog boxes are closed and the **Point_Feature** layer is added to the **Display Manager** tab of the **TASK PANE**.

Adding Features to the Point_Feature Layer

The created shape file (data store) is added as a feature layer in the map. In this section of the tutorial, you will create new features in this feature layer.

1. Select the **Point_Feature** layer in the **Display Manager** tab of the **TASK PANE**; the **Vector Layer** (contextual) tab is displayed in the Ribbon.

2. Choose the **Point** tool from **Vector Layer** (contextual) **> Create > New Feature** drop-down; you are prompted to specify a point. Notice that the display of the cursor changes to a crosshair in the drawing window.

3. Place the crosshair on the Building 1, refer to Figure 4-29, and then click on it; a point feature is placed on the building and the **DATA TABLE** is displayed.

4. In the **DATA TABLE**, click in the cell below the **Name** property and enter **Warehouse** in it and press ENTER.

5. Right-click on the **Point_Feature** layer in the **Display Manager** tab of the **TASK PANE**; a shortcut menu is displayed. In this shortcut menu, choose the **Check In Features** option; the editing is completed.

Tip
*You can also check in the created features by choosing the **Check In** tool from the **Edit Set** panel of the **Vector Layer** tab.*

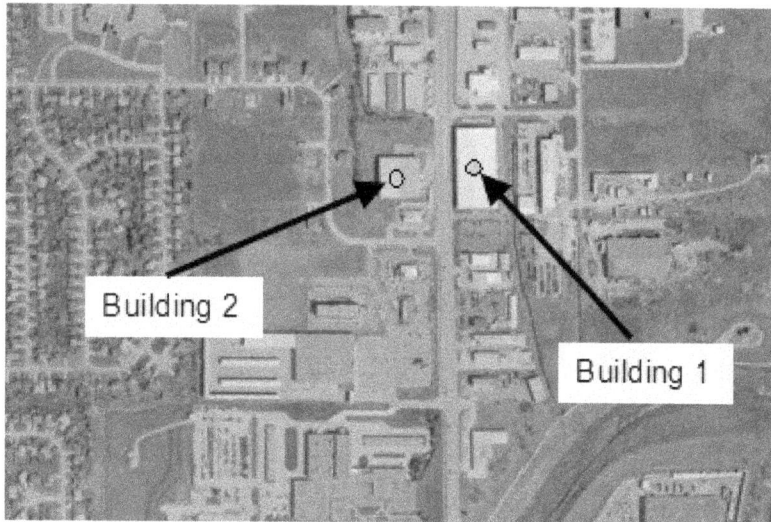

Figure 4-29 *Point features located in the raster image*

6. Repeat the procedure given in steps 1 through step 3 to create a point feature on Building 2, as shown in Figure 4-29.

7. Repeat the procedure followed in steps 4 and 5, and specify **Megastore** in the **Name** property corresponding to the new feature layer and close the **DATA TABLE**.

Create a Shape File for the Line Feature

In this section of the tutorial, you will create an SHP file with line geometry type and define attribute property for storing the attribute data.

1. Choose the **SHP** tool from the **Feature Data Store** panel in the **Create** tab; the **Choose Shape File** dialog box is displayed.

2. In the **Choose Shape File** dialog box, browse to the following location:

    C:\m3d_2017\c04_m3d_2017_tut\c04_tut01

3. Next, enter the name **Line_Feature** in the **File name** edit box.

4. Make sure that the **Shape File (*.shp)** option is selected in the **Save as type** drop-down list. Next, choose the **Save** button in the **Choose Shape File** dialog box; the dialog box is closed and the **Specify Coordinate System** dialog box is displayed, as shown in Figure 4-30.

5. In this dialog box, ensure that the **NE83** coordinate system is chosen in the **Coordinate System** edit box. Next, choose the **OK** button; the **Specify Coordinate System** dialog box is closed and the **Schema Editor** dialog box is displayed.

6. In the **Schema Editor** dialog box, expand the **SHP Schema** folder in the **Schema** list box and select **FeatureClass 1**; the **Logical Feature Class** tab is displayed in the right pane of the dialog box.

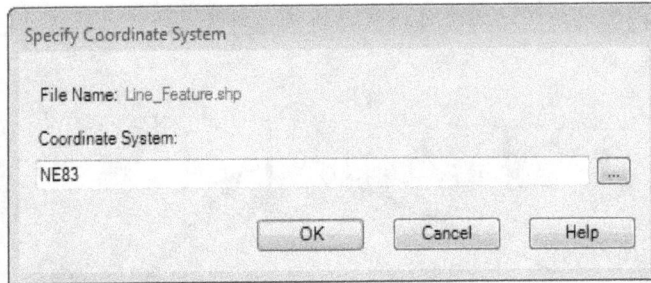

Figure 4-30 The *Specify Coordinate System* dialog box

7. Enter **Road_lines** in the **Name** edit box and **Line geometry representing line feature** in the **Description** text box. Next, ensure the **Feature Class** option is selected in the **Type** drop-down list.

8. Expand the **FeatureClass 1** node in the **Schema** list box and then select the **FeatId** property.

   Note that the name of the feature class is changed from **FeatureClass 1** to **Road_lines** in the **Schema** list box and the **Logical Property** tab is displayed in the right pane of the dialog box.

9. Next, select the **Geometry** property in the **Road_lines** feature class node; the **Logical Property** tab displays the options corresponding to the **Geometry** property.

10. In the **Data Attributes** area of this tab, ensure that the cell corresponding to the **Geometric Types** displays **2 (Curve)**. If not, click in the cell; a drop-down list is displayed in the cell. Select the **Curve** option from this drop-down list.

11. Next, select the **Road_lines** feature class in the **Schema** list box; the **Logical Feature Class** tab is displayed in the right pane of the dialog box and the **New Property** button is activated.

12. Choose the **New Property** button; a new property **Property1** is added to the **Road_lines** feature class node and the **Logical Property** tab is displayed in the right pane of the dialog box.

13. In the **Logical Property** tab, enter **Width** in the **Name** edit box and **Road width** in the **Description** edit box.

14. In the **Data Attributes** area, click in the cell corresponding to **Data Type**; a drop-down list is displayed in the cell. Select **Decimal** from this drop-down list. Next, specify **5** and **2** as the precision and scale values in the cells corresponding to **Precision** and **Scale** attributes, respectively.

15. Next, select the **Road_lines** feature class in the **Schema** list box of the **Schema Editor** dialog box; the name **Property1** changes to **Width** in the **Road_lines** feature class node.

16. Choose the **OK** button; the **Submit Changes** message box is displayed, as shown in Figure 4-31. Choose the **Yes** button from this dialog box; the message box is closed and the shape data store (shape file) **Line_Feature** layer is added to the **Display Manager** tab of the **TASK PANE**.

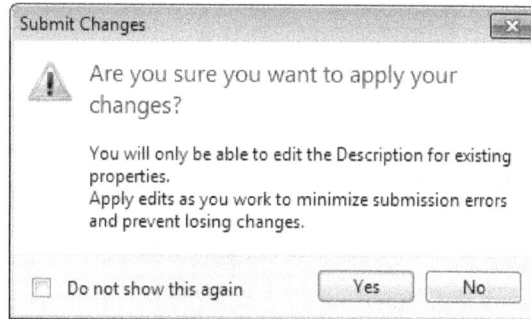

*Figure 4-31 The **Submit Changes** message box*

Adding Feature Data to the Line_Feature Layer

In this section of the tutorial, you will create features in the **Line_feature** layer and then specify attribute data for the created feature in the **DATA TABLE**.

1. Select the **Line_Feature** layer in the **Display Manager** tab of the **TASK PANE**; the **Vector Layer** (contextual) tab is displayed in the ribbon.

2. Choose the **Line** tool from **Vector Layer > Create > New Feature** drop-down; you are prompted to specify the start point of the line. Note that the cursor has changed to a crosshair in the drawing window.

3. Click on the north end of the North-South road (Line 1), refer to Figure 4-27; you are prompted to specify the next point. To specify the next point, click in the center of the road width; you will be prompted to specify the next point.

4. Continue specifying points in the center of the road until you reach the south end of the road. On reaching the south end of the road, press ENTER to end digitizing; AutoCAD Map 3D will create a line feature, refer to Figure 4-32. The **DATA TABLE** window will also be displayed with a data row corresponding to the created feature.

Note
*If the **DATA TABLE** is not displayed by default, choose the **Table** tool from the **View** panel of the Vector Layer tab to display the **DATA TABLE**.*

5. Next, click in the **Width** cell of the data row corresponding to the created feature. Clear the cell and enter **20** as its value and press ENTER.

6. Right-click on the feature layer in the **Display Manager** tab of the **TASK PANE**; a shortcut menu is displayed. In this shortcut menu, choose the **Check In Features** option; the process of editing is completed.

7. Repeat the procedure followed in step 1 through step 4 to create a feature for the road in East-west direction (Line 2), as shown in Figure 4-32.

Figure 4-32 Line features created in the raster image

8. Next, follow the steps 5 and 6, and specify **10** as the value in the **Width** cell.

Generating a Shape File for the Polygon Feature

In this section of the tutorial, you will create an SHP file with polygon geometry type and define attribute property for storing attribute data.

1. Choose the **SHP** tool from the **Feature Data Store** panel of the **Create** tab; the **Choose Shape File** dialog box is displayed.

2. In the **Choose Shape File** dialog box, browse to the following location:

    C:\m3d_2017\c04_m3d_2017_tut\c04_tut01

3. Enter **Polygon_Feature** in the **File name** edit box. Ensure that the **Shape File(*.shp)** option is selected in the **Save as type** drop-down list and then choose the **Save** button from the **Choose Shape File** dialog box; this dialog box is closed and the **Specify Coordinate System** dialog box is displayed with the **NE83** code in the **Coordinate System** edit box.

4. Choose the **OK** button in this dialog box; the dialog box is closed and the **Schema Editor** dialog box is displayed.

5. In the **Schema Editor** dialog box, expand the **SHP Schema** folder in the **Schema** list box and select the **FeatureClass 1** feature class; the **Logical Feature Class** tab is displayed in the right pane of the dialog box.

6. Enter the name **ConstructionSite** in the **Name** edit box and the text **Polygon feature representing construction site** in the **Description** text box. Next, ensure the **Feature Class** option is selected in the **Type** drop-down list.

7. Select the **FeatId** property in the **Schema** list box; the name of the **FeatureClass 1** feature class is updated to **ConstructionSite**, and the **Logical Property** tab is displayed in the right pane of the dialog box.

8. Select the **Geometry** property in the **ConstructionSite** feature class node of the **Schema** list box; the options corresponding to the **Geometry** property are displayed in the **Logical Property** tab of the dialog box.

9. In the **Data Attribute** area of this tab, specify the **Geometric Types** option as **4 (Surface)** by choosing the **Surface** option from the **Geometric Types** drop-down list.

10. Next, select the **ConstructionSite** feature class node in the **Schema** list box; the **New Property** button located above the **Logical Feature Class** area will be activated.

11. Choose the **New Property** button; a new property **Property1** is added to the **ConstructionSite** feature class node.

12. Enter **Type** in the **Name** edit box and **Building type** in the **Description** edit box. Next, select the **ConstructionSite** feature class; the **Property1** name is updated to **Type**.

13. Choose the **OK** button; the **Submit Changes** message box is displayed. In this window, choose the **Yes** button; all the dialog boxes are closed and the **Polygon_Feature** is added as feature layer to the **Display Manager** tab of the **TASK PANE**.

Adding the Polygon Feature to the Feature Layer

1. Select the **Polygon_Feature** layer in the **Display Manager** tab of the **TASK PANE**; the **Vector Layer** tab is displayed.

2. Choose the **Polygon** tool from **Vector Layer > Create > New Feature** drop-down; you will be prompted to specify the start point of the polygon. Notice that the cursor changes to a crosshair in the drawing window.

3. Click at any corner of the plot (shown as polygon in Figure 4-27); you will be prompted to specify the next point. Continue specifying the points along the perimeter of the plot until you close the polygon.

> **Tip**
> *To achieve precise closure of objects while digitization, activate object snapping with the **Endpoint** snap mode.*

4. After closing the polygon, press ENTER; you are prompted to specify the option to exit or to continue creating a ring. Choose **Exit** from the interactive menu displayed in the drawing window as you need to create a solid polygon.

    Note that, on choosing the **Exit** option, AutoCAD Map 3D creates a polygon feature and displays the **DATA TABLE**. If the **DATA TABLE** is not displayed, choose the **Table** tool from the **View** panel of the **Vector Layer** tab to display it.

5. In the **DATA TABLE**, enter **Commercial** in the **Type** cell corresponding to the polygon feature.

6. Next, right-click on the feature layer in the **Display Manager** tab of the **TASK PANE**; a shortcut menu is displayed. Choose the **Check In Features** option from the shortcut menu; the process of editing is completed. Figure 4-33 shows the created polygon.

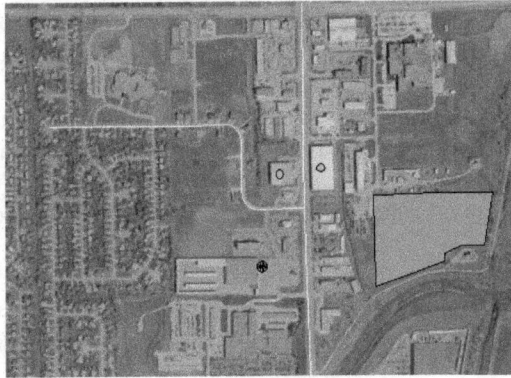

Figure 4-33 The polygon features created in the raster image

7. Close the **DATA TABLE**.

Saving the Drawing File

1. Choose the **Save As** option from the Application Menu; the **Save Drawing As** dialog box is displayed.

2. In the **Save Drawing As** dialog box, enter **c04_Tut01a** in the **File name** edit box and select the **AutoCAD 2013 Drawing (*.dwg)** option in the **Files of type** drop-down list if not selected by default. In the **Save Drawing As** dialog box, choose the **Save** button next to the **File name** edit box; the current drawing file is saved with the given name.

Tutorial 2   Creating SDF Feature Data Store

In this tutorial, you will create an SDF data store with multiple feature class and geometry type. You will also add properties in the feature class to store attribute data. Next, you will export the SDF schema for future references. Finally, you will create feature data using various routines in the COGO input tool, as shown in Figure 4-34.   **(Expected time: 1hr 15min)**

The following steps are required to complete this tutorial:

a. Open a new drawing.
b. Insert a raster image.
c. Create a SDF feature data store.
d. Create schema for feature data store.
e. Export the data store schema.
f. Create feature data.
g. Save the file.

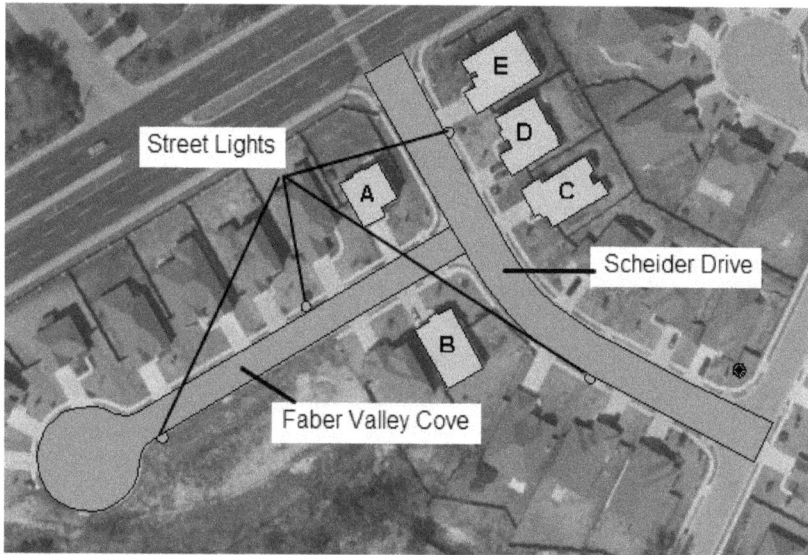

Figure 4-34 The point, line and polygon features created in the feature layers

Starting a New Drawing and Assigning a CRS

In this section of tutorial, you will start with a new drawing file and will assign a coordinate reference system to it.

1. Choose the **New** button from the Quick Access Toolbar; a new drawing file is created.

2. Choose the **Assign** tool from the **Coordinate System** panel in the **Map Setup** tab; the **Coordinate System - Assign** dialog box is displayed.

3. In the **Search** edit box of this dialog box, enter **TX83-CF**.

4. The list box will display a coordinate system with description **NAD83 Texas State Planes, Central Zone, US Foot**.

5. Select this coordinate system in the list box and choose the **Assign** button in the **Coordinate System - Assign** dialog box; the dialog box is closed and the selected coordinate system is assigned to the current Workspace.

Inserting the Raster Image

In this section of the tutorial, you will insert a georeferenced image using the **Image** tool.

1. Choose the **Image** tool from the **Image** panel in the **Insert** tab; the **Insert Image** dialog box is displayed.

2. In the **Insert Image** dialog box, browse to the following location

    C:\m3d_2017\c04_m3d_2017_tut\c04_tut02

3. Select the *AustinTX.jpg* file, and choose the **Open** button; the **Insert Image** dialog box is closed and the **Image Correlation** dialog box is displayed, as shown in Figure 4-35.

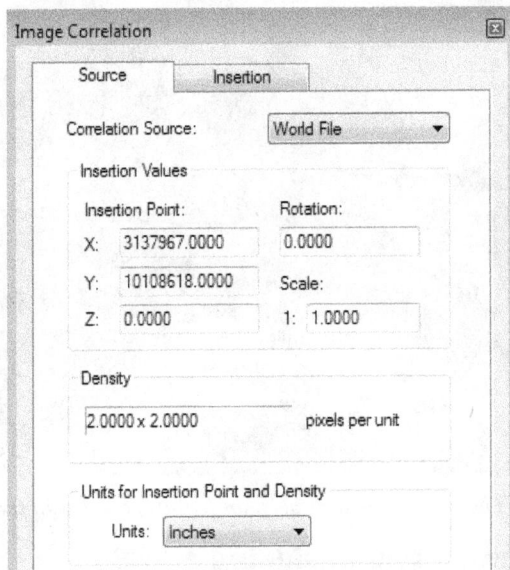

Figure 4-35 *Partial view of the* **Image Correlation** *dialog box*

4. Retain the default parameters in the **Image Correlation** dialog box and then choose the **OK** button; the dialog box is closed and the image is inserted in the drawing.

5. Choose the **Extents** tool from the **Navigate** panel of the **View** tab; the drawing will zoom to the extents and the image will be visible within the drawing window.

Creating SDF Feature Data Store

1. Choose the **SDF** tool from the **Feature Data Store** panel in the **Create** tab; the **Choose Spatial Database File** dialog box is displayed.

2. In the **Choose Spatial Database File** dialog box, browse to the following location:

   C:\m3d_2017\c04_m3d_2017_tut\c04_tut02

3. Enter **SDFstore** in the **File name** edit box and then choose the **Save** button from the **Choose Spatial Database File** dialog box; this dialog box is closed and the **Specify Coordinate System** dialog box with the **TX83-CF** code in the **Coordinate System** edit box is displayed.

4. Choose the **OK** button in this dialog box; the **Specify Coordinate System** dialog box is closed and the **Schema Editor** dialog box is displayed.

Creating Schema for Feature Data Store

1. In the **Schema Editor** dialog box, expand and select the **Schema1** node in the **Schema** list box.

2. In the right pane of the **Schema Editor** dialog box, enter **NeighbourhoodSchema** in the **Enter a schema name** edit box.

3. Next, expand and select the **FeatureClass 1** node in the **Schema** list box of the **Schema Editor** dialog box; the **Logical Feature Class** tab is displayed in the right pane of this dialog box. Note that the name **Schema 1** is now updated as **NeighbourhoodSchema**.

4. Enter **BuildingFC** in the **Name** edit box of the **Logical Feature Class** tab.

5. Specify **Building feature class (Geometry- Polygon)** in the **Description** edit box.

6. Next, select the **Geometry** property in the **FeatureClass1** node in the **Schema** list box; the **Logical Property** tab is displayed in the right pane of the **Schema Editor** dialog box. Also, note that the name **FeatureClass1** is updated to **BuildingFC** in the **Schema** list box.

7. Next, click in the cell corresponding to the **Geometric Types** in the **Data Attributes** area of the **Logical Property** tab; a drop-down list is displayed in the cell.

8. Select the **Surface** option from the displayed drop-down list; **4 (Surface)** is displayed corresponding to the **Geometric Types** attribute.

9. Next, choose the **BuildingFC** node in the **Schema** list box; the **New Property** New Property button is activated.

10. Choose the **New Property** button; a new property **Property1** is added to the **BuildingFC** node.

11. Choose **Property 1** in the **Schema** list box; the parameters corresponding to this property are displayed in the **Logical Property** tab of the **Schema Editor** dialog box.

12. Enter **Name** in the **Name** edit box of the **Logical Property** tab.

13. Next, enter **Building name** in the **Description** edit box.

14. Ensure that **String** is selected in the cell corresponding to **Data type** in the **Data Attributes** area.

15. Next, select the **NeighbourhoodSchema** node in the **Schema** list box; the New Feature Class **New Feature Class** button is activated in the **Schema Editor** dialog box.

16. Choose the **New Feature Class** button; **FeatureClass1** is created and is added in the **NeighbourhoodSchema** node in the **Schema** list box. Also, the right pane of the dialog box displays the **Logical Feature Class** tab.

17. Enter **RoadFC** in the **Name** edit box of this dialog box.

18. Enter **Road feature class (Geometry- Point, line, polygon)** in the **Description** edit box.

19. Next, choose the **FeatureClass1** node in the **Schema** list box; the **New Property** button is activated.

20. Choose the **New Property** button; the name of the feature class **FeatureClass1** is updated to **RoadFC** and a new property **Property1** is added to this feature class node.

21. Choose **Property1** in the **Schema** list box; the parameters corresponding to this property will be displayed in the **Logical Property** tab of the **Schema Editor** dialog box.

22. Enter **Geometry** and **Geometry- Point, line, and polygon** in the **Name** and **Description** edit boxes, respectively.

23. Next, select the **Geometry** option from the **Type** drop-down list.

24. Make sure that **7 (Point+Curve+Surface)** is displayed in the cell corresponding to the **Geometric Types** in the **Data Attributes** area.

25. Next, select the **RoadFC** node in the **Schema** list box; the **New Property** New Property button will be activated.

26. Repeat the procedure given in steps 10 to 12, and create three new properties, **FeatId**, **Name** and **Type** in the **RoadFC** feature class using the data given in the following table:

	Property 1	Property 2	Property 3
Name	FeatId	Name	Type
Type	Data	Data	Data
Description	Feature ID	Feature name	Feature type
Data Attributes			
Data Type	Int 32	String	String
Length		25	30
Auto Generated	True		

27. Next, choose the **RoadFC** node in the **Schema** list box; the **Logical Feature Class** tab is displayed in the right pane of the dialog box.

28. In the **Specify identifier propert(ies) and the order** area of this tab, a list of available properties in the **RoadFC** feature class is displayed. Select the check box corresponding to the **FeatId** property. Figure 4-36 shows the **FeatId** property selected as the identifier property.

   Next, you will save the data store schema as an xml file for future reference. To do so, you need to export the schema using the **Export Schema** tool in the **Schema Editor** tab.

29. Choose the **Schemas** node in the **Schema** list box of the **Schema Editor** dialog box; the **Export Schema** button in the dialog box is activated.

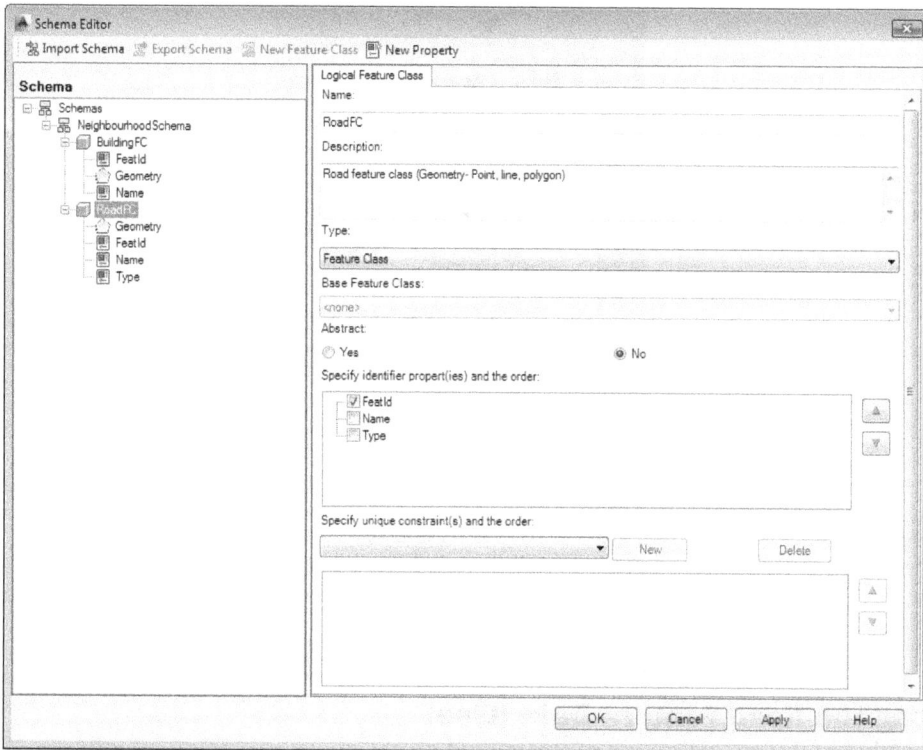

*Figure 4-36 The **FeatId** property selected as the identifier property in the **Schema Editor** dialog box*

30. Choose the **Export Schema** button; the **Save As** dialog box is displayed.

31. In this dialog box, browse to the location

    C:\m3d_2017\c04_m3d_2017_tut\c04_tut02

32. Enter **NeighbourhoodSchema** in the **File name** edit box.

33. Select the **XML files (*.xml)** option from the **Save as type** drop-down list and then choose the **Save** button in the dialog box; the **Save As** dialog box is closed and the schema is saved to the specified location with the specified name.

34. Next, choose the **OK** button in the **Schema Editor** dialog box; the **Submit Changes** message box is displayed.

35. Choose the **Yes** button in the **Submit Changes** message box; the message box and the **Schema Editor** dialog box are closed and the feature classes are added as feature layers to the drawing. The added feature layers are also displayed in the **Display Manager** tab of the **TASK PANE**, as shown in Figure 4-37.

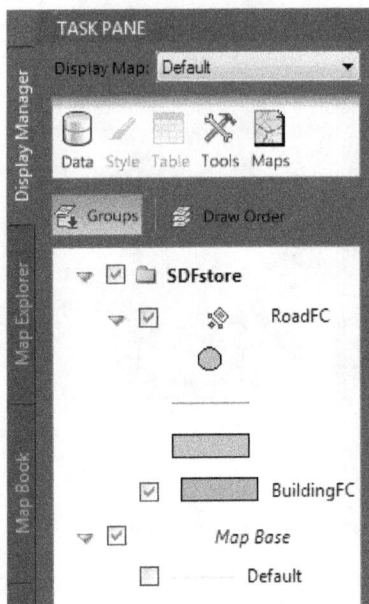

Figure 4-37 *Feature Layer displayed in the*
Display Manager tab of the Task Pane

Creating the Road Feature Data (Line)

In this section of the tutorial, you will create road center line connecting the points **A - F**, refer to Figure 4-38. For precise drafting of the center line vertices, you will use various routines available in the **COGO Input** dialog box.

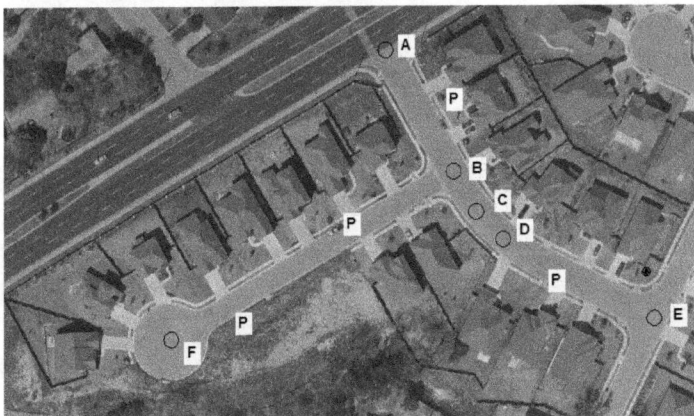

Figure 4-38 *Points representing the vertices of road center line and the street light poles*

1. Select the **RoadFC** feature layer from the **Display Manager** tab of the **TASK PANE**; the **Vector Layer** (contextual) tab is displayed in the Ribbon.

2. Choose the **Line** tool from **Vector Layer** (contextual) **> Create > New Feature** drop-down; an interactive input is displayed in the drawing area and you are prompted to specify the start point.

You will specify point **A** as the start point of the line by entering the coordinates of the point.

3. To specify the coordinates of point **A**, enter **3138614** in the interactive input as the X coordinate and then press TAB; the interactive input for the X coordinate will be locked and the interactive input for Y is activated. Now, you can specify the Y coordinate for the point.

4. Enter **10109774** and press ENTER; AutoCAD Map 3D will create a start point at **A** and you will be prompted to specify the next point.

   You will specify point **B** as the next point by using the **Bearing/Distance** routine from the **COGO Input** dialog box.

5. With the start point selected at **A**, choose the **COGO Input** tool from the **Vector Layer** (contextual) **> Create > COGO** drop-down; the **COGO Input** dialog box is displayed.

6. In this dialog box, choose the **Bearing/Distance** routine from the left pane; the edit boxes for specifying the parameters for the routine will be displayed in the right pane of the dialog box.

7. Specify **3138614** and **10109774** as the coordinates of the point in the **X** and **Y** edit boxes, respectively.

8. Select the **SE** option from the **Quadrant** drop-down list.

9. Enter **30** and **180** in the **Bearing** and **Distance** edit boxes, respectively.

10. Next, choose the **Calculate** button in the **COGO Input** dialog box; the coordinates of the resultant point are displayed as **3138704.000000** and **10109618.115427** in the **X** and **Y** edit boxes, respectively in the **Result** area, as shown in Figure 4-39.

11. Next, choose the **Create Point** button in the **COGO Input** dialog box; the **COGO Input** dialog box is closed and a line is displayed between points **A** and **B**.

12. Next, press ENTER; AutoCAD Map 3D processes and creates a line feature and displays the **DATA TABLE** containing a row corresponding to the created feature.

13. In the **DATA TABLE**, enter **Scheider** and **Drive** in the **Name** and **Type** cells, respectively.

14. Choose the **Check In** tool from the **Edit Set** panel of the **Vector Layer** tab; the created feature is checked in and the system generated ID for the feature is displayed in the **FeatId** cell of the **DATA TABLE**.

15. Choose the **Line** tool from **Vector Layer** (contextual) **> Create > New Feature** drop-down; you are prompted to specify the start point.

16. Press **F3** to switch on the object snap; open **Object Snap Settings** from the **Status Bar**, select the check box corresponding to **Object Snap on** and **Endpoint** node.

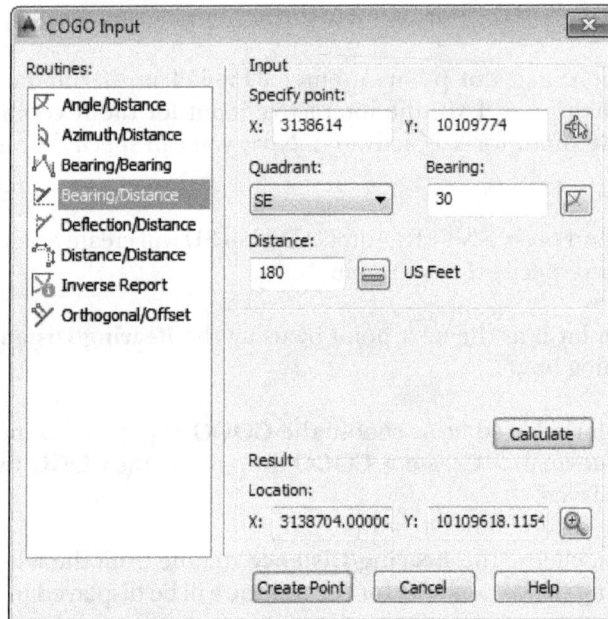

*Figure 4-39 The **COGO Input** dialog box displaying the result of the **Bearing/Distance** routine*

17. Choose the **OK** button in the **Drafting Settings** dialog box; the dialog box is closed and the object snap is switched on for the selected snap mode.

18. Select the end point **B**, you are prompted to specify the next point.

19. With the start point selected at **B**, choose the **COGO Input** tool from **Vector Layer** (contextual) > **Create** > **COGO** drop-down; the **COGO Input** dialog box is displayed.

20. In the **COGO Input** dialog box, choose the **Angle/Distance** routine from the left pane; the parameters corresponding to the selected routine are displayed in the right pane of the dialog box.

21. In this dialog box, choose the button to the right of the **Y** edit box to specify the start and end points; the dialog box is closed and you are prompted to select the line.

22. Click on line AB near point **B**; the **COGO Input** dialog box is displayed again with the coordinates of the points selected in the respective edit boxes.

23. Enter **180** and **60** in the **Angle** and **Distance** edit boxes, respectively.

24. Next, choose the **Calculate** button; the coordinates of the resultant point are displayed in the **X** and **Y** text boxes in the **Result** area. Figure 4-40 shows the **COGO Input** dialog box with the specified parameters and the coordinates of the resultant point for the **Angle/Distance** routine.

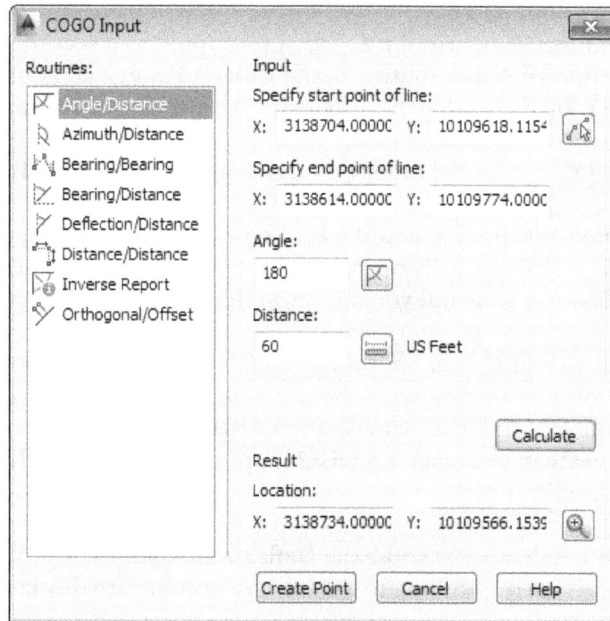

Figure 4-40 *The **COGO Input** dialog box displaying the input parameters and result of the* ***Angle/Distance*** *routine*

25. Choose the **Create Point** button; the point will be created at **C** and a line is displayed between **B** and **C**.

26. Press ENTER; AutoCAD Map 3D processes and creates a line feature and displays the **DATA TABLE** containing a row corresponding to the created feature.

27. Next, in the **DATA TABLE**, enter **Scheider** and **Drive** in the **Name** and **Type** cell of the row corresponding to the newly created line feature, respectively.

28. Choose the **Check In** tool from the **Edit Set** panel of the **Vector Layer**; the created feature is checked in.

29. Repeat the procedure given in steps 15 to 28 and create line features between **C** and **D**, and **D** and **E** using the information given below.

    For creating line features between points **C** and **D** follow the steps below:

    a.   Specify the start point at **C**
    b.   Click near the line BC near point **C**, when prompted to select the line.
    c.   Choose the **Angle/Distance** routine in the **COGO Input** dialog box.
    d.   In the **COGO Input** dialog box, specify **Angle** and **Distance** as **165** and **50**, respectively.
    e.   In **DATA TABLE**, specify **Scheider** and **Drive** in the **Name** and **Type** cells, respectively.

    For creating line features between points **D** and **E** follow the steps below:

    a.   Specify the start point at **D**

    b.   Click near the line CD near point **C**, when prompted to select the line.

    c.   Choose the **Angle/Distance** routine in the **COGO Input** dialog box.

    d.   In the **COGO Input** dialog box, specify **Angle** and **Distance** as **162** and **225**, respectively.

    e.   In **DATA TABLE**, specify **Scheider** and **Drive** in the **Name** and **Type** cell, respectively.

30. Next, to create a line feature between **B - F**, choose the **Line** tool from the **Vector Layer** (contextual) **> Create > New Feature** drop-down; an interactive input is displayed in the drawing area and you are prompted to specify the start point.

31. Next, choose the vertex at **B**; you are prompted to specify the next point.

32. With the start point selected at **B**, choose the **COGO Input** tool from the **Vector Layer** (Contextual) **> Create > COGO** drop-down; the **COGO Input** dialog box is displayed.

33. In the **COGO Input** dialog box, choose the **Deflection/Distance** routine from the left pane; the parameters corresponding to the selected routine are displayed in the right pane of the dialog box.

34. Choose the button at the right of the **Y** edit box. The **COGO Input** dialog box is closed and you are prompted to select the line.

35. Select the point on line AB near point **B**; the **COGO Input** dialog box is displayed again with the edit boxes displaying the coordinates of the start and end points of line AB.

36. Next, enter **90** and **430** in the **Deflection** and **Distance** edit boxes, respectively.

37. Choose the **Calculate** button; the coordinates of the resultant points are displayed in the text boxes of the **Result** area.

38. Next, choose the **Create Point** button in the **COGO Input** dialog box; the point **F** is created.

39. Press ENTER; AutoCAD Map 3D creates a line feature and displays the **DATA TABLE** containing a new row corresponding to the created feature.

40. Enter **Faber Valley** and **Cove** in the **Name** and **Type** cells of this row, respectively.

41. Choose the **Check In** tool from the **Edit Set** panel of the **Vector Layer**; the created feature is checked in and a system generated ID for the feature is displayed in the **FeatId** cell of **DATA TABLE**. Figure 4-41 shows the line feature created connecting points **A - F**.

Figure 4-41 *The line feature created between points* **A-F** *representing the center line of the road*

Creating the Road Feature Data (Point)

In this section of the tutorial, you will create multi-point data that represents the street lights.

1. Choose the **MultiPoint** tool from **Vector Layer** (contextual) **> Create > New Feature** drop-down; an interactive input is displayed in the drawing area and you are prompted to specify the point.

2. Click on the street lights that are visible in the raster along the Faber Valley Cove. Circle **P** in the Figure 4-38 shows the location of the street lamps along these roads.

3. After specifying the locations for the street lights (total of 2), press ENTER; the **DATA TABLE** is displayed with a new row.

4. Enter **Faber Valley Co** and **Street Light** in the **Name** and **Type** cells, respectively.

5. Next, check in the created feature by choosing the **Check In** tool from the **Edit Set** panel of the **Vector Later** (contextual) tab.

6. Repeat the procedure given in steps 1 to 5 and create street light feature along the Scheider Drive. Enter **Scheider Dr** and **Street Light** in the **Name** and **Type** cells in the new row of the **DATA TABLE**, respectively.

Creating the Road Feature Data (Polygon)

Next, you will create a polygon feature for the pavement along the Scheider Drive and Faber Valley Cove. Figure 4-42 shows the polygon feature created along the two roads.

Figure 4-42 *The polygon feature created along the Scheider Drive and Faber Valley Cove*

1. Choose the **Polygon** tool from **Vector Layer** (contextual) **> Create > New Feature**; an interactive input is displayed in the drawing area and you are prompted to specify the first point.

2. Click on the edge of the pavement of the **Scheider Drive**; you are prompted to specify the second point.

3. Digitize along the edge of the pavement to create a polygon.

    Note that the geometry must be closed to create a polygon. Use Object snap to achieve a precise closed geometry.

4. On closing the polygon geometry, press ENTER; you are prompted to specify whether you wish to exit or create a ring.

5. Choose the **eXit** option from the interactive options; the **DATA TABLE** is displayed with a new row corresponding to the new feature.

6. Enter **Scheider Dr** and **Pavement** in the **Name** and **Type** cells of this row, respectively.

7. Repeat the procedure given in steps 1 to 6 and create a polygon along the Faber Valley Cove. Enter **Faber Valley Co** and **Pavement** in the **Name** and **Type** cells of the new row in the **DATA TABLE**, respectively.

8. Choose the **Check In** tool from the **Edit Set** panel of the **Vector Layer**; the created feature is checked in. Figure 4-43 shows the **DATA TABLE** for the **RoadFC**.

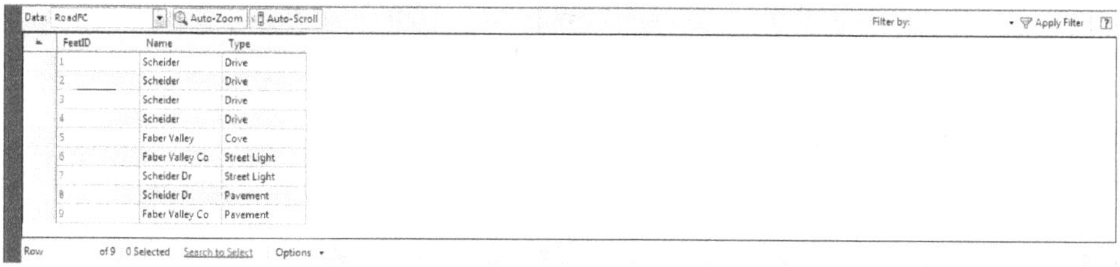

FeatID	Name	Type
1	Scheider	Drive
2	Scheider	Drive
3	Scheider	Drive
4	Scheider	Drive
5	Faber Valley	Cove
6	Faber Valley Co	Street Light
7	Scheider Dr	Street Light
8	Scheider Dr	Pavement
9	Faber Valley Co	Pavement

Figure 4-43 *The* ***DATA TABLE*** *for the* ***RoadFC***

Creating the Building Feature Data (Polygon)

Next, you will create polygon feature for the buildings along the Scheider Drive and Faber Valley Cove, as shown in Figure 4-44.

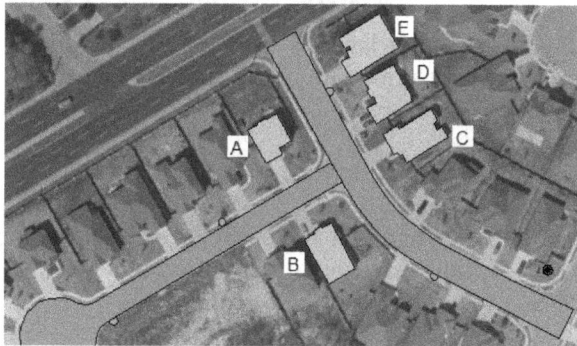

Figure 4-44 *The building features along the Scheider Drive and Faber Valley Cove*

1. Select the **BuildingFC** from the **Display Manager** of the **TASK PANE**.

2. Next, choose the **Polygon** tool from **Vector Layer** (contextual) **> Create > New Feature** drop-down; an interactive input is displayed in the drawing area and you are prompted to specify the first point.

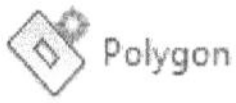

3. Click on the edge of the Building A; you are prompted to specify the second point. Digitize along the edge of the building to create a closed polygon.

4. On closing the polygon, press ENTER; you will be prompted to specify whether you wish to exit or create a ring.

5. Choose the **eXit** option from the interactive options; the **DATA TABLE** is displayed with a new row corresponding to the new feature.

6. Enter **White Rose Cottage** in the **Name** cell of the **DATA TABLE** and press ENTER.

7. Repeat the procedure given in steps 2 to 6 and create a polygon for building **B**, **C**, **D**, and **E**. Specify **Albany Cottage**, **The Lawn**, **The Retreat**, and **Red Cottage** as the name in the **Name** cell of the **DATA TABLE.**

8. Choose the **Check In** tool from the **Edit Set** panel of the **Vector Layer**; the created feature is checked in.

Saving the Drawing File

1. Choose the **Save As** option in the Application Menu; the **Save Drawing As** dialog box is displayed.

2. In the **Save Drawing As** dialog box, enter **c04_Tut02a** in the **File name** edit box and select the **AutoCAD 2013 Drawing (*.dwg)** option in the **Files of type** drop-down list, if it is not selected by default. In the **Save Drawing As** dialog box, choose the **Save** button corresponding to the **File name** edit box; the current drawing file is saved with the given name.

Tutorial 3   Editing Vector Layer and Data Table

In this tutorial, you will edit the vector layer by splitting and merging a polygon feature, and then you will edit the property or feature data in the **DATA TABLE**.

(Expected time: 1hr 30min)

The following steps are required to complete this tutorial:

a. Start a drawing file and then assign global coordinate system to the current drawing.
b. Load the data into the working space by using the **Connect** tool.
c. Split and merge polygon feature, as shown in Figure 4-45.
e. Save the drawing file.

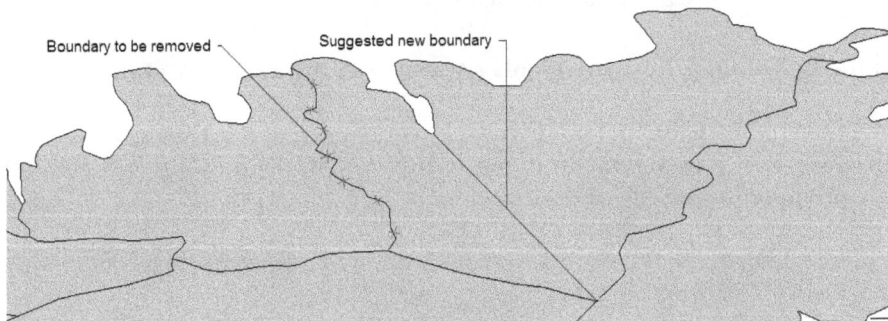

Figure 4-45 New boundary suggested between two feature polygons

Starting a Drawing and Assigning a Coordinate System to it

1. Open a new drawing file by choosing the **New** button from the Quick Access Toolbar.

2. Choose the **Assign** tool from the **Coordinate System** panel of the **Map Setup** tab; the **Coordinate System - Assign** dialog box is displayed.

3. In the dialog box, select the **Iceland** option from the **Category** drop-down list; the coordinate systems in the selected category are displayed in the list box in this dialog box.

4. Next, choose the **Hjorsey.IcelandGrid** code having the **Iceland Grid of 1955** description from the list box if not chosen by default.

5. Choose the **Assign** button in the dialog box; the dialog box is closed and the **Hjorsey.IcelandGrid** coordinate system is assigned to the current drawing.

Loading the Feature Data by Using the Connect Tool

1. Choose the **Connect** tool from the **Data** panel in the **Home** tab; the **DATA CONNECT** wizard is displayed.

2. In this wizard, choose the **Add SHP Connection** option from the **Data Connections by Provider** list box; the **OSGeo FDO Provider for SHP** page is displayed on the right of this wizard.

3. Choose the **SHP** button on the right of the **Source file or folder** edit box; the **Open** dialog box is displayed.

4. In this dialog box, browse to the following location:

    C:\m3d_2017\c04_m3d_2017_tut\c04_tut03

5. In the **Open** dialog box, select the *ic.shp* file and choose the **Open** button; the path of the *ic.shp* file is displayed in the **Source file or folder** edit box.

6. Choose the **Connect** button located below the **Source file or folder** edit box; the name of the *ic.shp* file is displayed in the list box in the **SHP** page of the **DATA CONNECT** wizard.

7. Choose the **Edit Coordinate Systems** button located above the list box in the **SHP** page; the **Edit Spatial Contexts** dialog box is displayed.

8. Select the first row in the **Spatial Contexts** list box and choose the **Edit** button; the **Coordinate System Library** dialog box is displayed.

9. In this dialog box, select the **Iceland** option from the **Category** drop-down list, and then select the **HJORSEY.LL** code with the **HJORSEY.LL Automatically generated LL system for WKT use** description from the list box.

10. Choose the **Select** button from the **Coordinate System Library** dialog box; the **HJORSEY.LL** code is displayed in the first row of the **Override** column. Next, choose the **OK** button from the **Edit Spatial Contexts** dialog box; the selected coordinate system is applied to the source file in the list box of the **SHP** page.

11. Choose the **Add to Map** button below the list box; the source file is added to the **Display Manager** tab in the **TASK PANE** and the feature layer is displayed in the drawing window. Now, close the **DATA CONNECT** wizard.

Splitting a Polygon into Two Feature Polygons

1. Select the **ic** feature layer in the **Display Manager** tab of the **TASK PANE**; the **Vector Layer** (contextual) tab is displayed.

2. Choose the **Zoom to Extents** tool from the **View** panel in the **Vector Layer** tab; the vector layer fits into the drawing window.

> **Note**
> *Choose the **Check Out** tool from the **Edit Set** panel in the **Vector Layer** tab; the cursor changes into a selection box in the drawing window and you are prompted to select an object in the drawing.*
>
> Check Out

3. Select the feature polygon with the attribute value **FeatId 7**, as shown in Figure 4-46, and then press ENTER; the selected polygon becomes editable.

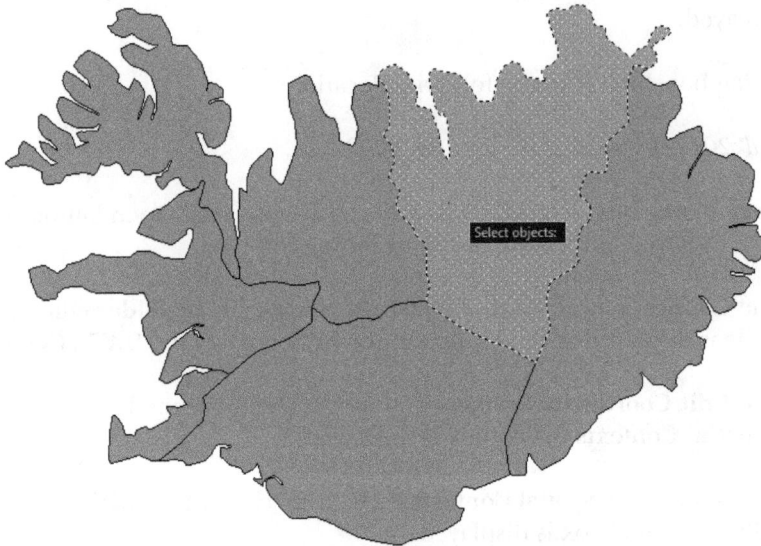

Figure 4-46 The selected feature polygon used for splitting

4. Choose the **Split Feature** tool from the **Split/Merge** panel in the **Vector Layer** tab; you are prompted to specify whether you want to create a new or a multi-part object.

    Split Feature

5. Type **N** for creating a new object and then press ENTER in the command line; you are prompted to choose the option to create new or use the existing ID.

6. Type **E** and then press ENTER; you are prompted to choose an option for selecting the method of specifying the splitting object.

7. To draw the splitting object, type **D** and then press ENTER; you enter the sketching mode and you are prompted to specify the start point.

8. Choose the **Object Snap** button in the Status Bar, if it is not turned on by default; the **Object Snap** is turned on.

9. Place the crosshair over the first corner point, as shown in Figure 4-47, and then click on it; the first point is selected.

10. Next, drag and place the cursor over the next corner, as shown in Figure 4-48, and then click to specify the end point of the line.

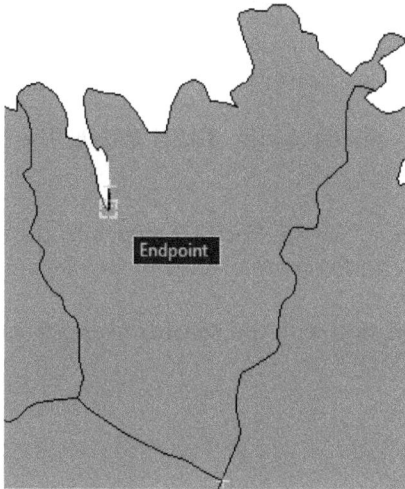

Figure 4-47 *Selecting the first corner point for splitting the polygon feature*

Figure 4-48 *Selecting the end corner point for splitting the polygon feature*

11. After specifying the end point, press ENTER; the selected polygon feature is split along the specified line and the **DATA TABLE** is displayed with two rows corresponding to each of the divided polygons.

12. Select the cell in the **FIPS_ADMIN** column of the first row corresponding to the split polygon (FeatID 7), and then change the name **IC** to **IC1**. Similarly, change the name **IC** to **IC2** in the other row corresponding to the split column. Figure 4-49 shows the **DATA TABLE** with the modified attribute values.

Data: ic Auto-Zoom Auto-Scroll

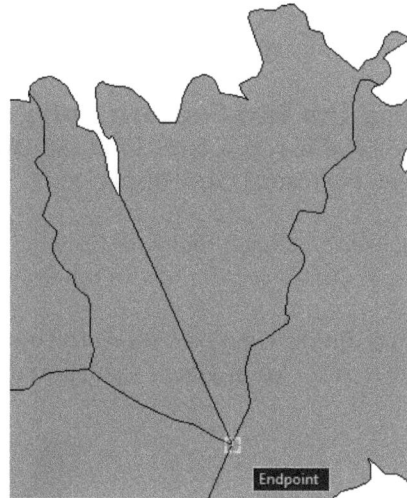

FeatId	FIPS_ADMIN	GMI_ADMIN	ADMIN_NAME	FIPS_CNTRY	GMI_CNTRY	CNTRY_NAME
1	IC	ISL-AST	Austurland	IC	ISL	Iceland
2	IC	ISL-NVS	Nordhurland V...	IC	ISL	Iceland
3	IC	ISL-SDH	Sudhurland	IC	ISL	Iceland
4	IC	ISL-VSL	Vesturland	IC	ISL	Iceland
5	IC	ISL-VST	Vestfirdhir	IC	ISL	Iceland
6	IC24	ISL-RYK	Reykjavik	IC	ISL	Iceland
7	IC1	ISL-NYS	Nordhurland Ey...	IC	ISL	Iceland
8	IC2	ISL-NYS	Nordhurland Ey...	IC	ISL	Iceland

Row 8 of 8 0 Selected Search to Select Options

Figure 4-49 *Modified attribute values of the generated features in the* ***DATA TABLE***

13. Choose the **Check In** tool from the **Edit Set** panel in the **Vector Layer** tab; the modification or splitting of the feature polygon is saved.

Note

*If you are unable to see the split polygons on checking in the features, right-click on the **ic** feature layer in the **Display Manager**; a shortcut menu is displayed. Choose the **Refresh** Refresh Layer **Layer** option from the displayed menu; the layer in the drawing will be refreshed and the objects in the drawing will now be visible.*

Merging the Two Feature Polygons

1. Select the **ic** vector layer from the **Display Manager** tab in the **TASK PANE**; the **Vector Layer** contextual tab is displayed.

2. Choose the **Check Out** tool from the **Edit Set** panel in the **Vector Layer** tab; the cursor changes to a selection box in the drawing window and you are prompted to select objects.

3. Using the selection box, select the two feature polygons with the **FeatIds** property values **2** and **7**, as shown in Figure 4-50, and then press ENTER.

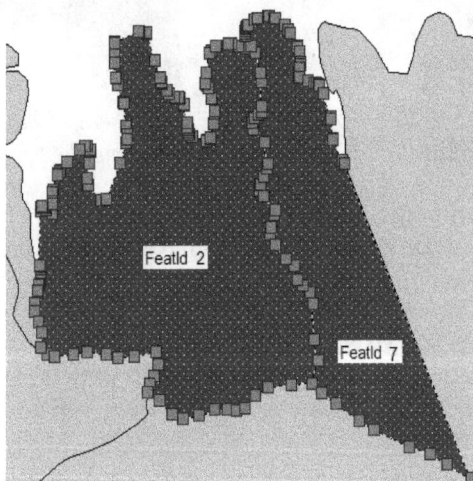

Figure 4-50 Two feature polygons selected for merging

4. Choose the **Merge Feature** tool from the **Split/Merge** panel in the **Vector Layer** (contextual) tab; the cursor changes into a crosshair in the drawing window and you are prompted to choose an option for specifying the feature ID.

    Merge
    Feature

5. Type **S** and press ENTER; the cursor changes into a selection box in the drawing window and you are prompted to specify the polygon feature from which the attribute values are to be assigned to the merged feature.

6. Select the larger polygon feature; the two polygon features are merged and the attributes of the large polygon are assigned to the resultant polygon.

7. Choose the **Check In** tool from the **Edit Set** panel in the **Vector Layer** tab; the process of editing is completed and the changes are saved to the **ic** vector file. The modified feature polygons are shown in Figure 4-51.

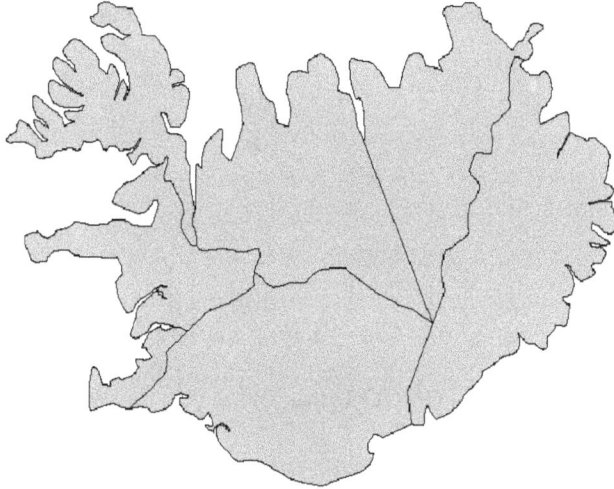

Figure 4-51 *The modified feature polygons*

Saving the Drawing File

1. Choose the **Save As** option in the Application Menu; the **Save Drawing As** dialog box is displayed.

2. In the **Save Drawing As** dialog box, enter **c04_Tut03a** in the **File name** edit box and select the **AutoCAD 2013 Drawing (*.dwg)** option in the **Files of type** drop-down list if not selected by default. In the **Save Drawing As** dialog box, choose the **Save** button corresponding to the **File name** edit box; the current drawing file is saved with the given name.

Self-Evaluation Test

Answer the following questions and then compare them to those given at the end of this chapter:

1. Which of the following tools is used to insert raster images using the **Image Correlation** dialog box?

    (a) **Image**   (b) **Insert**
    (c) **Connect**   (d) **Create**

2. Which of the following COGO routines uses azimuth and distance to locate a new point?

    (a) **Angle/Distance**   (b) **Deflection/Distance**
    (c) **Bearing/Bearing**   (d) **Azimuth/Distance**

3. How many types of geometries can a single shape file contain?

   (a) One   (b) Two
   (c) Three   (d) Four

4. Which of the following commands is used to locate a point based on the distances measured from two reference points?

   (a) **BB**   (b) **DD**
   (c) **DDIST**   (d) **BD**

5. Which of the following dialog boxes contain various routines for locating coordinate points geometrically?

   (a) **Choose Shape File**   (b) **Specify Coordinate System**
   (c) **Schema Editor**   (d) **COGO Input**

6. The _____ routine in the **COGO Input** dialog box is used to specify a point deflection and distance from a point.

7. While creating an SHP data store, you can customize the datasets and its properties by using the _____ dialog box.

8. You can split or merge the geometry of a feature polygon by using the tools in the _____ panel of the **Vector Layer** contextual tab.

9. The raster images need to be _____ for using them in the geospatial analysis.

10. The _____ data store can contain multiple geometry types in a single feature class.

11. The coordinate geometric functions (COGO routines) are used to locate a new point based on the specified geometric parameters. (T/F)

12. You can use any of the COGO methods to locate a new point and begin a new line in the drawing process. (T/F)

13. A shape file or an SDF file is not a vector data file. (T/F)

14. You cannot edit feature data in the **DATA TABLE**. (T/F)

15. You need to check in the edited feature to save the edits to the feature source. (T/F)

Review Questions

Answer the following questions:

1. Which of the following COGO routines is used to locate a point based on distance and angle given from a point?

        (a) **Angle/Distance**   (b) **Deflection/Distance**
        (c) **Distance/Distance**   (d) **Bearing/Bearing**

2. Which of the following tools is used to merge two polygons in a feature layer?

        (a) **Check In**   (b) **Merge Features**
        (c) **Select Layer**   (d) **Check Out**

3. Which of the following COGO routines is used to locate a new point using the bearing angle measured from two points?

        (a) **Angle/Distance**   (b) **Deflection/Distance**
        (c) **Bearing/Bearing**   (d) **Bearing/Distance**

4. Which of the following tools is used to create multiline string features?

        (a) **MultiLine**   (b) **Line**
        (c) **MultiPolygon**   (d) **Point**

5. Which of the following tabs is used to display the feature layers in the **TASK PANE**?

        (a) **Survey**   (b) **MapBook**
        (c) **Map Explorer**   (d) **Display Manager**

6. The _____ option in the **Routines** list box of the **COGO Input** dialog box is used to obtain the details of geometric alignment between two points.

7. If a raster image has already been georeferenced then the default value will appear in the _____ dialog box during the process of image insertion.

8. You can edit the feature attribute (data) of any feature using _____.

9. A fire hydrant is best represented by a feature of _____ geometry type.

10. The _____ type feature data is used to represent an area object.

11. You cannot edit the geometry of an object in a feature layer. (T/F)

12. You can define the schema in the **Schema Editor** dialog box by importing an xml file containing schema definition. (T/F)

13. In the **Schema Editor** dialog box, you can assign polyline geometry to a new shape file by selecting the **2(Curve)** option from the **Geometric Types** drop-down list in the **Data Attributes** area. (T/F)

14. You can select an existing line as a splitting object while using the **Split Features** tool. (T/F)

15. The **Check Out** tool is used to make a feature editable and the **Check In** tool is used to save
the edits made to the feature source. (T/F)

EXERCISES

Exercise 1

Download the folder **c04_ex01** from *www.cadcim.com*. Open a new drawing and assign the
UTM84-10N UTM-WGS 1984 datum, Zone 10 North, Meter;Cent, Meridian 123d W coordinate
system to the drawing. Next, insert the *c04-m3d-2017-exr01.jpg* image from the downloaded folder
into your drawing. Create an SHP data store using the following specifications:

(Expected time: 1 hr)

Data Store Name	Geometry Type	Coordinate System	Create New Property	Property type	Length
Road	Curve	UTM 84- 10 N	Name	String	20
Plot	Surface	UTM 84- 10 N	Name	String	30

Next, digitize the feature shown in Figure 4-52 using the data given below:

Feature	Data Store	Name
Line_Feature_AB	Road	North Anderson Drive
Boundary01	Plot	Park view complex
Boundary01	Plot	Park

Figure 4-52 Locations of the feature data to be created

Exercise 2

Download the *ic.shp* file from *www.cadcim.com*. Next, merge the two polygon features by removing
the boundary between them, refer to Figure 4-53.   **(Expected time: 45 min)**

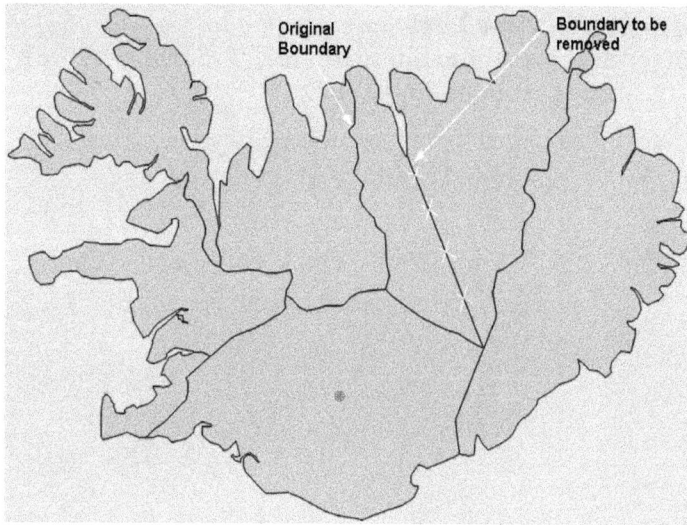

Figure 4-53 *Modifications to be performed on the feature data*

Answers to Self-Evaluation Test

1. a, **2.** d, **3.** a, **4.** c, **5.** d, **6. Deflection/Distance**, **7. Schema Editor**, **8. Split/Merge**, **9.** georeferenced, **10.** SDF, **11.** T, **12.** T, **13.** F, **14.** F, **15.** T

Chapter 5

Styling and
Querying Feature Data

Learning Objectives

After completing this chapter, you will be able to:
- *Use the thematic layer in a map*
- *Style the feature data*
- *Query the feature data by using different types of filters*
- *Search and select the required data from single or multiple vector layers*

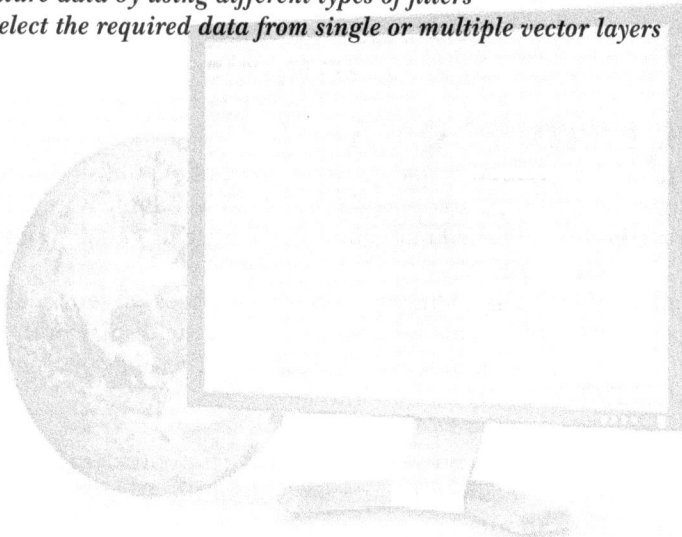

INTRODUCTION

In the previous chapter, you learned about the feature data store and the method to define a data store. You also learned the procedure to create, edit, and save feature data in the feature class of the data store.

In this chapter, you will learn to define the display style for representing the feature data in the map. You will also learn techniques of identifying, selecting, and finding features from huge database. Moreover, in this chapter, you will learn to define a query for filtering feature data based on various conditions.

THEMATIC LAYERS

A thematic layer is defined as a layer with a set of logical feature data placed under single theme such as soil type, road, river, wells, and buildings. The features in these layers represent the real-world objects and are created using geometries such as point, line, and polygon.

Figure 5-1 shows different types of thematic layers such as Text layer, Point feature layer, Line feature layer, Polygon feature layer, and Multi Polygon feature layer lying over one another. These layers are spatially referenced to a geographical area and each layer contains geographical data corresponding to a theme. For example, you can use a text layer that contains names, labels, and text annotations for displaying the names of the geographical features. A Point feature layer is a vector layer that contains point geometry and is used for representing the real world objects such as fire hydrants, hospitals, and bus stops as points. Similarly, you can use the Line feature layer containing line geometries representing the linear objects such as roads or rivers, while a Polygon feature layer containing polygon geometry can be used to represent area features such as district and county.

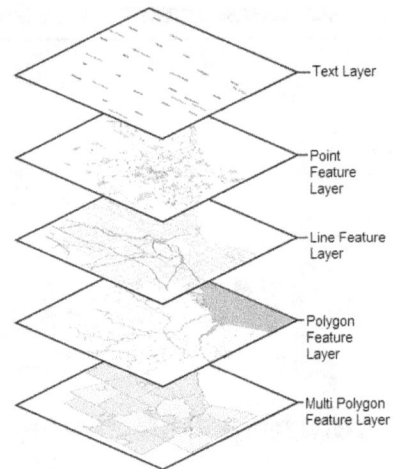

Figure 5-1 *Various thematic layers placed over one another*

STYLING THE FEATURE DATA

Once you have generated the required map data, you may want to publish the final map. While publishing it, you can apply different colors and symbols to its map features. This will help in understanding the data in the published map and will enhance its visual appearance.

AutoCAD Map 3D allows users to implement various techniques of geovisualization, such as creating class interval and assigning scale ranges that control the display of the feature data at various scale (zoom) levels. You can define these visualization settings by using the options in the **Style** tab of the AutoCAD Map 3D Ribbon interface.

You can also style the feature data and define scale ranges using the options in the **STYLE EDITOR** dialog box. To do so, select the required feature layer from the **Display Manager** tab of the **TASK PANE** and right-click; a shortcut menu is displayed. Choose the **Edit Style** option from

the shortcut menu; the **STYLE EDITOR** dialog box for the selected feature layer will be
displayed. Alternatively, to invoke the **STYLE EDITOR** dialog box, choose the **Style** button
from the **Display Manager** tab of the **TASK PANE**. Figure 5-2 shows the **STYLE EDITOR**
dialog box invoked for a polygon feature. The areas in this dialog box are discussed next.

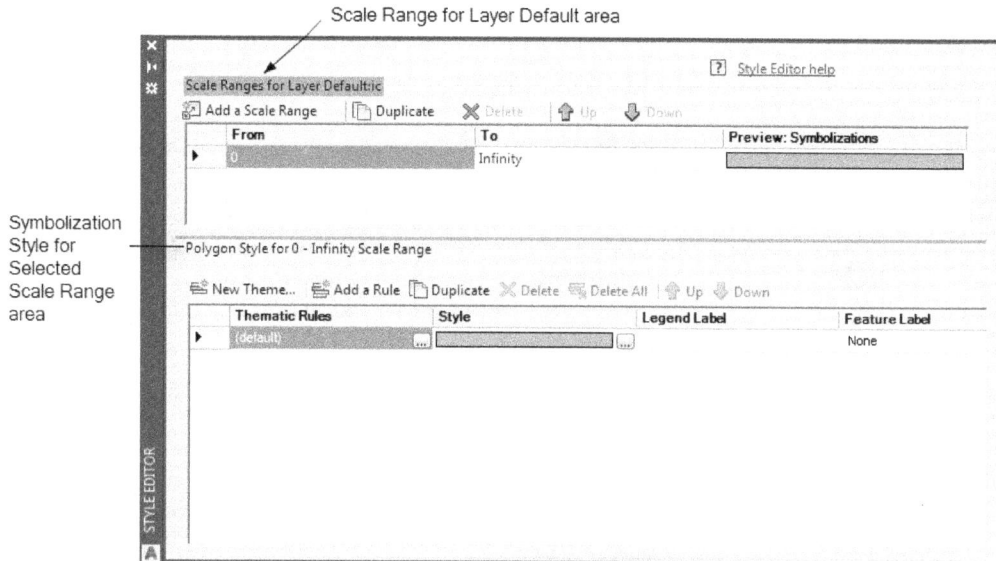

Figure 5-2 *The STYLE EDITOR dialog box*

Scale Ranges for Layer Default Area

The options in the **Scale Ranges for Layer Default** area are used to create, edit, and delete
scale ranges of the selected feature layer. The method to use different options for customizing
the scale ranges in this area are discussed next.

Adding a New Scale Range

To add a new scale range to the existing list of scale ranges, choose the **Add a Scale Range**
button in the **Scale Ranges for Layer Default** area; a copy of the existing row will be added
to the list box. In this added row, you can edit the range values in the **From** and **To** columns.

Duplicating an Existing Scale Range

To duplicate an existing scale range row, select a row in the list box and then choose the **Duplicate**
button located above the list box; a copy of the selected row will be placed below the selected
row in the list box. By default, this duplicate scale range will have all the display settings of the
original scale range.

Editing a Scale Range

To edit a scale range in the list box, select a row and click on the cell to be edited; the cell will
become editable. Now, you can enter a scale value in the cell.

Arranging the Scale Ranges in the List Box

To arrange the scale ranges, select the required row in the first column in the list box of the
Scale Ranges for Layer Default: <name> area; an arrow will appear in the first column of the

selected row. Now, choose the **Up** button to move the row upward and choose the **Down** button to move the row downward.

Deleting the Scale Range

To remove or delete an unwanted scale range row, select the row from the scale range list box in the **Scale Range for Layer Default** area, and then choose the **Delete** button at the top of the scale range list box; the selected row will be deleted.

Symbolization Style for Selected Scale Range Area

AutoCAD Map 3D allows users to define the display style of the data in the feature layers based on its display scale. To style the data based on the scale range, you can use the options in the **Symbolization Style for Selected Scale Range** area of the **STYLE EDITOR** dialog box, refer to Figure 5-2. Figure 5-3 shows the illustration where the symbolization style is applied to the data at varying scale ranges. Each scale range applied to the feature data can have different styles and can be classified into multiple classes.

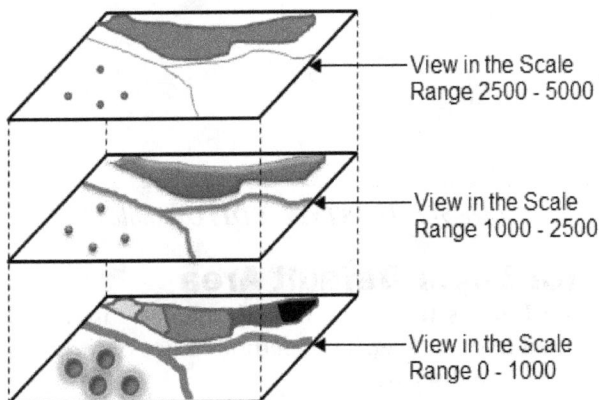

View in the Scale
Range 2500 - 5000

View in the Scale
Range 1000 - 2500

View in the Scale
Range 0 - 1000

Figure 5-3 Styles applied to the data on varying scale ranges

Using the options in the **Symbolization Style for Selected Scale Range** area, you can apply new visual theme settings to an existing feature layer or can modify the existing visual theme settings as required. You can create new thematic rules to classify data into different groups. These thematic layers are used to classify and compare data with different attributes or properties. You can also create thematic layers by classifying the dataset based on the predefined statistical classification methods such as Equal, Standard Deviation, Quantile, Natural Breaks, and Individual Values.

The methods of creating thematic rules, styling the thematic rules, and editing the legend and feature labels of the feature dataset by using different options in the **Symbolization Style for Selected Scale Range** area are discussed next.

Creating Thematic Rules for a Feature Data

You can create a thematic rule for a feature dataset in a selected scale range. To do so, choose the **New Theme** button from the **Symbolization Style for Selected Scale Range** area in the **STYLE EDITOR** dialog box; the **Theme Layer** dialog box will be displayed, as shown in Figure 5-4. Different areas in the **Theme Layer** dialog box are discussed next.

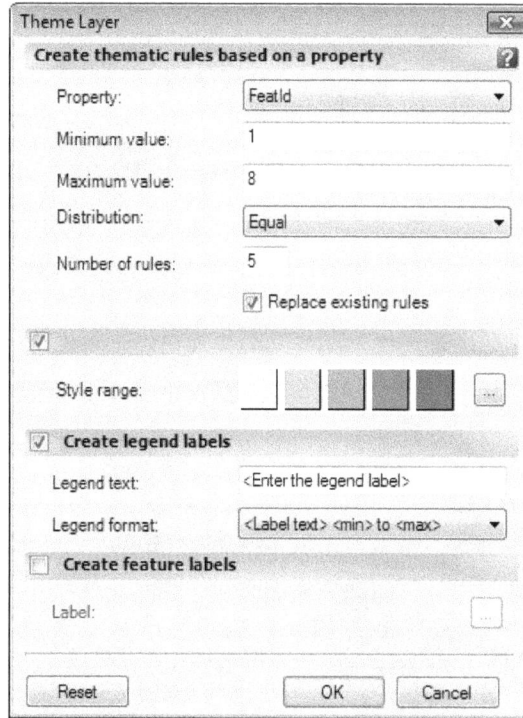

Figure 5-4   *The* **Theme Layer** *dialog box*

Create thematic rules based on a property Area

In this area, you can classify the given feature data into the required number of ranges. To create a thematic rule based on a property of the feature data, select the required property from the **Property** drop-down list. The **Minimum value** and **Maximum value** edit boxes will show the minimum and maximum values, respectively, of the property selected in the **Property** drop-down list. You can also set the minimum and maximum values manually for the selected range in the **Minimum value** and **Maximum value** edit boxes.

Next, select an option from the **Distribution** drop-down list to specify the classification method for the feature dataset. The options available in this drop-down list are: **Equal**, **Standard Deviation**, **Quantile**, **Jenks (Natural Breaks)**, and **Individual Values**. The selected layer will be divided into class intervals. The number of class interval created depends on the options specified in the **Number of rules** edit box and the **Distribution** drop-down list.

Note

*The **Number of rules** edit box will be disabled on selecting the **Individual Values** option from the **Distribution** drop-down list of the **Theme Layer** dialog box.*

Style range Option

The check box above the **Style range** option is selected by default. As a result, the **Style range** option is activated. You can clear the check box to ignore the styling of theme. To apply a style range to a theme, choose the Browse button next to the **Style range** option; the **Style and Label Editor** dialog box will be displayed, as shown in Figure 5-5. The methods of specifying options in different tabs of this dialog box are discussed next.

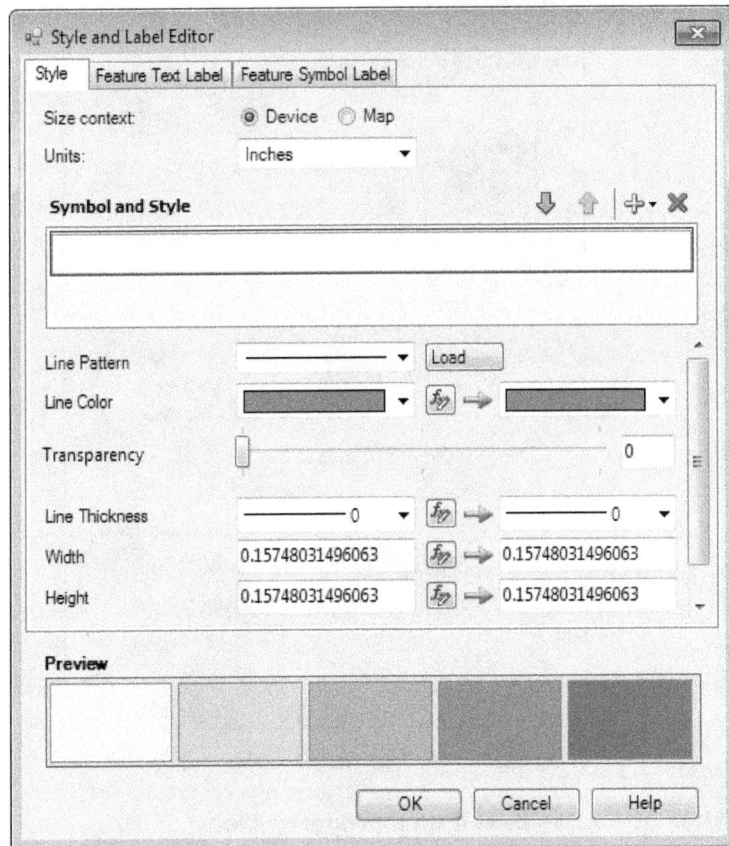

*Figure 5-5 The **Style and Label Editor** dialog box*

Note

*The options in the **Style and Label Editor** dialog box are displayed based on the feature class selected. If you are working in the line or polygon feature class, the options in this dialog box will be displayed accordingly.*

Style Tab: The options in the **Style** tab are used to specify the fill pattern or the symbol for the feature geometry. Using the options in this tab, you can specify the fill pattern for a polygon geometry or the line pattern for the line geometry. For the point feature, the options in the **Style** tab will be used to specify the point symbol. The **Device** radio button corresponding to the **Size context** parameter is selected by default. As a result, the size of the symbol and text of the feature data is measured with respect to drawing units. If you select the **Map** radio button, the size of the symbol and text will be measured with respect to the units used in the map. Next, select an option from the **Units** drop-down list to specify the measuring unit for symbol style.

You can combine multiple symbol styles to generate a style for the current feature class. To do so, choose the **Add** button located above the **Symbol and Style** display box; a drop-down list will be displayed. You can choose the required option from the drop-down list to add a border or fill a symbol. On choosing the required option, a new symbol with style will be added to the display box. To modify a symbol, select it from the

Symbol and Style display box; the settings of the selected style will be displayed below the display box.

For example, to specify a display pattern for the fill symbol, choose it from the **Symbol and Style** display box; the predefined settings of the selected symbol will be displayed below the display box. Next, select an option from the **Fill Pattern** drop-down list to apply a symbol style. Also, you can load a new symbol style into the **Fill Pattern** drop-down list by using the **Load** button next to it. The **Ramp** radio button corresponding to the **Fill color** option is selected by default. As a result, you can specify a color range by choosing two colors. If you select the **Palette** radio button, a drop-down list with predefined color palettes will be displayed below this radio button. Next, select the required palette from the drop-down list displayed. Then, specify all values for the other parameters such as **Line color**, **Line thickness**, **Rotation**, and so on.

Tip
*You can use the **Edit Expression** button corresponding to a property to invoke the* [fx] *Create/Modify Expressions window. You can use this window to define an expression to specify styling options such as width, height, or repeat interval.*

Feature Text Label Tab: The options in the **Feature Text Label** tab are used to specify the display properties of the feature label. To set the measuring unit for the thickness of the feature label, select a radio button corresponding to the **Size context** option and then select an option from the **Units** drop-down list. Next, add a new text style by choosing the **Add** button. On doing so, a new text style will be added to the **Symbol and Style** display box. Then, set various parameters displayed below the display box based on your requirement.

Feature Symbol Label Tab: The options in this tab are used to apply a symbol style to the feature class. To specify a measuring unit to the symbol style of the selected feature class, select the **Device** or **Map** radio button and then select an option from the **Units** drop-down list. Then, set various properties of the symbol style by using different options displayed below the **Symbol and Style** display box.

Tip
*The changes made to the default style are displayed in the **Preview** display box of the **Style and Label Editor** dialog box.*

After specifying all display properties of the feature class, choose the **OK** button from the **Style and Label Editor** dialog box; the dialog box will be closed and the preview of the style specified will be displayed in the **Style range** preview of the **Theme Layer** dialog box.

Create legend labels Area
The options in the **Create legend labels** area in the **Theme Layer** dialog box are used to apply common label to each range in the theme. To do so, enter the label text in the **Legend text** edit box in this area. Next, specify the display format for the legend by choosing the required option from the **Legend format** drop-down list.

Create feature labels Area

Select the check box in the **Create feature labels** area to label the features in the selected feature layer. The preview in this area displays the label preview. To define the style and syntax of the label, choose the Browse button corresponding to the **Label** option; the **Feature Text Label** tab of the **Style and Label Editor** dialog box will be displayed. In the dialog box, choose the **Add** button; a new text style will be added to the **Symbol and Style** display box and the properties of the new text symbol will be displayed below this display box, refer to Figure 5-6.

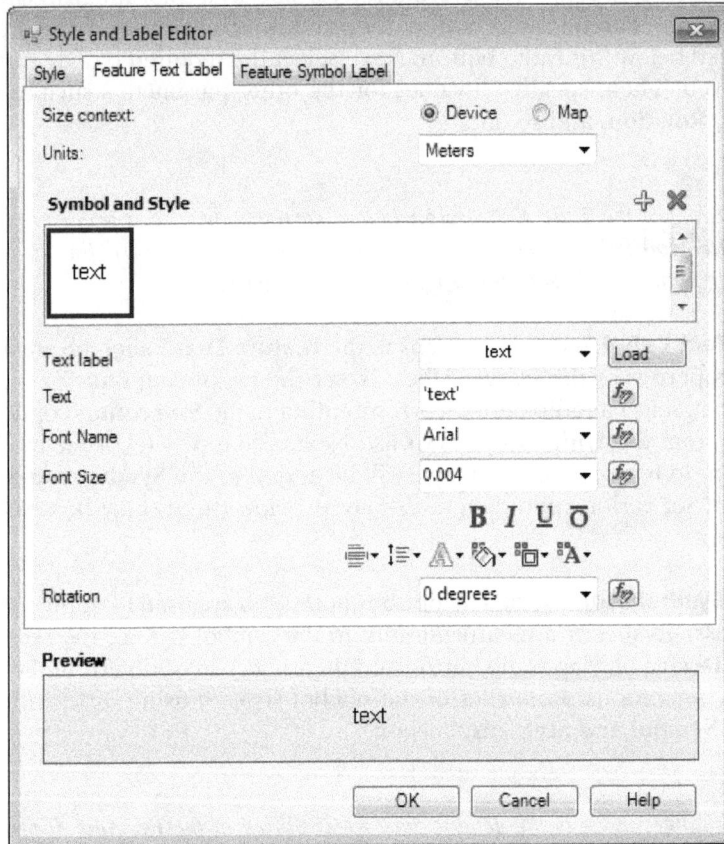

*Figure 5-6 The **Feature Text Label** tab of the **Style and Label Editor** dialog box*

In the **Feature Text Label** tab, you can set the appearance of the feature label. Specify all the required options in the tab and then choose the **OK** button; the **Style and Label Editor** dialog box will be closed and the modified label text will be displayed in the **Label** preview of the **Theme Layer** dialog box.

After specifying all the required parameters in the **Theme Layer** dialog box, choose the **OK** button; the dialog box will be closed and the defined thematic rules will be added to the **STYLE EDITOR** dialog box. These thematic rules are also applied to the corresponding feature layer in the drawing window. Figure 5-7 shows the **STYLE EDITOR** dialog box with the thematic rules for a polygon feature layer.

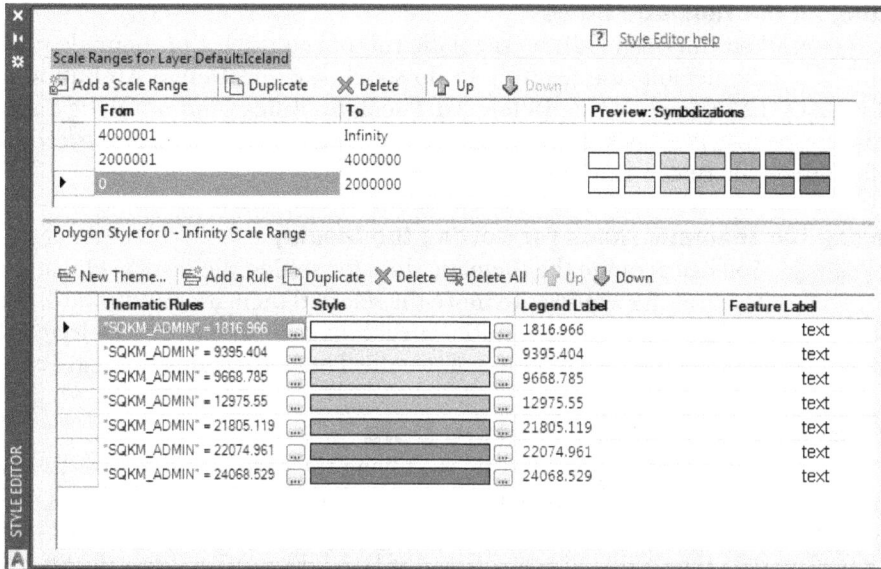

Figure 5-7 The **STYLE EDITOR** *dialog box with the thematic rules added*

Editing the Thematic Rules

To edit the expression of a thematic rule, choose the Browse button corresponding to the required rule in the list box; the **Create/Modify Expression** window will be displayed. Modify the thematic expression in this window and choose the **OK** button; the modified expression will be applied to the selected rule.

You can also add a new rule or duplicate an existing thematic rule by using the options in the **STYLE EDITOR** dialog box. The methods of creating, editing, and arranging thematic rules for the current themes are discussed next.

Adding a New Thematic Rule

You can add a new rule into the current themes list box in the **STYLE EDITOR** dialog box. To do so, choose the **Add a Rule** button from the **STYLE EDITOR** dialog box, refer to Figure 5-2; a new rule will be added into this list box.

Duplicating an Existing Thematic Rule

You can duplicate an existing theme and then apply rules to it. To do so, choose the **Duplicate** button in the **STYLE EDITOR** dialog box; a copy of the thematic rule will be added to the list box.

Deleting a Thematic Rule

You can delete a thematic rule from the current list of thematic rules. To do so, select the thematic rule to be removed and then choose the **Delete** button from the **STYLE EDITOR** dialog box; the selected thematic rule will be deleted from the **Thematic Rules** list.

Deleting All the Thematic Rules

Delete All
You can delete all the thematic rules from the list of thematic rules except the default line feature. To do so, choose the **Delete All** button from the **STYLE EDITOR** dialog box; the **Delete All Thematic Rules** window will be displayed. In this window, choose the **Yes** button; all the thematic rules will be deleted, except the basic or default thematic rule in the **Thematic Rules** list box.

Arranging the Thematic Rules for Setting the Display

Up   Down
You can arrange the thematic rules for setting the display of feature data in the drawing window. To move the selected thematic rule upward, choose the **Up** button located above the list box in this area. Similarly, to move the selected thematic rule downward, choose the **Down** button above the **Thematic Rules** list box.

Editing the Display Style of Feature Data

You can edit the display of the feature data by modifying the style of the theme. To modify the display properties or style of the feature data in a selected scale range, select the cell in the **Style** column corresponding to the required thematic layer row in the **STYLE EDITOR** dialog box; the **Style <feature data type>** dialog box will be displayed. Figure 5-8 shows the **Style Polygon** dialog box. You can edit the display style of the feature data by specifying options in this dialog box. The options in this dialog box are discussed next.

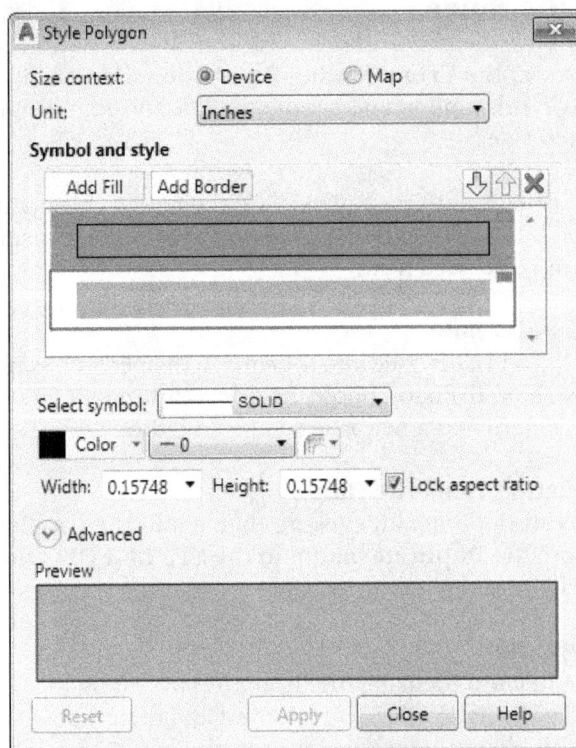

Figure 5-8 The Style Polygon dialog box

Note

You can modify the style for the point, line, and polygon features by setting the parameters in the respective style editing dialog boxes. The styling parameters in the dialog box will be displayed based on the geometry type of the feature.

The options in the **Style** dialog box are used to specify the parameters for the feature display in a selected style range. To specify a measuring unit for the current style, select the **Device** or **Map** radio button corresponding to the **Size context** and then select the required option from the **Units** drop-down list.

In the **Symbol and style** area, for the polygon feature class, you can add two types of styles: fill and border. You can fill a color into the polygon feature class by choosing the **Add Fill** button. On doing so, a new fill style will be added to the list box. Similarly, you can add a border to the polygon feature class by choosing the **Add Border** button in the **Symbol and style** area. On doing so, a new border style will be added to the display box in this area.

For the line and point feature class, you can add symbols by choosing the **Add Symbol** button in the **Symbol and style** area. On doing so, a new fill symbol will be added to the list box in the area.

After adding the required number of fill, border, and symbol styles, select the style to be modified from the display box in the **Symbol and style** area; the predefined property settings of the selected style will be displayed below this area. Next, set the property value of each option based on your requirement. After specifying all property settings, choose the **Apply** button in the dialog box; the modified style will be applied to the feature data and the feature data in the drawing window will change accordingly. Next, choose the **Close** button to close the dialog box.

Applying the Composite Style to a Line Feature

A composite style is made by combining two or more display symbols (types) to form a single display style. You can use a composite style to represent features such as the road with a median (divider) or road with multiple lanes. To apply a composite style to a line feature, select the line feature and then choose the **Style** button from the **Display Manager** tab in the **TASK PANE**; the **STYLE EDITOR** dialog box will be displayed. To style a line feature in the selected scale range, click in the **Style** column on the cell corresponding to the required theme; the **Style Line** dialog box will be displayed, as shown in Figure 5-9.

In the **Style Line** dialog box, choose the **Add Symbol** button from the **Symbol and Style** area; a new line style will be added to the display box. Change the | Add Symbol | thickness of the underlying line by choosing the required option in the **Line Thickness** drop-down list so that the thickness of this layer is greater than the thickness of the line over it. Figure 5-10 shows the composite style or overlaying of two line styles applied to a line feature. To apply the required settings to the line style, select the line style from the display box. Next, to specify a line type for the line, select an option from the **Select symbol** drop-down list. Also, specify other options based on your project requirement.

To change the color of the line, select an option from the **Color** drop-down list. To apply line thickness to the line feature, select an option from the drop-down list next to the **Color** drop-down list. Any change made to the line feature will be displayed in the **Preview** display box. After setting the required parameters, choose the **Apply** button; the line style will be modified based on the parameters specified. Next, choose the **Close** button to close the dialog box.

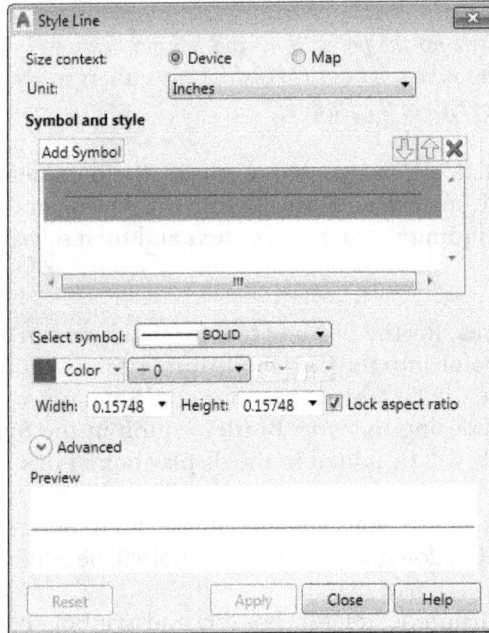

Figure 5-9 The **Style Line** *dialog box*

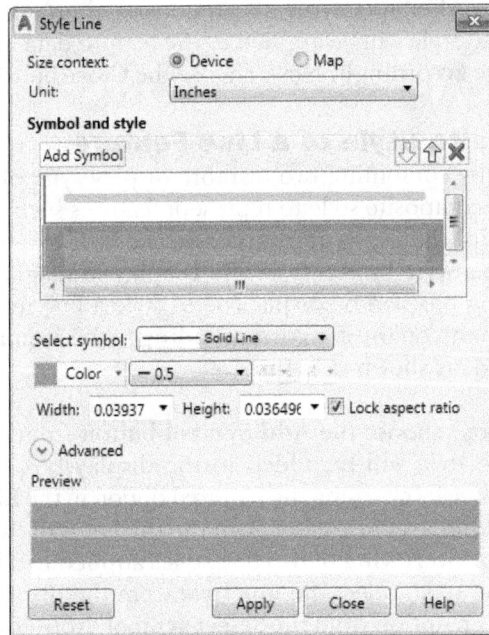

Figure 5-10 The **Style Line** *dialog box with the composite line style*

Note

*You can choose the **Reset** button located on the lower left corner of the **Style Line** dialog box to apply default AutoCAD settings to the selected line style.*

Editing the Labels of Layers

You can edit the labels of a layer as required. To edit the label of the feature data, you need to either edit the **Legend Label** cell or the **Feature Label** cell in the **STYLE EDITOR** dialog box. Figure 5-11 shows an example of the legend and feature labels representing feature class data in a map. The procedures of editing the legend and feature labels are discussed next.

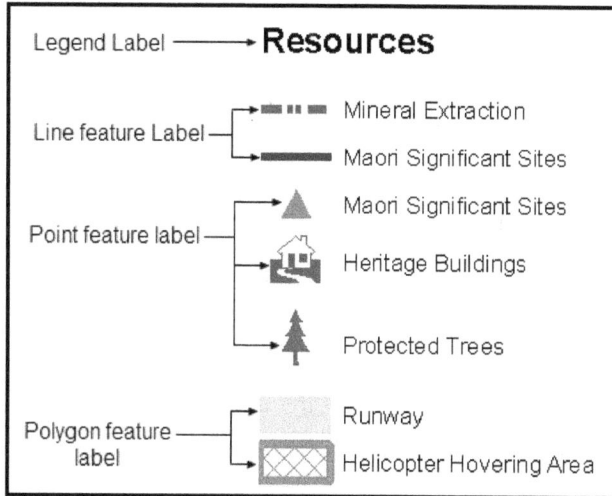

Figure 5-11 The legend and feature labels

Editing the Legend Label

You can edit a legend label of the feature dataset in the **STYLE EDITOR** dialog box. To do so, click in the **Legend Label** cell that you want to edit; the cell becomes editable. Enter the desired text for the legend label and then click outside the selected cell; the specified name will be saved.

Editing the Feature Label

You can edit the label of the feature dataset falling in the corresponding class-interval for the required vector data. To edit the feature label of a dataset, click in the cell corresponding to the required feature or range of feature class in the **Feature Label** column; the **Style Label** dialog box will be displayed, as shown in Figure 5-12.

In this dialog box, you can alter the settings of the feature label as required. To specify screen units for the width and height of the label, retain the default selection for the **Size context** property. If you select the **Map** radio button for the **Size context** property, the width and height measurements of the label are measured with respect to map units. Next, to apply measuring units, select an option from the **Units** drop-down list. In the **Label and style** area, you can add multiple label styles to the feature class data. To add a new label style, choose the **Add Label** button from the **Label and style** area; the predefined settings of the added label style will be displayed below the area. The options displayed below the area are based on the symbol or text style selected in the display box. Specify the values for these options and then choose the **Apply** button; the settings will be applied to the feature label. Next, choose the **Close** button in the **Style Label** dialog box; the dialog box will be closed.

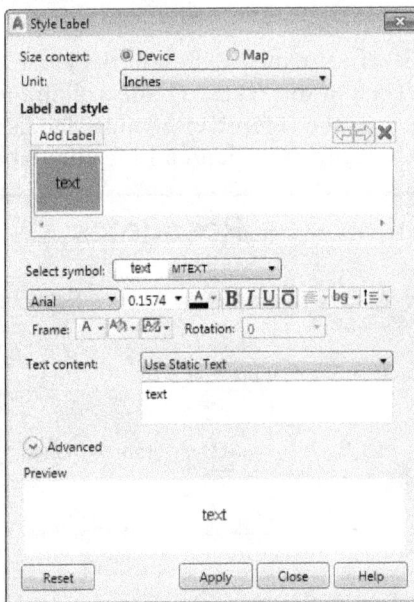

*Figure 5-12 The **Style Label** dialog box*

USING FILTERS TO QUERY THE DATA

A spatial dataset contains data that may cover a large geographic region such as a country or state. These datasets can also contain a huge volume of attribute data corresponding to the geographical features within the region.

While working with spatial dataset in GIS, you may come across a situation where the area of interest is limited to a small geographical area, such as city or county. Similarly, while performing spatial analysis, you may need to use a specific type of feature from the entire dataset. In such a situation, the entire dataset consumes a large computing power for performing spatial analysis. Also, it increases the processing time thereby lowering the performance of the entire system.

To deal with such situations, you can create a subset of the entire dataset. To create a subset, you can apply filters (spatial or non spatial) to select the required features from the dataset. You can apply such filters by using queries to select the required data for your analysis.

Figure 5-13 shows the schematic diagram of the filter process for selecting the required data for the analysis. Filtering the data will limit the volume of data required for performing the data analysis resulting in increased performance during the process of data analysis.

Data filters are the conditions or criteria that are applied for the data selection. These conditions can be applied to the database by using logical expressions. The steps involved in querying feature data using data filter are discussed next.

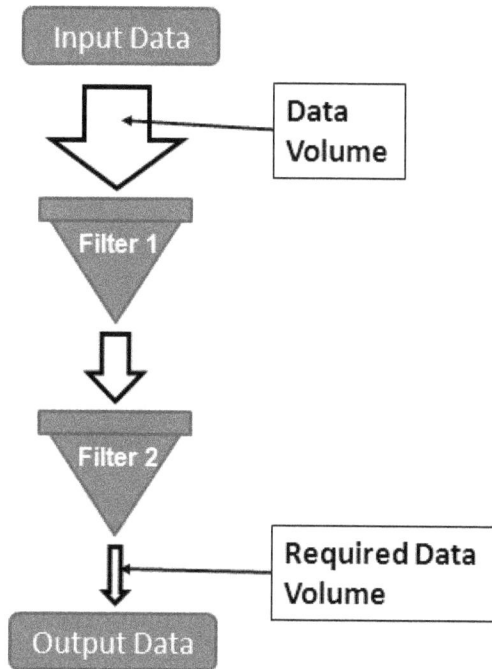

Figure 5-13 Schematic diagram of the data filter process showing the data flow through multiple filters

Step 1 - Invoking the Create Query Window

Ribbon: Vector Layer (Contextual) > View > Query to Filter

AutoCAD Map 3D provides the **Query to Filter** tool for constructing the query expression. This query expression (statement) contains the logical conditions that are applied to a dataset for extracting the required data from the entire dataset. This filtered (subset) data can be used further in the analysis. By using this subset you can limit the volume of data required to be processed for performing a spatial analysis and this will inturn result in an efficient and quick analysis.

To invoke the **Create Query** window, select a feature data layer from the **Display Manager** tab of the **TASK PANE**; the **Vector Layer** contextual tab will be displayed. Next, choose the **Query to Filter** tool from the **View** panel of this tab; the **Create Query** window with the **Getting Started with Filters** page will be displayed, as shown in Figure 5-14. You can write a query expression in the editing area of the **Create Query** window. For users who are new to the technique of query writing, AutoCAD Map 3D provides basic help in building the query expression. The three options in the **Getting Started with Filters** page provide syntax for a query expression. You can edit the query syntax to create a required expression. The options in the **Getting Started with Filters** page are discussed next.

Note
*You cannot invoke the **Query to Filter** or **Search** tool without adding a vector layer to the workspace. So, before invoking these tools, ensure that all the features to be queried are checked-in. The **Search** tool is discussed later in this chapter.*

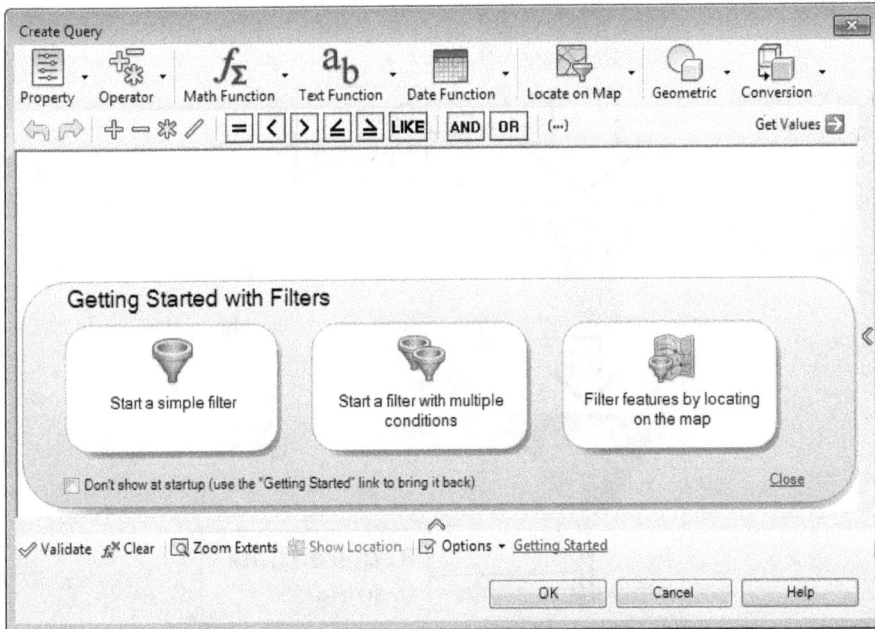

*Figure 5-14 The **Create Query** window with the **Getting Started with Filters** page displayed*

Start a simple filter

You can use the **Start a simple filter** option to build a simple query expression. A simple expression is the one where you filter data by applying a single condition on a dataset. To create a simple query statement, choose the **Start a simple filter** button from the **Getting Started with Filters** page; the syntax for the query expression **[property] > [value]** will be displayed in the **Create Query** window. Next, you need to edit this syntax to form a query expression. An example of a simple query expression is given next.

BUILTYPE = 'SHOPPING MALL'

In this query, all the features in the selected feature dataset, having the attribute SHOPPING MALL in the BUILTYPE field will get filtered. This filtered data is a subset of the entire dataset and has the same structure as that of the original dataset.

Another example of a simple query expression is also given next.

RESTAURANT = 'YES'

In this query, all the features in the selected feature dataset, having the attribute YES in the RESTAURANT field will get filtered.

Start a filter with multiple conditions

The **Start a filter with multiple conditions** option provides syntax to create a complex query. A complex query is a combination of two or more simple queries put together. It is used to check multiple criteria. For example, in the query BUILTYPE = 'SHOPPING MALL' AND RESTAURANT = 'YES', only those features in the selected feature layer will be filtered for which the SHOPPING MALL attribute is in the BUILTYPE field and the YES attribute is available in the RESTAURANT field.

To create a query with multiple conditions, choose the **Start a filter with multiple conditions** button from the **Getting Started with Filters** page; the **[property] > [value] AND [property] < [value]** syntax will be displayed in the **Create Query** window. Next, you need to edit these expressions to form a query statement.

Filter features by locating on the map

AutoCAD Map 3D allows the use of spatial filters to select data within the required area. The spatial filters are created by specifying the geometry of the desired area in the drawing window. You can also apply additional non-spatial filters to refine the data. To select data by applying spatial filter, choose the **Filter features by locating on the map** button in the **Getting Started with Filters** page; the **Locate on Map** drop-down list will be displayed. You can select the shape option from the list which is suitable for object selection in your project. On selecting it, you will be prompted to choose an option to define the spatial filter by creating geometry or by selecting an existing geometry from the drawing. Define the geometry of the spatial filter in the drawing; the query statement with the spatial query expression will be displayed in the editor area of the **Create Query** window. You can use this query expression to filter the data using the defined spatial filter.

Consider the query, [LOCATION:INSIDE.POLYGON.ID1] AND BUILTYPE = 'SHOPPING MALL' AND RESTAURANT = 'Y'. This is a complex query expression that defines spatial and non spatial filters. This query will search for features within the defined geographic area that have the attributes SHOPPING MALL and Y in the BUILTYPE and RESTAURANT fields, respectively.

After choosing the required option from the **Getting Started with Filters** page, the query syntax will be displayed in the editing area of the **Create Query** window. You can edit this syntax to create the required query expression. The method to write a query expression is discussed in the next section.

Note

*If you select the **Don't show at startup (use the "Getting Started" link to bring it back)** check box, the **Getting Started with Filters** page will not be displayed on invoking the **Create Query** window the next time.*

Step 2 - Writing Query Expressions

Expressions are the statements or arguments that contain filters to search data within a dataset. The data is filtered by applying the conditions that are specified within the query expression. You can create a simple query for checking the data that satisfies a simple condition or you can build a complex query that verifies the data for multiple conditions using various logical, conditional, mathematical, and relative operators.

The process of query writing can be broadly classified into two steps: developing the logic (concept) for the query and writing expression. These steps are discussed next.

Developing Concept for Querying

A query expression consists of identifiers, operators, functions and arguments, constants and values. Before writing a query expression, a user must know the objective (what/which) and condition (where) for the query. In addition, the user must understand the structure of the data model.

Let us take an example where you are required to find regions in Iceland where population is less than 20000, and area is less than 10000 sqkm.

Question:	What (object)
Answer:	Regions (areas)
Question:	Where (condition)
Answer:	Population less than 20000 and area less than 10000 sqkm
	From feature layer Iceland

Consider a feature layer **ic** that shows various regions of Iceland, as shown in Figure 5-15. The attribute data corresponding to this vector layer is displayed in the **DATA TABLE** window, as shown in Figure 5-16. The fields ADMIN_NAME, POP_ADMIN, and SQKM_ADMIN in the data table contain attribute information for the region name, population, and area, respectively.

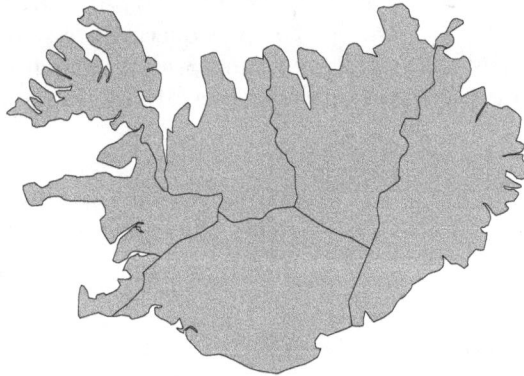

Figure 5-15 *Spatial Map of Iceland*

Figure 5-16 *The **DATA TABLE** window displaying the attribute data of Iceland*

Note
*To display the **DATA TABLE** window, select the required feature class in the **Display Manager** of the **TASK PANE** and then choose the **Table** tool.*

Figure 5-17 shows the flow diagram displaying the data filtering process in this example. First, you must have access to the ICELAND dataset. On establishing the connection with the dataset, you can perform the query process to search the entire attribute data of the feature layer. Next, you will apply the first condition (Population is less than 20000); a subset of attributes that satisfy the condition will be created.

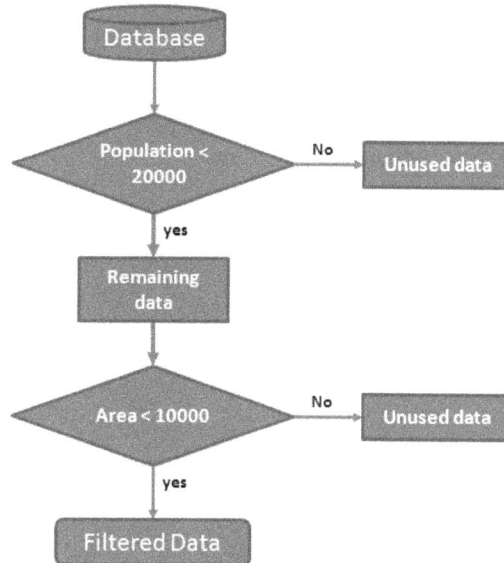

Figure 5-17 Flow diagram depicting the data flow in the process of using filters

Next, the second condition will refine this subset by retaining the attributes for the features that satisfy it. As a result, the data subset will now contain the records for the features that satisfy both conditions in the query.

Writing an expression in the **Create Query** window using the various options are discussed next.

Writing Query Expression

After developing the concept for the query, you need to write the query expression in editor area of the **Create Query** window. A query expression has three basic components: Identifiers (property), Operator to perform operation or check condition (such as <, =, >, LIKE, OR, *, +), and Value. The structure (syntax) of a simple query is given below.

[Property] operator [Value]

To write an expression in the above syntax, substitute the [Property], operator, and [Value] with the property name, operator, and value, respectively. An operator can be logical (AND, OR, NOT), arithmetic (+, -, *), or conditional (=, <, >).

Writing a query expression for searching areas in the feature layer for Iceland where population less than 20000 and area less than 10000 sq km is discussed next.

To write a query expression for the feature layer for Iceland, first invoke the **Create Query** window. Next, to write an expression for population less than 20000, choose the **Property** drop-down list from the **Create Query** window; the list of all the properties in the attribute table of the feature layer will be displayed. From this list, select the property that represents the population of the regions in the feature class (**POP_ADMIN**) option, as shown in Figure 5-18. On doing so, the selected property name (**POP_ADMIN**) will be displayed in the editor area of the **Create Query** window. Next, type **< 20000** (less than 20000).

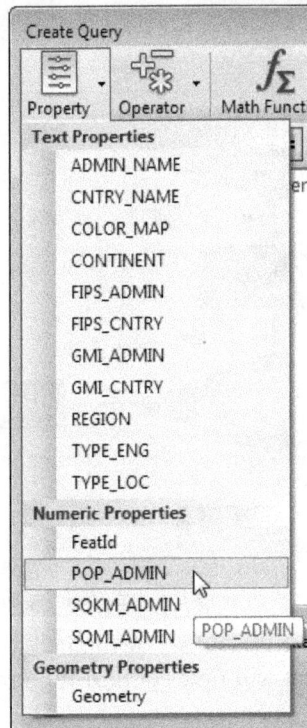

Figure 5-18 *Selecting the* **POP_ADMIN** *option from the* **Property** *drop-down list*

The editor area will now display the following text:

POP_ADMIN < 20000

This is a simple query that searches the first condition for the area with population less than 20000.

Similarly, write the second condition as follows:

SQKM_ADMIN < 10000

This expression searches regions in the feature layer for Iceland with area less than 10000 sq km.

Now, there are two simple expressions in the editor window. To find the region with population less than 20000 and area less than 10000, you need both the conditions to be true. Therefore, you need to add the AND operator between the two simple expressions. After adding this operator, the query expression will be as follows:

POP_ADMIN < 20000 AND SQKM_ADMIN < 10000

The query expression for the required condition is now complete. Before executing the query, a validation check for examining the correctness of the expression is recommended.

Step 3 - Validating the Query Expression

⊘ Validate   After writing the expressions, choose the **Validate** button. If the expression is valid, the message **The expression is valid** will be displayed in the text box of the **Create Query** window, as shown Figure 5-19.

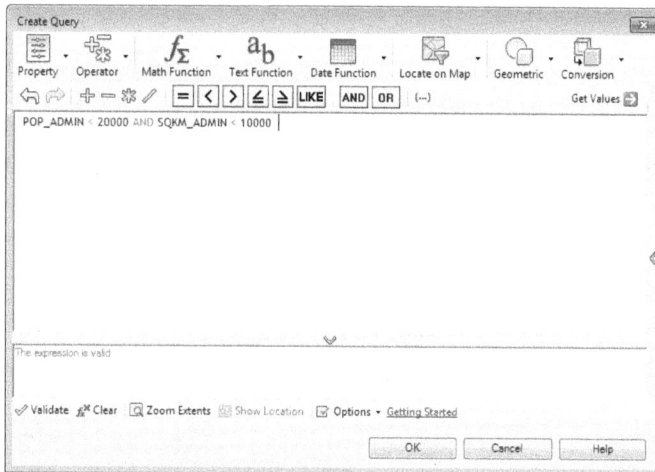

Figure 5-19 *The text box in the* **Create Query** *window displaying the query validation message*

Next, choose the **OK** button to execute the query; the **Create Query** dialog box will be closed and the filtered vector layer will be displayed in the drawing window, as shown in Figure 5-20.

Figure 5-20 *The regions displayed in the drawing window (after Query operation)*

Figure 5-21 shows the **DATA TABLE** window displaying the attribute data for the corresponding queried feature the filtered data is displayed in tabular format.

Figure 5-21 *The **DATA TABLE** window displaying the attribute data for the queried feature*

USING FILTERS FOR SEARCHING DATA

Ribbon: Home > Data > Search
Command: MAPSEARCH

The **Search** tool in AutoCAD Map 3D allows you to search and select the features from the feature layer using the specified search criteria. The process of searching data is similar to that of querying data. To search the required data, choose the **Search** tool from the **Data** panel of the **Home** tab; the **Search for Features Across Multiple Layers** dialog box will be displayed, as shown in Figure 5-22.

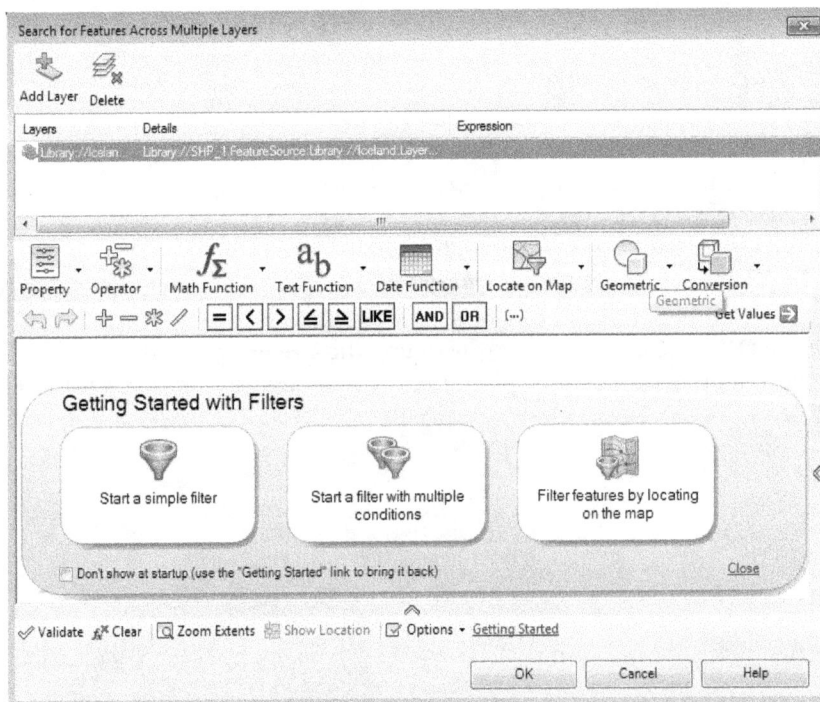

Figure 5-22 *The **Search for Features Across Multiple Layers** dialog box*

In this dialog box, choose the **Add Layer** button located at the upper left corner; a drop-down list containing the available feature layers will be displayed. From this list, select the feature layers for which you wish to conduct the search operation. On doing so, the selected feature layer will be added to the list box below the **Add Layer** button. When you add more than two layers, the **Common Filter** option will be displayed below the layers in the **Layers** list. Click on the down arrow in the **Common Filter** option; the list of added layers will be displayed with their corresponding check boxes selected, as shown in Figure 5-23. To ignore a layer during the filtering process, clear the check box corresponding to that layer.

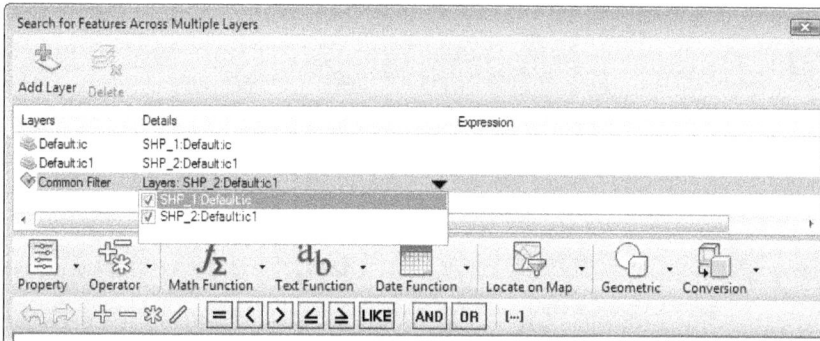

Figure 5-23 *The partial view of the* ***Search for Features Across Multiple Layers*** *dialog box showing the layers in the* ***Common Filter*** *option*

After selecting the required layer/s, write and validate the query expression for searching the data in the editor region of this dialog box, as explained previously. Next, choose the **OK** button to execute the query; the dialog box will be closed and the result of the query will be displayed in the drawing window. Figure 5-24 shows a feature object selected after performing a search operation.

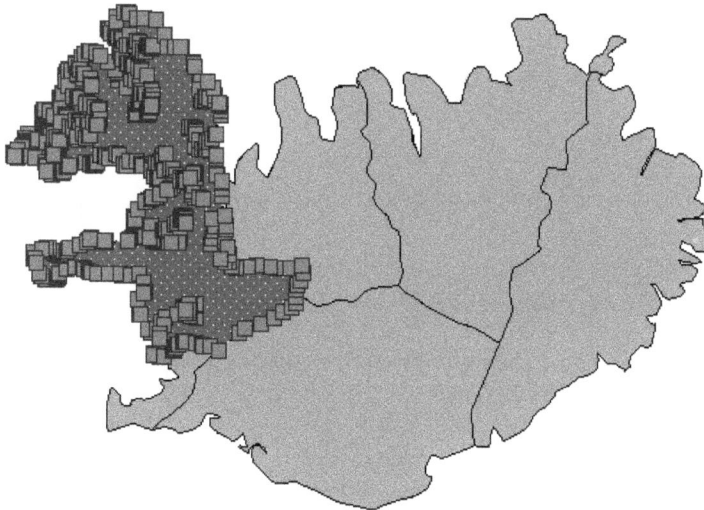

Figure 5-24 *The selected feature data after performing search operations*

TUTORIALS
General instructions for downloading tutorial files:

Before starting the tutorials, you need to download the tutorial data to your computer. To do so, follow the steps given below:

1. Log on to *www.cadcim.com* and browse to *Textbooks > Civil/GIS > Map 3D > Exploring AutoCAD Map 3D 2017*. Next, select the *c05_m3d_2017_tut.zip* file from the **Tutorial Files** drop-down list. Next, choose the corresponding **Download** button to download the data file.

2. Extract the contents of the zip file to the following location:

    C:\m3d_2017

Notice that the *c05_m3d_2017_tut* folder is created within the *m3d_2017* folder.

Tutorial 1   Styling the Feature Data

In this tutorial, you will apply visualization techniques for representing the geospatial data.
(Expected time: 1hr)

The following steps are required to complete this tutorial:

a. Load the *Iceland.shp* file into the workspace using the **Connect** tool.
b. Create scale ranges for the ***Iceland.shp*** feature data.
c. Define display themes for the scale ranges.
d. View the applied themes.
e. Save the file.

Loading the Shape File
In this section of the tutorial, you will open a new drawing file and assign a coordinate reference system to the drawing. Next, you will load the Iceland SHP file into the drawing using the **DATA CONNECT** wizard.

1. Choose the **New** button from the **Quick Access Toolbar**; a new drawing file is opened.

2. Choose the **Assign** tool from the **Coordinate System** panel in the **Map Setup** tab; the **Coordinate System - Assign** dialog box is displayed.

3. In this dialog box, select the **Iceland** option from the **Category** drop-down list; the coordinate systems in this category are displayed in the list box.

4. Select the option with the **Hjorsey.IcelandGrid** code from the list box and then choose the **Assign** button; the **Coordinate System - Assign** dialog box is closed and the code of the selected coordinate system is assigned to the current drawing.

5. Next, choose the **Connect** tool from the **Data** panel in the **Home** tab; the **DATA CONNECT** wizard is displayed.

6. In the **DATA CONNECT** wizard, choose the **Add SHP Connection** option from the **Data Connections by Provider** page, if it is not chosen by default; the **OSGeo FDO Provider for SHP** page is displayed in the right pane of this wizard.

7. In the **OSGeo FDO Provider for SHP** area of the **DATA CONNECT** wizard, choose the **SHP** button; the **Open** dialog box is displayed.

8. In the **Open** dialog box, browse to the following location:

    C:\m3d_2017\c05_m3d_2017_tut\c05_tut01

9. In the **c05_tut01** folder, select the *Iceland.shp* file; the shape file name is displayed in the **File name** edit box. Next, choose the **Open** button; the shape file path is displayed in the **Source file or folder** edit box.

10. Choose the **Connect** button in the **OSGeo FDO Provider for SHP** page; the **SHP** page with source data is displayed.

11. Select **Iceland < unknown >** from the list box in the **Add Data to Map** area, and then choose the **Edit Coordinate System** button located above the list box; the **Edit Spatial Contexts** dialog box is displayed.

12. In the **Edit Spatial Contexts** dialog box, select the first row and then choose the **Edit** button; the **Coordinate System Library** dialog box is displayed.

13. In the **Coordinate System Library** dialog box, select the **Iceland** option from the **Category** drop-down list; a list of coordinate systems in the selected category is displayed in the list box.

14. Next, choose the option with the **HJORSEY.LL** code and then choose the **Select** button; the dialog box is closed and the selected coordinate system is displayed in the **Override** column in the **Edit Spatial Contexts** dialog box.

15. Next, choose the **OK** button in the **Edit Spatial Contexts** dialog box; the **HJORSEY.LL** coordinate system is applied to the source shape file in the **Add Data to Map** list box.

16. Choose the **Add to Map** button located below the source file list box; the regional boundary of the **Iceland** feature polygon data appears in the drawing window. Also, the vector data is loaded into the working space.

17. Next, close the **DATA CONNECT** wizard.

Creating the Scale Ranges for the Vector Data

In this section of the tutorial, you will create and specify various scale ranges for the Iceland feature layer.

1. Select the **Iceland** feature layer in the **Display Manager** tab of the **TASK PANE**; the **Vector Layer** contextual tab is displayed. Also, the **Style** and **Table** buttons in the **Display Manager** tab are activated.

2. Next, choose the **Style** button from the **Display Manager** tab of the **TASK PANE**; the **STYLE EDITOR** dialog box is displayed with default settings.

3. In the **STYLE EDITOR** dialog box, choose the **Add a Scale Range** button in the **Scale Ranges for Layer Default:Iceland** area; a new scale range is added to the scale range list box. Repeat this step to create another scale range in the list.

4. Now, click in the **From** cell of the first row in the **Scale Ranges for Layer Default:Iceland** area; the cell becomes editable.

5. Clear the content of the cell and then specify **4000001** in the cell.

6. Next, specify **2000001** in the **From** cell and **4000000** in the **To** cell of the second row.

7. Specify **2000000** in the **To** cell corresponding to the third row. Figure 5-25 shows the three rows with the specified scale range.

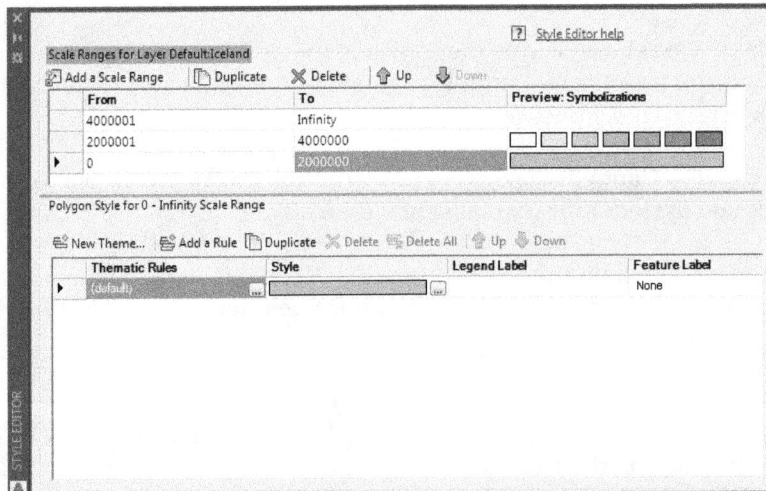

*Figure 5-25 The **STYLE EDITOR** dialog box with the added scale ranges*

Applying Theme to the Feature Data for Scale Range 4000001 - Infinity

In this part of the tutorial, you will create a theme and then define its settings for displaying the feature data in the scale range of 4000001 - Infinity.

1. Select the first row that defines the scale range from 4000001 - Infinity in the **Scale Ranges for Layer Default:Iceland** area of the **STYLE EDITOR** dialog box; the **Polygon Style for 4000001 - Infinity Scale Range** area is displayed at the bottom of this dialog box.

2. Choose the **New Theme** button in the **Polygon Style for 4000001 - Infinity Scale Range** area; the **Theme Layer** dialog box is displayed.

3. In the **Create thematic rules based on a property** area of this dialog box, select the **CNTRY_NAME** option from the **Property** drop-down list; **Iceland** is displayed in the **Minimum value** and **Maximum value** edit boxes and **1** is displayed in the **Number of rules** (uneditable) text box.

4. Choose the Browse button corresponding to the **Style range**; the **Style and Label Editor** dialog box with the **Style** tab selected is displayed.

Note that the line symbol is selected by default in the **Symbol and Style** list box of the **Style** tab, also the options below the list box display the parameters such as **Line Pattern** and **Line Color** corresponding to the selected line symbol.

5. Next, choose the **Delete** button above the **Symbol and Style** list box; the selected line symbol is deleted and the list box now displays a single symbol corresponding to the **Fill Pattern**, as shown in Figure 5-26.

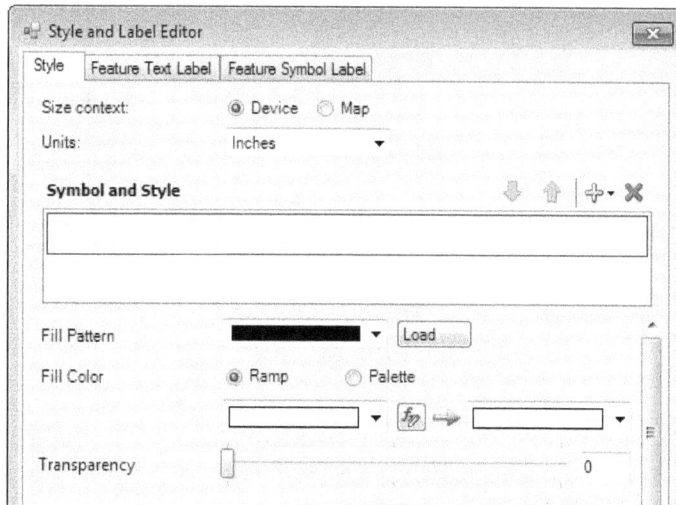

*Figure 5-26 Partial view of the **Style and Label Editor** dialog box showing the selected symbol and style*

6. Now, choose the **OK** button in the **Style and Label Editor** dialog box; the dialog box is closed and the selected style is set and displayed in the preview area corresponding to the **Style range** in the **Theme Layer** dialog box.

7. In the **Theme Layer** dialog box, ensure that the check box corresponding to the **Create feature labels** is cleared. Figure 5-27 shows the **Theme Layer** dialog box with the specified settings.

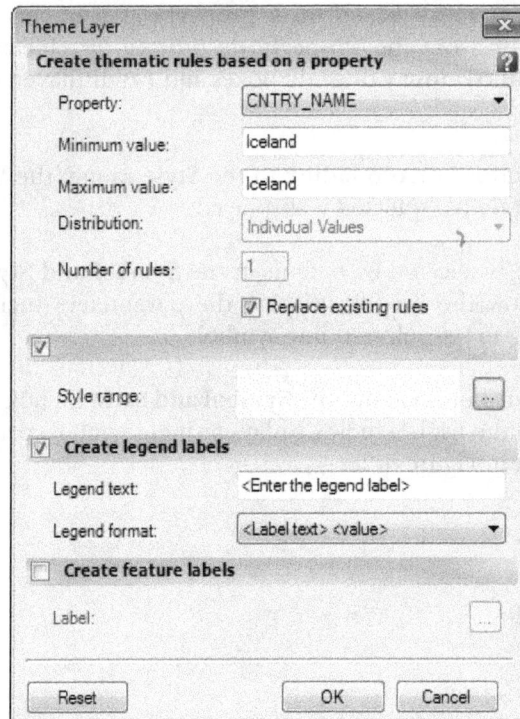

*Figure 5-27 The **Theme Layer** dialog box with specified settings*

8. Next, choose the **OK** button in the **Theme Layer** dialog box; the thematic rule is created and displayed in the **Polygon Style for 4000001 - Infinity Scale Range** area of the **STYLE EDITOR** dialog box.

9. Next, select the row corresponding to the **(default)** value in the **Thematic Rules** column and choose the **Delete** button in the **Polygon Style for 4000001 - Infinity Scale Range** ✕ Delete area; the selected rule is deleted.

Applying Theme to the Feature Data for Scale Range 2000001 - 4000000

In this part of the tutorial, you will create the theme to display the administrative areas of Iceland for the scale range of 2000001 - 4000000. You will then define the settings for this theme.

1. Select the second row (scale range from 2000001 to 4000000) from the **Scale Ranges for Layer Default:Iceland** area of the **STYLE EDITOR** dialog box; the **Polygon Style for 2000001 - Infinity Scale Range** area is displayed in the dialog box.

2. Next, choose the **New Theme** button in the **Polygon Style for 2000000 - 4000001 Scale Range** area; the **Theme Layer** dialog box is displayed.

3. In this dialog box, select the **ADMIN_NAME** option from the **Property** drop-down list.

    Note that the **Distribution** and **Number of rules** options are now uneditable.

4. Next, select the check box corresponding to **Create feature labels**; the **Label** option is activated.

5. Choose the Browse button corresponding to the **Label** option; the **Feature Text Label** tab of the **Style and Label Editor** dialog box is displayed.

6. In the **Feature Text Label** tab, choose the **Edit Expression** button corresponding to the **Text** property; the **Create/Modify Expressions** dialog box is displayed.

7. In the **Create/Modify Expressions** dialog box, clear the content in the edit area.

8. Next, choose the **ADMIN_NAME** option from the **Property** drop-down list, as shown in Figure 5-28; **ADMIN_NAME** is displayed in the edit area of the **Create/Modify Expressions** dialog box.

9. Now, choose the **OK** button in the **Create/Modify Expressions** dialog box; the dialog box is closed. Also, in the **Style and Label Editor** dialog box, the **Text** option of the **Feature Text Label** tab displays the expression created in the **Create/Modify Expressions** dialog box.

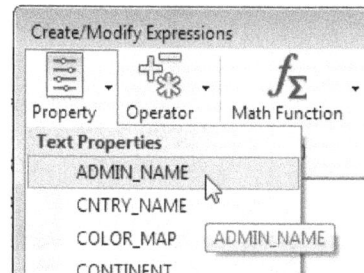

Figure 5-28 Selecting the ADMIN_NAME option from the Property drop-down list

10. Retain other settings in the **Style and Label Editor** dialog box, and then choose the **OK** button; the dialog box is closed and the specified settings for the feature label text style are set in the **Theme Layer** dialog box.

11. Next, choose the **OK** button in the **Theme Layer** dialog box; the data is classified based on the specified settings and the thematic rules are created and displayed in the **Polygon Style for 2000001- Infinity Scale Range** area of the **STYLE EDITOR** dialog box.

12. Next, select the row corresponding to the **(default)** value in the **Thematic Rules** column and choose the **Delete** button in the **Polygon Style for 2000001 - Infinity Scale Range** area; the selected rule is deleted. Figure 5-29 shows the **STYLE EDITOR** dialog box with the displaying theme for the scale range 2000001 - 4000000.

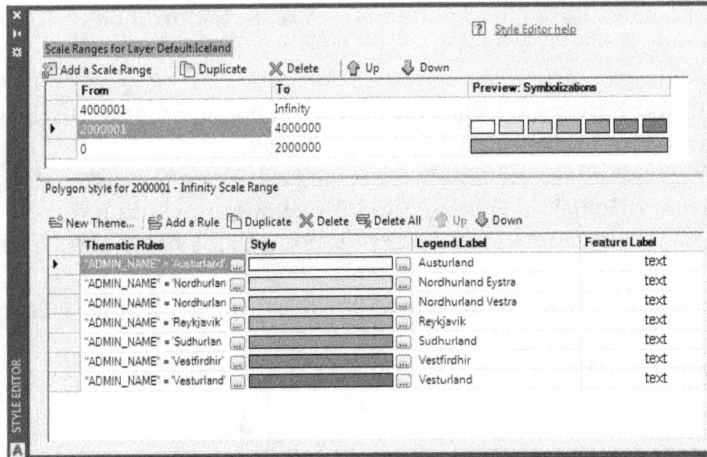

Figure 5-29 The *STYLE EDITOR* dialog box displaying the theme for the scale range 2000001 - 4000000

Applying Theme to Feature Data for Scale Range 0 - 2000000

1. Select the third row (Scale range 0 - 2000000) from the scale ranges list box in the **Scale Ranges for Layer Default:Iceland** area, if it is not selected by default; the **Polygon Style for 0 - Infinity Scale Range** area is displayed.

2. Choose the **New Theme** button in the **Polygon Style for 0 - Infinity Scale Range** area; the **Theme Layer** dialog box is displayed.

3. In this dialog box, select the **SQKM_ADMIN** option from the **Property** drop-down list; the minimum and maximum values corresponding to this property are displayed in the **Minimum value** and **Maximum value** edit boxes, respectively.

4. Select the **Individual Values** option from the **Distribution** drop-down list; the **Number of rules** edit box becomes uneditable and displays **7**.

5. Choose the Browse button corresponding to the **Style range**; the **Style and Label Editor** dialog box is displayed.

6. In this dialog box, select the fill symbol in the **Symbol and Style** list box; options such as **Fill Pattern** and **Fill Color** corresponding to the selected fill symbol are displayed.

7. Ensure that the **Ramp** radio button for the **Fill Color** option is selected.

8. Next, select the Yellow and Red color from the two color drop-down list displayed below the **Fill Color** option; the **Preview** area displays the color for the different classes, refer to Figure 5-30.

9. Now, choose the **OK** button in the **Style and Label Editor** dialog box; the dialog box is closed and the selected color and symbol is displayed in the preview corresponding to the **Style range** option in the **Theme Layer** dialog box.

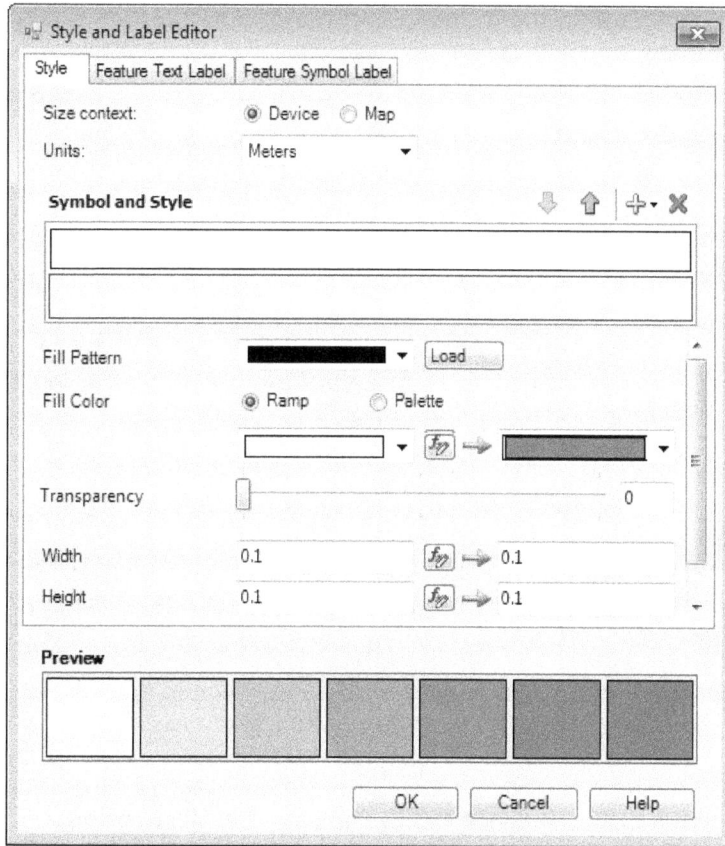

Figure 5-30 *The* **Style and Label Editor** *dialog box displaying the selected color ramp*

10. Next, select the **Create feature labels** check box in the **Theme Layer** dialog box; the **Label** option is activated.

11. Choose the Browse button corresponding to the **Label** option; the **Style and Label Editor** dialog box is displayed with the **Feature Text Label** tab chosen.

12. In the **Feature Text Label** tab, choose the **Edit Expression** button corresponding to the **Text** option; the **Create/Modify Expressions** dialog box is displayed.

13. In the **Create/Modify Expressions** dialog box, clear the contents in the editor area.

14. Enter **CONCAT (ADMIN_NAME, '\n', 'Area: ' , SQKM_ADMIN)** in the editor area and choose the **OK** button; the **Create/Modify Expressions** dialog box is closed. The expression written in the **Create/Modify Expressions** dialog box is displayed in the text box corresponding to the **Text** option in the **Feature Text Label** tab of the **Style and Label Editor** dialog box.

15. Choose the **OK** button in the **Style and Label Editor** dialog box; the dialog box is closed and the **Theme Layer** dialog box is displayed.

16. Choose the **OK** button in the **Theme Layer** dialog box; the data is classified into seven class-intervals, as shown in Figure 5-31.

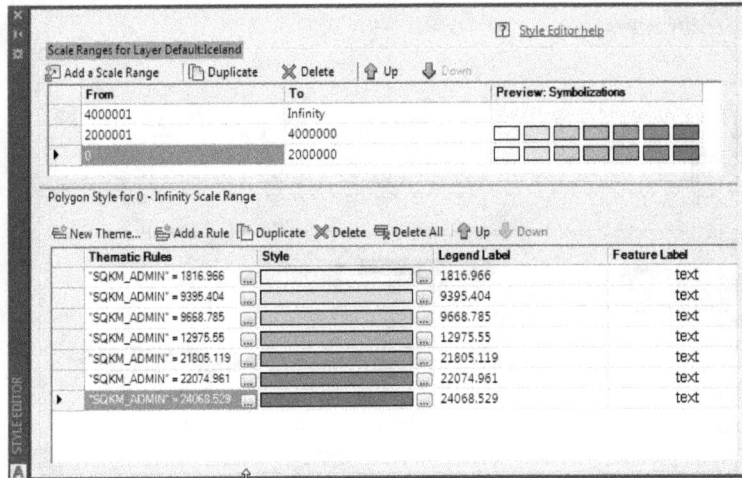

Figure 5-31 *The STYLE EDITOR dialog box showing the thematic rules created*

17. Next, select the row corresponding to the **(default)** value in the **Thematic Rules** column and choose the **Delete** button from the **Thematic Rules** area; the selected rule is ✖ Delete deleted.

18. Close the **STYLE EDITOR** dialog box by choosing the **Close** button.

Checking the Applied Themes

1. Choose the **Extents** tool from the **Navigate** panel of the **View** tab; the drawing is zoomed to its extent and displayed in the drawing window. Notice the theme of the displayed data. Also, notice the feature label in the drawing and in the **Display Manager**.

2. Next, zoom in and zoom out the drawing at various scales. Notice the change in the display of the data in the drawing window. Figure 5-32 shows the feature data for Iceland at the zoom scale between 2000001 to 4000000.

Saving the Drawing File

1. Choose the **Save As** tool from the Application Menu; the **Save Drawing As** dialog box is displayed.

2. In the **Save Drawing As** dialog box, enter **c05_Tut01a.dwg** in the **File name** edit box.

3. Select the **AutoCAD 2013 Drawing (*.dwg)** option in the **Files of type** drop-down list located at the bottom of the **Save Drawing As** dialog box, if it is not selected by default.

4. Choose the **Save** button; the drawing file is saved.

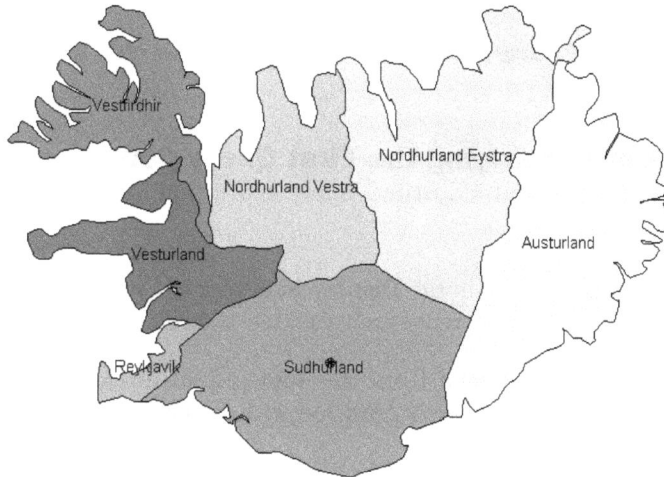

Figure 5-32 The vector data displayed at a zoom scale between 2000001 - 4000000

Tutorial 2   Querying the Feature Data

In this tutorial, you will write query expressions to search the data for the conditions specified below.   **(Expected time: 30 min)**

You need to apply the following conditions to query the feature data:

- **Area** (SQKM_ADMIN) of the selected region(s) should be more than 20000.
- **Population** (POP_ADMIN) of the selected region(s) should be more than 10000 and less than 15000.

The following steps are required to complete this tutorial:

a. Open the drawing file.
b. Perform query to filter the data by applying the condition.
c. Save the file.

Opening the Drawing File

1. Choose the **Open** button from the Quick Access Toolbar; the **Select File** dialog box is displayed.

2. In this dialog box, browse to the following location:

    C:\m3d_2017\c05_m3d_2017_tut\c05_tut02

3. Select the **m3d_c05_Tut02.dwg** option from the **c05_tut02** folder and then choose the **Open** button in this dialog box; the geometry of the selected drawing is displayed in the drawing window.

Note
*If the **m3d_c05_Tut02.dwg** file shows error message, open the **c05_Tut01a.dwg** file to perform query.*

Query Feature Data - Applying the First Query Condition

In this section of the tutorial, you will write a query expression to find the regions in Iceland with area greater than 20000 sq km.

1. Select the **Iceland** feature layer in the **Display Manager** tab of the **TASK PANE**, if it is not selected by default; the **Vector Layer** contextual tab is displayed.

2. Choose the **Query to Filter** tool from the **View** panel in the **Vector Layer** tab; the **Create Query** window with the **Getting Started with Filters** page is displayed.

3. In this page, choose the **Start a simple filter** button; the **[property] > [value]** expression is displayed in the edit area of this dialog box.

4. Next, select the **[property]** text in the displayed expression, if it is not selected by default, and then choose the **Property** button; the **Property** drop-down list is displayed.

5. In this drop-down list, select the **SQKM_ADMIN** option; the **[property]** text is replaced by **SQKM_ADMIN** in the edit area of the **Create Query** window.

6. Select the **[value]** text in the expression and replace this text with the numeric value **20000**. The edit area in the **Create Query** window will now contain the following expression:

   SQKM_ADMIN > 20000

7. Choose the **Validate** button at the lower left portion of the **Create Query** dialog box; the statement **The expression is valid** is displayed, confirming that the written expression is a valid query statement.

8. Choose the **OK** button; the polygon features with the **Area** property more than 20000 are displayed in the drawing window, as shown in Figure 5-33.

Figure 5-33 Feature polygons with area greater than 20000

Query the Feature Data - Applying the Second Query Condition

In this section of the tutorial, you will write a query expression to find the regions in Iceland with population between 10000 and 15000.

Note that this query requires to check for the condition where:

Population is less than 15000 and greater than 10000.

1. Select the **Iceland** vector layer in the **Display Manager** tab of the **TASK PANE**; the **Vector Layer** contextual tab is displayed.

2. Choose the **Query to Filter** tool from the **View** panel in the **Vector Layer** tab; the **Modify Query** dialog box with the previous expression is displayed.

3. In this dialog box, choose the **fx Clear** button displayed below the editor area; f_x^{x} Clear the expression used in the previous filter is cleared.

4. Click on the **Getting Started** link; the **Getting Started with Filters** page is displayed.

5. Choose the **Start a filter with multiple conditions** button in the **Getting Started with Filters** wizard; the following query syntax is displayed in the editor area of the **Modify Query** dialog box:

    [property] > [value] AND [property] < [value]

6. Select the **[property]** text in the first condition, if it is not selected by default.

7. Next, select the **POP_ADMIN** option from the **Property** drop-down list; the **[Property]** text is replaced by **POP_ADMIN**.

8. Select the **[value]** text in the first condition and then replace it with **10000**.

9. Similarly select the **[property]** text in the second condition, if it is not selected by default, and select the **POP_ADMIN** option from the **Property** drop-down; **[property]** will be replaced by **POP_ADMIN.**

10. Select the **[value]** text in the second condition and then replace it with **15000**. The editor area now displays the following query expression as follows:

    POP_ADMIN > 10000 AND POP_ADMIN < 15000

11. Choose the **Validate** button at the lower left corner of the **Modify Query** window; **The expression is valid** statement is displayed confirming that the expression is in a readable format.

Tip
*1. Instead of using the predefined query syntax, you can formulate the query expression in the editor area of the **Create Query / Modify Query** window by typing the required expression.*

*2. For the condition **Population of the region should be between 10000 and 150000**, write the following expression in the edit area of the **Modify Query** window:*

POP_ADMIN > 10000 AND POP_ADMIN < 15000

12. Choose the **OK** button; the polygon features with the **Population** property range between 10000 and 15000 are displayed in the drawing window, as shown in Figure 5-34.

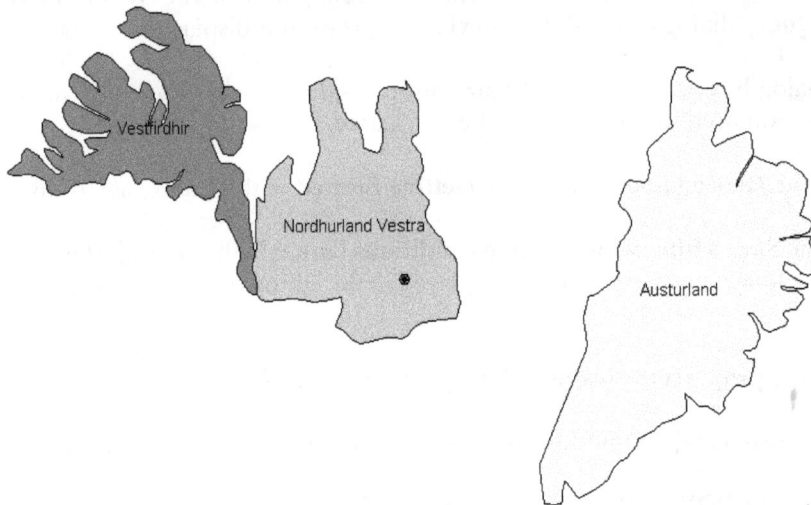

Figure 5-34 Feature polygons with population between 10000 and 15000

Saving the Drawing File

1. Choose **Save As > AutoCAD Drawing** from the Application Menu; the **Save Drawing As** dialog box is displayed.

2. In the **Save Drawing As** dialog box, enter the text **c05_Tut02a.dwg** in the **File name** edit box, and then choose the **Save** button; the drawing file is saved at the specified location.

Self-Evaluation Test

Answer the following questions and then compare them to those given at the end of this chapter:

1. Which of the following options is used to define a new rule to an existing theme?

    (a) **New Theme**   (b) **Add a Rule**
    (c) **Add a Scale Range**   (d) **Duplicate**

2. Which of the following options in the **Theme Layer** dialog box is used to define the number ranges to be created when using the **Quantile** distribution method?

    (a) **Property**   (b) **Minimum value**
    (c) **Maximum value**   (d) **Number of rules**

3. Which of the following tabs in the **Style and Label Editor** dialog box is used to define the display symbol and style of a feature?

    (a) **Style**   (b) **Feature Text Label**
    (c) **Feature Symbol Label**   (d) None of the above

4. Which of the following is a comparison operator?

    (a) **>**   (b) **+**
    (c) **-**   (d) **AND**

5. Which of the following drop-down lists in the **Create Query** window contains the options to define the geometry of the spatial filter?

    (a) **Property**   (b) **Operator**
    (c) **Locate on Map**   (d) **Math Function**

6. The _____ option in the **Create Query** window is used to validate the expression written in the text box.

7. The _____ filter option in the **Getting Started with Filters** wizard is used to filter the feature data from a particular geometrical section of a map.

8. The _____ drop-down in the **Create Query** window contains a list of attributes in the attribute table of the feature layer.

9. To invoke the **STYLE EDITOR** dialog box, choose the _____ tool from the **Display Manager** tab of the **TASK PANE**.

10. To search and select the features from the feature layer using the specified search criteria, invoke the _____ tool from the **Data** panel of the **Home** tab.

11. You can apply styling to a feature data in any scale range. (T/F)

12. Different scale ranges cannot have different themes. (T/F)

13. You can apply composite style to a line feature. (T/F)

14. You can use multiple conditions in a query expression to filter the data. (T/F)

15. The symbol **+** is a logical operator. (T/F)

Review Questions

Answer the following questions:

1. Which of the following options is used to clear the expressions in the text box in the **Create Query** window?

   (a) **Validate**   (b) **fx Clear**
   (c) **Zoom Extents**   (d) **Getting Started**

2. Which of the following options is used to control the display of the feature data at a given scale?

   (a) **Distribution**   (b) **Label**
   (c) **Scale Range**   (d) **Theme**

3. Which of the following expressions represents the statement '[Property] is greater than or equal to [value]'?

   (a) **[Property] > [value]**   (b) **[Property] = [value]**
   (c) **[Property] <= [value]**   (d) **[Property] >= [value]**

4. Which of following logical operators is used to check that atleast one of the two conditions in a query expression is true?

   (a) **OR**   (b) **AND**
   (c) **NOT**   (d) None of the above

5. Which of the following operators is not a comparison operator?

   (a) **+**   (b) **OR**
   (c) **NOT**   (d) All of the above

6. In a query expression, you can perform a comparison check by using the _____ type of operator.

7. You cannot specify the number of rules, when you select the _____ option from the **Distribution** drop-down list in the **Theme Layer** dialog box.

8. Select the _____ check box in the **Theme Layer** dialog box to display the labels for features.

9. You can edit the scale range in the **Scale Ranges for Layers** area in the **STYLE EDITOR** dialog box. (T/F)

10. You can create and edit the theme of a feature layer using the options in the **STYLE EDITOR** dialog box. (T/F)

11. Applying the scale range of 0 - infinity to a feature layer will display its content in the drawing window at all viewing scale. (T/F)

12. In the **Create Query** window, you can write a query expression without using the options in the **Getting Started with Filters** page. (T/F)

13. You can use the built-in mathematical and text functions to write the expressions in the **Create Query** window. (T/F)

14. You can define the geometry of a spatial filter by choosing an option from the **Locate on Map** drop-down list in the **Create Query** window. (T/F)

15. Arithmetic (Math) operators are used to perform mathematical operations. (T/F)

EXERCISE

Download *c05_m3d_2017_exe.zip* from *www.cadcim.com* and extract it for the following exercise.

Exercise 1

Extract the **c05_exr01** folder from *c05_m3d_2017_exe.zip*. Using the **Iceland.shp** file given in the **extracted** folder and the data given below, create style for scale range 0 - 3000000 and 3000001 - infinity.   (**Expected time: 45 min**)

   Drawing Coordinate System: **Hjorsey.IcelandGrid**
   Feature Layer: **Iceland** (Coordinate System: **HJORSEY.LL**)

   For scale range 3000001- infinity, refer to Figure 5-35
       Create theme rule based on **ADMIN_NAME** property of the feature class
       Feature label to display- Region name and area in sqkm

   For scale range 0 - 3000000, refer to Figure 5-36
       Create theme rule based on **POP_ADMIN** property of the feature class
       Distribution Method- **Standard Deviation**
       Number of rules- **5**
       Style range
       Use Fill Color or your choice
       Feature label to display- Region name and area and population density

Hint: Use the syntax given below
CONCAT ([property], '\n', 'Label: ', [property] , '\n', 'Label: ', ([property] / [property]))

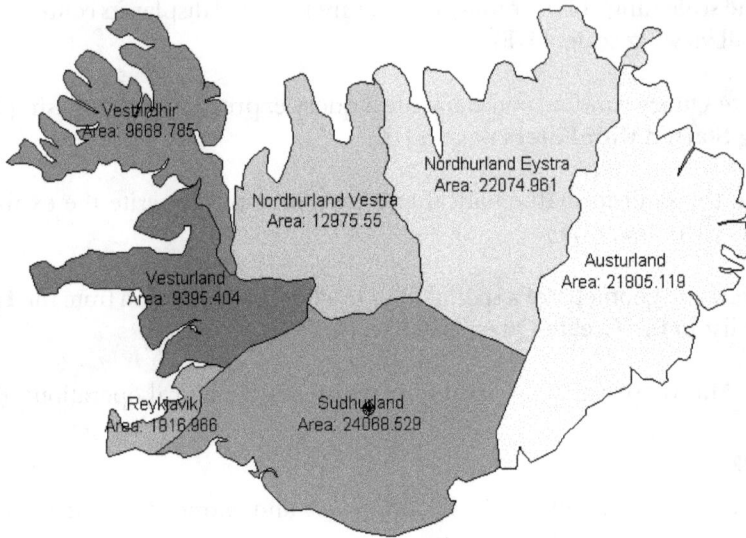

Figure 5-35 *Vector map displayed at scale range 3000001 - infinity*

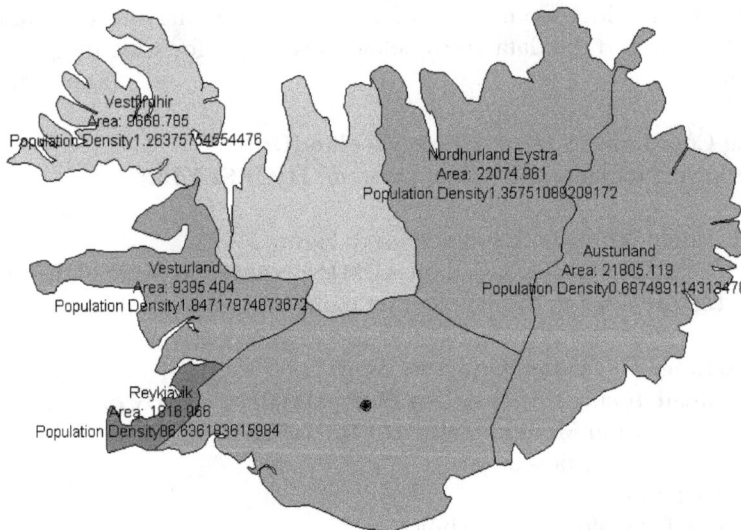

Figure 5-36 *Vector map displayed at scale range 0 - 3000000*

Answers to Self-Evaluation Test

1. b, **2.** d, **3.** a, **4.** a, **5.** d, **6. Validate, 7. Filter features by locating on the map, 8. Property, 9. Style, 10. Search, 11.** T, **12.** F, **13.** T, **14.** T, **15.** F

Chapter 6

Creating Object Data and Attaching External Database

Learning Objectives

After completing this chapter, you will be able to:
• *Create a drawing geometry and attach data fields to it*
• *Edit and modify the data fields of drawing objects*
• *Attach and detach data fields from drawing objects*
• *Incorporate drawing objects from external drawing files*
• *Create a query category*
• *Define a query process*
• *Execute query in different query modes*

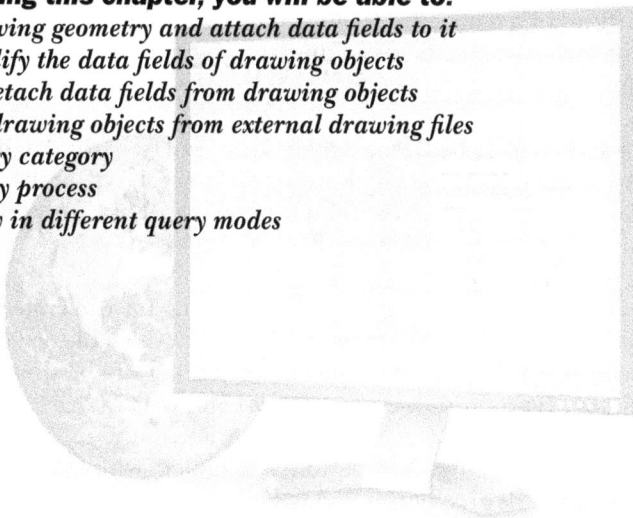

INTRODUCTION

In the previous chapters, you learned to use the **DATA CONNECT** wizard to connect to the various types of data sets. You also learned to create, edit, and query the feature data in the feature layer. In this chapter, you will learn to attach CAD data from other drawing files to the current drawing. You will also learn to query, edit, and update the drawing objects in the attached CAD drawing.

In this chapter, you will learn to work with the object data table. You will learn the procedure to create an object data table by defining the data fields. After creating the object data table, you will learn to insert data into the table and then attach it to the drawing objects using various methods in AutoCAD Map 3D.

This chapter also explains the procedure for attaching an external database to the drawing. After attaching an external database, you will use the link templates to link the data from the attached database to the drawing objects.

OBJECT DATA

The spatial object represents the geographical location of the real world features. Apart from the geographic location, a feature may have nonspatial information associated with it. For example, a post box may have data such as post code or time of collection associated with it. You can store such non-geographic data corresponding to the feature data in an object data table.

In AutoCAD Map 3D, you can associate nonspatial data to a spatial object by linking the object to a data table stored in an external file or by creating an object data table within the drawing, which will store the attribute information corresponding to the object.

Defining the Object Data Table

Ribbon:	Map Setup > Attribute Data > Define Object Data

In AutoCAD Map 3D, you can create an object data table for storing the nonspatial data. The structure of a data table is defined by the table definition. This definition includes the columns (attribute fields) of the data table and the data type corresponding to each column.

Define Object Data

To define an object data table, choose the **Define Object Data** tool from the **Attribute Data** panel of the **Map Setup** tab; the **Define Object Data** dialog box will be displayed, as shown in Figure 6-1. Next, to create a new data table, choose the **New Table** button; the **Define New Object Data Table** dialog box will be displayed, as shown in Figure 6-2.

In the **Define New Object Data Table** dialog box, enter a name for the table to be created in the **Table Name** edit box. To define fields (attribute columns) in the table, specify the required name in the **Field Name** edit box. Next, specify the data type that will be stored in the field by selecting an option from the **Type** drop-down list. Optionally, write a brief description of the data field in the **Description** text box. To assign a default value to the data field, enter a value in the **Default** edit box. Next, choose the **Add** button; a data field with the specified name will be added in the **Object Data Fields** area of this dialog box. Repeat this process to add required fields to the data table.

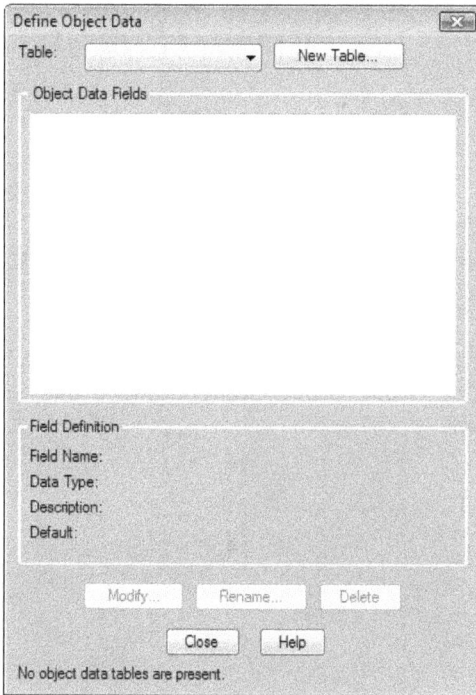

Figure 6-1 The *Define Object Data* dialog box

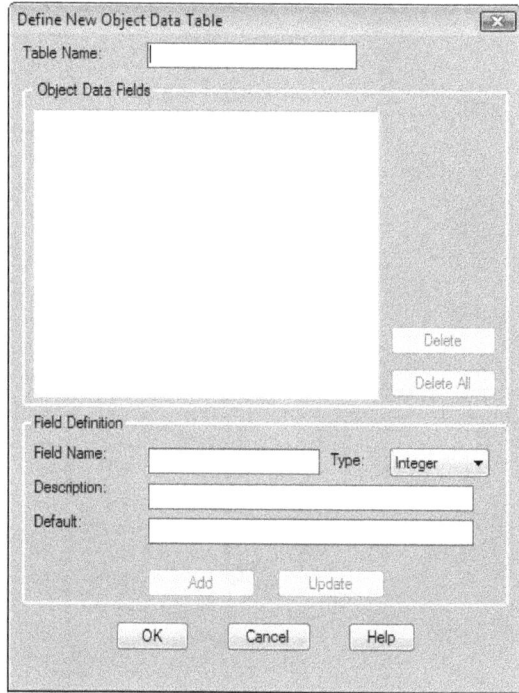

Figure 6-2 The *Define New Object Data Table* dialog box

Note

A field name cannot contain spaces and special characters. Use only alphanumeric characters in the **Field Name** *edit box to define the name of the fields in the data table.*

After specifying the fields in the **Define New Object Data Table** dialog box, choose the **OK** button; the dialog box will be closed and the name of the created table will be added to the options list of the **Table** drop-down list in the **Define Object Data** dialog box. The **Object Data Fields** area of this dialog box displays the fields of the object data table selected in the **Table** drop-down list. To view the definition parameters of the field in the data table, select the field from the **Object Data Fields** area; the definition parameters of the field will be displayed in the **Field Definition** area. Choose the **Close** button to close the **Define Object Data** dialog box.

Attaching the Object Data to an Existing Drawing Object

Ribbon: Create > Drawing Object > Attach/Detach Object Data

After creating an object data table, you can use it to store and attach object data to the drawing objects in the current drawing. To attach object data to a drawing object, specify the attribute data for each field in the selected object data table and then attach the record to the drawing object.

To attach object data, choose the **Attach/Detach Object Data** tool from the **Drawing Object** panel of the **Create** tab; the **Attach/Detach Object Data** dialog box will be displayed. Next, select the required object data table from the **Table** drop-down list; the fields in the selected data table will be displayed in the **Object Data Field** area of the dialog box, refer to Figure 6-3.

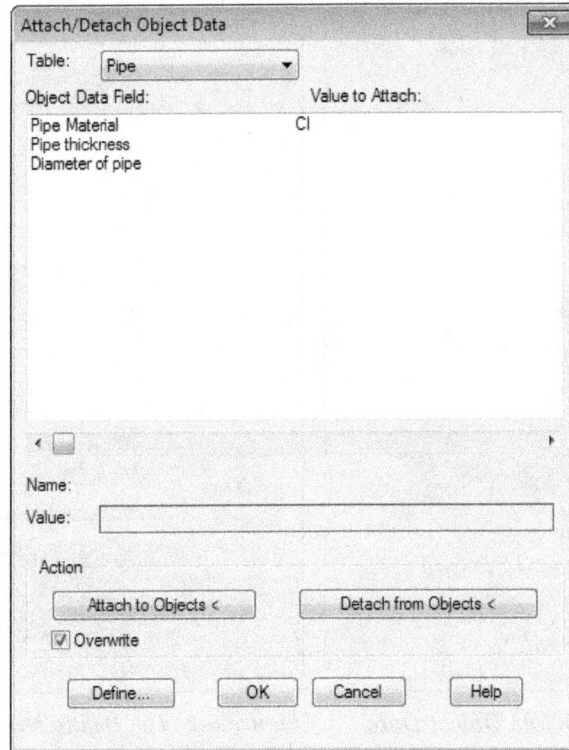

*Figure 6-3 The **Attach/Detach Object Data** dialog box*

After choosing the required data table in the **Attach/Detach Object Data** dialog box, you will specify the data in the fields of the object data table. To specify the data, select the first row in the **Object Data Field** area; the name of the selected row (field) will be displayed below this area. Next, specify the data for the selected field by entering the data value in the **Value** edit box. Next, press ENTER; the specified data will be displayed in the **Object Data Field** area next to the row selected earlier, and the next row in this area will be selected.

Specify the data value for all the fields in the **Attach/Detach Object Data** dialog box. Next, select the **Overwrite** check box to overwrite the existing values (object data) for this object data table. If you clear this check box, another record in addition to the existing record defined for the selected object data will be created. It is highly recommended that you avoid the situations where a single spatial object has multiple records from a single object data table.

A data record has now been created in the object data table. To attach this row to a drawing feature, choose the **Attach to Objects<** button in the **Action** area; you will be prompted to select the objects in the drawing area. Select the required drawing objects by clicking on individual objects or by creating a selection box. Next, press ENTER; the value specified will be attached to the drawing object.

Tip

*You can verify the object data attached to a drawing object. To do so, select the drawing object to which the object data has been attached and then right-click; a shortcut menu will be displayed. In this shortcut menu, choose the **Properties** option; the **Properties** palette will be displayed. In the **Properties** palette, the attached object data and the corresponding data fields will be displayed in the table below the **OD :<Object Data table name>** heading.*

Note

You can attach object data from two or more data object tables to a single spatial object.

Attaching Object Data While Digitizing a Drawing Object

In AutoCAD Map 3D, you can attach data while digitizing a drawing object (point or line) using the **Digitize** tool. The **Digitize** tool digitizes the objects with the settings that are specified by the user. To digitize a point or line, you need to specify its settings in the **Digitize Setup** dialog box and then use the **Digitize** tool. The various options in the **Digitize Setup** dialog box and the method of digitizing drawing objects are discussed next.

Specifying the Options in the Digitize Setup Dialog Box

While creating the drawing objects, the **Digitize** tool uses the settings that are specified in the **Digitize Setup** dialog box. It is therefore necessary to specify the settings in this dialog box before using the **Digitize** tool. To invoke the **Digitize Setup** dialog box, choose **Digitize Setup** from **Create > Drawing Object >Digitize** drop-down; the **Digitize Setup** dialog box will be displayed, as shown in Figure 6-4. The various areas and options in the dialog box are discussed next.

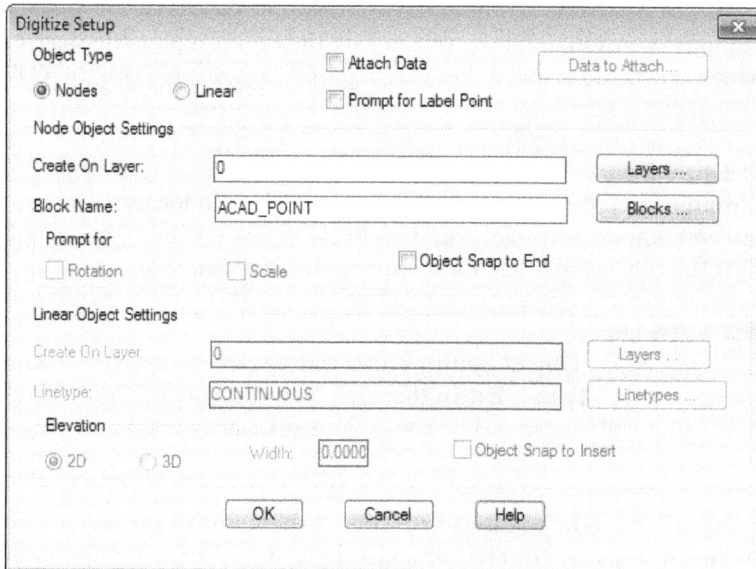

*Figure 6-4 The **Digitize Setup** dialog box*

Object Type

You can select the type of object that you want to create by selecting the radio button in this area. Select the **Nodes** radio button to create a node object. Alternatively, you can select the **Linear** radio button to create a linear object.

Attach Data

Select the **Attach Data** check box for enabling data attachment to the created object. On selecting this check box, the **Data to Attach** button will be activated. Next, choose this button; the **Data to Attach** dialog box will be displayed, as shown in Figure 6-5.

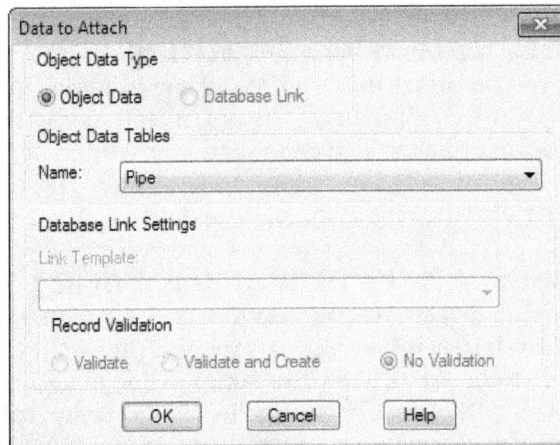

*Figure 6-5 The **Data to Attach** dialog box*

Using the options in this dialog box, you can set the object data table that is to be attached to the drawing object. After specifying the required options, choose the **OK** button in the dialog box to set the object data.

Prompt for Label Point

Select the **Prompt for Label Point** check box to specify the location of the text associated with the digitized objects. On selecting this check box, you will be prompted to specify the label point for the object while creating it using the **Digitize** tool.

Node Object Settings

The options in the **Node Object Settings** area will be activated if the **Nodes** radio button is selected in the **Object Type** area. In this area, you can specify the settings for the node drawing objects that will be created by using the **Digitize** tool. Various options in this area are discussed next.

Create On Layer: This edit box is used to specify the name of the layer on which you want to create the linear objects. To select an existing layer for creating new nodes, choose the **Layers** button displayed on the right of the **Create On Layer** edit box. On choosing this button, the **Select** dialog box containing the list of layers in the drawing will be displayed. Select the required layer in this dialog box and then choose the **OK** button; the dialog box will be closed and the selected layer will be set as the layer for new node generation.

Block Name: This edit box is used to specify the name of the block which is to be used to display the point object. To select the required block from the list of available blocks, choose the **Blocks** button displayed on the right of the **Block Name** edit box; the Select dialog box with list of the blocks in the drawing will be displayed.

| Blocks ... |

Select the required block from the list in this dialog box and choose the **OK** button; the dialog box will be closed and the selected block will be used for representing the created nodes.

Rotation: Select the **Rotation** check box to specify the angle of rotation of a block that is being digitized.

Scale: Select the **Scale** check box to specify the scale for the block while digitizing the point object. Note that if this check box is not selected, the scale for the block is set to 1.

Object Snap to End: Selecting the **Object Snap to End** check box will allow you to snap to the endpoint of an object such as the arc, elliptical arc, line, and polyline.

Linear Object Settings

The options in the **Linear Object Settings** area will be activated if the **Linear** radio button is selected in the **Object Type** area. In the **Linear Object Settings** area, you can specify the settings for the linear drawing objects that will be created by using the **Digitize** tool. The various options in this area are discussed next.

Create On Layer: In this edit box, specify the name of the layer on which you want to create the linear objects, as explained earlier.

Linetypes: In this edit box, you can specify the line type for the linear geometry. To specify the line type, choose the **Linetypes** button displayed on the right of the edit box; the **Select** dialog box will be displayed. Choose the required linetype from the list of linetypes displayed in the dialog box. Next, choose the **OK** button; the **Select** dialog box will be closed and selected linetype will be set for the linear object.

After specifying the options in the **Digitize Setup** dialog box, choose the **OK** button; the dialog box will be closed and the options specified in the dialog will be used to create objects.

Creating the Drawing Object and Attaching the Object Data

After specifying various options in the **Digitize Setup** dialog box, choose the **Digitize** tool from the **Create > Drawing Object > Digitize** drop-down. Depending on the object type selected in the **Digitize Setup** dialog box, you will be prompted to create a line or node object. Digitize the object and press ENTER; the **Attach Object Data** dialog box will be displayed. Figure 6-6 shows the **Attach Object Data** dialog box for the Pipe object data.

This dialog box contains the list of all object data fields (columns/ attribute fields) of the object data table that has been selected in the **Digitize Setup** dialog box. To add the field data, choose the required field name in the dialog box. Next, enter the required attribute value in the **Value** edit box and press ENTER; the specified value will be displayed in the list box of the dialog box. Specify the attribute data for all the required fields and then choose the **OK** button; the specified attributes will be attached to the drawing object.

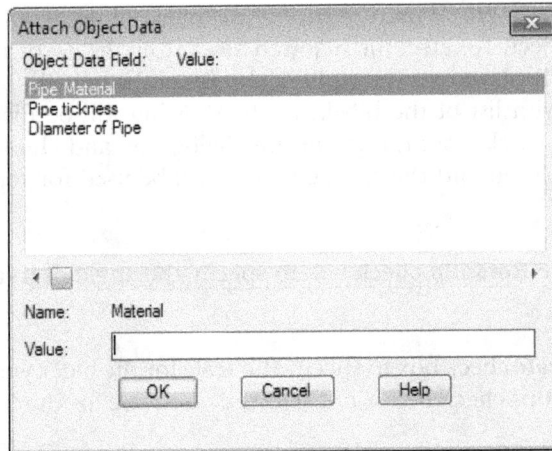

Figure 6-6 The **Attach Object Data** *dialog box*

Editing the Object Data

Ribbon: Tools > Map Edit > Edit Object Data

You can edit attribute values in the data field of a drawing object after attaching it to a drawing object. To do so, choose the **Edit Object Data** tool from the **Map Edit** panel; you will be prompted to select object in the drawing. Click on the object in the drawing area; the **Edit Object Data** dialog box will be displayed with the data table attached to it, refer to Figure 6-7.

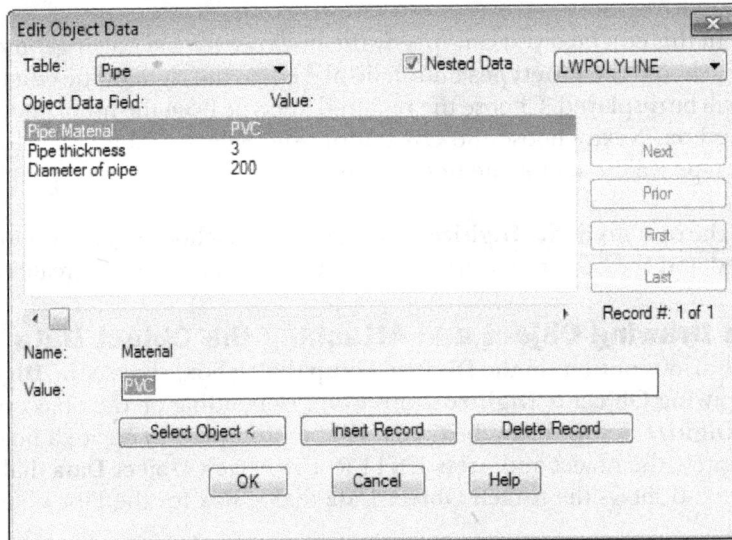

Figure 6-7 The **Edit Object Data** *dialog box*

In the **Edit Object Data** dialog box, the name of all the object data tables attached to the selected object will be listed in the **Table** drop-down list. Select the required object data table from this drop-down list. The list of data fields in the selected table and the corresponding attribute values will be displayed in the **Object Data Field** list box. The different ways of editing the data field values of drawing objects are discussed next.

Editing the Values in the Data Field

To edit the value of a data field in the **Edit Object Data** dialog box, select the required data field in the **Object Data Field** list box; the name of the selected data field will be displayed next to the **Name** parameter and the value of the selected data field will be displayed in the **Value** edit box. Next, enter a new value in the **Value** edit box and then press ENTER; the value of the selected data field in the **Object Data Field** list box will be modified accordingly.

Inserting a New Data Record

You can insert a new attribute data record into the object data table for the selected drawing object. To do so, choose the **Insert Record** button located below the **Value** edit box; a new record will be created with the set of default data fields and will be displayed in the **Object Data Field** list box. Next, insert the data into this row as explained earlier. By inserting a record for a feature, you will have a situation where the selected feature will have multiple records in a single object data table. Note that when a feature has multiple records in single object data table, the **Next**, **Prior**, **First**, and **Last** buttons are activated. You can use these buttons to browse, display, and edit records.

Deleting a Data Record

In the **Edit Object Data** dialog box, you can delete a data record that is not required. Select the required records of a data field by using the **Next**, **Prior**, **First**, and **Last** buttons located on the right of the **Object Data Field** list box. Next, choose the **Delete Record** button; the entire set of data field values will be deleted from the selected table.

Editing the Object Data Table

AutoCAD Map 3D allows you to edit an object data table upon its creation. To edit an existing object data table, you need to invoke the **Define Object Data** dialog box. To do so, choose the **Define Object Data** tool from the **Attribute Data** panel of the **Map Setup** tab; the **Define Object Data** dialog box will be displayed.

Renaming an Object Data Table

To change the name of the object data table, choose the **Rename** button from the **Define Object Data** dialog box; the **Rename Table** dialog box will be displayed, refer to Figure 6-8.

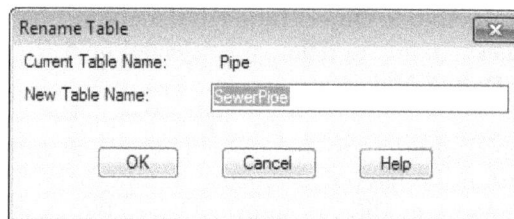

Figure 6-8 The **Rename Table** dialog box

In this dialog box, enter the required table name in the **New Table Name** edit box and choose the **OK** button; the **Rename Table** dialog box will be closed and the object data table name will be updated.

Adding, Deleting, and Updating the Fields in the Object Data Table

To edit the fields in the selected object data table, choose the **Modify** button in the **Define Object Data** dialog box; the **Modify Object Data Table** dialog box will be displayed, refer to Figure 6-9.

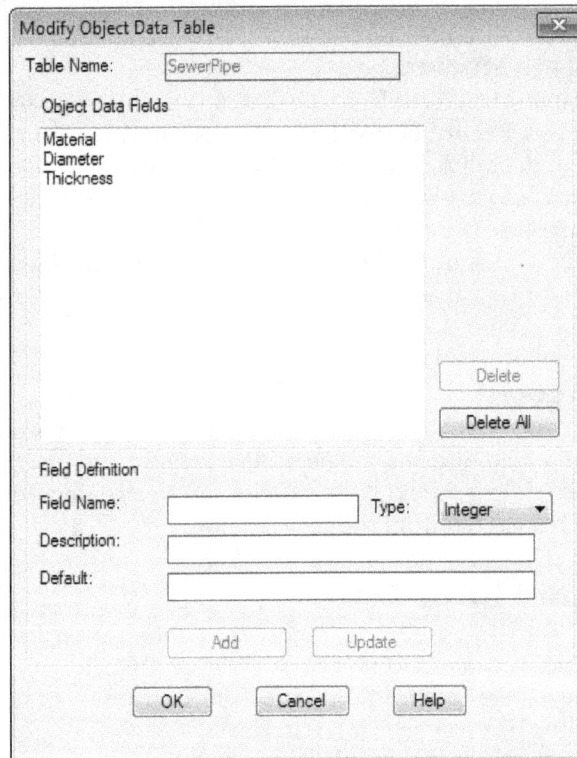

*Figure 6-9 The **Modify Object Data Table** dialog box*

To delete a field from the object table, select the required field from the **Object Data Fields** area of the **Modify Object Data Table** dialog box; the **Delete** button in this area will be activated. Choose the **Delete** button; the **Delete Object Data Field** message box will be displayed. Choose the **Yes** button from the message box to delete the selected field. To delete all the fields from the data table, choose the **Delete All** button.

To modify the information (Field description and default value) and the data type of a field, select the required field from the list box; the existing field parameters will be displayed in the **Field Definition** area. Next, modify the information in the **Description** and **Default** edit boxes. If required, change the data type of the field by selecting an option from the **Type** drop-down list. Next, to save the modifications made to the field, choose the **Update** button in the **Field Definition** area; the modifications made to the field will be saved in the table.To add a new field to an object data table, specify all the parameters in the **Field Definition** area. Next, choose the **Add** button to create a new field in the object data table.

Note

Whenever you specify or edit the name of a field in the Field Name edit box, the Add button is activated.

Detaching the Object Data Table

Ribbon:	Create > Drawing Object > Attach/Detach Object Data

You can detach or disconnect the object data table pertaining to a drawing object by using the options in the **Attach/Detach Object Data** dialog box. To do so, choose the **Attach/Detach Object Data** tool from the **Drawing Object** panel of the **Create** tab; the **Attach/Detach Object Data** dialog box will be displayed. In this dialog box, select the required object data table from the **Table** drop-down list; the data fields in the selected table will be displayed in the **Object Data Field** list box. Next, to detach the selected object data table, choose the **Detach from Objects** button in the **Action** area; you will be prompted to select the required object. Click on the required object and press ENTER; the object data table will be detached from the selected drawing object.

Note

It is recommended that you always select the correct drawing object while performing the delete or detach operation, as you cannot retrieve the data once it is detached or deleted.

ATTACHING AN EXTERNAL DATABASE

Attaching the object data to a drawing object using the object data table is a simple and flexible way of creating attribute data within a drawing environment. However, in some cases, the data that you may want to include in your project is available in data files/database. AutoCAD Map 3D allows you to attach such external data to your drawing objects. The process of attaching external database is discussed next.

Creating a Link between Map 3D Drawing and Database

To attach data from an external file/database, you first need to create a link that provides access to the table within the database. The link (connection) to the database is defined in the file known as the Universal Data Link (UDL) file.

To attach an external database, right-click on the **Data Sources** node in the **Map Explorer** tab of the **TASK PANE**; a shortcut menu will be displayed. Next, choose the **Attach** option from the displayed menu, as shown in Figure 6-10. On doing so, the **Attach Data Source** dialog box will be displayed, as shown in Figure 6-11.

In this dialog box, select the data source type from the **Files of type** drop-down list. Next, browse to the location of the required data and select it. Next, choose the **Attach** button; the selected data file / database will now be connected to the drawing file. The connected database is displayed in the **Data Sources** node of the **TASK PANE**. Figure 6-12 shows a database connected to the drawing file.

Note

On attaching a database, AutoCAD Map 3D creates a UDL for the attached database and this link is then stored in the default Data Links folder.

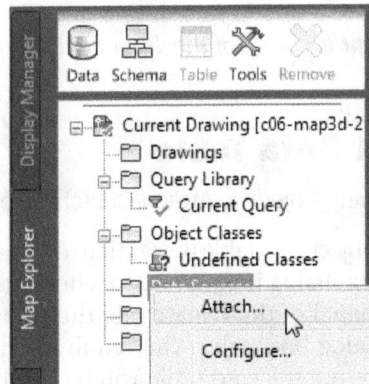

Figure 6-10 *Choosing the **Attach** option from the shortcut menu*

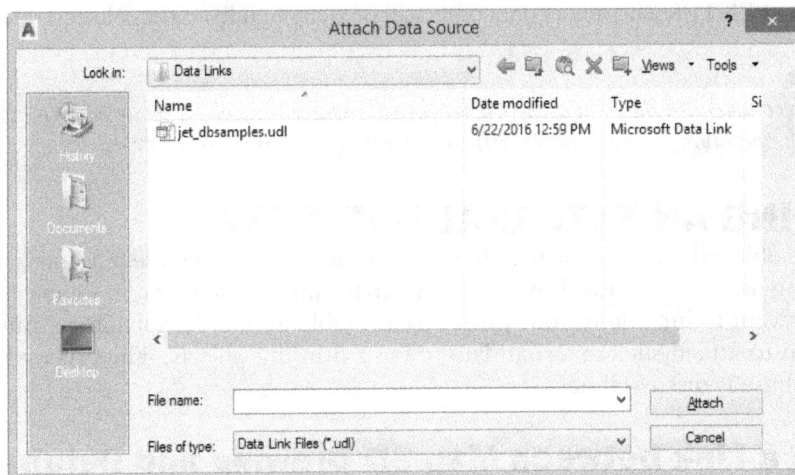

Figure 6-11 *The **Attach Data Source** dialog box*

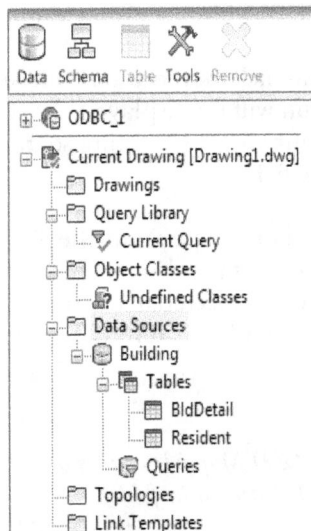

Figure 6-12 *The external database added to the **Data Sources** node*

Defining Link Templates

After attaching a database to a drawing it becomes accessible. Note that at this stage, though you can view and edit the data in the tables of the attached database, there is still no link between the drawing object and the data in the tables. To create a link between the data and the drawing object, you need to define a link template. A link template defines the parameters such as the data source, data table, and key field in the data table. On creating a link template, you can link the drawing objects to the data records. To define a link template, choose the **Define Link Template** tool from the **Attribute Data** panel of the **Map Setup** tab; the **Define Link Template** dialog box will be displayed, refer to Figure 6-13.

Define Link
Template

Note
*You can also invoke the **Define Link Template** dialog box by using the options in the **TASK PANE**. To do so, right-click on the required table name in the **Data Sources** node of the **Map Explorer** tab in the **TASK PANE**; a shortcut menu will be displayed. Next, choose the **Define Link Template** option from the shortcut menu; the **Define Link Template** dialog box will be displayed.*

*However, when you invoke the **Define Link Template** dialog box using the options in the **TASK PANE**, the **Data Source** and **Table Name** options in this dialog box are uneditable.*

In this dialog box, choose the data source by selecting an option from the **Data Source** drop-down list. Next, choose the table from the selected data source by selecting an option from the **Table Name** drop-down list. Next, specify the name for the template in the **Link Template** text box. Next, in the **Key Selection** area of the dialog box, select the check box corresponding to the required column name to identify it as a key column. Next, choose the **OK** button; the **Define Link Template** dialog box will be closed and the defined link template's name will be displayed under the **Link Template** node in the **Map Explorer** tab of the **TASK PANE**.

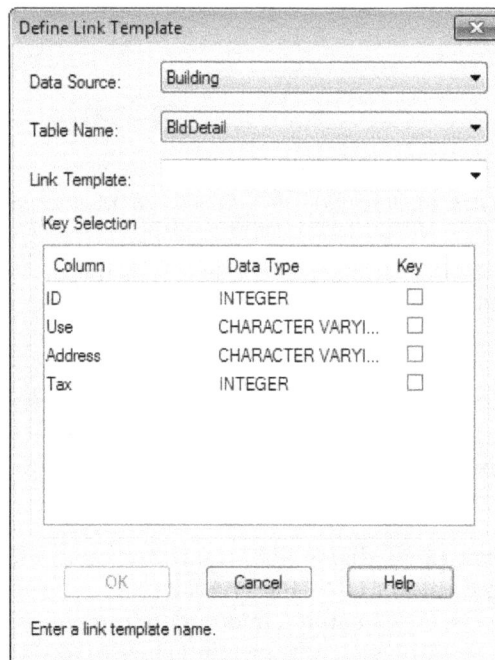

*Figure 6-13 The **Define Link Template** dialog box*

Linking Data from the Data Table to the Drawing Objects

After creating a link template, you can generate links to attach the data records from the table to the drawing object. You can create link to attach the data to the object using one of the following methods:

Linking Records Manually

The manual data linking method is slow and lengthy way to link data. This method should be used when the drawing objects have no attribute data associated with them. To create a link manually, you will have to select a data row individually and then attach it to the required drawing feature. The procedure to generate links manually is described next.

After defining the link template, right-click on the table name (for which you have created the link template) in the **Map Explorer** tab of the **TASK PANE**; a shortcut menu will be displayed. Choose the **View Table** option from the displayed menu; the **Data View <Table Name>** window will be displayed. The **Data View** window will display the records in the selected data table. Next, right-click in the gray box of the required row; a shortcut menu will be displayed. Choose the **Link to Object(s)** option from the displayed menu, refer to Figure 6-14.

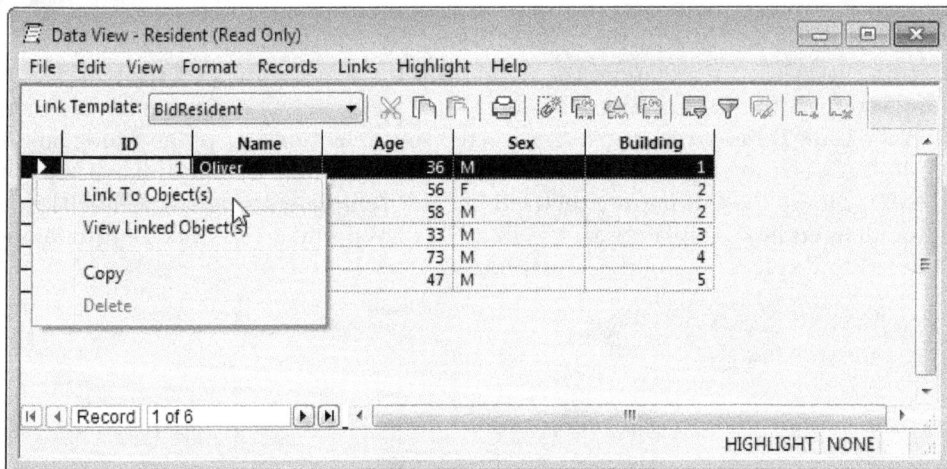

*Figure 6-14 Choosing the **Link To Object(s)** option from the shortcut menu in the **Data View** window*

On doing so, you will be prompted to select the object in the drawing window. Click on the required object to select it and then press ENTER. The selected record in the data table will be linked to the selected drawing object.

Linking Records Automatically Based on Text or Blocks in the Drawing

If the drawing objects have associated object data (or text) that matches with the attribute data in the data table, you can use the **Generate Data Links** dialog box to create links automatically based on the matching information.

To invoke the **Generate Data Links** dialog box, right-click on the link template in the **Links Templates** node of the **Map Explorer** tab of the **TASK PANE**; a shortcut menu will be displayed, as shown in Figure 6-15. Choose the **Generate Links** option from the shortcut menu; the **Generate Data Links** dialog box will be displayed, as shown in Figure 6-16.

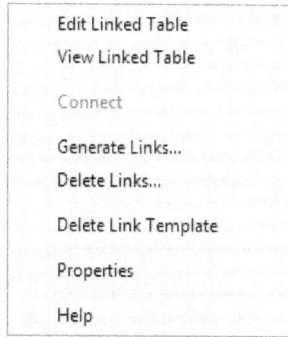

Figure 6-15 The link template shortcut menu

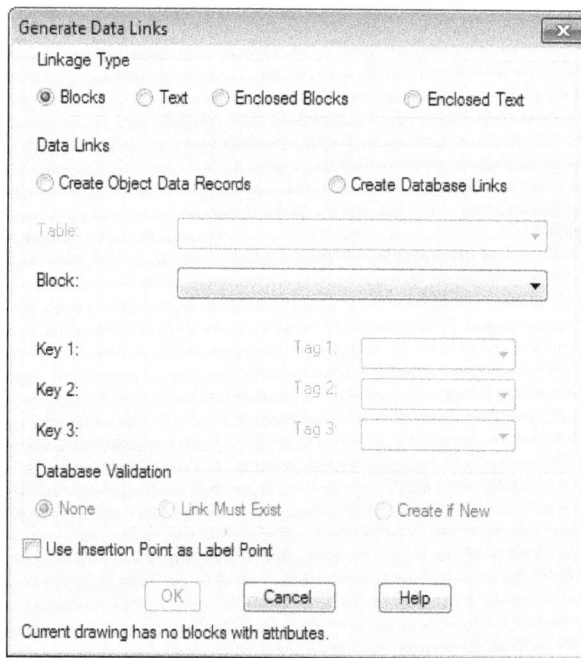

Figure 6-16 The **Generate Data Links** dialog box

From the **Linkage Type** area of this dialog box, select the required radio button to specify the type of link. You can select the **Blocks** radio button to create links from block attribute data. To create a link from the text in the drawing, select the **Text** radio button. Alternatively, you can select the **Enclosed Blocks** or the **Enclosed Text** radio buttons to create links from block attribute data and from the text that lies within a closed (enclosed) polyline, respectively.

Note

*On selecting the **Enclosed Blocks** radio button, the links are created on the polyline that encloses the block. In case the block is not enclosed by a polyline, the link will not be created.*

Next, ensure that the **Create Database Links** radio button is selected and then select a radio button from the **Database Validation** area to specify the method of validation. After specifying all options in the dialog box, choose the **OK** button; the dialog box will be closed and you will be prompted to select the required objects. Select the required objects and press ENTER; the links to the selected objects will be created.

Linking Records Automatically to the Drawing Objects Using Object Data / Converting Existing Object Data to the Linked Database Table

After creating the link template, enter the command **MAP0D2ASE** at the Command prompt; the **Convert Object Data to Database Links** dialog box will be displayed, refer to Figure 6-17. In the **Source Object Data Table** area of this dialog box, select the object data table by selecting the required option from the **Name** drop-down list. Next, you can select the **Remove Data from Objects Processed** check box to delete the object data after creating the link.

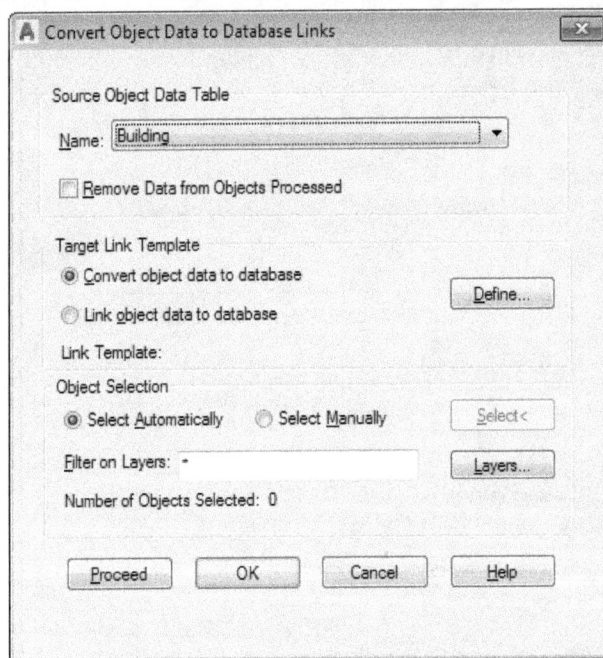

*Figure 6-17 The **Convert Object Data to Database Links** dialog box*

To convert object data to a new table in the linked database, select the **Convert object data to database** radio button in the **Target Link Template** area. Next, choose the **Define** button; the **Define Link Template** dialog box will be displayed. In this dialog box, select the database in which the table is to be created. To do so, select an option from the **Data Source** drop-down list and then choose the **Connect** button. On connecting to the database, the other options in the dialog box will be activated. Define a link template using the options in the dialog box and then choose the **OK** button; the **Define Link Template** dialog box will be closed and the name of the defined template will be displayed in the **Target Link Template** area of the **Convert Object Data to Database Links** dialog box (corresponding to the **Link Template** text).

To link records from the data table of the attached database to the drawing objects using object data, select the **Link object data to database** radio button in the **Target Link Template** area of the **Convert Object Data to Database Links** dialog box. Next, choose the **Define** button; the **Select Existing Link Template** dialog box will be displayed, refer to Figure 6-18.

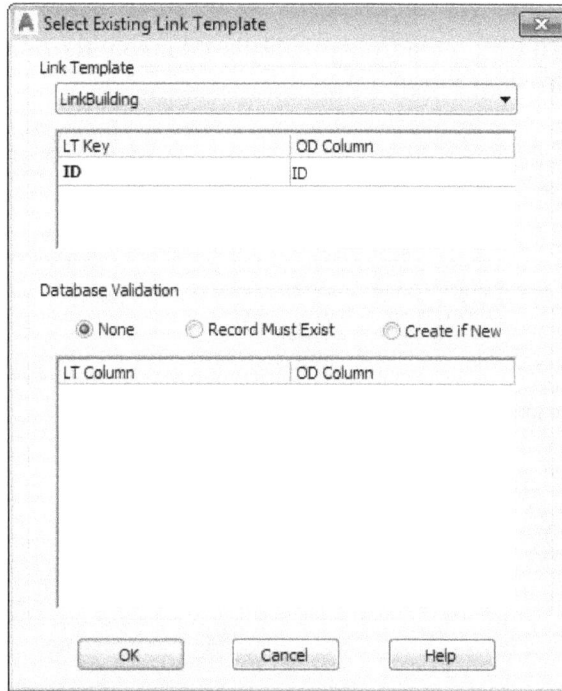

*Figure 6-18 The **Select Existing Link Template** dialog box*

In this dialog box, select the required link template that defines the connection between the required object data table and the external database (linked) table; the key column in the database table will be displayed in the cell of the **LT Key** column. Next, choose the key column in the object data table. To do so, click in the **OD Column** cell; a drop-down list containing the names of object data fields will be displayed. Select the required field as the key column in the object data table from this drop-down list. Next, select the required radio button from the **Database Validation** area. Select the **None** radio button to create links without checking the database. To create a link between feature and data table where the attribute value of the object data table matches the key field value of the table in the attached database, select the **Record Must Exist** radio button. In situations where no match is found, you can create a new record by selecting the **Create if New** radio button. After specifying the options, choose the **OK** button; the **Select Existing Link Template** dialog box will be closed.

Next, in the **Convert Object Data to Database Links** dialog box, specify the options in the **Object Selection** area for selecting the object and then choose the **OK** button; the dialog box will be closed and the data will be linked based on the options specified.

ATTACHING A DRAWING FILE

In AutoCAD Map 3D, you can attach a drawing file to the current drawing. After attaching the drawing file, you can display, query, and edit the drawing objects in the attached file. You can also save the edits to the attached drawing file.

You can select the file/s to attach using the **Select drawings to attach** dialog box. To invoke this dialog box, choose the **Attach** tool from the **Data** panel of the **Home** tab. Alternatively, right-click on the **Drawings** node in the **Map Explorer** tab of the **TASK PANE**; a shortcut menu will be displayed. Choose the **Attach** option from the displayed menu; the **Select drawings to attach** dialog box will be displayed, as shown in Figure 6-19.

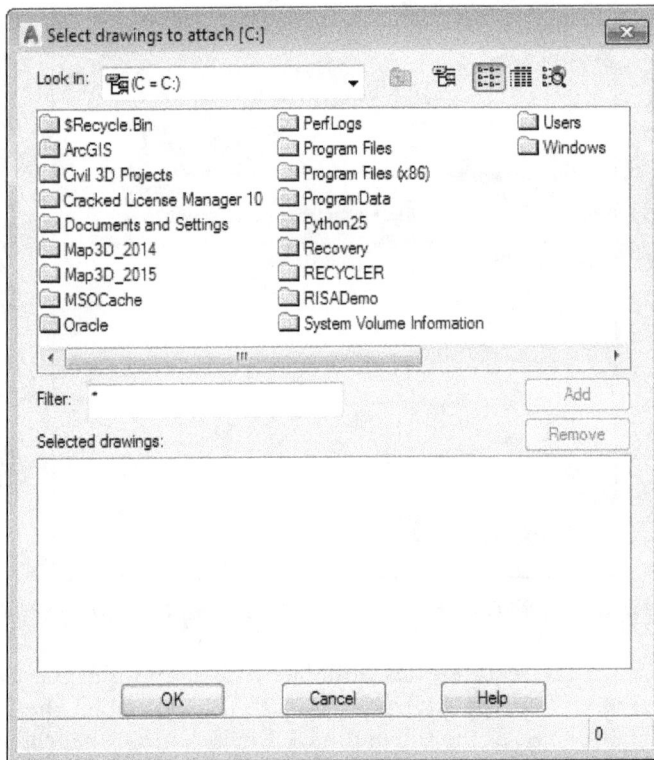

*Figure 6-19 The **Select drawings to attach** dialog box*

In the **Select drawings to attach** dialog box, browse to the folder containing the drawing file/s and then select the required files. You can also select multiple files for attaching them to the current drawing. Next, choose the **Add** button; the selected drawing file will be added in the **Selected drawings** area of the dialog box.

Note
*To remove a file from the **Selected drawings** list, select the file from the list and then choose the **Remove** button; the selected file will be removed from the list.*

Next, choose the **OK** button; the drawing file will be added to the **Current Drawing** node. If the drawing is not visible in the drawing window, right-click on the **Drawings** folder; a shortcut menu will be displayed. In this shortcut menu, choose the **Quick View** option; the objects in the attached drawing file will be displayed in the drawing window. If multiple files are attached to the drawing, then on choosing the **Quick View** option from the shortcut menu, the **Quick View Drawings** dialog box will be displayed, as shown in Figure 6-20.

In this dialog box, select the required drawing file in the **Select Active Drawings to Quick View** area. Next, select the **Zoom to the Extents of Selected Drawings** check box to zoom all contents of the selected drawings to fit in the drawing window. Next, choose the **OK** button; the drawing objects in the selected drawing will be displayed.

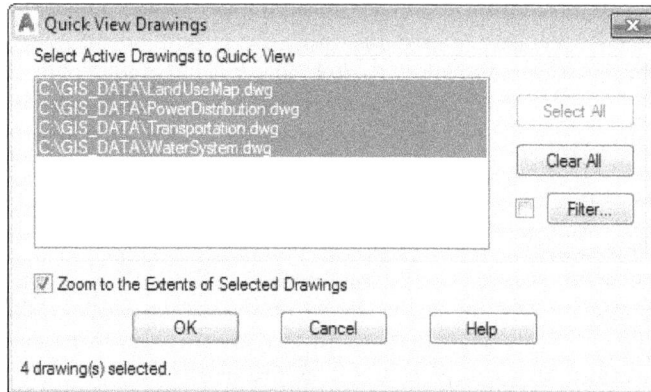

*Figure 6-20 The **Quick View Drawings** dialog box*

QUERYING AN ATTACHED DRAWING

You can define a query to filter the drawing objects from the attached drawings. You can create a query template and store it in your drawing for further reference. To create a query template, select the **Query Library** node in the **Map Explorer** tab of the **TASK PANE** and then right-click; a shortcut menu will be displayed. In this shortcut menu, choose the **Administration** option; the **Query Library Administration** dialog box will be displayed, as shown in Figure 6-21.

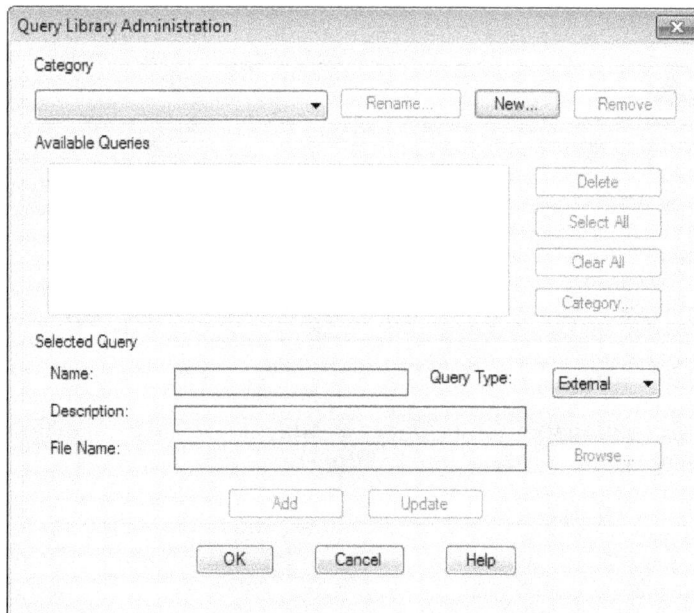

*Figure 6-21 The **Query Library Administration** dialog box*

The **Available Queries** area of the **Query Library Administration** dialog box will display a list of available queries in the selected category. Note that if no queries have been defined in a category, the list in the **Available Queries** area will be empty and all the buttons will be deactivated.

In the **Category** area of this dialog box, choose the **New** button; the **Define New Category** dialog box will be displayed. In this dialog box, enter a name in the **New category name** edit box and then choose the **OK** button; the new category will be displayed in the drop-down list in the **Category** area. Next, choose the **OK** button in the **Query Library Administration** dialog box; the **Query Library Administration** dialog box will close and a new category with the specified name will be added to the **Query Library** folder in the **Map Explorer** tab. Next, you can define and save queries in the created query category. The methods for defining queries by using various types of query filters are discussed next.

Location Query Type

You can create and execute a query to filter data in the attached drawing. To define a query, select the required query category from the **Query Library** folder and then right-click; a shortcut menu will be displayed. In this shortcut menu, choose the **Define** option; the **Define Query of Attached Drawing(s)** dialog box will be displayed, as shown in Figure 6-22. You can use the options in this dialog box to define a query.

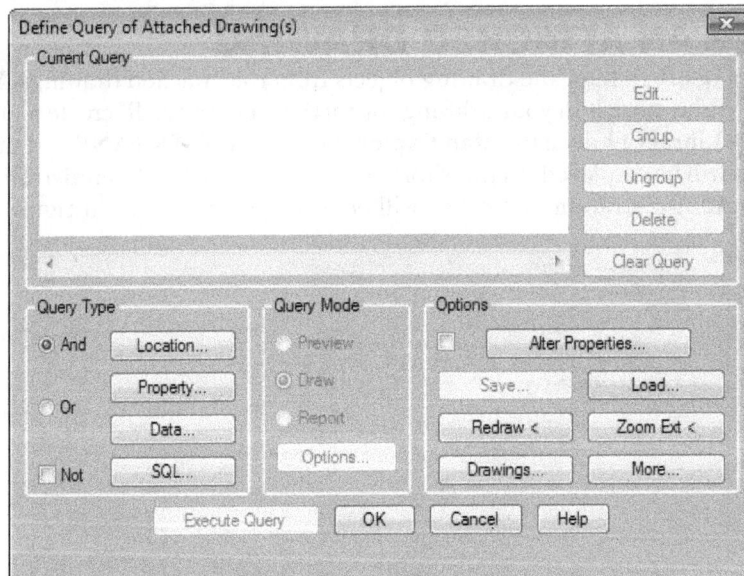

*Figure 6-22 The **Define Query of Attached Drawing(s)** dialog box*

Note that you can also invoke the **Define Query of Attached Drawing(s)** dialog box by choosing the **Define Query** tool from the **Data** panel of the **Home** tab.

To filter the drawing objects based on location, choose the **Location** button in the **Query Type** area; the **Location Condition** dialog box will be displayed, as shown in Figure 6-23.

In this dialog box, you can select the geometry type for defining the boundary. To do so, select the radio button corresponding to the required option in the **Boundary Type** area; the **Selection Type** area and the **Define** button located at the lower left corner of this dialog box will be activated. To draw the boundary, choose the **Define** button in the **Location Condition** dialog box; the dialog box will be closed and you will be prompted to draw the boundary.

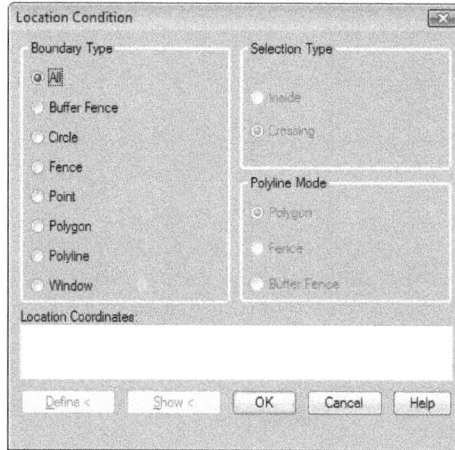

Figure 6-23 *The* **Location Condition** *dialog box*

In the **Section Type** area, the **Crossing** radio button is selected by default. As a result, the drawing objects lying across the drawn boundary will be selected. If you select the **Inside** radio button, the drawing objects lying inside the defined boundary in the drawing window will be selected. Next, choose the **Define** button; the **Location Condition** dialog box will close and the cursor will change to a crosshair in the drawing window. Define the boundary for querying drawing objects. After you have drawn the boundary, press ENTER; the selected region will be represented with a red-dotted boundary. Figure 6-24 shows a circular boundary that is used to define a query boundary.

The region selected for querying

Figure 6-24 *The circular boundary of the region selected for querying*

Also, in the list box, the specified query parameters will be displayed in the display box in the **Current Query** area of the **Define Query of Attached Drawing(s)** dialog box, as shown in Figure 6-25. After selecting the region for querying, you will notice that most of the options are activated in this dialog box. You can edit the parameters in the display box in the **Current Query** area. To do so, select the required query parameter/s and then choose the **Edit**, **Group**, **Ungroup**, **Delete**, or **Clear Query** button on the right of the display box.

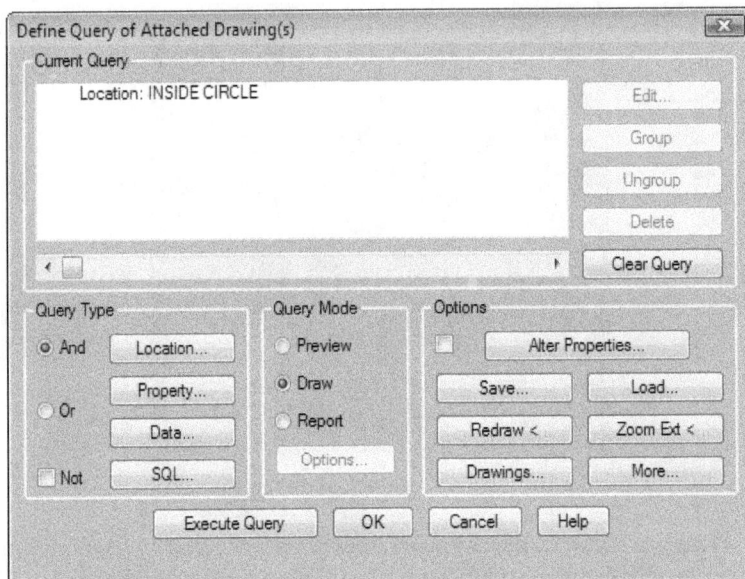

Figure 6-25 *The query type parameter displayed in the display box in the **Current Query** area*

After defining the query, you can save the query in the current category. To do so, choose the **Save** button in the **Options** area; the **Save Current Query** dialog box will be displayed. In this dialog box, select a category option in the **Category** drop-down list. To name the query process, enter the desired name in the **Name** edit box and then enter details in the **Description** edit box. Next, choose the **OK** button in the **Save Current Query** dialog box; the dialog box will be closed. Next, choose the **Execute Query** button in the **Define Query of Attached Drawing(s)** dialog box; this dialog box will be closed and the query will be executed with the options specified. Also, the query with the specified name will be added to the specified query category in the **Map Explorer** tab of the **TASK PANE**. To view the added query, expand the category node by clicking on the [+] symbol corresponding to the category name in the **Map Explorer** tab of the **TASK PANE**.

Property Query Type

You can assign property query type to the current query process for filtering out the data based on the required properties. To do so, in the **Define Query of Attached Drawing(s)** dialog box, choose the **Property** button in the **Query Type** area; the **Property Condition** dialog box will be displayed, as shown in Figure 6-26.

In the **Property Condition** dialog box, select the desired radio button in the **Select Property** area. To specify the query condition, select the operator from the **Operator** drop-down list. Next, specify the value in the **Value** edit box. You can also specify the value in the **Value** edit box from the list of existing values for the selected property. To do so, choose the **Values** button; the **Select** dialog box will be displayed. This list box will display the list of existing values. Select the required option from the list box; the **OK** button in this dialog box will be activated. Choose the **OK** button; the **Select** dialog box will be closed and the selected value will be displayed in the **Value** edit box in the **Property Condition** dialog box.

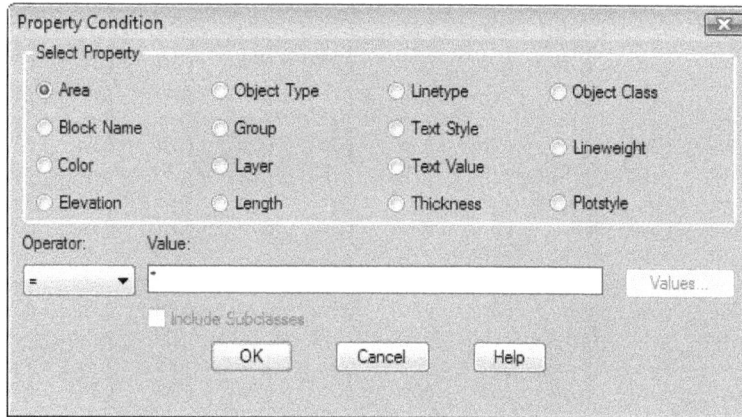

*Figure 6-26 The **Property Condition** dialog box*

After specifying the required parameters, choose the **OK** button in the **Property Condition** dialog box; this dialog box will be closed and the property-based filter command will be displayed in the **Current Query** area of the **Define Query of Attached Drawing(s)** dialog box. Figure 6-27 shows a query with statement for the property condition in the **Current Query** area of this dialog box.

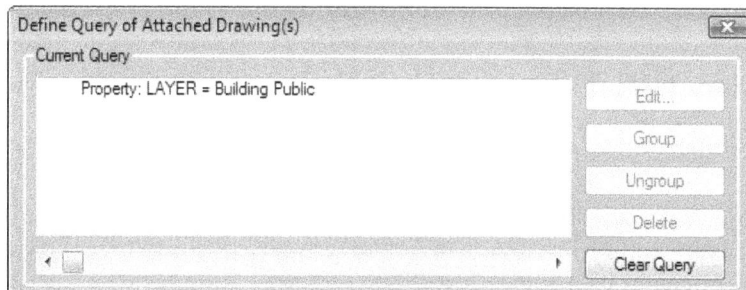

*Figure 6-27 The partial view of the **Define Query of Attached Drawing(s)** dialog box*

Specify other parameters in the **Define Query of Attached Drawing(s)** dialog box. On specifying all the parameters, you can save the current query in a category. Next, choose the **Execute Query** button in the **Define Query of Attached Drawing(s)** dialog box; this dialog box will be closed and the query will be executed with the options specified.

Data Query Type

You can define a query to filter the drawing objects based on the object data attached to the drawing objects in the external (attached) drawing file. To do so, choose the **Data** button in the **Query Type** area of the **Define Query of Attached Drawing(s)** dialog box; the **Data Condition** dialog box will be displayed, as shown in Figure 6-28.

In the **Data Condition** dialog box, select the required radio button to select the data to query; the attributes corresponding to the selected radio button will be displayed in the dialog box. For example, if you select the **Object Data** radio button, the list of object data tables in the attached drawing file will be displayed in the **Tables** drop-down list. Choose the required object data table by selecting the option in the **Tables** drop-down list; the fields in the selected data table will be displayed in the **Object Data Fields** list box, refer to Figure 6-28.

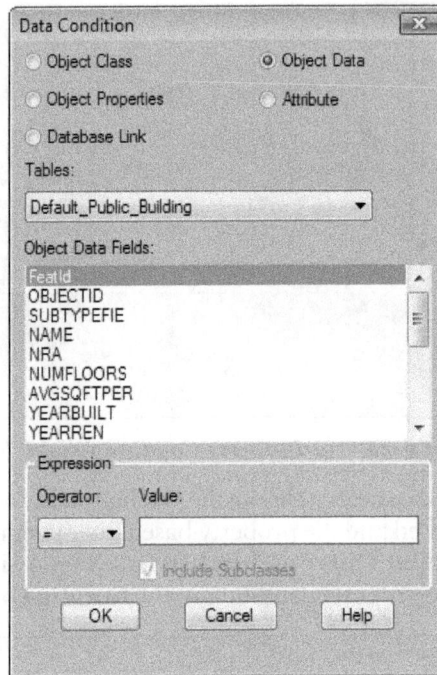

*Figure 6-28 The **Data Condition** dialog box with an example dataset*

Next, define the query condition by specifying the required object data field and operator from the **Object Data Fields** and **Operator** drop-down lists, respectively. Also, specify the query value in the **Value** text box. Next, choose the **OK** button in the **Data Condition** dialog box; the dialog box will be closed and the defined query condition will be displayed in the **Define Query of Attached Drawing(s)** dialog box. Figure 6-29 shows a query statement displayed in the **Current Query** area of the **Define Query of Attached Drawing(s)** dialog box.

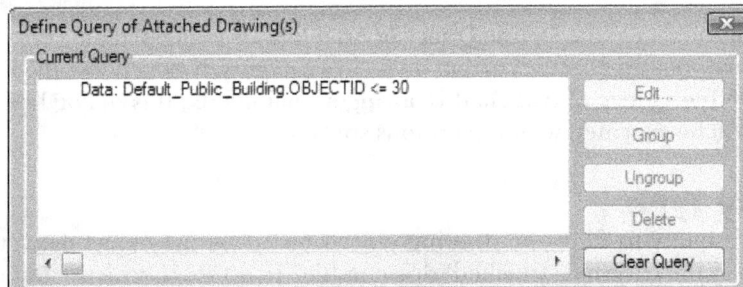

*Figure 6-29 Partial view of the **Define Query of Attached Drawing(s)** dialog box with the selected statement*

In the **Define Query of Attached Drawing(s)** dialog box, select the statement in the display box in the **Current Query** area; the buttons on the right of the display box will be activated. Using these buttons, you can edit, group, ungroup, or delete an expression. You can also save the defined query by choosing the **Save** button. Next, choose the **Execute Query** button in the **Define Query of Attached Drawing(s)** dialog box; this dialog box will be closed and the query will be executed with the options specified.

Note
*The options in the **SQL** query type will be activated only if some external database is attached to the drawing data.*

Defining a Complex Query for Checking Multiple Conditions

You can define a complex query that validates multiple conditions. You can write a complex query statement by using the logical operators such as the **And**, **Or**, and **Not** between two query conditions. These logical operators can be included in the query statement by selecting the radio button corresponding to the required operator in the **Query Type** area of the **Define Query of Attached Drawing(s)** dialog box.

Specifying the Query Mode

After defining the query using the options in the **Query Type** area in the **Define Query of Attached Drawing(s)** dialog box, you can specify the desired mode for the output data for the query process. To do so, you need to select an option from the **Query Mode** area. The different options available in this area are discussed next.

Preview

You can select the **Preview** radio button in the **Query Mode** area to display the queried data. Note that the displayed objects are not actually filtered but are merely displayed and will disappear when you redraw (regenerate) the screen.

Draw

To filter the drawing objects using the query condition and bring them into the current drawing, select the **Draw** radio button in the **Query Mode** area.

Report

The **Report** radio button in the **Query Mode** area is used to get output data in the form of a text file. On selecting this radio button, the **Options** button below this radio button will be activated. To specify parameters or apply constraints on the output data, choose the **Options** button; the **Output Report Options** dialog box will be displayed, as shown in Figure 6-30.

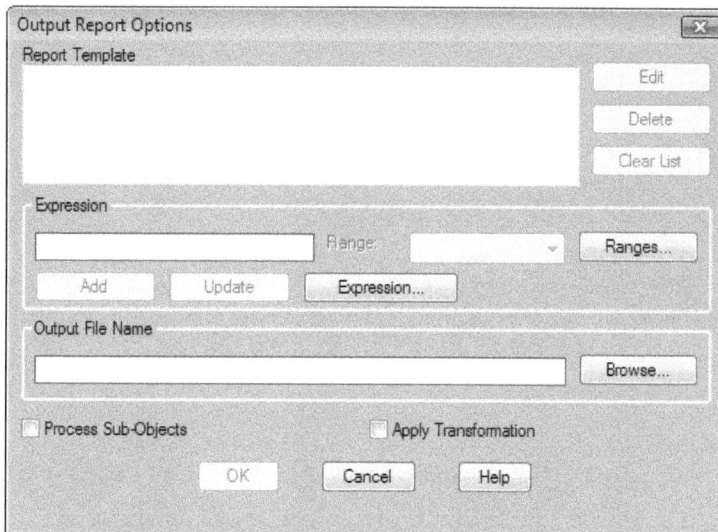

*Figure 6-30 The **Output Report Options** dialog box*

In the **Output Report Options** dialog box, you can write expressions and apply filters to query the drawing object data. To do so, choose the **Expression** button in the **Expression** area; the **Report Template Expression** dialog box will be displayed, as shown in Figure 6-31.

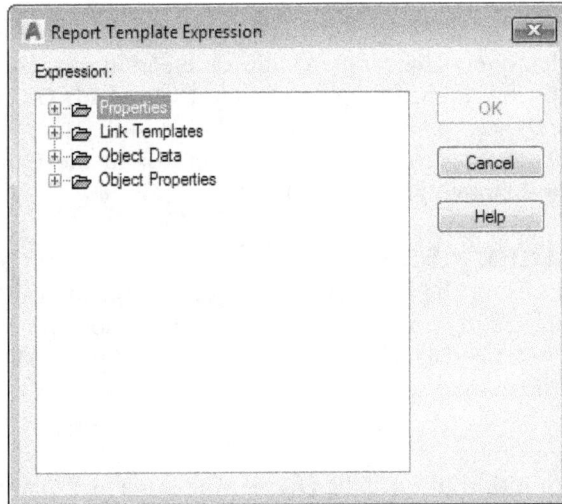

*Figure 6-31 The **Report Template Expression** dialog box*

Various folders such as the **Properties**, **Object Data**, and **Object Properties** are displayed in the **Expression** list box. Expand the required folder in the **Expression** list box, and then select the required data file; the **OK** button will be activated. Then, choose the **OK** button; the **Report Template Expressions** dialog box will be closed and the selected data table name will be displayed in the edit box in the **Expression** area of the **Output Report Options** dialog box. To set the limit for the data range to be included in the report, you can specify the data range. To do so, choose the **Ranges** button in the **Output Report Options** dialog box; the **Define Range Table** dialog box will be displayed, as shown in Figure 6-32.

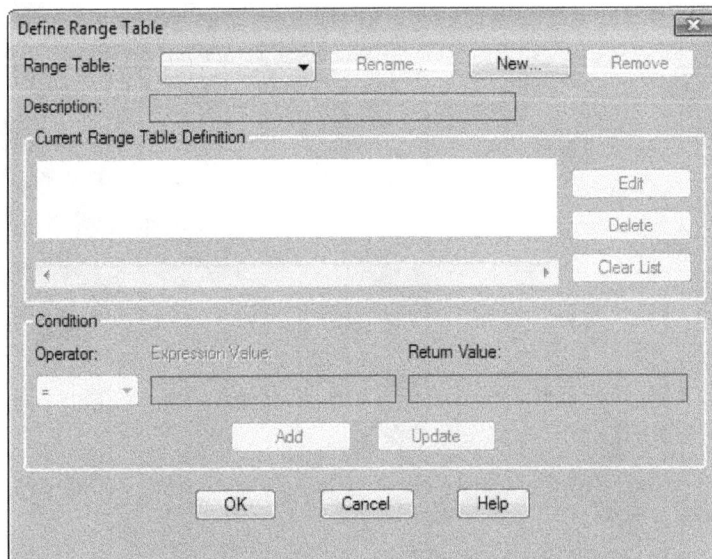

*Figure 6-32 The **Define Range Table** dialog box*

In the **Define Range Table** dialog box, choose the **New** button; the **New Range Table** dialog box will be displayed. Enter the name of the table in the **New Range Table** edit box and then choose the **OK** button; the **New Range Table** will be closed and the name specified will be displayed in the **Range Table** drop-down list. Enter the details of the table in the **Description** edit box. To apply the logical operator in the expression, select an option from the **Operator** drop-down list in the **Condition** area. To specify the limiting value of the data, enter the limiting value in the **Expression Value** edit box. To return a required value when the condition is satisfied, enter the value in the **Return value** edit box. Then, choose the **Add** button; the expression will be added to the display box in the **Current Range Table Definition** area, as shown in Figure 6-33. Next, choose the **OK** button; the **Define Range Table** dialog box will be closed and the **Output Report Options** dialog box will be displayed again.

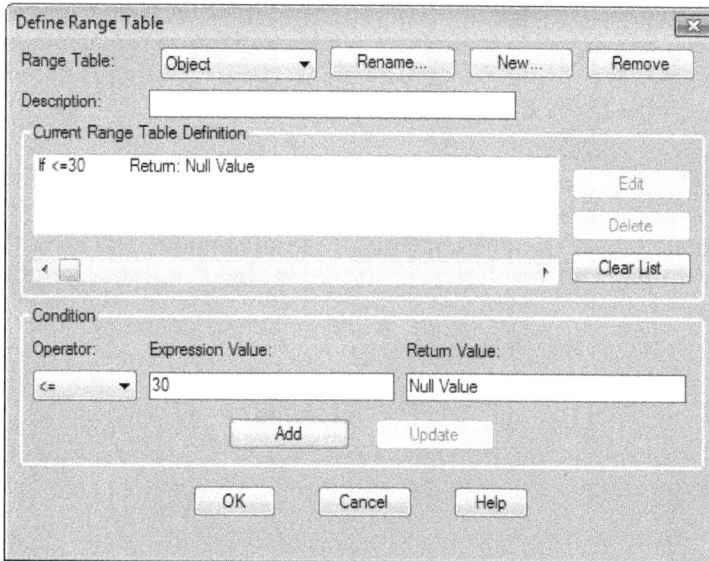

*Figure 6-33 The **Define Range Table** dialog box displaying the added expression*

In the **Output Report Options** dialog box, select a table option from the **Range** drop-down list in the **Expression** area and then choose the **Add** button below it; the newly created range with the expression of the created query will be displayed. Figure 6-34 shows the **Output Report Options** dialog box with the new range added to the **Report Template** list box.

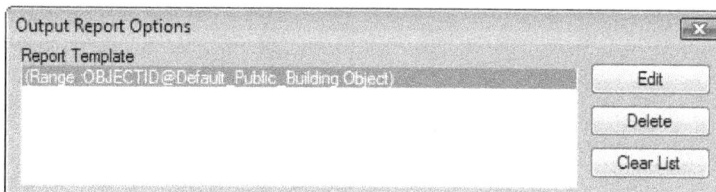

*Figure 6-34 The partial view of the **Output Report Options** dialog box with the added report template*

Next, choose the **Browse** button in the **Output File Name** area of the **Output Report Options** dialog box; the **Select Report File** dialog box will be displayed. Browse to the required location and enter the desired name in the **File name** edit box and then choose the **Save** button in the

Select Report File dialog box; this dialog box will be closed and the path of the text file will be displayed in the **Output File Name** edit box of the **Output Report Options** dialog box. The **Process Sub-Objects** check box is clear by default. As a result, the output report file will contain information about the selected objects but not its components. To get the report about the objects and classified objects in the object drawing, select the **Process Sub-Objects** check box. To find out the information about the object data that has been modified by coordinate transformation, scaling, or rotation, select the **Apply Transformation** check box. Next, choose the **OK** button; the **Output Report Options** dialog box will be closed.

Setting the Options for Query Output Data

You can specify the property options for query output data in the **Options** area in the **Define Query of Attached Drawing(s)** dialog box. To apply the changes to the query output data, select the check box at the upper left corner in the **Options** area. You can use various options in the **Options** area to alter or change the appearance of output data resulting after the query process. These options are discussed next.

Alter Properties

You can modify the drawing objects that are filtered by executing a query. To specify the changes to the queried objects, choose the **Alter Properties** button from the **Options** area of the **Define Query of Attached Drawing(s)** dialog box; the **Set Property Alterations** dialog box will be displayed, as shown in Figure 6-35.

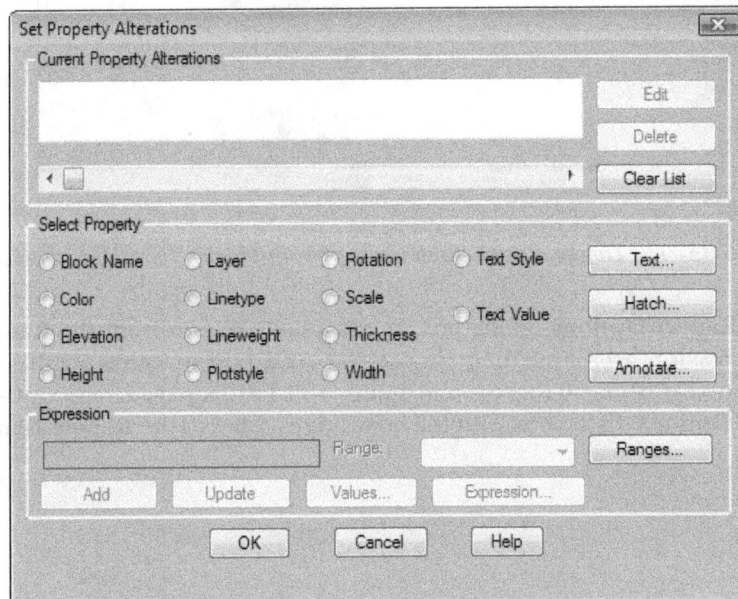

Figure 6-35 The Set Property Alterations dialog box

In the **Set Property Alterations** dialog box, you can select the required property from the **Select Property** area. To select a property, select the radio button corresponding to the required property in the **Select Property** area; the **Expression** area will be activated. You can specify the appearance for the output data by using the **Text**, **Hatch**, and **Annotate** buttons in the

Select Property area. To apply the property settings to the text labels in the selected property in the attached drawing layer, choose the **Text** button in the **Select Property** area; the **Define Text** dialog box will be displayed, as shown in Figure 6-36.

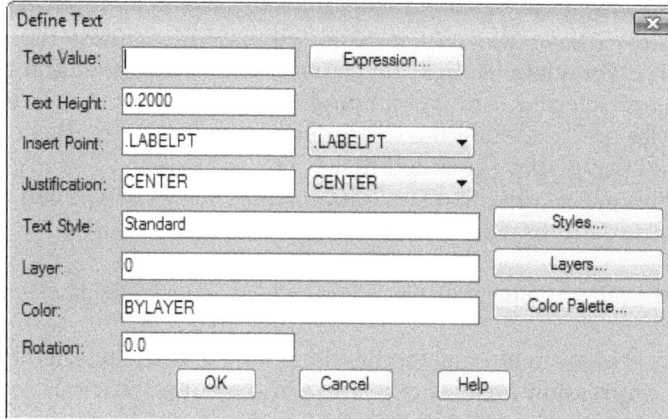

*Figure 6-36 The **Define Text** dialog box*

In this dialog box, specify the appearance of the text by using the parameters provided. To add a text to each of the retrieved object data, enter the text in the **Text Value** edit box. Alternatively, you can select a text from the attribute list. To do so, choose the **Expression** button in the right of the **Text Value** edit box; the **Text Value Expression** dialog box will be displayed. In this dialog box, expand the folder to select the required property from the table. Next, choose the **OK** button; the **Text Value Expression** dialog box will be closed and the selected expression will be displayed in the **Text Value** edit box.

Then, to specify the text height, enter a value in the **Text Height** edit box. To set the location point of a label, select the required option from the drop-down list corresponding to the **Insert Point** parameter. To apply justification to the label text, select the required option from the drop-down list corresponding to the **Justification** parameter. To apply a text style to the label text, enter the style name in the **Text Style** edit box or choose the **Styles** button on the right of this edit box; the **Select** dialog box will be displayed. In the **Select** dialog box, select an option from the list of text styles. Next, choose the **OK** button; the **Select** dialog box will close and the selected text style will be displayed in the **Text Style** edit box. To apply the settings only to the required layer, enter the name of the layer in the **Layer** edit box or choose the **Layers** button on the right of the **Layer** edit box; the **Select** dialog box will be displayed again. In this dialog box, select a layer from the list and then choose the **OK** button; the **Select** dialog box will be closed and the name of the selected layer will be displayed in the **Layer** edit box. To apply a color to the selected layer, choose the **Color Palette** button; the **Select Color** dialog box will be displayed. Select a color from the dialog box. To specify the angle of rotation for the base of the label text, enter the required angle in the **Rotation** edit box. Next, choose the **OK** button in the **Define Text** dialog box; this dialog box will be closed.

To hatch the required drawing object after querying, choose the **Hatch** button in the **Select Property** area of the **Set Property Alterations** dialog box; the **Hatch Options** dialog box will be displayed. In this dialog box, you can define the parameters for hatching the object data. To annotate the queried object data, choose the **Annotate** button in the **Select Property** area;

the **Insert Annotation** dialog box will be displayed. To apply an annotation template, select the required template from the **Annotation Template** list box and then choose the **Insert** button; the annotation template will be applied to query the output data.

If you want to edit the parameters of a selected template file, choose the **Advanced** button below the **Annotation Template** list box; the expanded **Insert Annotation** dialog box will be displayed. To change a selected annotation template file, set properties in the **Insert options** and **Insert properties** areas. To hide the expanded areas, choose the **Basic** button below the **Annotate Template** list box; the expanded dialog box will return to its original shape. Next, choose the **Insert** button; the modified annotation template will be applied to the query data.

Note

You cannot use the annotation function until you have created a template for annotating objects.

You can also use expressions to alter or modify the property of the queried data. To do so, use the options in the **Expression** area, as explained in the earlier section. After specifying the required properties, choose the **OK** button; the **Set Property Alterations** dialog box will be closed and the check box at the upper left corner in the **Options** area of the **Define Query of Attached Drawing(s)** dialog box will be selected.

Save

You can save a query in the required category as per project requirement. To set the options for saving a query, choose the **Save** button in the **Options** area; the **Save Current Query** dialog box will be displayed. To save the current query settings in the existing query categories, select an option from the **Category** drop-down list. Also, you can save the current query in a new category. To create a category, choose the **New Category** button; the **Define New Category** dialog box will be displayed. In this dialog box, enter a name for the new category in the **New Category Name** edit box. Then, choose the **OK** button; the **Define New Category** dialog box will be closed and the name entered will be displayed in the **Category** drop-down list. Also, the **Name** and **Description** edit boxes will be activated.

After creating a category, you need to specify a name for the current query. To do so, enter a name for the current query in the **Name** edit box. Then, enter the details about the query process in the **Description** edit box. To save the current query in a category located at an external location, select the **Save to External File** check box in this dialog box; the options in the **Save Options** area will be activated. In the **Save Options** area, the **Save List of Active Drawings** check box is selected by default. As a result, the list of drawing files selected and activated will be saved after the query has been executed. You can select the **Save Alter Properties** check box to save the changes made in the **Set Property Alterations** dialog box. You can select or clear the **Save Location Coordinates** check box to save or remove location coordinates from the query process. If you select the **Auto Execute** check box, the current query will be executed each time you run a query. To save the description and details about the query in the query library, select the **Keep Reference in Library** check box. You can use the default file location path for saving a query file or save the file in the required folder. To do so, choose the **Browse** button next to the **File Name** edit box; the **Create File** dialog box will be displayed. In this dialog box, browse to the required folder and enter a name in the **File name** edit box. Then, choose the **Save** button; the **Create File** dialog box will close and the query file will be saved.

Load

You can load a previously saved query file from a query category into the current query process. To select the required query category, choose the **Load** button; the **Load Internal Query** dialog box will be displayed, as shown in Figure 6-37.

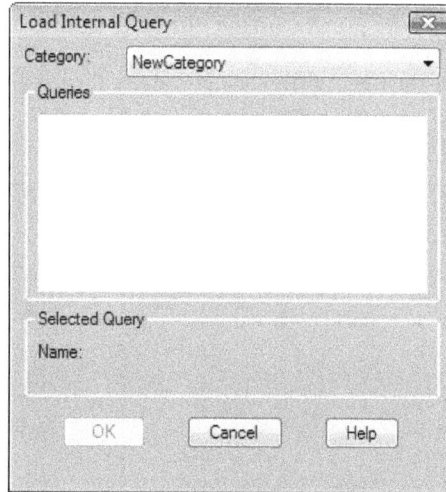

*Figure 6-37 The **Load Internal Query** dialog box*

In this dialog box, select the required option from the **Category** drop-down list; the saved query files in this category will be displayed in the list box in the **Queries** area. Also, the name of the selected query file will be displayed in the **Selected Query** area. Next, choose the **OK** button; the selected query file will be loaded into the query process.

Redraw

After altering the queried drawing, you can redraw the drawing object data in the queried file. To do so, choose the **Redraw** button; the preview of the modified drawing objects will be displayed.

Zoom Ext <

You can zoom to the limits of a drawing by using the **Zoom Ext <** button. On choosing this button in the **Options** area, the **Zoom Drawing Extents** dialog box will be displayed. In this dialog box, the **Select Active Drawings to Zoom** area displays the list of drawing files attached to the current drawing. To select the required drawing file from this list, click on its name. You can also choose the **Select All** button to select all drawings in the list. You can choose the **Clear All** button to clear all the selected drawings in the list. Moreover, you can use a drawing file as filter to select the data to be included in the query process.

You can use the **Filter** button to zoom the content in a particular drawing. To use the **Filter** button, select the check box on the left of this button; the selected drawings in the list box will be cleared and the **Select All** button will be activated. Next, choose the **Filter** button; the **Drawing Set Display Filter** dialog box will be displayed.

In this dialog box, you can select the required drawing file by using the starting word in the search command in the form of ***\XXXXX*.DWG**, where **XXXXX** represents the starting word of the required drawing file name. To filter the drawing file, enter the text ***\XXXXX*.DWG** in the **File name** edit box, and then choose the **OK** button; the drawing file/s that start with the word **XXXXX** will be displayed in the **Select Active Drawings to Zoom** area. Next, choose the **OK** button in the following dialog boxes; the settings applied for the zoom extent option will be saved. The process of using the **Filter** button to select the required drawing files is illustrated by using an example given next.

Consider an example where two drawing files are attached before defining the query process. To specify the zooming extents for the queried data, choose the **Zoom Ext** button in the **Options** area; the **Zoom Drawing Extents** dialog box will be displayed, as shown in Figure 6-38.

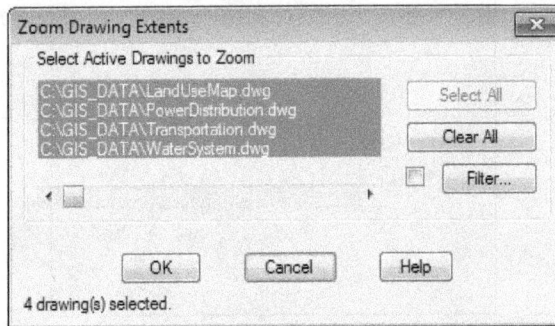

Figure 6-38   *The **Zoom Drawing Extents** dialog box*

In this dialog box, you can filter drawing files by using the starting word of the file name. By default, all attached drawing files are displayed in the list box in the **Select Active Drawings to Zoom** area. To filter the selected drawing files, select the check box beside the **Filter** button and then choose this button; the **Drawing Set Display Filter** dialog box will be displayed, as shown in Figure 6-39.

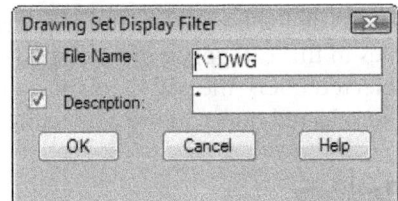

Figure 6-39   *The **Drawing Set Display Filter** dialog box*

Tip
*To zoom drawing, it is recommended to retain the default symbol provided in the **Description** edit box in the **Drawing Set Display Filter** dialog box.*

In this dialog box, select the check box corresponding to the **File Name** edit box, if it is not selected by default. Enter the text ***\Building*.DWG** in the **File name** edit box and choose the **OK** button; the **Drawing Set Display Filter** dialog box will be closed and the name of the drawing file starting with the text **Building** will be retained in the list box in the **Select Active Drawings to Zoom** area of the **Zoom Drawing Extents** dialog box. Select the drawing file in the list box; the **OK** button will be activated. Choose the **OK** button; the **Zoom Drawing Extents** dialog box will be closed and the zoom settings will be applied to the current query file.

Drawings

You can specify the drawings to be used in the current query process. To do so, choose the **Drawings** button in the **Options** area; the **Define/Modify Drawing Set** dialog box will be displayed, as shown in Figure 6-40.

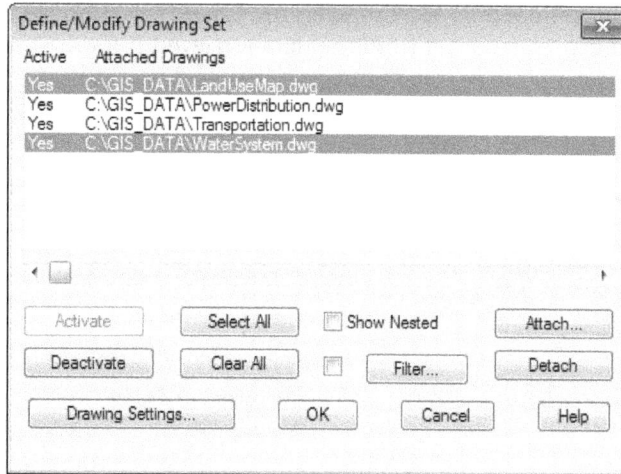

*Figure 6-40 The **Define/Modify Drawing Set** dialog box*

In the **Define/Modify Drawing Set** dialog box, you can attach new drawing files to the query process. To do so, choose the **Attach** button; the **Select Drawings to Attach** dialog box will be displayed. Browse to the required folder in this dialog box and then select the drawing files for the current query process.

You can specify the object data from the drawing files that will be used for the process of querying. In order to use the object data from a drawing file, you need to activate this file in the **Attached Drawings** list box, if it is not activated by default. To activate files, select the required drawing file/s in the list, and then choose the **Activate** button; the selected drawings in the **Attached Drawings** list box will be activated and the **No** status under the **Active** column in the list box will change to the **Yes** status. You can deactivate the drawing files that are not required for the current query process. To do so, select the unwanted drawing file/s from the **Attached Drawings** list box and then choose the **Deactivate** button; the **Yes** status corresponding to the selected drawing file will change to the **No** status in the **Active** column. Also, you can use the **Filter** button to search and filter the drawing files based on their file name, as explained in the previous section. After you have selected the drawings for the current query process, choose the **OK** button; the **Define/Modify Drawing Set** dialog box will be closed and the specified drawing file/s will be used for the query process.

More

You can choose the **More** button in the **Options** area to invoke the **Query** page of the **AutoCAD Map 3D Options** dialog box. Using the options in this page, you can specify the query options such as using the case sensitivity of text in the query syntax and saving query with drawing. Figure 6-41 displays the Query page of the **AutoCAD Map 3D Options** dialog box.

Query options Area

You can use the options in the **Query options** area to specify the property and default settings of the queries. You can select the **Save current query with drawing** check box to save a query in the current drawing even if the query is not saved as a separate query file. To match the case sensitivity of the text values in the query process, select the **Use case sensitivity when**

matching text values check box. You can select the **Create selection set from queried objects** check box to create a selection set from the data retrieved from the query process.

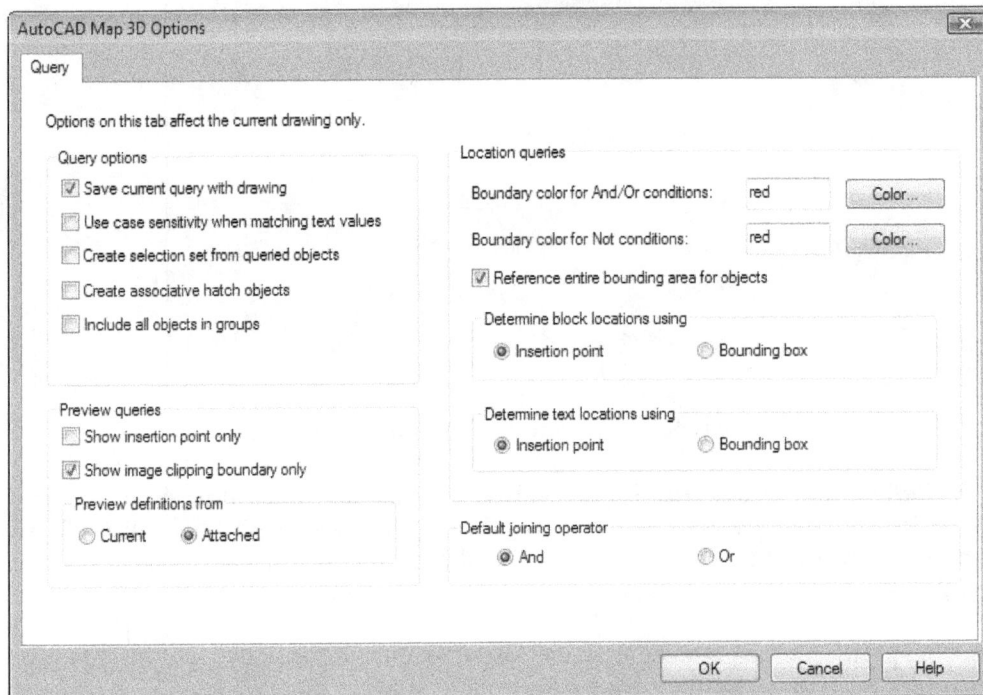

*Figure 6-41 The **AutoCAD Map 3D Options** dialog box*

To retain the hatching property of the object data retrieved by using the **ADEFILLPOLYG**, **ADEQUERY**, **MAPTOPOQUERY**, and **MAPTHEMATIC** commands, select the **Create associative hatch objects** check box. You can include or avoid all objects in the groups by selecting or clearing the **Include all objects in groups** check box.

Preview queries Area
In the **Preview queries** area, you can set properties for the preview of the queried data. You can select the **Show insertion point only** check box to display the insertion point of inserted blocks using the **X** symbol. On clearing this check box, the drawing object block will be displayed in the drawing. To show only the raster image boundary in the **Preview query** mode instead of the entire raster image, select the **Show image clipping boundary only** check box. Clear this check box to display the entire raster image in the **Preview query** mode. You can specify the location for the preview definition of the current drawing or of the attached drawings. To do so, select the **Current** or **Attached** radio button in the **Preview definitions from** area.

Location queries Area
In the **Location queries** area, you can set the appearance of the data retrieved from a query. To set color for a boundary while using the **And** or **Or** operator, choose the **Color** button next to the **Boundary color of And/Or conditions** edit box; the **Color Palette** will be displayed. In the **Color Palette**, select the required color and then choose the

OK button; the number corresponding to the selected color will be displayed in the **Boundary color of And/Or conditions** edit box. Similarly, you can select the required color for the boundary while using the **Not** operator by specifying the required color in the **Boundary color for Not conditions** edit box.

You can select the **Reference entire bounding area objects** check box, if it is not selected by default, to treat the hatch or hatch objects in the drawing area. As a result, if the query location lies inside or touches the boundary of hatch, the entire hatch object will be selected. If you clear this check box, the hatch boundary will be treated as an edge. As a result, the query location must enclose or intersect the boundary edge of the hatch object to include the object in the selection.

You can use the **Insertion point** or **Bounding box** radio button to specify the insertion point or the bounding box of the drawing block, respectively to determine whether the block meets the **Location** condition, as shown in Figures 6-42(a) and 6-42(b).

Figure 6-42(a) shows an instance of the drawing object referenced by using the **Insertion Point** parameter. Since the insertion point of the drawing object lies outside the location boundary, the drawing object will not be selected for the query process.

Figure 6-42(b) shows an instance of the drawing object referenced by using the **Bounding box** parameter. In this case, the location boundary intersects the bounding box of the drawing object. Therefore, the drawing object will be selected for the query process. Similarly, you can select the **Insertion Point** or **Bounding box** radio button to specify the location of text by using the insertion point or bounding box of the drawing object.

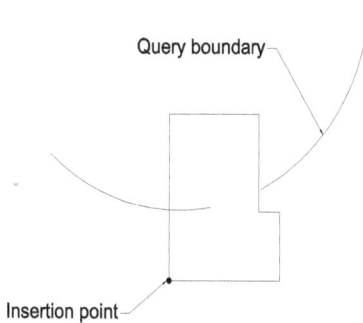

Figure 6-42(a) *The insertion point lying outside the boundary*

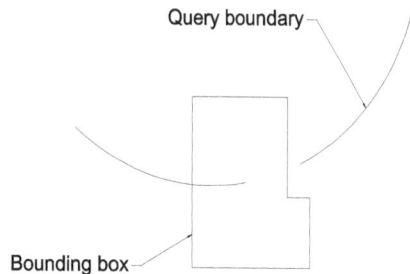

Figure 6-42(b) *The boundary intersecting the bounding box*

Default joining operator Area

The options in the **Default joining operator** area are used to specify the default relational operator in the **Query Type** area of the **Define Query of the Attached Drawings** dialog box. In the **Default joining operator** area, the **And** radio button is selected by default. As a result, each time you start defining a query type, the **And** radio button will be selected in the **Query Type** area. You can select the **Or** radio button in the **Default joining operator** area to set this operator as the default joining operator between two query types.

After specifying all parameters in the **AutoCAD Map 3D Options** dialog box, choose the **OK** button; this dialog will be closed and the specifications will be applied to the current query process.

Executing a Query

After specifying the query parameters (conditions) and the query result options in the **Define Query of Attached Drawing(s)** dialog box, you can execute the query. To do so, choose the **Execute Query** button at the bottom of the **Define Query of Attached Drawing(s)** dialog box; the defined query will be executed.

> **Note**
> *The output of an executed query depends on the options selected while defining the query process in the **Query Mode** area of the **Define Query of Attached Drawing(s)** dialog box.*

EDITING AN OBJECT IN AN ATTACHED DRAWING

AutoCAD Map 3D allows the user to edit the drawing objects in the attached drawing. To do so, filter the drawing objects to be edited by executing a query on the attached drawing as explained earlier. On doing so, AutoCAD Map 3D will find the drawing objects from the attached drawing and copy them into the current drawing. Now, you can perform edit operations on these filtered objects. On modifying the drawing object, the **Confirm Save Back** message box will be displayed. If you want to save the modified drawing object to the source (attached) drawing, choose the **Yes** button in this message box; the modified object will be added to the save set. Next, save the current drawing by choosing the **Save** option from the Application Menu. On saving the current drawing file, the **Save Objects to Source Drawing** dialog box will be displayed. In this dialog box, ensure that the **Save Queried Object** check box is selected. Next, choose the **OK** button; the edited drawing objects will be saved in the source file.

SETTING USER RIGHTS FOR DRAWING

In AutoCAD Map 3D, you need to set certain user privileges such that most of the functions available for the administrator are restricted for other users. You can set the administrator control over the object classification by logging in as an administrator. To do so, choose the **Map Explorer** tab from the **TASK PANE**; the list of folders in the **Current Drawing** node will be displayed. Then, choose the **Current Drawing** node in the **Map Explorer** tab and then right-click; a shortcut menu will be displayed. In this shortcut menu, choose the **User Login** option, as shown in Figure 6-43; the **User Login** window will be displayed, as shown in Figure 6-44. In this window, enter the registered user login name in the **Login Name** edit box and then enter the password in the **Password** edit box. Specify the names in both the edit boxes and then choose the **OK** button; you will log in to the current drawing space.

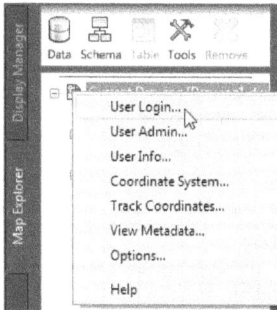

Figure 6-43 Choosing the **User Login** option from the shortcut menu

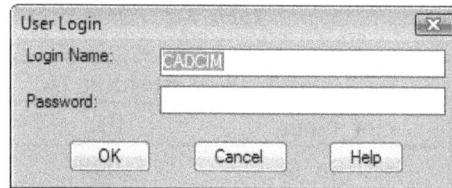

Figure 6-44 The **User Login** window

Note

*If you do not have a registered login user name and password, enter **SUPERUSER** in the **Login Name** edit box, and then enter **SUPERUSER** in the **Password** edit box. It is recommended that you use exact texts for login name and password, as they are case-sensitive.*

TUTORIALS

General instructions for downloading tutorial files:

Before starting the tutorials, you need to download the tutorial data to your computer. To do so, follow the steps given below:

1. Log on to *www.cadcim.com* and browse to *Textbooks > Civil/GIS > Map 3D > Exploring AutoCAD Map 3D 2017*. Next, select *c06_m3d_2017_tut.zip* file from the **Tutorial Files** drop-down list. Next, choose the corresponding **Download** button to download the data file.

2. Extract the contents of the zip file to the following location:

    C:\m3d_2017

    Notice that the *c06_m3d_2017_tut* folder is created within the *m3d_2017* folder.

Tutorial 1   Creating Drawing Objects

In this tutorial, you will create drawing objects and object data using the downloaded raster image. Then, you will attach the object data created to the drawing objects.

(Expected time: 1hr 30 min)

The following steps are required to complete this tutorial:

a. Open the tutorial file.
b. Create polygon geometries for the buildings labeled as **(A)**, **(B)**, and **(C)** in Figure 6-45.
c. Generate the object data table.

d. Insert data record, given in the Table 1-1, into the object data table and then attach the object data to the polygon geometries created for the buildings (A), (B), and (C).

e. Verify and edit the object data.

f. Save the file.

Table 1-1 Data record for creating drawing objects

	Building (A)	Building (B)	Building (C)
Plot_Num	101	102	103
Rent(in_$)	1000	1500	1750
Address	Marvel_St01	Marvel_St02	Marvel_St03

Figure 6-45 The buildings labeled as (A), (B), and (C) for creating drawing objects

Opening the Tutorial Drawing

1. Choose the **Open** button from the Quick Access Toolbar; the **Select File** dialog box is displayed.

2. In this dialog box, browse to the following location:

    C:\m3d_2017\c06_m3d_2017_tut\c06_tut01

3. Next, select the **c06_Tut01.dwg** file; the preview of the selected drawing file is displayed in the **Preview** area and the text **c06_Tut01.dwg** is displayed in the **File name** edit box.

4. Choose the **Open** button on the right of the **File name** edit box; the **Select File** dialog box is closed and the drawing is displayed in the drawing window.

Creating Polygon Geometries

1. Zoom into the raster image so that the building (A) is clearly visible in the drawing window, refer to Figure 6-45.

2. Choose the **Polyline** tool from the **Draw** panel in the **Home** tab; you are prompted to specify the start point of the line and the cursor changes to a crosshair in the drawing window.

Polyline

3. Place the cursor at the lower-left corner of the building **(A)**, as shown in Figure 6-46(a), and then click; the first corner of the building is selected.

4. Digitize the building by using the **Polyline** tool **(A)**. Next, press C and then press ENTER; a polygon is created, as shown in Figure 6-46(b).

Figure 6-46(a) *The crosshair placed at the lower-left corner of the building (A)*

Figure 6-46(b) *The polygon geometry drawn around the edge of the building (A)*

5. Repeat the procedure given in steps 1 to 4 and create polygon geometries for the buildings **(B)** and **(C)**. Figure 6-47 shows the polygon geometries created for the buildings **(A)**, **(B)**, and **(C)**.

Figure 6-47 *The polygon geometries drawn for the buildings (A), (B), and (C)*

Generating the Data Field Object Table

1. Choose the **Define Object Data** tool from the **Attribute Data** panel in the **Map Setup** tab; the **Define Object Data** dialog box is displayed.

2. In this dialog box, choose the **New Table** button next to the **Table** drop-down list; the **Define New Object Data Table** dialog box is displayed.

3. In this dialog box, enter **Building** in the **Table Name** edit box.

4. Next, enter **Plot_Num** and **Plot number of the building** in the **Field Name** and **Description** edit boxes, respectively. Note that the **Add** button in the **Field Definition** area of this dialog box is activated.

5. Ensure that the **Integer** option is selected in the **Type** drop-down list.

6. Next, choose the **Add** button; the defined data field **Plot_Num** is added to the list box in the **Object Data Fields** area.

7. In the **Field Definition** area, enter **Rent** in the **Field Name** edit box and then enter **Rent of the building in dollars** in the **Description** edit box; the **Add** button is activated.

8. Ensure that the **Integer** option is selected in the **Type** drop-down list. Next, choose the **Add** button in the **Field Definition** area; a data field with the name **Rent** is added to the list box in the **Object Data Fields** area.

9. In the **Field Definition** area, enter **Address** in the **Field Name** edit box, and then select the **Character** option from the **Type** drop-down list.

10. Enter **Address of the building** in the **Description** edit box; the **Add** button is activated.

11. Choose the **Add** button in the **Field Definition** area; a data field with the name **Address** is added to the list box in the **Object Data Fields** area.

12. Next, choose the **OK** button in the **Define New Object Data Field Table** dialog box; the dialog box is closed and the defined **Building** object data table is added to the **Table** drop-down list in the **Define Object Data** dialog box.

    Note that the **Building** object data table is selected in the **Table** drop-down list of the **Define Object Data** dialog box and the data fields in this table are displayed in the **Object Data Fields** list box.

> **Tip**
> *To view the details of data field parameters, select the required option in the list box in the* **Object Data Fields** *area; the* **Field Name**, **Data Type**, **Description**, *and* **Default** *parameters associated with the selected data field are displayed in the* **Field Definition** *area.*

13. Choose the **Close** button in the **Define Object Data** dialog box; the dialog box is closed.

Attaching the Object data to the Polygon Geometries

1. Choose the **Attach/Detach Object Data** tool from the **Drawing Object** panel in the **Create** tab; the **Attach/Detach Object Data** dialog box is displayed, as shown in Figure 6-48.

2. In this dialog box, select the **Plot number of the building** option in the **Object Data Field** list box; the name of the data field **Plot_Num** is displayed next to the **Name** parameter.

3. In the **Value** edit box for **Plot_Num**, specify the value **101** and then press ENTER; the specified value is displayed next to the **Plot number of the building** option in the

Object Data Field list box, and the **Rent of the building in dollars** option in the list box is selected.

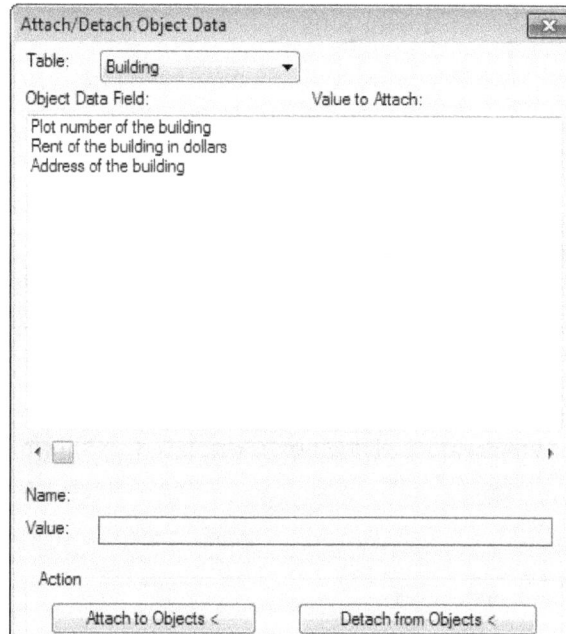

Figure 6-48 *The partial view of **Attach/Detach Object Data** dialog box displaying the data fields in the selected table*

4. Next, in the **Value** edit box corresponding to the **Rent** data field, specify the value **1000** and then press ENTER; the specified value is displayed next to the **Rent of the building in dollars** option in the **Object Data Field** list box, and the **Address of the building** option is selected.

5. Specify **Marvel_St01** in the **Value** edit box and then press ENTER; the specified value is displayed next to the **Address of the building** option in the **Attach/Detach Object data** dialog box.

6. Next, choose the **Attach to Objects** button in the **Action** area of the **Attach/Detach Object Data** dialog box; the dialog box closes and you are prompted to select the object in the drawing area. Also, note that the cursor changes into a selection box in the drawing window.

7. Place the cursor over the polygon drawn along the edge of the building **(A)**, refer to Figure 6-45; the polygon drawn is highlighted.

8. Click on the polygon and then press ENTER; the record created in the object data table is attached to the selected drawing object.

9. Repeat steps 1 to 8 to create and attach attribute data to the building (drawing object **(B)** and **(C)**). Use the data given below:

Building B
        Plot_Num -   102
        Rent(in_$) -   1500
        Address -   Marvel_St02

Building C
        Plot_Num -   103
        Rent(in_$) -   1750
        Address -   Marvel_St03

Verifying and Editing the Object Data

1. Choose the **Edit Object Data** tool from the **Map Edit** panel in the **Tools** tab; you are prompted to select the required drawing feature. Note that the cursor has changed to a selection box in the drawing window.

2. Place the selection box over the polygon drawn around the edge of the building (**A**), and then click; the **Edit Object Data** dialog box is displayed, as shown in Figure 6-49.

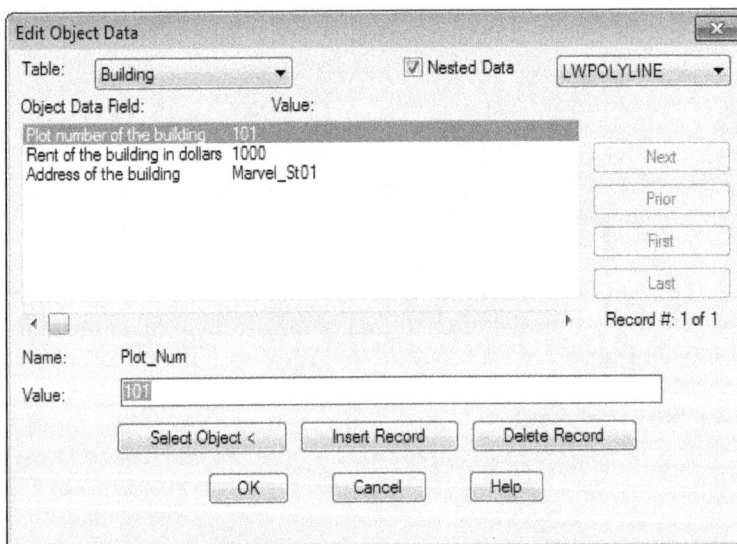

Figure 6-49 The **Edit Object Data** dialog box displaying the data fields and values of the building (**A**)

3. To verify the data fields and values of the building (**B**), choose the **Select Object** button in the **Edit Object Data** dialog box; the dialog box is closed and you are prompted to select the required drawing object.

4. Place the selection cursor on the polygon drawn over the building (**B**) and then click on it; the **Edit Object Data** dialog box is displayed with the data fields and values corresponding to the building (**B**).

5. Compare the values in each data field with the values given in the tutorial data table. If any of the values does not match with the value given in the table, repeat the steps in previous section and define and attach the data to the object.

Note

*In case you are attaching the data again, make sure that the **Overwrite** check box in the **Attach/Detach Object Data** dialog box is selected. In this way, you can avoid creating multiple records for the selected drawing object.*

Saving the Drawing File

1. Choose the **Save As** option in the Application Menu; the **Save Drawing As** dialog box is displayed.

2. In the **Save Drawing As** dialog box, enter **c06_Tut01a** in the **File name** edit box and select the **AutoCAD 2013 Drawing (*.dwg)** option in the **Files of type** drop-down list, if it is not selected by default. In the **Save Drawing As** dialog box, choose the **Save** button next to the **File name** edit box; the current drawing file is saved with the given name.

Tutorial 2   Using the Location and Property Query Types

In this tutorial, you will incorporate the required drawing objects into the current drawing by applying the **Location** and **Property** filters.   **(Expected time: 1 hr 30 min)**

The following steps are required to complete this tutorial:

a. Create a new drawing file using the **map2d** template.
b. Load the **GIS_ADMIN_BM_Subdiv** shape file.
c. Log in as an administrator.
d. Attach a drawing file to the current drawing.
e. Define and execute the **Location** query by drawing the polygon with **FeatId 647**.
f. Add property constraint of **Area < 2000** (Building area less than 2000 sqft) for the **Property** based query.
g. Execute the query in the **Preview** query mode.
h. Save the drawing file.

Starting a New Drawing File

1. Choose **New > Drawing** from the Application Menu; the **Select template** dialog box is displayed.

2. In the **Select template** dialog box, select the *map2d.dwt* template file and then choose the **Open** button; a new drawing is created by applying the settings defined in the **map2d** template.

Loading the Shape File

1. Choose the **Connect** tool from the **Data** panel in the **Home** tab; the **DATA CONNECT** wizard is displayed.

2. In this wizard, select the **Add SHP Connection** option from the **Data Connections by Provider** list box; the **OSGeo FDO Provider for SHP** page is displayed on the right side of this list box.

3. In this page, choose the **SHP** button next to the **Source file or folder** edit box; the **Open** dialog box is displayed.

4. In this dialog box, browse to the following location:

    C:\m3d_2017\c06_m3d_2017_tut\c06_tut02

5. Select the **GIS_ADMIN_BM_Subdiv** file in the list box and then choose the **Open** button; the **Open** dialog box is closed and the path of the selected file is displayed in the **Source file or folder** edit box of the **DATA CONNECT** wizard.

6. In the **OSGeo FDO Provider for SHP** page, choose the **Connect** button; the **GIS_ADMIN_BM_Subdiv** file is added to the list box below the **Edit Coordinate Systems** button in the **SHP** page.

7. Choose the **Add to Map** button in the **SHP** page; the **GIS_ADMIN_BM_Subdiv** feature layer is added to the list box in the **Display Manager** tab of the **TASK PANE**. Also, the geometry of the feature layer is displayed in the drawing window, as shown in Figure 6-50. Next, close the **DATA CONNECT** wizard.

Figure 6-50 The vector layer displayed in the drawing window

Logging in as an Administrator

In this section of the tutorial, you will login as a superuser (default user created by AutoCAD Map 3D) using the credentials given in the steps of this section.

1. Choose the **Map Explorer** tab in the **TASK PANE**, if it is not chosen by default; the folders in this tab are displayed in a tree view.

2. In the **Map Explorer** tab, right-click on the **Current Drawing** node; a shortcut menu is displayed.

3. In this shortcut menu, choose the **User Login** option, as shown in Figure 6-51; the **User Login** dialog box is displayed.

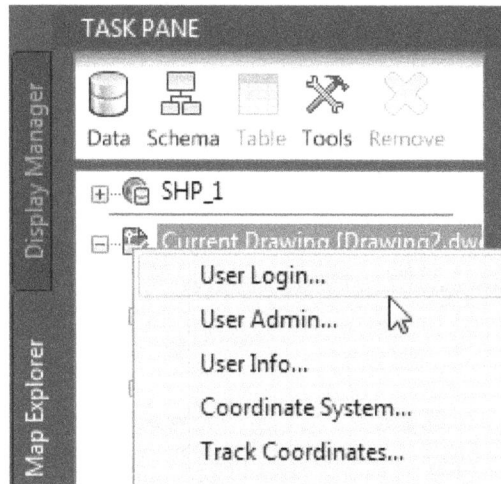

Figure 6-51 *Choosing the **User Login** option from the shortcut menu*

4. In the **User Login** dialog box, enter **SUPERUSER** in the **Login Name** edit box and in the **Password** edit box. Next, choose the **OK** button; now you are logged in as SUPERUSER. Note that the login password is case sensitive and is required to be entered as specified in the step.

Attaching the External Drawing File to the Current Drawing File

1. In the **Map Explorer** tab of the **TASK PANE**, right-click on the **Drawings** folder; a shortcut menu is displayed.

2. In this shortcut menu, choose the **Attach** option; the **Select drawings to attach [C:]** dialog box is displayed.

3. In this dialog box, browse to the following location:

   C:\m3d_2017\c06_m3d_2017_tut\c06_tut02

4. Select the **Building 101409 DWG** option in the drawing file list box; the **Add** button below the list box is activated.

5. Choose the **Add** button; the path of the **Building 101409 DWG** file is added to the **Selected drawings** list box, as shown in Figure 6-52.

6. Next, choose the **OK** button in the **Select drawings to attach [C:]** dialog box; the dialog box is closed and the path of the attached drawing is displayed below the **Drawings** folder in the **Map Explorer** tab.

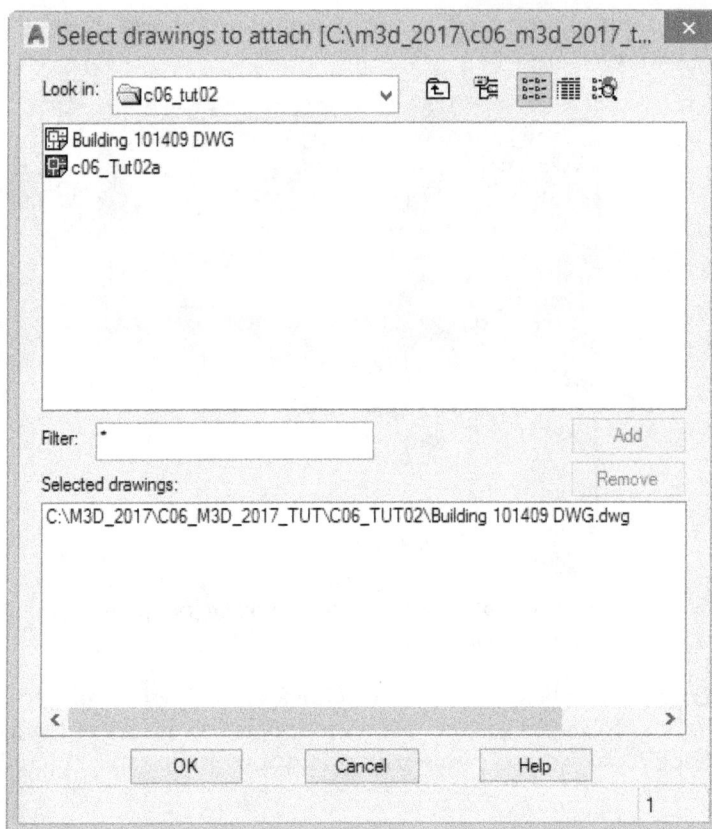

Figure 6-52 The Select drawings to attach [C:] dialog box showing the added drawing location in the Selected Drawings list box

7. Right-click on the attached drawing location path in the **Drawings** folder in the **Map Explorer** tab; a shortcut menu is displayed. In this shortcut menu, choose the **Quick View** option; the geometry associated with the attached drawing is displayed in the drawing window.

Defining and Executing the Query Based on the Location Query Type

In this section of the tutorial, you will create a query category. Next, you will define and execute a location based query on the attached drawing.

1. Right-click on the **Query Library** node in the **Map Explorer** tab; a shortcut menu is displayed.

2. In this shortcut menu, choose the **New Category** option; a new query category is added to the **Query Library** node. Enter the name **Map3D_C06_Tut_02** for the new category and then press ENTER; a new query category is created with the specified name.

3. Right-click on the **Map3D_C06_Tut_02** category in the **Query Library** node; a shortcut menu is displayed.

4. In this shortcut menu, choose the **Define** option; the **Define Query of Attached Drawing(s)** dialog box is displayed.

5. In the **Query Type** area of this dialog box, select the **And** radio button, if it is not selected by default. Next, choose the **Location** button; the **Location Condition** dialog box is displayed, as shown in Figure 6-53.

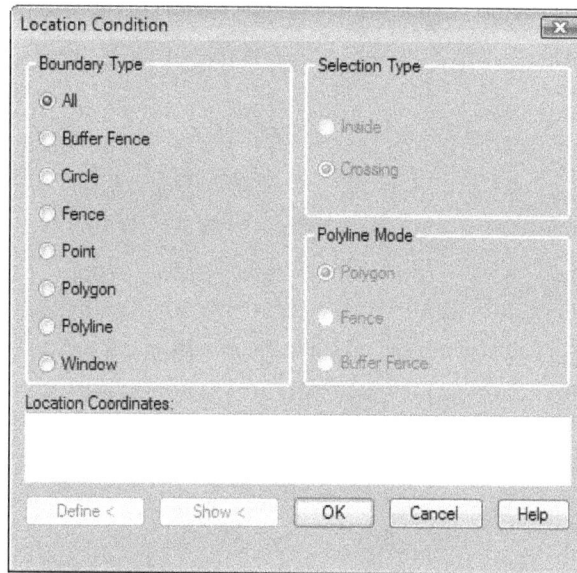

*Figure 6-53 The **Location Condition** dialog box*

6. In the **Boundary Type** area of the **Location Condition** dialog box, select the **Polygon** radio button; the options in the **Selection Type** area are activated. Notice that the **Define** button at the bottom of the **Location Condition** dialog box is also activated.

7. Select the **Inside** radio button in the **Selection Type** area.

8. Next, choose the **Define** button in this dialog box; you are prompted to specify the first point and the cursor changes to a crosshair in the drawing window.

9. Zoom into the attached drawing so that the boundary of the polygon with **FeatId 647** is clearly visible, refer to Figure 6-54. Next, turn on the object snapping by choosing the **Object Snap** button from the Status Bar, if it is turned off.

10. Place the crosshair over the lower left vertex of the polygon with **FeatId 647**, refer to Figure 6-55, and then click; the first vertex of the polygon is selected. Next, trace the drawing line along each vertex of this polygon by left-clicking until the entire polygon boundary has been selected.

11. After tracing the entire polygon boundary, right-click; the **Define Query of Attached Drawing(s)** dialog box is displayed. Also, the query **Location: INSIDE POLYGON** is displayed in the **Current Query** list box.

12. In the **Query Mode** area of the **Define Query of Attached Drawing(s)** dialog box, select the **Preview** radio button, if it is not selected by default.

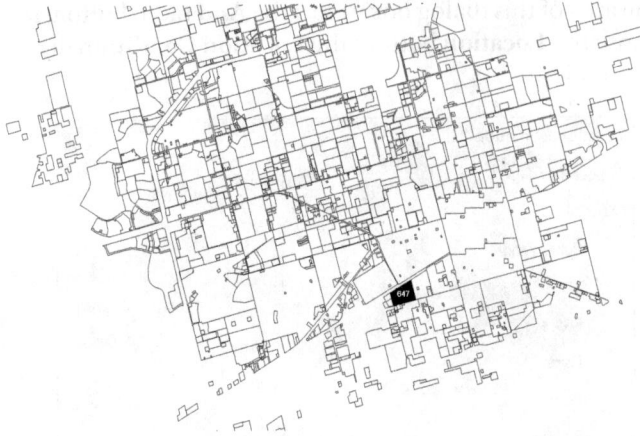

*Figure 6-54 The polygon used to define the **Location** query type boundary*

Figure 6-55 The crosshair placed over the starting point of the polygon boundary

13. Next, choose the **Execute Query** button at the bottom of this dialog box; the **Location** based query is executed and the drawing objects in the attached drawing are filtered based on the query expression. Also, the filtered drawing objects are displayed in the drawing window.

Defining and Performing the Query Based on Property Query Type

In this section, you will create a complex query to filter drawing objects using the location and property query type.

1. In the **Map Explorer** tab of the **TASK PANE**, right-click on the **Map3D_C06_Tut_02** category in the **Query Library** node; a shortcut menu is displayed. In this shortcut menu, choose the **Define** option; the **Define Query of Attached Drawing(s)** dialog box is displayed with the previous query statement.

2. In the **Query Type** area of the **Define Query of Attached Drawing(s)** dialog box, ensure that the **And** radio button is selected.

3. Next, choose the **Property** button in this area; the **Property Condition** dialog box is displayed, as shown in Figure 6-56.

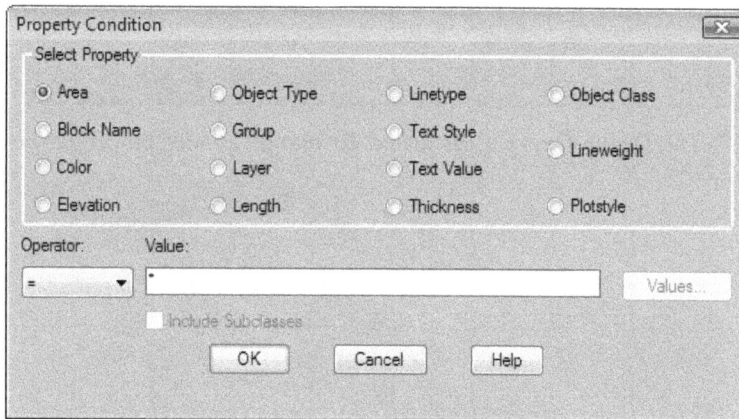

*Figure 6-56 The **Property Condition** dialog box*

4. In the **Select Property** area of the **Property Condition** dialog box, select the **Area** radio button, if it is not selected by default.

5. Next, select the **<** (less than) option from the **Operator** drop-down list, and then enter **2000** in the **Value** edit box next to the **Operator** drop-down list.

6. Choose the **OK** button; the **Property Condition** dialog box is closed and the **AND Property: AREA < 2000** statement is added to the existing query expression in the **Current Query** list box of the **Define Query of Attached Drawing(s)** dialog box, as shown in Figure 6-57.

7. Next, choose the **Execute Query** button from the **Define Query of Attached Drawing(s)** dialog box; the dialog box is closed and the filtered data is displayed in the drawing window. Figure 6-58 shows the filtered data after executing the query statement.

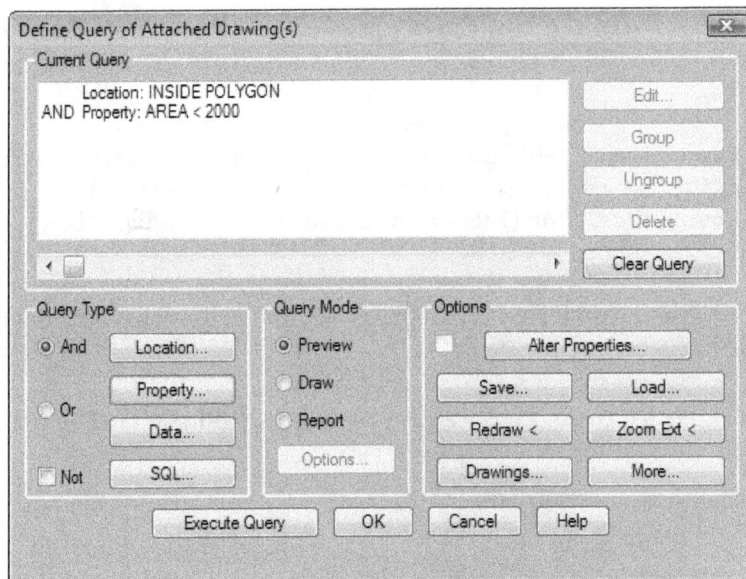

Figure 6-57 The **Define Query of Attached Drawing(s)** *dialog box displaying multiple query expressions*

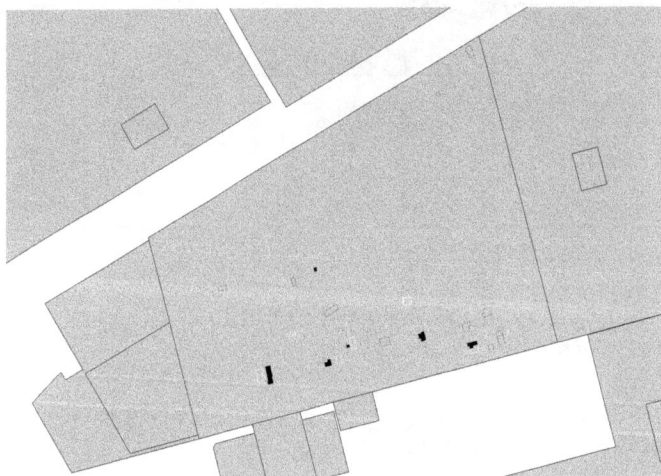

Figure 6-58 Filtered data after executing the query

Saving the Drawing File

1. Choose the **Save As** option in the Application Menu; the **Save Drawing As** dialog box is displayed.

2. In the **Save Drawing As** dialog box, enter **c06_Tut02a** in the **File name** edit box and select the **AutoCAD 2013 Drawing (*.dwg)** option in the **Files of type** drop-down list, if it is not selected by default. Next, choose the **Save** button; the current drawing file is saved with the given name.

Self-Evaluation Test

Answer the following questions and then compare them to those given at the end of this chapter:

1. Which of the following files stores the connection links to the external database?

   (a) **.mdb**   (b) **.udl**
   (c) **.dgn**   (d) **.shp**

2. Which of the following tools is used to attach the object data to a drawing object while digitizing?

   (a) **Digitize**   (b) **Line**
   (c) **Polyline**   (d) **Polygon**

3. Which of the following options/tools is used to attach an external CAD (.dwg) file to the current drawing?

   (a) **Insert**   (b) **Connect**
   (c) **Map Import**   (d) **Attach**

4. Which of the following tools is used to modify the object data in the object data table of a drawing object?

   (a) **Digitize**   (b) **Define Object Data**
   (c) **Edit Object Data**   (d) **Attach/Detach Object Data**

5. Which of the following nodes in the **Map Explorer** tab of the **TASK PANE** displays the external database that is attached to the drawing?

   (a) **Data Sources**   (b) **Drawings**
   (c) **Link Templates**   (d) **Query Library**

6. The _____ tool in the **Attribute Data** panel of the **Map Setup** tab is used to define a new object data table.

7. You can use the _____ option in the **Query Type** area of the **Define Query of Attached Drawing(s)** dialog box to filter drawing objects based on its object data.

8. Using the _____ option in the **Query Mode** area of the **Define Query of Attached Drawing(s)** dialog box, you can produce the result of the query in a text file.

9. The _____ button in the **Define Query of Attached Drawing(s)** dialog box is used to invoke the **Set Property Alterations** dialog box for specifying the changes to be applied to the drawing objects that are filtered by a query.

10. You can attach only one external CAD drawing file to the current drawing. (T/F)

11. You cannot attach attribute data to the drawing object. (T/F)

12. The drawing geometry or the attribute data field can be edited any number of times after it is created. (T/F)

13. You can attach or detach an object data table from a drawing object in the drawing file. (T/F)

14. For a drawing object, you cannot add more than one data record in a single object data table. (T/F)

15. You cannot import drawing objects from an external drawing file into the current drawing file. (T/F)

Review Questions

Answer the following questions:

1. Which of the following options in the **Query Type** area is used to filter drawing objects based on the object properties?

    (a) **Location**   (b) **Property**
    (c) **Data**   (d) **SQL**

2. Which of the following buttons in the **Define Query of Attached Drawing(s)** dialog box is used to perform or process a defined query?

    (a) **Redraw**   (b) **OK**
    (c) **Execute Query**   (d) **Cancel**

3. You can set up the parameters for the **Digitize** tool in the _____ dialog box.

4. The _____ radio button in the **Query Mode** area is used to filter drawing objects from the attached drawing files and make them editable in the drawing window.

5. Link templates are not required to attach data from an external database table to the drawing object in the drawing. (T/F)

6. You cannot create or define a new attribute table for geometries. (T/F)

7. You can define a new object data table by using the options in the **Define Object Data** dialog box. (T/F)

8. You cannot modify an object data table once it has been defined. (T/F)

9. The **Link Template** node in the **Map Explorer** tab displays the list of external data sources that are connected to the drawing. (T/F)

10. The operators such as AND, OR, NOT in the **Query type** area are used to create a query with multiple (complex) conditions. (T/F)

EXERCISES

Exercise 1

Download the *c06-m3d-2017-exr01.jpg* raster image from *http://www.cadcim.com*. Insert the downloaded image into your drawing and then digitize a road between point **A** and **B** using the raster image. Next, create an object data table using the data given in Table1-2 and then attach the object data to the digitized road object. **(Expected time: 45 min)**

Table 1-2 *Data record for creating drawing objects*

	Line AB
Description	Motor way
Field Name	Nat_Highway32
Value	9

Exercise 2

Download the *c06-m3d-2017-exr02.dwg* file from *http://www.cadcim.com*. Then, attach this drawing file to your current drawing. Next, filter the data from the attached drawing file by applying a property type query (**Property: Layer= Driveway**). **(Expected time: 30 min)**

Answers to Self-Evaluation Test

1. b, **2.** a, **3.** d, **4.** c, **5.** a, **6.** Define Object Data, **7.** Data, **8.** Report, **9.** Alter Properties, **10.** F, **11.** F, **12.** T, 1

Chapter 7

Classifying Objects and Working with Classified Objects

Learning Objectives

After completing this chapter, you will be able to:

• *Set the basic environment for object classification*
• *Create and attach a definition file*
• *Define an object class by specifying parameters*
• *Select and classify drawing objects*
• *Create objects and maps by using an object class*
• *Generate and display the metadata file*
• *Customize the metadata file*
• *Publish and share the metadata file*

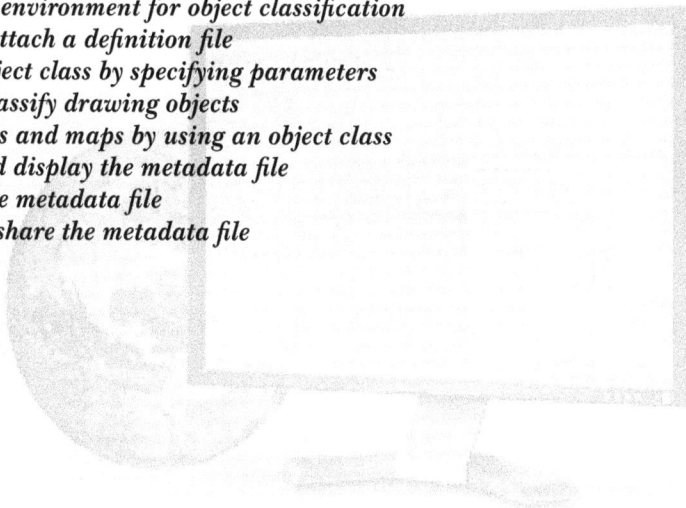

INTRODUCTION

In the previous chapter, you learned to work with external reference files and object data. In this chapter, you will learn to define the object class in the drawing and then classify the drawing objects into different object classes based on their properties and attributes.

To define an object class, the user must login as an AutoCAD Map 3D user having privileges to perform such operations. This chapter contains a detailed description for creating Map 3D users with administrative privileges. The procedure to create users with privileges which grants them access to commands such as creating definition file and defining an object class has also been explained in the chapter.

This chapter explains the procedure of creating an object class by using the tools in the **Task-Based Ribbon** and the **TASK PANE**. Next, you will learn to define the parameters in an object class. After creating an object class and defining its parameters, you will also learn to classify the drawing objects into defined object class.

Later in this chapter, you will learn the procedure to generate a metadata file for the current drawing file. The tools that are used for generating and sharing metadata are also described briefly in this chapter.

CLASSIFYING DRAWING OBJECTS

Using the drawing classification in AutoCAD Map 3D, you can classify the CAD objects into object classes that are defined by a set of parameters. Classifying drawing objects helps you to standardize the objects in the drawing. The objects in the drawing can be standardized by enforcing the parameters (such as color, layer, and object type) defined for the object class.

The process of classifying drawing objects in a CAD drawing involves creating the definition file, creating object classes, and classifying existing objects. For creating a definition file, first you need to login as a user with appropriate user privileges. The process of classifying objects is discussed in the forthcoming sections.

After defining the object class in the drawing, you can also use it to create new objects. This process ensures that the objects created conform to the parameters that have been set for the object class.

Note
You can classify an object by using the tools available in the Ribbon interface. These object classification tools can also be invoked from the TASK PANE. In the tutorial section of this chapter, the classification tools have been invoked using the options in the TASK PANE.

Logging in as Map 3D User with Appropriate User Privileges

Ribbon:	Map Setup > Map > User Login
Command:	MAPLOGIN

User Login   In AutoCAD Map 3D, some commands are restricted based on user privileges. To use these commands, you must log in to your AutoCAD Map 3D project with appropriate user privileges.

The superuser in AutoCAD Map 3D can perform administrative functions such as creating new users. The superuser can create a user by assigning login name and passwords. Using the login credentials, a user can login to the project. To control the access of a user for performing actions such as classifying objects, defining an object class, and modifying an existing object class in the drawing, the administrator can set user privileges for an individual login name.

To create a new user, you need to login as a superuser. To login, choose the **User Login** tool from the **Map** panel of the **Map Setup** tab; the **User Login** window will be displayed, as shown in Figure 7-1. In the **User Login** window, enter the user name and password in the **Login Name** and **Password** edit boxes, respectively. Next, choose the **OK** button; the text **Login Successful as <user name>** will be displayed in the Command prompt.

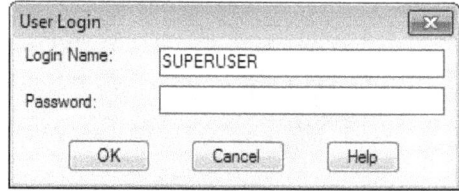

Figure 7-1 The *User Login* window

Tip
*If you do not have a registered login name and password, enter **SUPERUSER** in both the **Login Name** and **Password** edit boxes. Note that the password is case-sensitive and is required to be entered precisely as specified.*

Once you login as a superuser, you can create a new user and set the user privileges by using the options in the **User Administration** dialog box. To invoke this dialog box, choose the **User Administration** tool from the **Map** panel of the **Map Setup** tab; the dialog box will be displayed, as shown in Figure 7-2.

Figure 7-2 The *User Administration* dialog box

Next, to create object classes, you can use tools available in the ribbon interface or the **TASK PANE**. In the ribbon interface of the **Planning and Analysis Workspace**, the tools for object classification are available in the **Object Class** and **Drawing Object** panels of the **Map Setup** and **Create** tabs, respectively.

The object classification tools are also available in the **Classification** contextual tab. To invoke the **Classification** contextual tab, click on the **Object Classes** node in the **Map Explorer** tab of the **TASK PANE**; the **Classification** contextual tab will be displayed, as shown in Figure 7-3.

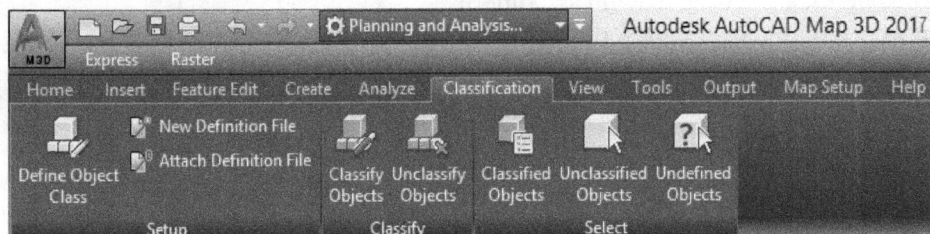

*Figure 7-3   The **Classification** contextual tab in the AutoCAD Map 3D Ribbon interface*

Specifying a Definition File

After the successful login with required user privileges, you can create a definition file. A definition file is a text file in XML format, which contains all the parameters used for the object classification procedure. You can specify a definition file for object classification by creating a new definition file or using an existing definition file. In the next section, you will learn how to create a new file or attach an existing definition file.

Creating a New Definition File

Ribbon:	Map Setup > Object Class > New Definition
	Or Classification (Contextual tab) > Setup > New Definition File
Command:	NEWDEF

You can create a new definition file that contains all properties and parameters of the current object class. To create a new definition file, choose the **New Definition** tool from the **Object Class** panel of the **Map Setup** tab; the **New Object Class Definition File** dialog box will be displayed. In this dialog box, enter a name for the object definition class file in the **File name** edit box and then choose the **Save** button; the **New Object Class Definition File** dialog box will be closed and a new object class definition file will be created with the specified name.

Attaching a Definition File

Ribbon:	Map Setup > Object Class > Attach Definition
	Or Classification (Contextual tab) > Setup > Attach Definition File
Command:	ATTACHDEF

You can attach a new definition file or use an existing one to apply the object classification parameters to the current object classification. To attach a definition file, choose the **Attach Definition** tool from the **Object Class** panel of the **Map Setup** tab; the **Attach Object Class Definition File** dialog box will be displayed. In this dialog box, browse to the folder containing the required definition file and then select it; the name of the selected file will be displayed in the **File name** edit box located below the list box. Next, choose the **Open** button; the selected definition file will be opened and can be used for the object classification process.

Note
The object class definition file will not be displayed as it is opened in the virtual mode.

Defining an Object Class

Ribbon:	Map Setup > Object Class > Define
	Or Classification (Contextual tab) > Setup > Define Object Class
Command:	FEATUREDEF

Define   You can define an object class based on the sample object selected in the current drawing. After choosing the sample object, you need to specify which of its parameters (properties) are to be included in the object class definition.

Note
If you do not have a sample object available in your drawing, you need to create an object with required parameters (properties) and then select it to include in the object class definition.

To define an object class, choose the **Define** tool from the **Object Class** panel; the cursor will change to a selection box in the drawing window and you are prompted to select the required sample object. Select the required drawing object as a model object class for further classification. Next, press ENTER; the **Define Object Classification** dialog box will be displayed, as shown in Figure 7-4.

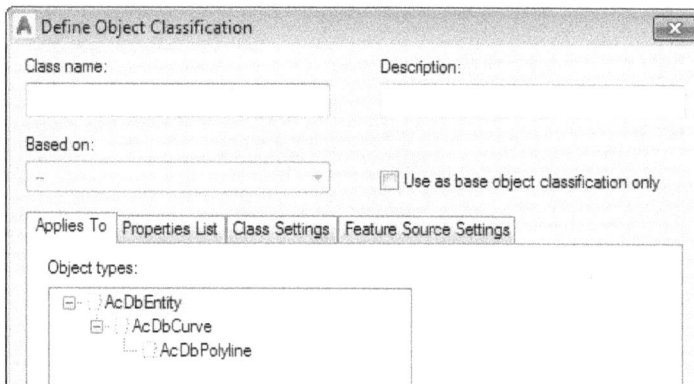

*Figure 7-4 Partial view of the **Define Object Classification** dialog box*

Note
*In the **Define Object Classification** dialog box, the hierarchy of nodes in the **Object types** area in the **Applies To** tab is displayed depending on the drawing object selected.*

In the **Define Object Classification** dialog box, you can specify various parameters of the object class. To specify a name for the object class, enter a name in the **Class name** text box and then enter details of the object class in the **Description** text box. The **Use as base object classification only** check box is clear by default. On selecting this check box, the current object class will be considered as the base class and you will not be able to create an object by using this object class. Next, specify the required object classification parameters of the object class in the **Applies To**, **Properties List**, **Class Settings**, and **Feature Source Settings** tabs. These tabs are discussed next.

Applies To Tab

The options in the **Applies To** tab are used to specify the object types to be used for the object classification process. To select all object types in the current drawing for classification, select the check box corresponding to the **AcDbEntity** node. Alternatively, to select a specific type of object for classification, select the check box corresponding to the required object type node. For example, to select all the polyline objects, select the check box corresponding to the **AcDbPolyline** node.

Properties List Tab

The options in the **Properties List** tab are used to specify the properties and data range values of the drawing objects to be included in the object classification. To specify properties for object classification, choose the **Properties List** tab; different options in this tab will be displayed, as shown in Figure 7-5.

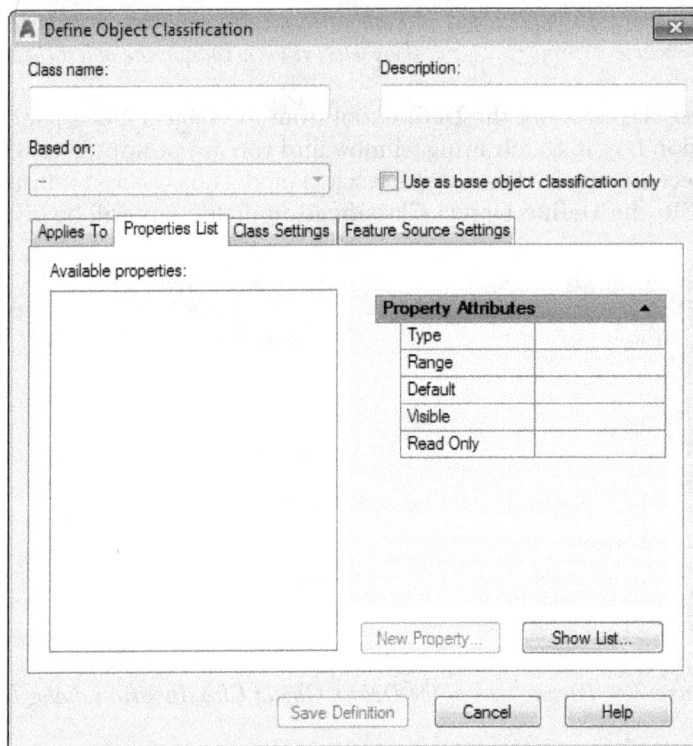

Figure 7-5 *The **Define Object Classification** dialog box with the **Properties List** tab chosen*

Note
*If you have already selected the check box corresponding to the **AcDbEntity** node or sub-nodes in the **Applies To** tab, a list of properties corresponding to the selected node will be displayed in the **Available properties** list box in the **Properties List** tab.*

To include the properties of a drawing object in the current object classification, select the check box(es) corresponding to the required property nodes in the **Available properties** list box. On doing so, the attributes of the selected property will be displayed in the **Property Attributes** rollout located on the right of the **Available properties** list box. Attributes such as **Type**, **Range**, **Default**, **Visible**, and **Read Only** are displayed in this rollout. These attributes are discussed next.

Type

On selecting a property in the **Available properties** list box, its data type will be displayed in the **Type** cell of the **Property Attributes** rollout. For properties that are not user defined, the data type is set automatically and cannot be changed. The process of defining a custom (user defined) property is discussed later in this chapter.

Range

The **Range** option in the **Property Attributes** rollout is used to specify the permissible data values of the selected property. The Table 7-1 shows several examples of how to specify the data range of the property of an object selected for classification. Also, this table indicates how to specify the range condition and format for writing data range in the **Range** edit box.

Table 7-1 *Description of the data range of the property of an object selected for classification*

Entity	Required Range Condition	Specify the Range Condition as
Materials	sand, gravel, & limestone	Sand,Gravel,Limestone
Door Number	D1 to D3	D1,D2,D3
Tiles	1 to 125	[1,125]
Lineweight	Lineweight of 0.25	25
To delete a range		--

Tip
*The data range for some of the properties in the **General** category such as **Color**, **Layer**, **Linetype**, and so on can be specified by using the built-in data type range editor functions. To specify these properties, select the check box for the required property node in the **Available properties** list box; the properties corresponding to the selected property will be displayed in the **Property Attributes** rollout. In this rollout, click in the cell for the **Range** property; the **Range** property will be selected and a Browse button will be displayed on the right of the cell. You can use this button to specify the predefined range for the selected property value.*

Default

The **Default** option is used to specify the default value for the objects if the value of the object data is out of range. To specify the default value for a property such as color, lineweight, and layer, click in the default value cell; a drop-down list will be displayed. Select the required default value from the drop-down list.

Visible

The **Visible** option is used to display or hide the selected property in the **Object Class** tab of the **PROPERTIES** palette after the object has been classified. To set the visibility, click in the corresponding value cell; a drop-down list will be displayed in the cell. In this drop-down list, you can select the **Yes** option to display the selected property, or select the **No** option to hide it.

Read Only
The **Read Only** option is used to specify if the properties of a classified object are editable in the **Object Class** tab of the **PROPERTIES** palette. To specify the parameter for this option, click on it; a drop-down list will be displayed on its right. Select the **Yes** option from this list to make the selected property only readable or select the **No** option to make it editable in the **Object Class** tab.

In addition to an existing category and properties, you can also add custom properties in the **Properties** tab. To do so, choose the **New Property** button below the **Property Attributes** rollout; the **New Property** window will be displayed, as shown in Figure 7-6.

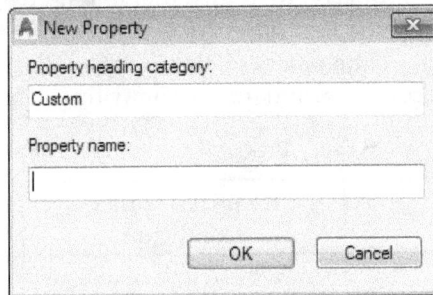

*Figure 7-6 The **New Property** window*

In the **New Property** window, you can specify the property category in the **Property heading category** edit box and the name of the new property in the **Property name** edit box. After specifying these options, choose the **OK** button; the **New Property** window will be closed and the newly created property category and the new property will be added to the **Available properties** list box. Also, the property values corresponding to the newly added property will be displayed in the **Property Attributes** rollout. In this rollout, you can specify the data type for the new property by clicking on the **Type** property. On doing so, the **Type** property will be selected and a drop-down list will be displayed on the right. Select the required option from this drop-down list. Now, you can specify other property options in the **Property Attribute** drop-down list, as explained in the previous sections.

Class Settings Tab
The options in the **Class Settings** tab are used to specify class display settings and geometry for creating objects in an object class. To specify these settings, choose the **Class Settings** tab in the **Define Object Classification** dialog box; the options in this tab will be displayed, as shown in Figure 7-7. These options are discussed next.

Show object class in Map Explorer
The **Show object class in Map Explorer** check box is selected by default. As a result, on saving the definition file, the current object class will be displayed in the **Map Explorer** tab. You can clear this check box if you want to hide the current object class in the **Map Explorer** tab after the definition file has been saved.

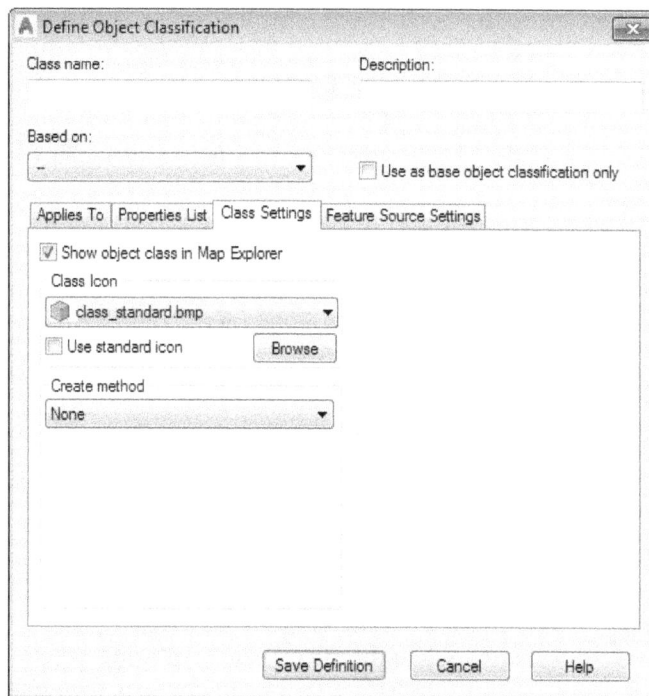

Figure 7-7 The **Define Object Classification** *dialog box with the* **Class Settings** *tab chosen*

Class Icon Area

The options in the **Class Icon** area are used to specify a standard cubical icon or a Bitmap (**.bmp*) image for the current object class. This icon will be displayed with the object class name in the **Map Explorer** tab. To apply the standard icon to the current object class, select the **<Standard Icon>** option from the drop-down list in the **Class Icon** area. Alternatively, select the **Use standard icon** check box in the **Class Icon** area. You can also use an image from an external location as an icon for the object class. To do so, use the **Browse** button to select the image; the selected image will be added in the drop-down list. You can use this image to use as a class icon.

Create method Area

The drop-down list in the **Create method** area is used to specify a geometry for creating objects by using the current object class. To specify a geometry for creating an object in the current object class, select the required option from the drop-down list in this area. If you do not want to create any object by using the current object class, select the **None** option from the drop-down list.

Feature Source Settings Tab

The options in the **Feature Source Settings** tab are used to specify whether to copy the linked data to the feature source or to copy the link of the connected data to the feature source. You can select the **Move my linked data to Feature Source** radio button, if it is not selected by default. If it is selected, then on adding an object with linked data, the data in the linked data source will be copied to the feature source. You can select the **Keep my data linked in Feature Source** radio button to copy the link to the feature source when an object with linked data is added to it.

Saving the Definition File

After specifying the properties in various tabs, you can save the modified settings in the definition file. To do so, choose the **Save Definition** button located at the bottom of the **Define Object Classification** dialog box; the specified classification properties will be saved in the definition file and the object class created will be added to the **Object Classes** folder in the **Map Explorer** tab.

Selecting Drawing Objects for Classification

You can select the classified, unclassified, or undefined drawing objects in the current drawing file for classification. The different methods of selecting the drawing objects are discussed next.

Selecting Classified Objects

Ribbon:	Create > Drawing Object > Select Classified
	Or Classification (Contextual tab) > Select > Classified Objects
Command:	MAPSELECTCLASSIFIED

You can select the objects that have already been classified in the current drawing file for classification. To do so, choose the **Select Classified** tool from the **Drawing Object** panel; the cursor will change into a crosshair in the drawing window. Now, you can select all drawing objects in the current drawing or select the preclassified drawing objects. To select the entire drawing object data in the current drawing file, specify * at the Command prompt and then press ENTER; all drawing objects will be selected.

You can also select the drawing objects pertaining to a specific object class by using the wildcard characters. For example, consider a drawing file with the object class **XYZ** predefined in it. To select this object class, enter the text **XYZ*** in the Command Line and then press ENTER; all drawing objects in this object class will be selected for the current classification process.

Selecting Unclassified Objects

Ribbon:	Create > Drawing Object > Select Unclassified
	Or Classification (Contextual tab) > Select > Unclassified Objects
Command:	MAPSELECTUNCLASSIFIED

You can select the drawing objects that are unclassified but are defined previously during the object classification process. To do so, choose the **Select Unclassified** tool from the **Drawing Object** panel; all drawing objects that are not subjected to any object classification will be selected.

Selecting Undefined Objects

Ribbon:	Create > Drawing Object > Select Undefined
	Or Classification (Contextual tab) > Select > Undefined Objects
Command:	MAPSELECTUNDEFINED

You can select the drawing objects that have been classified but are not defined in the current definition file. To select the undefined drawing objects, choose the **Select Undefined** tool from the **Drawing Object** panel of the **Create** tab; all drawing objects that are not defined in the current definition file will be selected.

If the current drawing file contains drawing objects that have been classified and saved with a different object class definition file, then on choosing the **Select Undefined** tool from the **Drawing Object** panel, the **Select Undefined** dialog box will be displayed. Figure 7-8 shows the **Select Undefined** dialog box with the description of various statements displayed in it. Next, choose the **OK** button in this dialog box; the **Select Undefined** dialog box will be closed and all drawing objects except the drawing objects used in different definition files will be selected. To select the drawing objects that have been defined in different definition files, add their respective definition files.

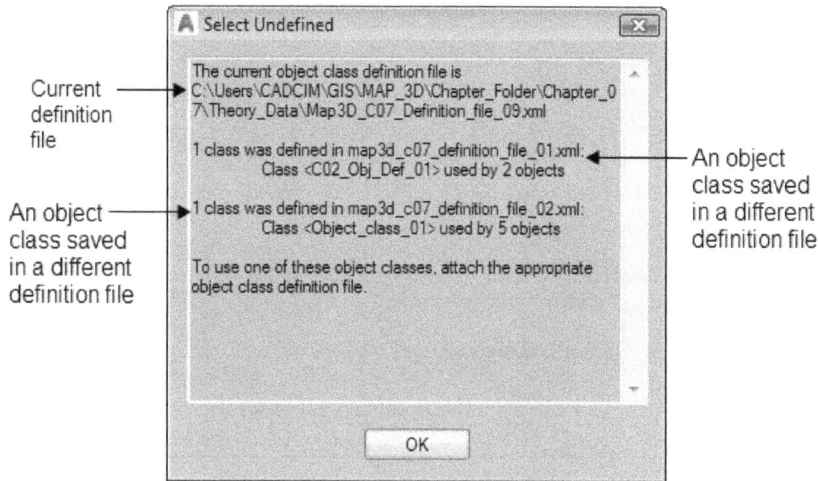

*Figure 7-8 The **Select Undefined** dialog box displaying the details of the statements*

Selecting the Drawing Objects Quickly Based on Their Property Values

You can select the drawing objects based on their property in the drawing window. Before proceeding with the quick selection, make sure that the basic environment for the object classification is set, as explained in the earlier section of this chapter. To perform the quick selection of the drawing objects, right-click in the drawing window; a shortcut menu will be displayed. In this shortcut menu, choose the **Quick Select** option; the **Quick Select** dialog box will be displayed, as shown in Figure 7-9.

In the **Quick Select** dialog box, you can specify properties, conditions, values, and so on for the selection criteria, based on the requirements of the object selection. In the **Apply to** drop-down list, the **Entire drawing** option is selected by default. As a result, the selection condition will be applied to the entire drawing.

Besides checking the selection condition for the entire drawing, you can select objects and then specify the filter (selection) condition. To use this filtering process, invoke the **Quick Select** dialog box, as explained in the previous section. Next, choose the **Select Objects** button displayed on the right of the **Apply to** drop-down list in this dialog box. On doing so, the cursor will change into a selection box in the drawing window and you will be prompted to select the objects in the drawing. Select the required objects and then press ENTER. Note that the **Current Selection** option is selected in the **Apply to** drop-down list of the **Quick Select** window.

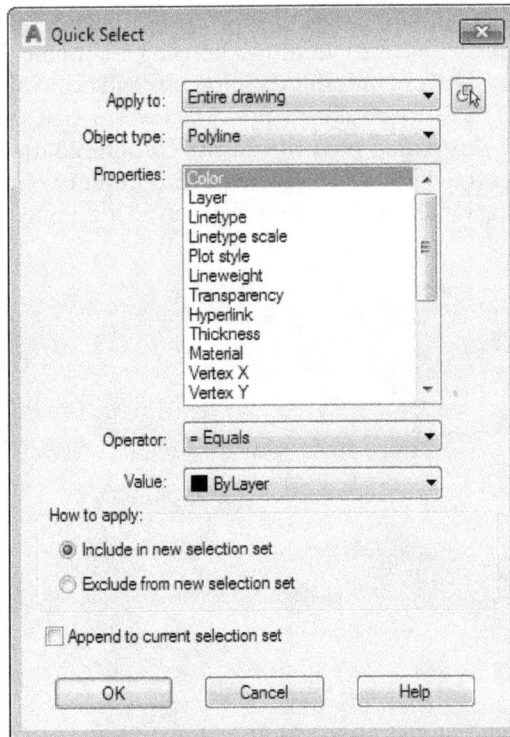

*Figure 7-9 The **Quick Select** dialog box*

After selecting the drawing objects, you will select an object type for this selection and then impose a filter condition on it. To select a specific type of object, select an option from the **Object Type** drop-down list; the properties corresponding to the selected option will be displayed in the **Properties** list box.

To impose a condition on a property of the object type, select the required property option in the **Properties** list box; the **Value** field below the **Operator** drop-down list will change accordingly. Next, select a suitable operator from the **Operator** drop-down list and then specify the required option in the **Value** field.

In the **How to apply** area, you can select the **Include in new selection set** radio button to create a drawing object selection set that satisfies the condition imposed. You can select the **Exclude from new selection set** radio button to remove that set from the selection set, which satisfies the condition imposed. The **Append to current selection set** check box is clear by default. As a result, the current selection set replaces the existing selection set. To attach the current selection set to the existing selection set, select the **Append to current selection set** check box.

After specifying all parameters required for the current selection set, choose the **OK** button from the **Quick Select** dialog box; the dialog box will be closed and the drawing objects that satisfy the imposed condition will be selected.

Classifying Drawing Objects

Ribbon: Create > Drawing Object > Classify
Or Classification (Contextual tab) > Classify > Classify Objects
Command: CLASSIFY

After selecting the required drawing objects for the object classification process, you can classify them using the object class definition file. To classify the selected drawing objects, choose the **Classify** tool from the **Drawing Object** panel of the **Create** tab; the **Classify** dialog box will be displayed.

In this dialog box, the **Include objects with missing or out of range property values** check box is selected by default. Clear this check box to exclude classification of the objects with out-of-range values. The **Exclude objects already tagged with a class name** check box in the **Classify** dialog box is also selected by default. As a result, objects that have an already assigned object class are not assigned the new class.

After specifying the required options, choose the **OK** button; the selected drawing objects will be classified, based on the properties defined in the attached definition file.

Unclassifying Drawing Objects

Ribbon: Classification (Contextual tab) > Classify > Unclassify Objects
Command: UNCLASSIFY

Unclassifying the drawing objects means removing the object class assigned to them. You can unclassify the classified drawing objects in the current drawing file. To do so, choose the **Unclassify Objects** tool from the **Classify** panel of the **Classification** contextual tab; you will be prompted to select objects. Note that the cursor will change into a selection box in the drawing window. Using this selection box, select the required drawing objects in the drawing window and press ENTER; the **AutoCAD** message box will be displayed, as shown in Figure 7-10. In this message box, choose the **OK** button to confirm the objects to be unclassified; the object class assigned to the selected drawing objects will be removed.

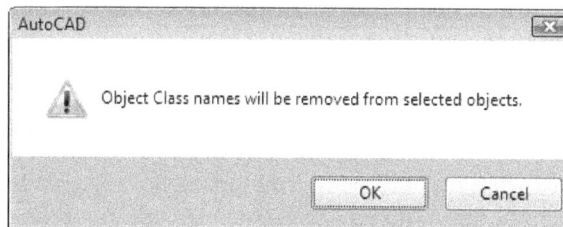

Figure 7-10 The AutoCAD message box

Verifying the Details of the Classified Data

You can verify the details of the classified data after the object classification has been performed on drawing objects. To verify the classification details, select a drawing object in the drawing window; the selected drawing object will be highlighted with points at each vertex. Then, right-click on the selected drawing object; a shortcut menu will be displayed. In this shortcut menu, choose the **Properties** option; the **PROPERTIES** palette will be displayed.

In the **PROPERTIES** palette, the **Design** tab is chosen by default. As a result, the properties of the drawing objects will be displayed. To display the object class assigned to the selected drawing object, choose the **Object Class** tab; the property categories with their respective properties will be displayed. If no data is displayed in the **Object Class** tab, then it implies that the drawing object is not classified.

WORKING WITH CLASSIFIED OBJECTS

An object class created by the object classification method is used to prepare a map for specific purpose. Using this object class, you can create and edit objects as well as create map and files with *.sdf* extension. In addition, you can generate the metadata file. Various methods of working with classified objects are discussed next.

Creating Objects by Using Object Classes

You can create objects by using the existing object classes, as explained in the previous section. To create objects by using an existing object class, select the required object class sub-node in the **Object Classes** node in the **Map Explorer** tab of the **TASK PANE**, and then right-click on it; a shortcut menu will be displayed. In this shortcut menu, choose the **Create Classified Objects** option; the cursor will change into a crosshair in the drawing window. Using this crosshair, you can create drawing objects as discussed earlier.

Note

While creating objects by using an object class, the **Create Classified Objects** *option will be activated only if you have selected an option other than* **None** *from the* **Create method** *drop-down list in the* **Class settings** *tab of the* **Define Object Classification** *dialog box.*

Editing an Existing Object Class

You can edit an existing object class by specifying new values for the object classification. To do so, select an object class sub-node from the **Object Classes** node in the **Map Explorer** tab of the **TASK PANE** and then right-click on it; a shortcut menu will be displayed. In this shortcut menu, choose the **New Object Class** option; the crosshair will change into a selection box in the drawing window. Using this selection box, select an object in the drawing window, and then press ENTER; the **Define Object Classification** dialog box will be displayed. In this dialog box, you can specify various parameters for customizing the object class, as discussed in the previous section.

DISPLAYING AND SHARING THE METADATA FILE

Map 3D software automatically generates the metadata file for the current drawing and its object class definition. Moreover, you can share this file with any other user. In the metadata file, you can read information about the object classification as well as the coordinate systems assigned to the current drawing. The metadata file is very helpful to learn about the source, type of classification, coordinate system, and general properties of the drawing file. The methods of generating the metadata and sharing it are discussed next.

Including and Sharing the Metadata File

The metadata file is automatically generated by the AutoCAD Map 3D software and can be read by using the **METADATA VIEWER** window. To invoke this window, right-click on the **Current Drawing** node in the **Map Explorer** tab of the **TASK PANE**; a shortcut menu will be displayed. In this menu, choose the **View Metadata** option; the **METADATA VIEWER** window will be displayed. You can include a metadata file from another drawing file or export it to other drawing system by using the tools displayed in the top row of the **METADATA VIEWER** window. Figure 7-11 shows the tools in the **METADATA VIEWER** window. These tools are discussed next.

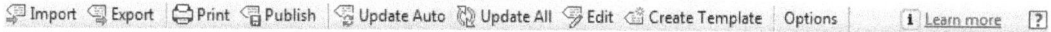

Figure 7-11 *The tools in the **METADATA VIEWER** window used for sharing the metadata file*

Import

The **Import** tool is used to import a metadata file prepared for different drawing files. To import a metadata file, choose the **Import** tool; the **Open** dialog box will be displayed. In this dialog box, browse to the folder containing the required metadata file and then select the file in it. Next, choose the **OK** button; the selected metadata file will be displayed in the **Metadata** tab.

Export

The **Export** tool is used to share the metadata file of the current drawing with other drawing files. The exported metadata file can be reused by other drawing files, and a new drawing file can be defined based on the current drawing file.

Print

The **Print** tool is used to print the content of the metadata file on a paper source.

Publish

The **Publish** tool is used to save the metadata information in the *.xml*, *.html*, or *.txt* format.

Update Auto

The **Update Auto** tool is used to update fields automatically.

Update All

The **Update All** tool is used to update all the fields automatically or manually.

Edit

The **Edit** tool is used to edit the content of a metadata file. To edit the content of the metadata file, choose the **Edit** tool; the **Metadata Editor (FGDC) - Current DWG** dialog box will be displayed, as shown in Figure 7-12.

In the **Metadata Editor (FGDC) - Current DWG** dialog box, the **Identification** tab is chosen by default. As a result, the general properties in this tab will be displayed. In this tab, you can specify various properties of the metadata file. To modify the existing value of a property, click in the cell corresponding to that property; the cell will become active. In this cell, enter the required parameters of the metadata file.

In this dialog box, there are other tabs such as **Data Quality**, **Data Organization**, **Spatial Reference**, and so on. You can also modify the properties in these tabs. To do so, choose the required tab in this dialog box; the properties in the chosen tab will be displayed. Then, click in the cell corresponding to the property to be edited; the cell will become editable. In this cell, enter the text or value. Note that when you select a property for editing by clicking in the cell, its description will be displayed in the area below the tabs of this dialog box.

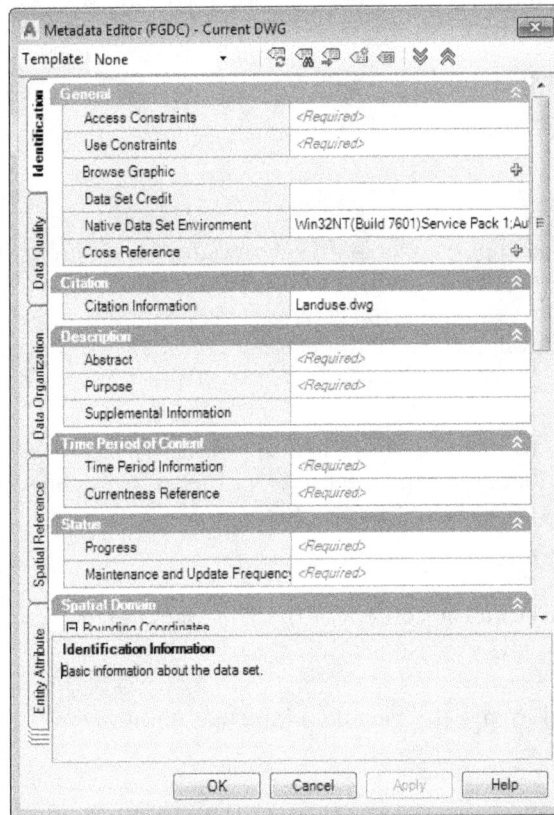

Figure 7-12 The Metadata Editor (FGDC) - Current DWG dialog box

After specifying the values for all the required properties in the **Metadata Editor** dialog box, choose the **OK** button; the **Metadata Editor** dialog box will be closed and the added or modified property values will be updated in the **Metadata** tab.

Create Template

The **Create Template** tool is used to save the current metadata settings as a template file. To do so, choose the **Create Template** tool from the top row; the **Create Metadata Template** window will be displayed, as shown in Figure 7-13. By default, the **Current DWG** text is displayed in the **Template** edit box of this window. Specify a name for the template to be created and then choose the **OK** button; the window will be closed and a template file with the specified name will be saved.

Options

The **Options** tool is used to specify the parameters of the current metadata file. To do so, choose this tool from the **METADATA VIEWER** window; the **Metadata Options** dialog box will be displayed, as shown in Figure 7-14.

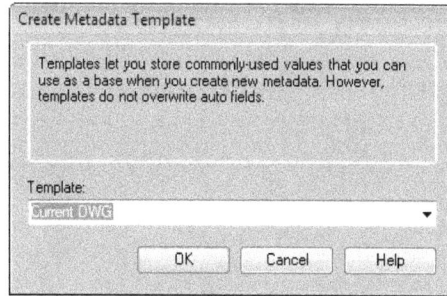

Figure 7-13 The **Create Metadata Template** *window*

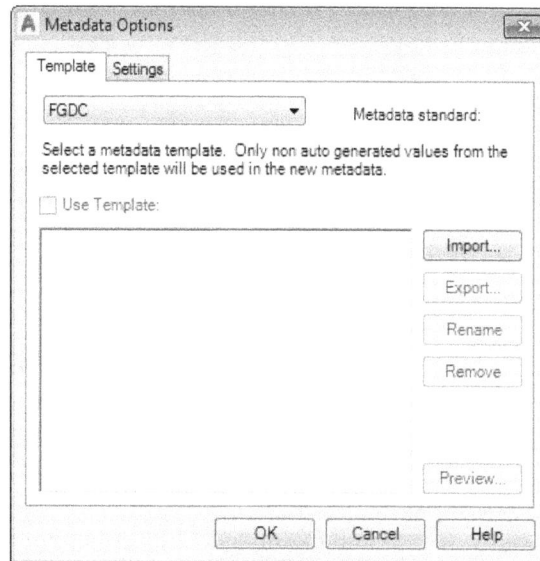

Figure 7-14 The **Metadata Options** *dialog box*

In the **Metadata Options** dialog box, the **Template** tab is chosen by default and the options in this tab are displayed. You can specify the format for the default metadata representation by selecting an option from the **Metadata standard** drop-down list in this tab.

To select an existing template from the list box, select the **Use Template** check box. If there is an existing template, then it will be displayed with a radio button in the list box below the **Use Template** check box. You can select that radio button to use the corresponding template.

In the **Settings** tab of the **Metadata Options** dialog box, you can specify the **Latitude** and **Longitude** precision of the metadata. You can also specify whether or not to update the metadata while selecting the data source.

After specifying all required parameters for the metadata, choose the **OK** button; the settings of the selected template or the modified settings will be applied to the current metadata.

Displaying the Metadata Information

You can display the metadata information of an entire drawing by using the **METADATA VIEWER** window. To do so, invoke the **METADATA VIEWER** window, as explained earlier; the information about the entire drawing will be displayed in this window, refer to Figure 7-15.

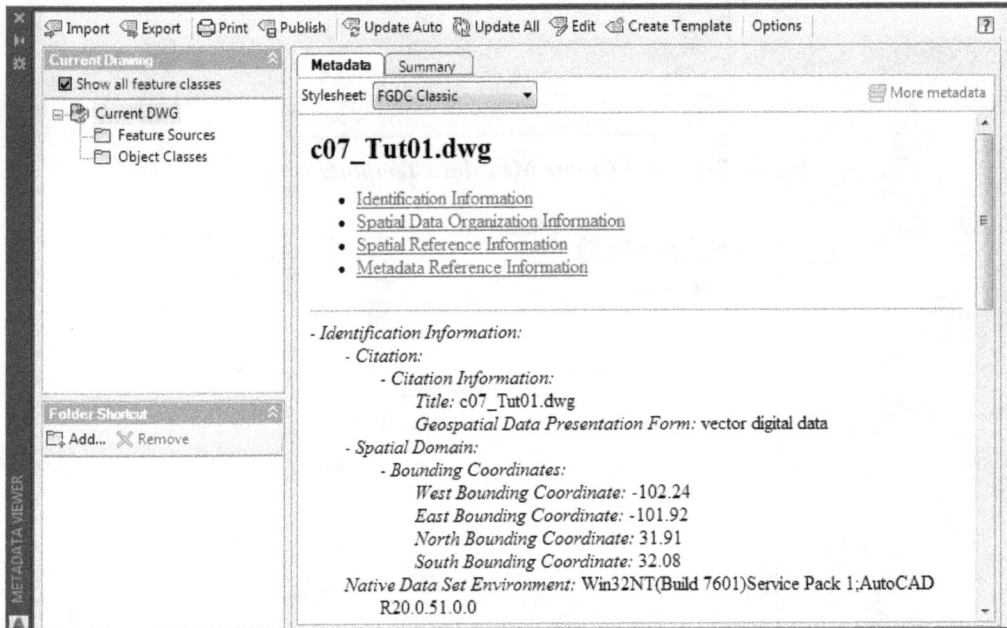

*Figure 7-15 The **METADATA VIEWER** window displaying the information of a sample drawing*

Metadata Tab

The **Metadata** tab is chosen by default and the metadata information of the current drawing is displayed in this tab. To display the metadata in the FGDC Classic, XML, or text format, select an option from the **Stylesheet** drop-down list located on the top row of the **Metadata** tab. On doing so, the metadata information will be displayed in the selected file format.

Summary Tab

When you choose the **Summary** tab, the general information regarding the current drawing file will be displayed in this tab.

Current Drawing Area

The options in the **Current Drawing** area are used to display information about a unique entity such as drawing, object class, feature, and so on in the current drawing file. To display the metadata pertaining to a unique item in the **Feature Source** or **Object Classes** node in this area, expand the node by clicking on the corresponding [**+**] symbol; a list of the feature class or object class data in the expanded node will be displayed. In this expanded node, select the required option; the metadata information corresponding to the selected entity will be displayed on the right of the **Metadata** tab.

Folder Shortcut Area

In the **Folder Shortcut** area, you can add folders that contain the required drawing entity, and then select the required item in the added folder to read information about its metadata. To add a folder containing a drawing entity, choose the **Add** button in the **Folder Shortcut** area; the **Browse For Folder** dialog box will be displayed. In this dialog box, browse to a folder location and then select the required folder from the **Select the folder that you want to connect** list box. Next, choose the **OK** button from the **Browse For Folder** dialog box; the dialog box will be closed and the selected folder will be added to the **Folder Shortcut** area. Expand the added folder and then select the required file by clicking on it; the metadata information of the selected drawing will be displayed on the right of the **Metadata** tab.

TUTORIALS

General instructions for downloading tutorial files:

Before starting the tutorials, you need to download the tutorial data to your computer. To do so, follow the steps given below:

1. Log on to *www.cadcim.com* and browse to *Textbooks > Civil/GIS > Map 3D > Exploring AutoCAD Map 3D 2017*. Next, select *c07_m3d_2017_tut.zip* file from the **Tutorial Files** drop-down list. Next, choose the corresponding **Download** button to download the data file.

2. Extract the content of the zip file to the following location:

    C:\m3d_2017

Notice that the *c07_m3d_2017_tut* folder is created within the *m3d_2017* folder.

Tutorial 1   Classifying Drawing Objects

In this tutorial, you will classify the drawing objects into different object classes.

(Expected time: 1 hr)

The following steps are required to complete this tutorial:

a. Open the tutorial drawing file and assign the **TX83-CF NAD83 Texas State Planes, Central Zone, US Feet** global coordinate system to the current drawing.
b. Login as a SUPERUSER.
c. Create and attach a new definition file for object classification.
d. Define the object class for the object classification by specifying the constraints as follows:

Property	Constraints to be imposed on the Object Class
Layer	Road_Edge_Unpaved
Color	Brown (32)
Length	1 to 11500

e.   Classify objects.
f.   Verify the results.
g.   Save the drawing file.

Opening the Tutorial Drawing File

In this section of the tutorial, you will open the required drawing file. Next, you will verify the geographic coordinate system of the drawing.

1.   Choose the **Open** button from the Quick Access Toolbar; the **Select File** dialog box is displayed.

2.   In the **Select File** dialog box, browse to the following location:

     C:\m3d_2017\c07_m3d_2017_tut\c07_tut01

3.   Select the **c07_Tut01.dwg** file from the list box; preview of the selected drawing is displayed in the **Preview** area.

4.   Choose the **Open** button in the **Select File** dialog box; the dialog box is closed and the drawing is displayed in the drawing window, as shown in Figure 7-16.

Figure 7-16 The street drawing displayed in the drawing window

5.   Make sure that the drawing has been assigned **TX83-CF** as its geographic coordinate system. The code of the assigned coordinate system is displayed in the **Drawing Status Bar**.

Tip
*You can assign the required coordinate system to the drawing by using the options in the **Coordinate System - Assign** dialog box. To invoke this dialog box, choose the **Assign** tool from the **Coordinate System** panel of the **Map Setup** tab.*

Logging in as a SUPERUSER

1.   Choose the **Map Explorer** tab in the **TASK PANE**; various options in this tab are displayed.

2. In this tab, right-click on the **Current Drawing** node; a shortcut menu is displayed. In this menu, choose the **User Login** option; the **User Login** window is displayed.

3. In the **User Login** window, enter **SUPERUSER** in both the **Login Name** and **Password** text boxes. Note that the password is case sensitive. It is therefore required to be entered as specified.

4. Choose the **OK** button from this window; the window is closed and you are logged in as a **SUPERUSER**.

Creating and Attaching a Definition File

1. In the **Map Explorer** tab of the **TASK PANE**, right-click on the **Object Classes** node; a shortcut menu is displayed. In this menu, choose the **New Definition File** option; the **New Object Class Definition File** dialog box is displayed.

2. In this dialog box, enter **c07-m3d-2017-tut01** in the **File name** edit box and ensure that the **Object Definition File (*.xml)** option is selected in the **Files of type** drop-down list.

3. Make sure that **c07_tut01** (tutorial data) folder is selected in the **Save in** drop-down list. Next, choose the **Save** button in the **New Object Class Definition File** dialog box; the dialog box is closed and a new object class definition file is created with the name **c07-m3d-2017-tut01.xml** at the specified location.

4. In the **Map Explorer** tab of the **TASK PANE**, right-click on the **Object Classes** node; a shortcut menu is displayed. In this shortcut menu, choose the **Attach Definition File** option; the **Attach Object Class Definition File** dialog box is displayed.

5. In the **Attach Object Class Definition File** dialog box, browse to the location *C:\m3d_2017\ c07_m3d_2017_tut\c07_tut01* and select the **c07-m3d-2017-tut01.xml** file.

6. Next, choose the **Open** button; the **Attach Object Class Definition File** dialog box is closed and the *c07-m3d-2017-tut01.xml* definition file is used for the current object classification process.

Defining the Object Class

1. In the **Map Explorer** tab of the **TASK PANE**, right-click on the **Object Classes** node; a shortcut menu is displayed. In this menu, choose the **Define Object Class** option; you are prompted to select the example object. Also, note that the cursor changes into a selection box in the drawing window.

2. Click on one of the brown colored polyline in the drawing window; the drawing object is highlighted for selection, as shown in Figure 7-17.

Tip
Figure 7-17 shows only an example of how to select a drawing object for classification. You can select any of the drawing objects in the current drawing for the object classification. You can also select more than one object as sample object.

Figure 7-17 Selecting an example object

3. Next, press ENTER; the **Define Object Classification** dialog box is displayed, as shown in Figure 7-18.

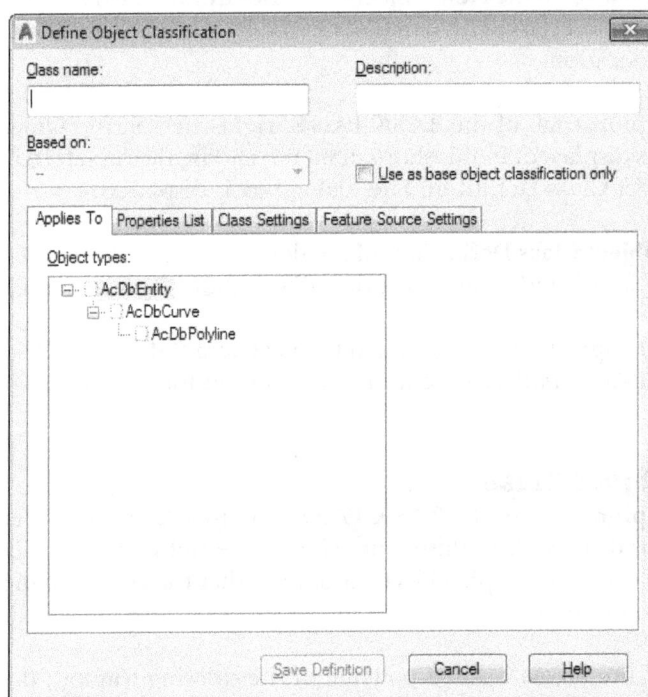

*Figure 7-18 The **Define Object Classification** dialog box*

4. In the **Define Object Classification** dialog box, enter **Object-Class-01** in the **Class name** edit box and select the check box corresponding to the **AcDbEntity** node in the **Object types** area of the **Applies To** tab.

5. Next, choose the **Properties List** tab from the dialog box; the options corresponding to the specified node are displayed.

6. In the **Properties List** tab, select the check box corresponding to the **Color** node in the **Available properties** list box; the attributes of the **Color** property are displayed in the **Property Attributes** rollout.

7. In this rollout, click in the cell corresponding to the **Range** option; the **Range** option is selected and the Browse button is displayed in the cell corresponding to it. Choose the Browse button; the **Color Range Editor** dialog box is displayed, as shown in Figure 7-19.

Figure 7-19   *The* **Color Range Editor** *dialog box*

8. In the **Color range** area of this dialog box, select the **Add a specific color** radio button; the **Color** drop-down list below this radio button is activated.

9. Select the **Select Color** option from the **Color** drop-down list; the **Select Color** palette is displayed.

10. In the **Select Color** palette, enter **32** in the **Color** text box.

11. Next, choose the **OK** button; the **Select Color** palette is closed and the **Color 32** option is displayed in the **Color** drop-down list of the **Color Range Editor** dialog box.

12. Next, choose the **Add** button from the **Color Range Editor** dialog box; the **All Colors** option in the **List of colors** list box is replaced with the **Color 32** option.

13. Next, choose the **OK** button in the **Color Range Editor** dialog box; the dialog box is closed and the value of the selected color is displayed in the **Range** property in the **Property Attributes** rollout.

14. In the **Define Object Classification** dialog box, select the **Layer** check box in the **Available properties** list box; the attributes of the **Layer** property are displayed in the **Property Attributes** rollout.

15. Click in the cell corresponding to the **Range** option in the **Property Attributes** rollout; the **Range** option is selected and the Browse button is displayed in the cell. Choose the Browse button; the **Layer Range Editor** dialog box is displayed, as shown in Figure 7-20.

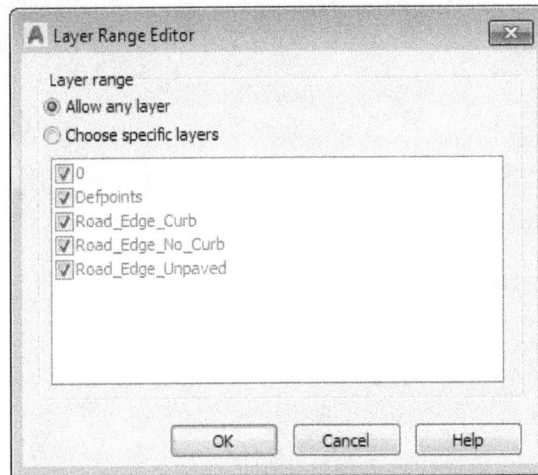

*Figure 7-20 The **Layer Range Editor** dialog box*

16. In the **Layer range** area of the **Layer Range Editor** dialog box, select the **Choose specific layers** radio button; the list box below this radio button is activated.

17. Clear all the check boxes in the list box except the **Road_Edge_Unpaved** check box.

18. Next, choose the **OK** button in the **Layer Range Editor** dialog box; the dialog box is closed and the selected **Road_Edge_Unpaved** option is displayed in the **Range** option of the **Property Attributes** rollout of the **Define Object Classification** dialog box.

19. In the **Define Object Classification** dialog box, expand the **Geometry** category in the **Available properties** list box by clicking on the [+] symbol; the properties in the **Geometry** category are displayed.

20. In the **Geometry** category, select the check box corresponding to the **Length** node; the properties of this node are displayed in the **Property Attributes** rollout on the right side.

21. In the **Property Attributes** rollout, click in the cell corresponding to the **Range** option; the **Range** option is selected and the cell becomes editable.

22. Enter [1,1500] in this cell and then press ENTER; the range for the **Length** property is set.

23. Choose the **Class Settings** tab; the options in this tab are displayed.

24. In the **Create method** area of the **Class Settings** tab, select the **Polyline** option from the drop-down list.

25. Retain the default settings of all other options and then choose the **Save Definition** button in the **Define Object Classification** dialog box; the dialog box is closed and the **Object-Class-01** object class is added to the **Object Classes** node in the **Map Explorer** tab of the **TASK PANE**.

Classifying Drawing Objects

1. In the **Map Explorer** tab of the **TASK PANE**, right-click on the **Object-Class-01** object class; a shortcut menu is displayed. In this menu, choose the **Classify Objects** option; the **Classify Objects** window is displayed, as shown in Figure 7-21.

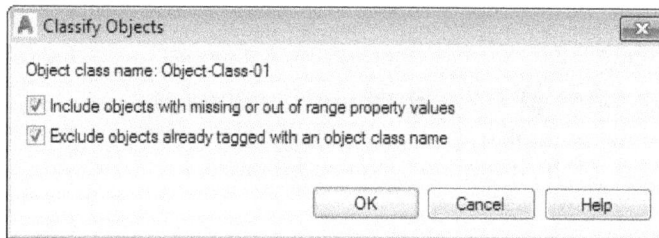

Figure 7-21 The Classify Objects window

2. In the **Classify Objects** window, keep the default settings and choose the **OK** button; the **Classify Objects** window is closed and the cursor changes into a selection box in the drawing window. Using this selection box, select the drawing objects shown in the zoomed portion of Figure 7-22.

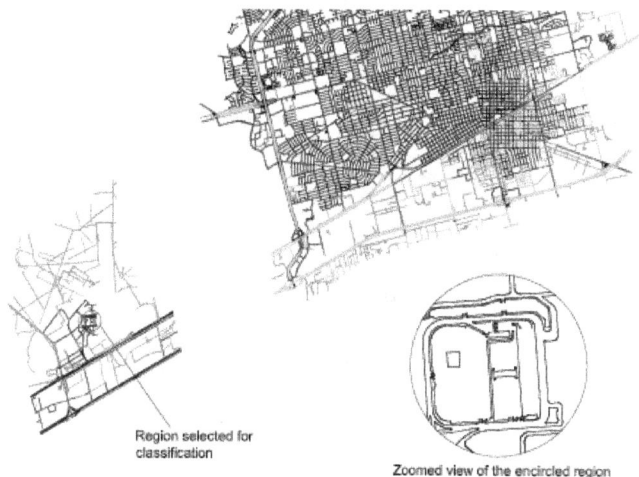

Figure 7-22 The region selected for object classification

3. After selecting all drawing objects, right-click in the drawing window; the selected objects are classified based on the **Object-Class-01** object class. Also, the color of the selected drawing objects changes from black to brown, refer to Figure 7-23.

Figure 7-23 Classified drawing objects displayed in brown color

Verifying the Results of Object Classification

1. In the classified object region (object represented by brown color, refer to Figure 7-23), place the cursor over one of the drawing objects and then click; the drawing object gets selected and highlighted by blue vertices.

2. With the object selected, right-click; a shortcut menu is displayed. In this shortcut menu, choose the **Properties** option; the **PROPERTIES** palette is displayed with the **Design** tab chosen, as shown in Figure 7-24.

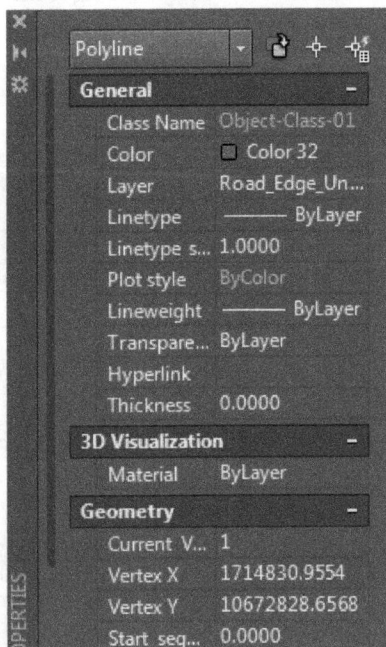

*Figure 7-24 The **PROPERTIES** palette with the **Design** tab chosen*

3. In the **PROPERTIES** palette, choose the **Object Class** tab; the object properties of the selected drawing are displayed, as shown in Figure 7-25.

    Note that the value of the length will vary depending upon the object you will be selecting.

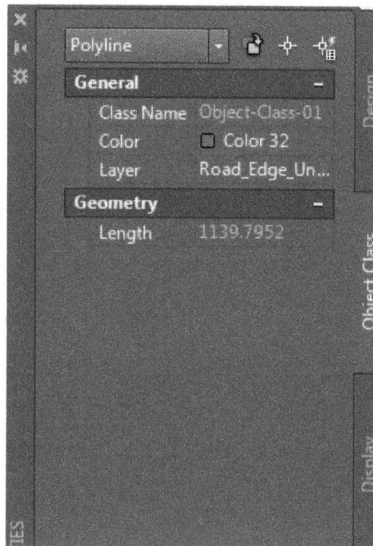

*Figure 7-25 The **PROPERTIES** palette with the **Object Class** tab chosen*

In order to verify that no other drawing object is selected outside the classified region, select a drawing object from the region outside the classified object region. You will notice that the **Object Class** tab in the **PROPERTIES** palette is empty. This shows that no object has been classified outside the classified region.

Saving the Drawing File

1. Choose the **Save As** option from the Application Menu; the **Save Drawing As** dialog box is displayed.

2. In the **Save Drawing As** dialog box, enter **c07_Tut01a.dwg** in the **File name** edit box and select the **AutoCAD 2013 Drawing (*.dwg)** option from the **Files of type** drop-down list, if it is not selected by default. Next, choose the **Save** button next to the **File name** edit box; the current drawing file is saved with the given name.

Tutorial 2   Creating Classified Objects

In this tutorial, you will create classified objects in an existing object class.

(Expected time: 30 min)

The following steps are required to complete this tutorial:

a. Open the drawing file
b. Log in as a SUPERUSER.

c.   Create an object, as indicated by the thick lines **AB**, **BC**, and **CD** in Figure 7-26, by using the **Object-Class-01** object class.

d.   Save the drawing file.

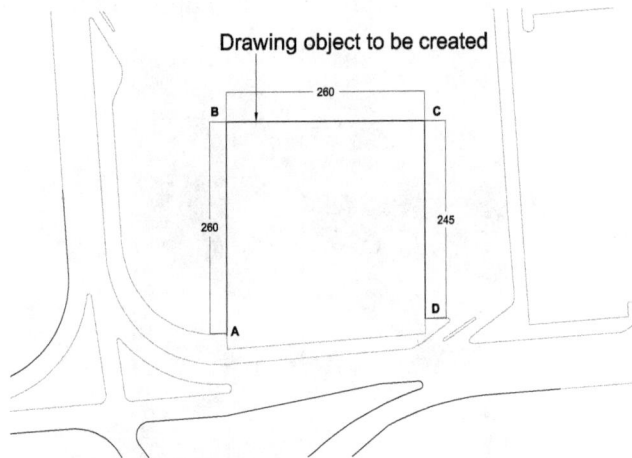

*Figure 7-26 The drawing object to be created by using the **Object-Class-01** object class*

Opening the Drawing File

1.   Choose the **Open** button from the Quick Access Toolbar; the **Select File** dialog box is displayed.

2.   In this dialog box, browse to the location:

     C:\m3d_2017\c07_m3d_2017_tut\c07_tut02

3.   Select the **c07_Tut02.dwg** file from the list box; preview of the selected drawing is displayed in the **Preview** area.

4.   Choose the **Open** button from this dialog box; the dialog box is closed and the drawing is displayed in the drawing window. Note that the **Object-Class-01** object class is displayed in the **Object Classes** node in the **Map Explorer** tab of the **TASK PANE**.

Logging in as a SUPERUSER

1.   In the **TASK PANE**, choose the **Map Explorer** tab, if not chosen by default; various options in this tab are displayed.

2.   Right-click on the **Current Drawing** node in this tab; a shortcut menu is displayed. In this shortcut menu, choose the **User Login** option; the **User Login** window is displayed.

3.   In the **User Login** window, enter the text **SUPERUSER** in the **Login Name** and **Password** edit boxes. Note that the password is case sensitive and should be entered as specified in the step.

4. Choose the **OK** button in this window; the **User Login** window is closed and the command prompt displays the message that you have successfully logged in as the **SUPERUSER**.

Creating a New Drawing Object by Using the Object Class

1. In the **Map Explorer** tab of the **TASK PANE**, right-click on **Object-Class-01** in the **Object Classes** node; a shortcut menu is displayed. In this shortcut menu, choose the **Create Classified Objects** option; you are prompted to specify the start point of the object to be created. Also, the cursor changes into a crosshair in the drawing window.

2. Before proceeding further, make sure the **Ortho Mode** and **Object Snap** buttons in the Status Bar are active. You can use the F8 key to toggle the **Ortho Mode.**

> **Tip**
> *Use the F8 and F3 keys on the keyboard to toggle the **Ortho Mode** and the **Snap Mode**, respectively.*

3. Place the crosshair on the vertex **A**, refer to Figure 7-27, of the new drawing object to be created and then click on it; the first point of the line object is selected.

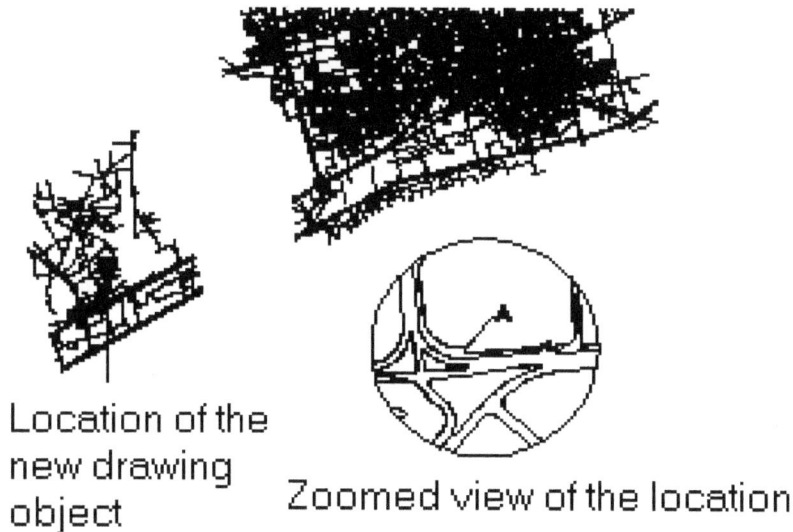

Location of the new drawing object   Zoomed view of the location

Figure 7-27 The region selected for object classification

4. Drag the crosshair vertical upward and then enter **260** in the Command prompt. Next, press ENTER; the **AB** line segment with the length **260** is drawn.

5. Move the cursor horizontally to the right, refer to Figure 7-28, and then enter **260** in the Command prompt. Next, press ENTER; the **BC** line segment is created.

6. Move the cursor vertically down and click on the vertex point, refer to Figure 7-29; the **CD** line segment is drawn. Next, right-click in the drawing area; a shortcut menu is displayed.

Choose the **Enter** option from the displayed menu; the object is created, as shown in Figure 7-30.

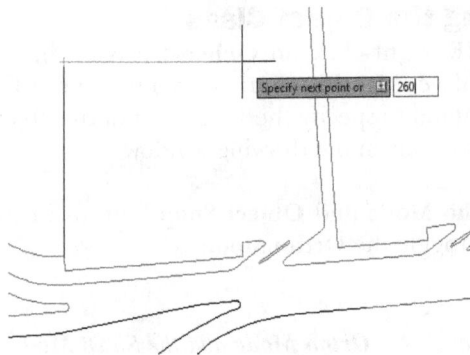

*Figure 7-28 Drawing the **BC** line segment*

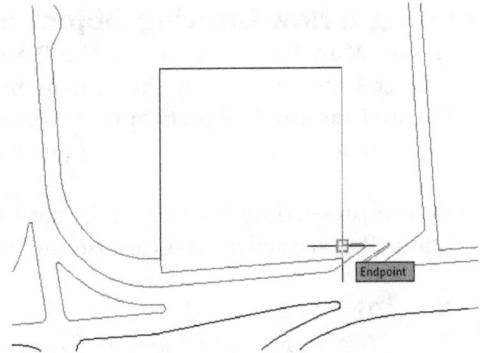

*Figure 7-29 Drawing the **CD** line segment*

Figure 7-30 A new drawing object created

7. Next, press ESC; you exit the drawing mode.

Note
*The color property of the **Object-Class-01** object class is specified as brown. The objects created in this object class will inherit the properties defined for this class and therefore will be displayed in brown color in the drawing window.*

Saving the Drawing File

1. Choose the **Save As** option from the Application Menu; the **Save Drawing As** dialog box is displayed.

2. In the **Save Drawing As** dialog box, enter **c07_Tut02a.dwg** in the **File name** edit box and select the **AutoCAD 2013 Drawing (*.dwg)** option from the **Files of type** drop-down list, if it is not selected by default. Next, choose the **Save** button next to the **File name** edit box; the current drawing file is saved with the given name.

Tutorial 3   Creating the Metadata File

In this tutorial, you will create the metadata file for the object class in the current drawing file.   **(Expected time: 30 min)**

The following steps are required to complete this tutorial:

a.   Open the drawing file.
b.   Log in as a SUPERUSER.
c.   Generate the metadata file for the drawing file in the .*XML* format
d.   Save the metadata file for the **Object-Class-01** object class in the .*TXT* format.

Opening and Saving the Tutorial Drawing File

1.   Choose the **Open** button from the Quick Access Toolbar; the **Select File** dialog box is displayed.

2.   In this dialog box, browse to the location:

     C:\m3d_2017\c07_m3d_2017_tut\c07_tut03

3.   Select the **c07_Tut03.dwg** file from the list box; preview of the selected drawing is displayed in the **Preview** area.

4.   Choose the **Open** button from this dialog box; the dialog box is closed and the drawing is displayed in the drawing window. Note that the **Object-Class-01** object class is displayed in the **Object Classes** node in the **Map Explorer** tab of the **TASK PANE**.

Logging in as a SUPERUSER

1.   In the **TASK PANE**, choose the **Map Explorer** tab if it is not chosen by default; various options in this tab are displayed.

2.   Right-click on the **Current Drawing** node in this tab; a shortcut menu is displayed. In this shortcut menu, choose the **User Login** option; the **User Login** window is displayed.

3.   In the **User Login** window, enter the text **SUPERUSER** in the **Login Name** edit box as well as in the **Password** text box.

4.   Choose the **OK** button in this window; the **User Login** window is closed and the command prompt displays the message that you have successfully logged in as the **SUPERUSER**.

Generating the Metadata File

1.   In the **Map Explorer** tab of the **TASK PANE**, right-click on the **Current Drawing** node; a shortcut menu is displayed. In this menu, choose the **View Metadata** option; the **METADATA VIEWER** window is displayed, as shown in Figure 7-31.

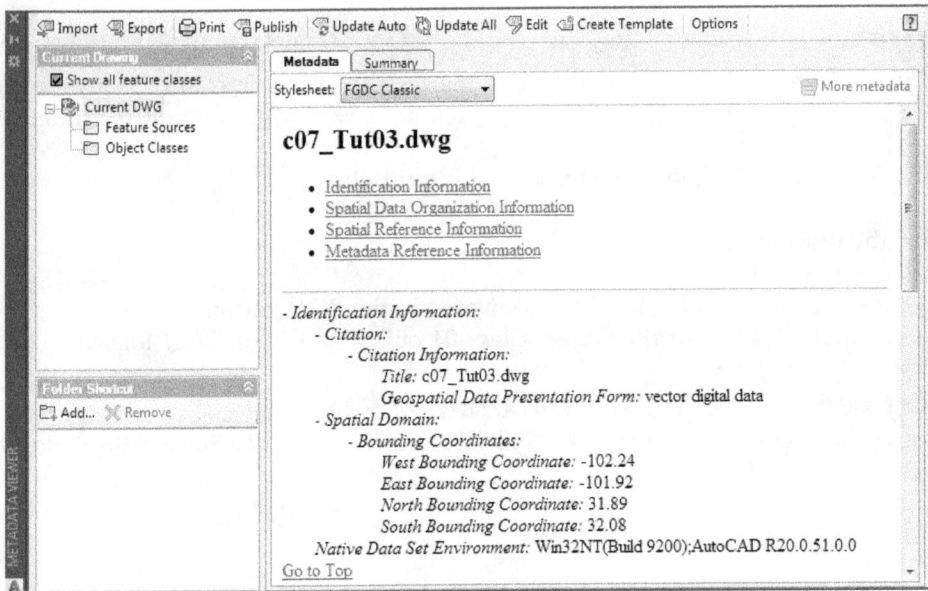

*Figure 7-31 The **METADATA VIEWER** window with the **Metadata** tab chosen*

2. In this window, choose the **Publish** tool; the **Save As** dialog box is displayed.

3. In this dialog box, browse to the following location:

    C:\m3d_2017\c07_m3d_2017_tut\c07_tut03

4. Enter **c07-m3d-2017-tut03** in the **File name** edit box and select the **XML File (*.xml)** option from the **Files of type** drop-down list in this dialog box. Next, choose the **Save** button; the metadata file is saved with the *.xml* format.

5. In the **Current Drawing** list box of the **METADATA VIEWER** window, left-click on the [**+**] symbol corresponding to the **Object Classes** node; the **Object-Class-01** sub-node is displayed below this node.

6. Select the **Object-Class-01** sub-node in the **Object Classes** node; the metadata information of the **Object-Class-01** object class is displayed in the **Metadata** tab.

7. In the **METADATA VIEWER** window, choose the **Publish** tool; the **Save As** dialog box is displayed.

8. In the **Save As** dialog box, enter **c07tut03-ObjectClass01** in the **File name** edit box and select the **Text File (*.txt)** option from the **Save as type** drop-down list.

9. Next, choose the **Save** button in the **Save As** dialog box; the metadata file for the selected object class is saved at the selected location in the *.txt* format and close the **METADATA VIEWER** window.

Self-Evaluation Test

Answer the following questions and then compare them to those given at the end of this chapter:

1. Which of the following options is used to bring in a metadata file into the **METADATA VIEWER** window?

    (a) **Import**   (b) **Publish**
    (c) **Export**   (d) **Update All**

2. Which of the following tabs in the **Define Object Classification** dialog box has the **New Property** button that is used to create a user defined property?

    (a) **Applies To**   (b) **Properties List**
    (c) **Class Settings**   (d) **Feature Source Settings**

3. Which of the following tools in the **Classification** (contextual) tab is used to select the objects that are not classified?

    (a) **Classify Objects**   (b) **Unclassified Objects**
    (c) **Unclassify Objects**   (d) **Classified Objects**

4. Which of the following tools in the **Map Setup** tab of the Ribbon is used to create an object class?

    (a) **New Definition**   (b) **Define Object Data**
    (c) **Define**   (d) **Attach Definition**

5. Which of the following user privileges can be assigned by an administrator to control the access to AutoCAD Map 3D?

    (a) **Superuser**   (b) **Edit Drawing**
    (c) **Alter Object Class**   (d) All of the above

6. You can perform the object classification procedure by using the tools in the _____ contextual tab.

7. You can specify the parameters of an object class in the _____ dialog box.

8. The **Define** tool in the _____ tab of the Ribbon is used to create a new object class.

9. In AutoCAD Map 3D, you can create a new user login name and set the user privileges by using the options in the _____ dialog box.

10. Object classification means classifying the drawing objects into sub-classes. (T/F)

11. The user does not need privileges to classify objects. (T/F)

12. The object classification procedure always requires a definition file. (T/F)

13. A Superuser in AutoCAD Map 3D has administration privileges. (T/F)

14. To create a classified object, you need to specify the object to be created by selecting the option in the **Create Method** drop-down list of the **Class Settings** tab in the **Define Object Classification** dialog box. (T/F)

15. You can customize a metadata file based on the requirements of a user. (T/F)

Review Questions

Answer the following questions:

1. Which of the following default administrator login name and password is used in AutoCAD Map 3D?

    (a) ADMINISTRATOR   (b) OBJECTCLASS
    (c) AutoCAD   (d) SUPERUSER

2. Which of the following tools is used to save the metadata file in the *.html* format?

    (a) **Import**   (b) **Print**
    (c) **Publish**   (d) **Edit**

3. In the _____ tab of the **Define Object Classification** dialog box, you can specify the icon to be displayed for the required object class. (T/F)

4. You can remove the object class attached to an object by using the _____ tool in the **Classify** panel of the **Classification** tab.

5. Classifying drawing objects will enforce the parameters defined for the object class on the classified objects. (T/F)

6. You do not need to attach the definition file to perform the object classification procedure. (T/F)

7. You can standardize the drawing by classifying the drawing objects into various object classes. (T/F)

8. You can reclassify an existing object class. (T/F)

9. You can specify the precision of latitude and longitude in the metadata file. (T/F)

10. The metadata file can be prepared for a specific entity. (T/F)

EXERCISES

Exercise 1

Download the *c07-m3d-2017-exr01.dwg* file from *www.cadcim.com*. Using this file, perform the object classification and generate the metadata file. To perform the object classification, use the drawing objects, refer to Figure 7-32, and set the following parameters:

Color: Blue
Total area range: 400000 to 500000

After performing the object classification, save the drawing with the name *c07-m3d-2017-exr01a. dwg*. **(Expected time: 1 hr)**

Zoomed view of the object
used for classification

Figure 7-32 *The object used for classification*

Exercise 2

Open the *c07-m3d-2017-exr02.dwg* file saved in the previous exercise and create a polyline drawing object, as shown in Figure 7-33, using the values for the coordinates of the vertices of the polyline object as given below: **(Expected time: 30 min)**

Points	X	Y
O	1728570.5363	10699718.1880
A	1728570.8244	10700003.5794
B	1728962.5710	10700003.5794
C	1728962.5710	10699628.7034
D	1728519.0719	10699628.5567

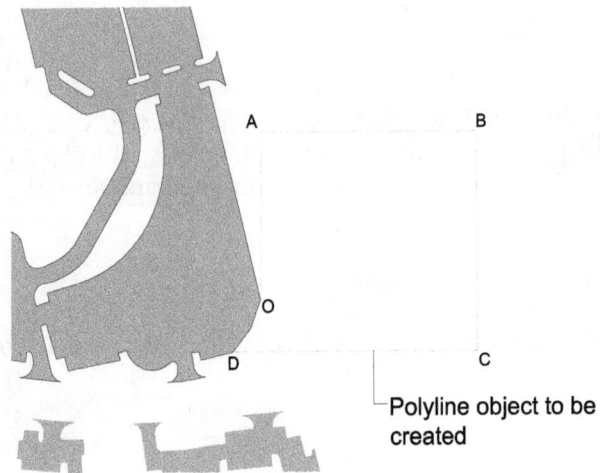

Figure 7-33 *Location of the polygon object to be created*

Chapter 8

Removing Digitization Errors and Working with Topologies

Learning Objectives

After completing this chapter, you will be able to:

- *Specify objects for drawing cleanup*
- *Select cleanup actions*
- *Set parameters of error marker*
- *Define topology rules*
- *Create topologies from drawing objects*
- *Define and perform a topology query*
- *Create an object thematic query*
- *Detect and rectify sliver polygons*

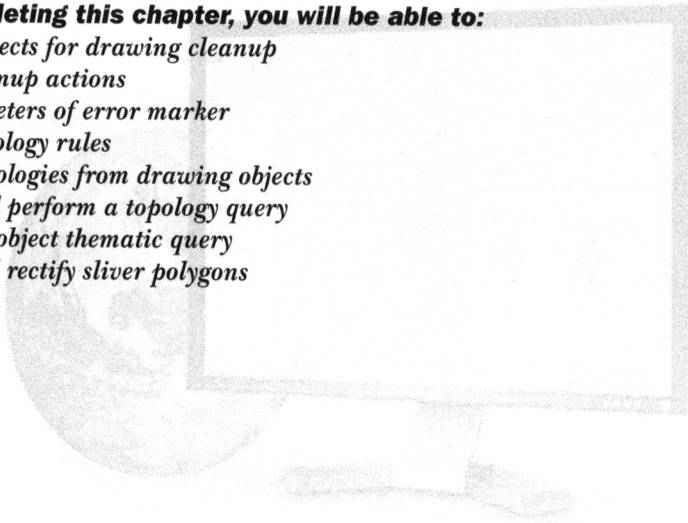

INTRODUCTION

In the previous chapter, you learned to classify the drawing objects into various object classes.

In this chapter, you will learn to identify the digitization errors in the drawing by performing the drawing cleanup using various drawing cleanup options available in AutoCAD Map 3D. Next, you will learn to fix the identified drawing errors by removing the anomalies. This chapter also explains the concept of creating and managing topologies in the drawing. The method to detect and rectify the sliver polygons, which may be generated while creating polygon topology, has also been explained in this chapter. Moreover, you will learn the procedure to define and execute a topology query.

DRAWING CLEANUP

In a project, there may be various types of errors that occur due to incorrect digitization of spatial objects. These errors can be in the form of duplicate objects, incorrect closure of area features, unwanted crossing objects, and many more. You need to rectify such errors in the drawing before proceeding with topology creation. The **Drawing Cleanup** wizard in AutoCAD Map 3D provides an easy-to-use interface to identify and rectify errors of the digitizing objects. By performing the cleanup operation on the drawing file, the flow of errors from AutoCAD environment to other geospatial file formats can be restricted. Moreover, the drawing cleanup helps in reducing the drawing file size by removing the drawing errors.

Note
It is always recommended that you save a copy as a back-up of the current drawing before performing the drawing cleanup process because the drawing may get altered after the cleanup process.

APPLYING CLEANUP TO DRAWING DATA

Ribbon:	Tools > Map Edit > Drawing Clean Up
Command:	MAPCLEAN

You can perform the drawing cleanup process to make the drawing of any error by using the options in the **Drawing Cleanup** wizard. To invoke this wizard, choose the **Drawing Clean Up** tool from the **Map Edit** panel of the **Tools** tab. On doing so, the **Select Objects** page of the **Drawing Cleanup** wizard will be displayed, as shown in Figure 8-1.

Drawing
Clean Up

The various pages in **Drawing Cleanup** wizard are discussed next.

Tip
*In the **Drawing Cleanup** wizard, you can also load an existing cleanup profile by using the **Load** button located at the lower left corner in this wizard.*

Select Objects Page

You can use the options in the **Select Objects** page to select the objects in the drawing for performing the cleanup. Using the options in this page, you can select the objects manually or automatically. In this page, you can also specify the objects to anchor. The methods of specifying the selection parameters in the **Objects to include in drawing cleanup** and **Object to anchor in drawing cleanup** areas of this page are discussed next.

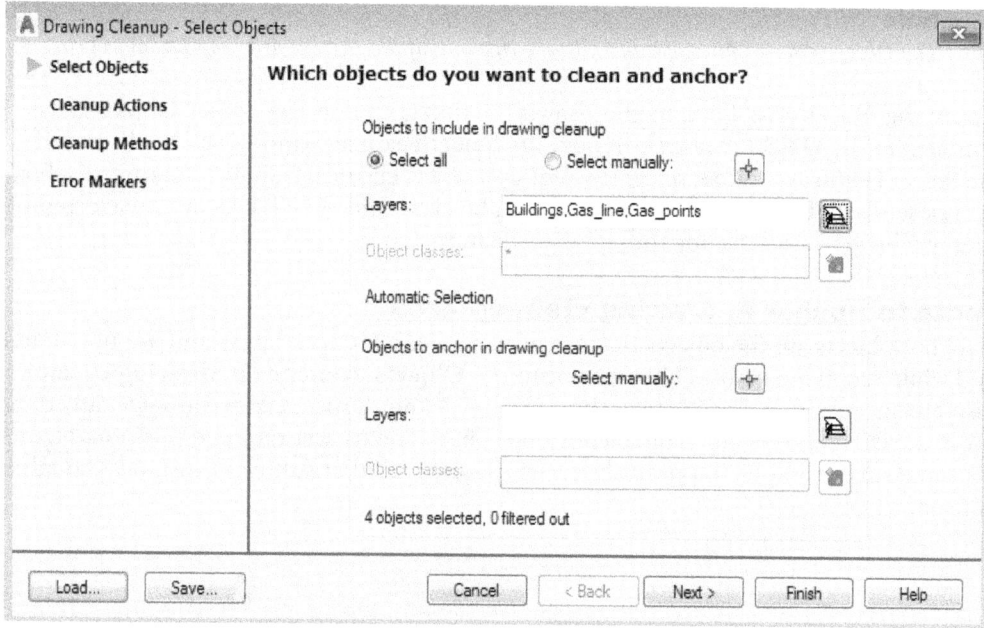

*Figure 8-1 The **Drawing Cleanup** wizard with the **Select Objects** page*

Objects to include in drawing cleanup Area

The options in the **Objects to include in drawing cleanup** area are used to select drawing objects for the cleanup process. The **Select all** radio button in this area is selected by default. As a result, all the drawing objects in the drawing will be included for the drawing cleanup.

> **Note**
> *You can perform the cleanup action only on Linear Objects, Points, Blocks, Text, and Mtext object types.*

To select the drawing objects manually from the drawing area, choose the **Select Objects** button next to the **Select manually** radio button; the **Drawing Cleanup** wizard will be close and you will be prompted to select the required drawing objects. Draw a selection box or click on the required drawing objects to select them in the drawing. After selecting the required objects, press ENTER; the **Drawing Cleanup** wizard will be displayed. Also, the number of selected objects will be displayed in the **Objects to include in drawing cleanup** area. The selected objects will be included in the cleanup process.

You can also apply the layer filter while selecting the drawing objects. To do so, choose the **Select Layers** button next to the **Layers** edit box; the **Select Layers** window will be displayed. In this window, select the required layers from the **Layers** list box and then choose the **Select** button; the **Select Layers** window will be closed and the names of the selected layer/s will be displayed in the **Layers** edit box. On specifying the layer/s, only the objects on the selected layer/s will be available for the cleanup process.

> **Tip**
> *In the **Select Layers** window, you can select multiple drawing layers for the cleanup process. To do so, press and hold CTRL and then select the required layer by clicking on it.*

Note
*The **Object classes** filter option will be active only if the current drawing has object classes.*

To apply the object class filter to the drawing objects, choose the **Select Object Class** button next to the **Object classes** edit box; the **Select Features** window will be displayed. In the **Select Features** window, select the required object class(es) from the **Features** list box and then choose the **Select** button; the window will be close and the objects in the selected object class/es will be now available for the drawing cleanup.

Objects to anchor in drawing cleanup Area

The anchor objects are the objects that are used as the reference objects and are not altered or moved while rectifying errors. The options in the **Objects to anchor in drawing cleanup** area are used to specify anchors objects in the drawing cleanup process. You can select any drawing object in the current drawing as an anchor object based on its accuracy and position. Figure 8-2 shows three survey points in a point layer, which are placed at sufficient distances suitable for anchoring.

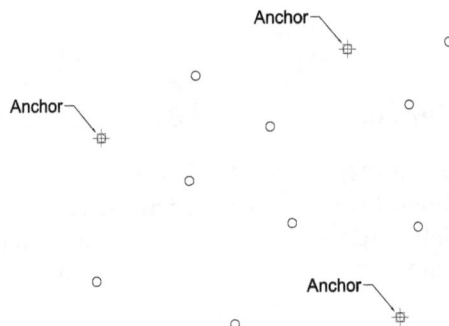

Figure 8-2 The surveyed points suitable for anchoring

Note
Anchoring is used to rectify the distortion and deformation of layer occurred during the cleanup process.

To specify an object in the current drawing as anchor, choose the **Select Objects** button in the **Objects to anchor in drawing cleanup** area; the **Drawing Cleanup** wizard will be closed and you will be prompted to select the required drawing objects. Note that the cursor will change to a selection box in the drawing window. By using this selection box, select the required drawing objects in the drawing area and then press ENTER; the number of objects selected for anchoring will be displayed at the bottom of this area. Moreover, you can apply the layer and object class filters to anchors, as discussed in the previous section.

After specifying the drawing objects and anchors for the cleanup process, choose the **Next** button at the bottom in **Drawing Cleanup** wizard; the **Select Actions** page will be displayed in the wizard, as shown in Figure 8-3.

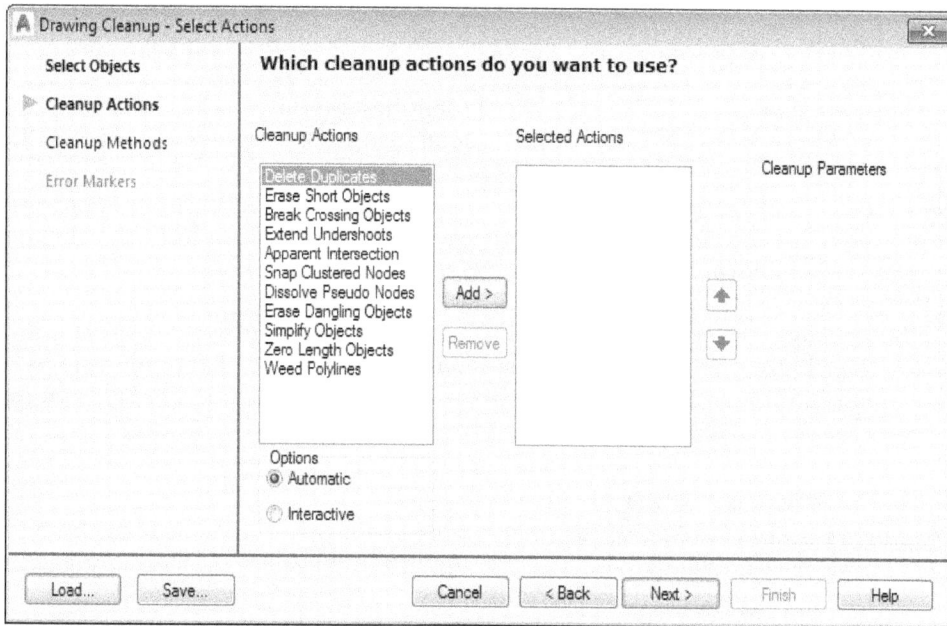

*Figure 8-3 The **Cleanup Actions** page of the **Drawing Cleanup** wizard*

Select Actions Page

In the **Select Actions** page, refer to Figure 8-3, you can specify the cleanup actions that you want to perform on the selected drawing objects. In this page, you can also specify the parameters required for performing a cleanup action in this page. The options in this page are discussed next.

Cleanup Actions and Selected Actions List Boxes

The **Cleanup Actions** list box displays all the available drawing cleanup actions available in AutoCAD Map 3D, while the cleanup actions selected by the user for the process of drawing cleanup are displayed in the **Selected Actions** list box. The table given next shows some of the cleanup actions and their usage.

Cleanup Action	Function
Delete Duplicates	Deletes objects that have same start and end points or have start and end points within defined tolerance
Erase Short Objects	Identifies and erases the objects whose length is shorter than the specified tolerance
Break Crossing Objects	Identifies the crossing objects and breaks them at the point of intersection
Erase Dangling Objects	Identifies and deletes those objects that have at least one end open (end not connected to other object)

| Dissolve Pseudo Nodes | Identifies and removes those nodes that are shared by two links |
| Snap Clustered Nodes | Identifies multiple nodes within a specified tolerance and clubs them into one node |

To select the required action for the cleanup process, select the cleanup action/s in the **Cleanup Actions** list box and then choose the **Add** button on the right of this list box; the selected cleanup action/s will be added to the **Selected Actions** list box.

If you want to remove any cleanup action from the **Selected Actions** list box, select it from the **Selected Actions** list box and then choose the **Remove** button on the left of this list box. If you want to arrange the selected actions, then use two arrows on the right of the **Selected Actions** list box in the **Select Actions** page.

Cleanup Parameters Area

The **Cleanup Parameters** area displays the parameters for the cleanup action selected in the **Selected Actions** list box. Figure 8-4 shows the options displayed in the **Cleanup Parameters** area for the **Delete Duplicates** cleanup action.

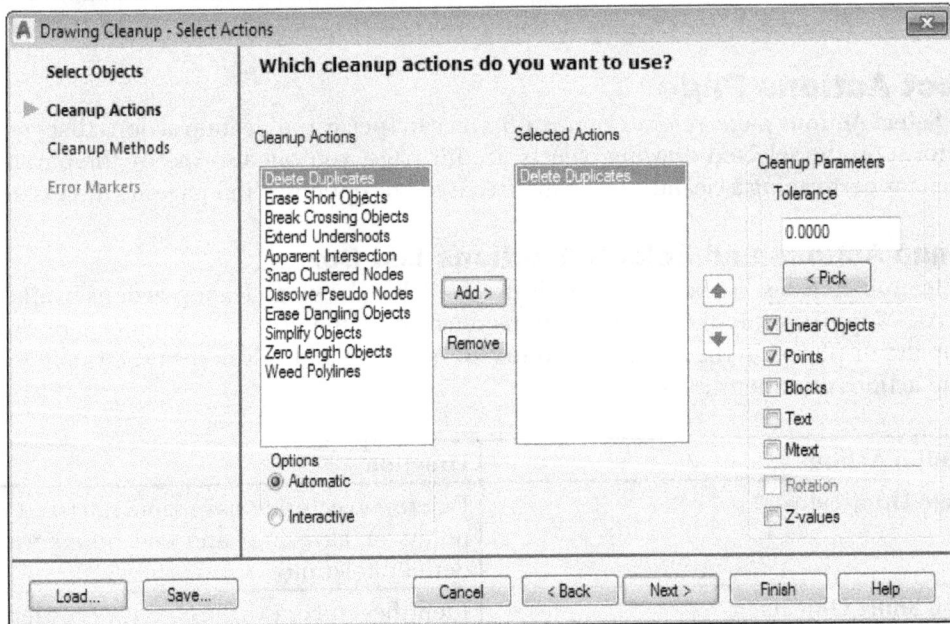

*Figure 8-4 Options in the **Cleanup Parameters** area displayed on selecting the **Delete Duplicates** action*

Note
*The options displayed in the **Cleanup Parameters** area will vary depending on the cleanup action selected.*

Options Area

In the **Options** area, you can specify the method for rectifying the drawing errors identified by the drawing cleanup process. Note that the **Automatic** radio button in this area is selected by default. As a result, all the errors found the drawing data are corrected automatically. To rectify the identified drawing errors manually, select the **Interactive** radio button in this area.

Note
*The **Error Markers** page in the **Drawing Cleanup** wizard will be activated only if you select the **Interactive** radio button in the **Options** area of the **Cleanup Actions** page.*

After selecting all the required cleanup actions and specifying their parameters, choose the **Next** button in the **Drawing Cleanup** wizard; the **Drawing Cleanup** wizard with the **Cleanup Methods** page will be displayed, as shown in Figure 8-5.

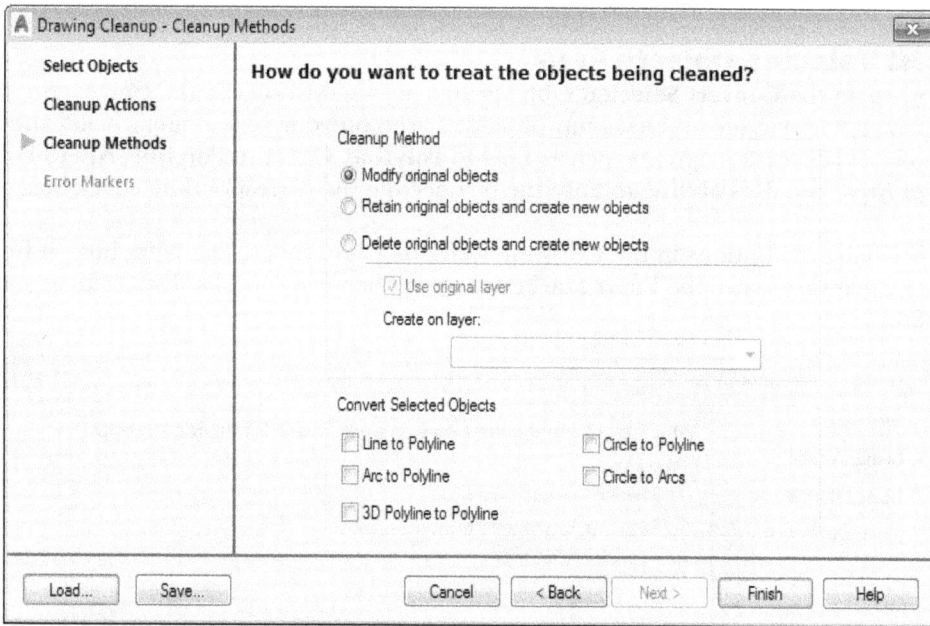

*Figure 8-5 The **Cleanup Methods** page of the **Drawing Cleanup** wizard*

Cleanup Methods Page

In the **Cleanup Methods** page, refer to Figure 8-5, you can specify the options for treating the objects used in the drawing cleanup process after the cleanup process is complete. The various output parameters in the **Cleanup Method** and **Convert Selected Objects** areas of this page are discussed next.

Cleanup Method Area

The options in the **Cleanup Method** area are used to specify whether to modify the original object or retain original object and create a new set of object data, or to delete the original data and create a new set of data after the cleanup process. In the **Cleanup Method** area, the **Modify original objects** radio button is selected by default. As a result, the selected original data will be modified according to the criteria specified in the cleanup process.

You can retain the original data as well as create a new set of data after the cleanup process. To do so, select the **Retain original objects and create new objects** radio button in this area; the **Use original layer** check box at the bottom of this area will be cleared and inactive. Also, the **Create on layer** edit box will be activated.

In the **Create on layer** edit box, you can either enter the name of the layer, or click on the down-arrow on the right of this edit box and then select the required option from the drop-down list displayed.

You can also delete the original data used for the drawing cleanup and create new objects after the cleanup process. To do so, select the **Delete original objects and create new objects** radio button; the **Use original layer** check box will be activated. If you want to create new objects on the original layer, then keep the **Use original layer** check box selected. Otherwise, clear this check box, and then select the required option from the **Create on layer** drop-down list.

Convert Selected Objects Area

The options in the **Convert Selected Objects** area are used to specify the conversion options for the objects selected for the cleanup process. Based on your requirement, you can use the various object conversion options such as **Line to Polyline**, **Circle to Polyline**, **Arc to Polyline**, **Circle to Arcs**, and **3D Polyline to Polyline** by selecting the corresponding check box.

After specifying the options in the **Cleanup Methods** page, choose the **Next** button from the **Drawing Cleanup** wizard; the **Error Markers** page of the wizard will be displayed, as shown in Figure 8-6.

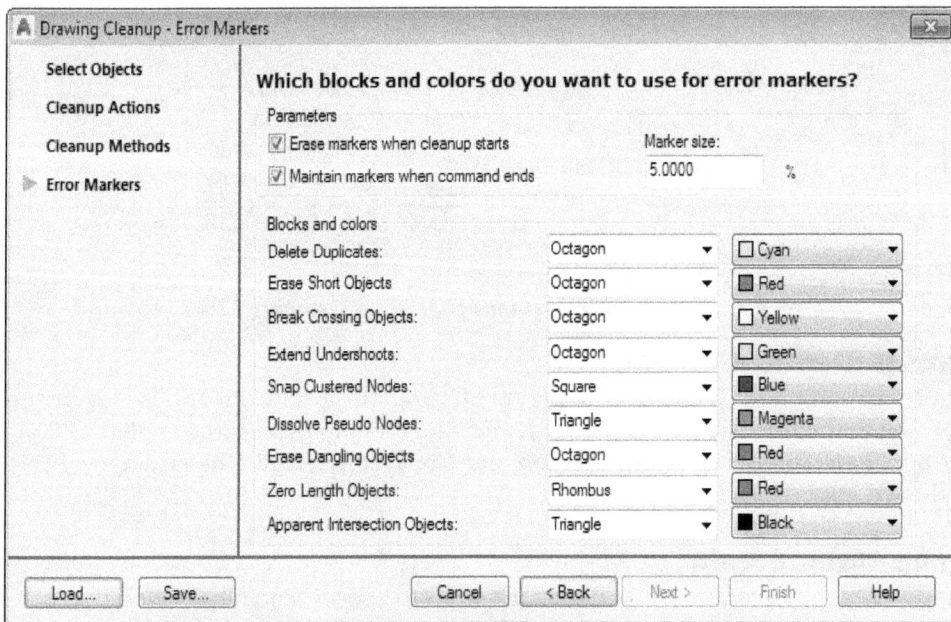

*Figure 8-6 The **Error Markers** page of the **Drawing Cleanup** wizard*

Error Markers Page

In the **Error Markers** page, refer to Figure 8-6, you can specify the parameters and properties of an error marker by using the options in the **Parameters** and **Blocks and colors** areas. The parameters and properties in these areas are discussed next.

Parameters Area

The options in the **Parameters** area are used to specify the parameters of error markers for the drawing cleanup process. The **Erase markers when cleanup starts** check box is selected by default. As a result, the error markers of the previous cleanup process will be removed from the drawing window before starting a new cleanup process. To retain the error markers of the previous cleanup process, clear the **Erase markers when cleanup starts** check box. The **Maintain markers when command ends** check box is also selected by default. As a result, the error markers will be retained after the cleanup process. To remove error markers after the cleanup process, clear the **Maintain markers when command ends** check box. You can specify the size of an error marker as a percentage of the screen area in the **Marker size** edit box. The default percentage value of the marker is set to **5%** of the screen area. To specify a different value, you can enter the required value in the **Marker size** edit box.

Blocks and colors Area

The options in the **Blocks and colors** area are used to specify the block type and color property of an error marker used during the cleanup action. To do so, select the required options from the drop-down lists in the first and second columns corresponding to the required cleanup action.

Performing Drawing Cleanup

After specifying the required parameters, you can start the drawing cleanup process by choosing the **Finish** button in the **Drawing Cleanup** wizard. On doing so, the drawing cleanup process will start.

If the **Automatic** radio button is selected from the **Options** area of the **Cleanup Actions** area in the **Cleanup Actions** page, the selected objects will be cleaned as per the options specified in the **Drawing Cleanup** wizard. But if the **Interactive** radio button is selected from the **Options** area of the **Cleanup Actions** page of the wizard, then on choosing the **Finish** button, the **Drawing Cleanup Errors** dialog box will be displayed, refer to Figure 8-7.

The **Cleanup action** area in the **Drawing Cleanup Errors** dialog box displays the list of all the cleanup actions that you have selected in the **Drawing Cleanup** wizard. You can expand the required cleanup action node in the **Drawing Cleanup Errors** dialog box by clicking on the corresponding [+] symbol. The number of errors will be displayed as an error statement in the expanded node. Next, click on this error statement; the error in the selected node will be marked and the drawing will zoom to show the erroneous object/s corresponding to the first error in the selected error statement.

To fix the error, choose the **Fix** button on the right of the **Cleanup action** list box; the error will be rectified and the marker will move to the next error. Alternatively, to fix all errors, select the parent node in the list box, and then choose the **Fix All** button; all errors will be rectified. To move to the next error without fixing it, choose the **Next** button. After fixing the errors, choose the **Close** button; the **Drawing Cleanup Errors** dialog box will be closed and the drawing cleanup procedure will be completed.

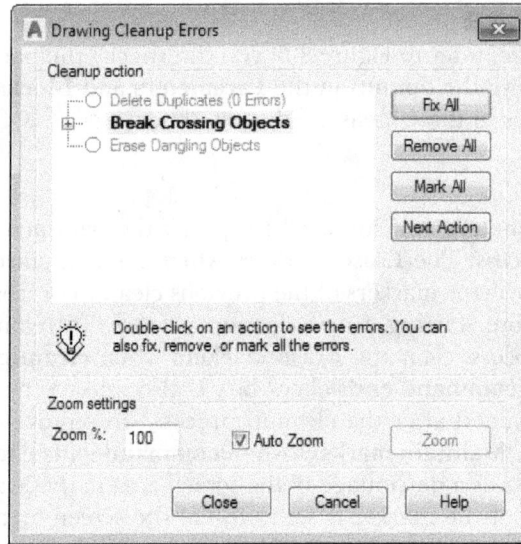

Figure 8-7 The **Drawing Cleanup Errors** *dialog box*

TOPOLOGY

Topology in GIS defines the relationships between the geographic features represented by point, line, and polygon geometries. In AutoCAD Map 3D, you can use a node topology to define the relationship between point features. Similarly, the network topology and polygon topology is used to define the relationship between the linear and area features respectively. Figure 8-8 shows an example of a combined map topology with three types of topological elements: node, lines, and polygon. Three types of topologies that are used for analysis are node, network, and polygon, as shown in Figures 8-9(a), (b), and (c).

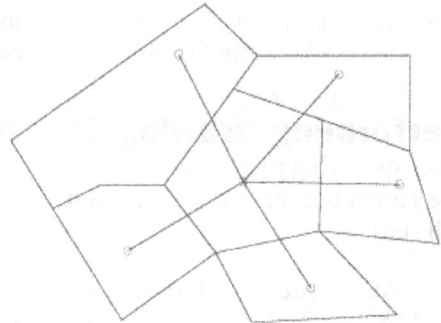

Figure 8-8 A combined model representing the nodes, links, and polygon elements in a map topology

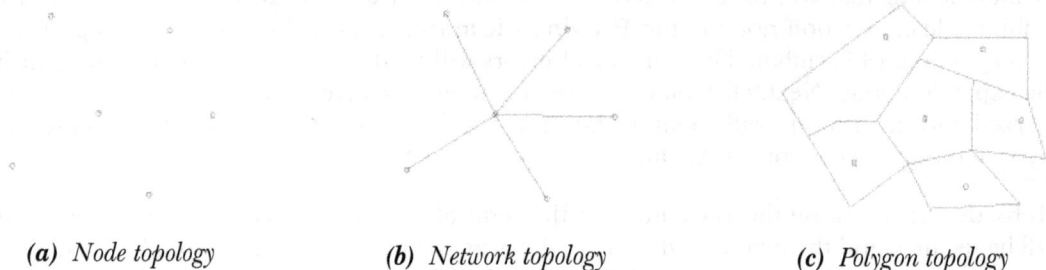

(a) Node topology

(b) Network topology

(c) Polygon topology

Figure 8-9 The topological elements represented separately

Creating a Topology

Ribbon:	Create > Topology > New
Command:	MAPTOPOCREATE

As discussed in the previous section, topology defines the relationship between spatial objects. These spatial relationships defined in the topology are useful for performing spatial operations such as dissolving boundaries between adjacent polygons or analyzing the network topology for finding the best route. To create a topology, you need to first create the base map containing drawing objects. To create a base map, you can import drawing objects from other drawings or digitize them by using any of the object creation method discussed previously.

After creating the base map, you will perform the drawing cleanup process to ensure that the drawing is free from drafting errors. Next, you will invoke the wizard to create topology.

To invoke the wizard for defining options for creating topology, choose the **New** tool from the **Topology** panel; the **Create Polygon Topology** wizard with the **Select Topology Type** page is displayed, as shown in Figure 8-10. Next, you will use various options in this wizard to create the required type of topology. The various pages available in the wizard are discussed next.

New

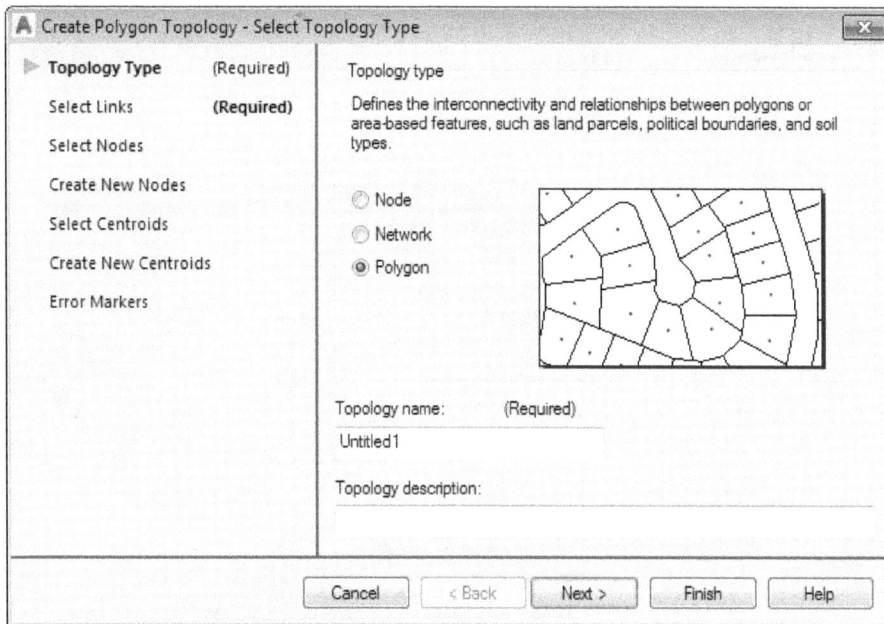

Figure 8-10 The Select Topology Type page of the Create Polygon Topology wizard

Note

In this chapter, you will learn how to create polygon topology only. Since the procedure to create the node and network topologies is similar to that of the polygon topology, the procedures to create the node and network topologies are omitted in this chapter. The procedure to create the network topology is discussed in the tutorial section of this chapter.

Select Topology Type Page

The **Select Topology Type** page of the wizard is displayed by default with the **Polygon** radio button selected in the **Topology type** area. As a result, the list of pages related to the creation of polygon topology will be displayed on the left pane in this wizard. In the **Select Topology Type** page, enter the label text in the **Topology name** text box below the **Topology type** area. Optionally, you can enter details about the topology to be created in the **Topology description** text box. Next, choose the type of topology that you want to create by choosing the radio button corresponding to the required topology type.

Notice that the links displayed in the left pane of the wizard will change based on the radio button selected in the **Topology type** area. For example, only two links are displayed in the left pane when you choose the **Node** radio button. On selecting the **Network** radio button, four links are displayed in the left pane of the wizard, including the two links that were displayed when the **Node** radio button was selected. On selecting the **Polygon** radio button, a full set of links is displayed in the left pane that includes all the links that were displayed when the **Network** radio button was selected.

After specifying various options in the **Select Topology Type** page, choose the **Next** button in the **Create Polygon Topology** wizard; the **Select Links** page will be displayed in the wizard, as shown in Figure 8-11.

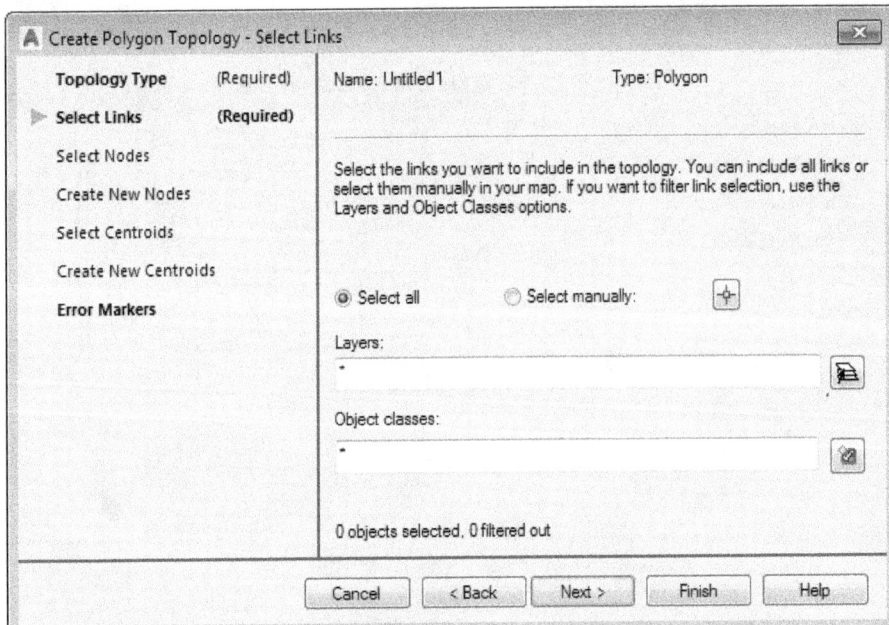

Figure 8-11   *The* **Select Links** *page of the* **Create Polygon Topology** *wizard*

Select Links Page

The options in the **Select Links** page are used to select the line objects for creating links. Both the name and type of topology are displayed on the top section of this page, refer to Figure 8-11. Moreover, you can specify the drawing objects to be used for creating topology in this page. To select all drawing objects in the current drawing, select the **Select all** radio button. To select drawing objects manually, select the **Select manually** radio button and then choose the **Select Objects** button next to this radio button; the cursor will change to a selection box in the drawing window. By using this selection box, select all the required drawing objects in the current drawing and then press ENTER; the number of objects selected will be displayed at the bottom of this page.

You can also filter the drawing objects in the current drawing based on layers and object classes. To select the drawing objects based on layer, choose the **Select Layer** button next to the **Layers** edit box below the **Select all** radio button; the **Select Layers** window will be displayed. In the **Select Layers** window, select the required layer from the list box and then choose the **Select** button; the name of the selected layer will appear in the **Layers** edit box. You can also select object classes corresponding to the selected layer.

To filter drawing objects in the selected layer based on an existing object class, choose the **Select Features** button next to the **Object classes** edit box; the **Select Features** window will be displayed. In this window, select the required object class from the list box and then choose the **Select** button; the name of the selected object class will be displayed in the **Object classes** edit box.

After specifying all parameters to create topologies in the **Select Links** page, choose the **Next** button; the **Create Polygon Topology** wizard with the **Select Nodes** page will be displayed.

Select Nodes Page

The options in the **Select Nodes** page are used to select the point objects for creating nodes. Select the required objects, layers, or block names in this page by using the specific tools explained in the earlier section. After specifying the required parameters in the **Layers**, **Block names**, and **Object classes** edit boxes, choose the **Next** button in this page; the **Create New Nodes** page of the **Create Polygon Topology** wizard will be displayed, as shown in Figure 8-12.

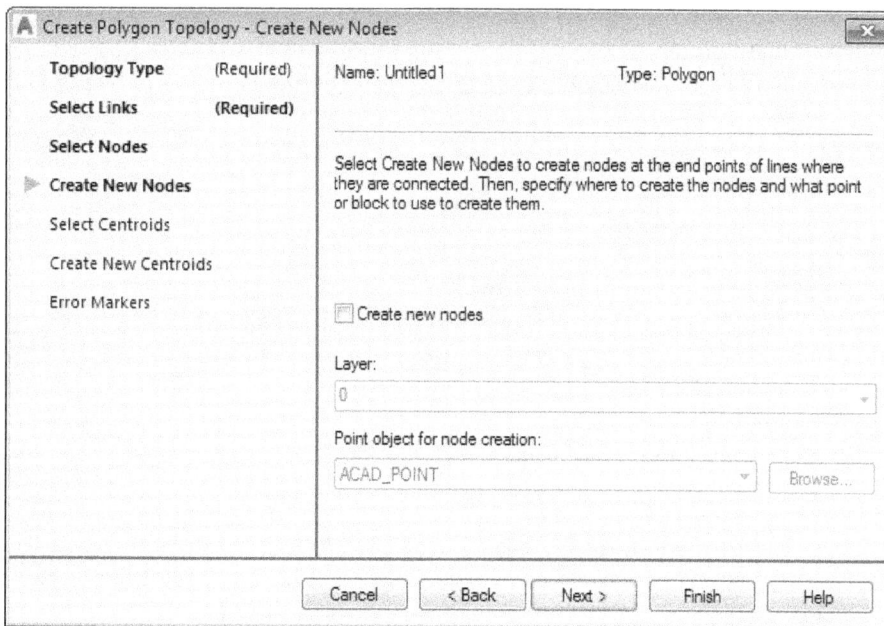

*Figure 8-12 The **Create New Nodes** page of the **Create Polygon Topology** wizard*

Create New Nodes Page

In the **Create New Nodes** page, the **Create new nodes** check box is clear by default, refer to Figure 8-12. As a result, the **Layer** and **Point object for node creation** options are inactive. In this page, you can specify parameters for creating nodes at the intersection (connection) of any two lines in the current drawing. To do so, select the **Create new nodes** check box; the **Layer** and **Point object for node creation** drop-down lists will be activated. In the **Layer** drop-down list, you can either enter the name of a layer, or select an option from the drop-down list by clicking on the down-arrow on the right of this edit box. Similarly, you can select the required option from the **Point object for node creation** drop-down list. Also, you can use the **Browse** button next to this drop-down list to load the required point object. After specifying these parameters, choose the **Next** button; the **Create Polygon Topology** wizard with the **Select Centroids** page will be displayed.

Select Centroids Page

In the **Select Centroids** page, you can select an existing point for creating centroids for the polygon objects in the current topology. To do so, select all object data in the current drawing file, or specify filters for object data by setting the required parameters in the **Layers**, **Block names**, and **Object classes** edit boxes. To select the required filters from the **Layers**, **Block names**, and **Object classes** edit boxes, follow the instructions given in the **Select Links Page** section of this chapter. After specifying the required parameters in this page, choose the **Next** button; the **Create Polygon Topology** wizard with the **Create New Centroids** page will be displayed.

Create New Centroids Page

In the **Create New Centroids** page, you can set the parameters to create new or missing centroids for object data. In this page, the **Create missing centroids** check box is selected by default. As a result, the **Layer** and **Point object for centroid creation** drop-down lists are activated. Specify the layer to be used for creating centroid in the **Layer** drop-down list and then select the required point object from the **Point object for centroid creation** drop-down list. If there is no need to create any new or missing centroids, clear the **Create missing centroids** check box. Then, choose the **Next** button; the **Create Polygon Topology** wizard with the **Set Error Markers** page will be displayed, as shown in Figure 8-13.

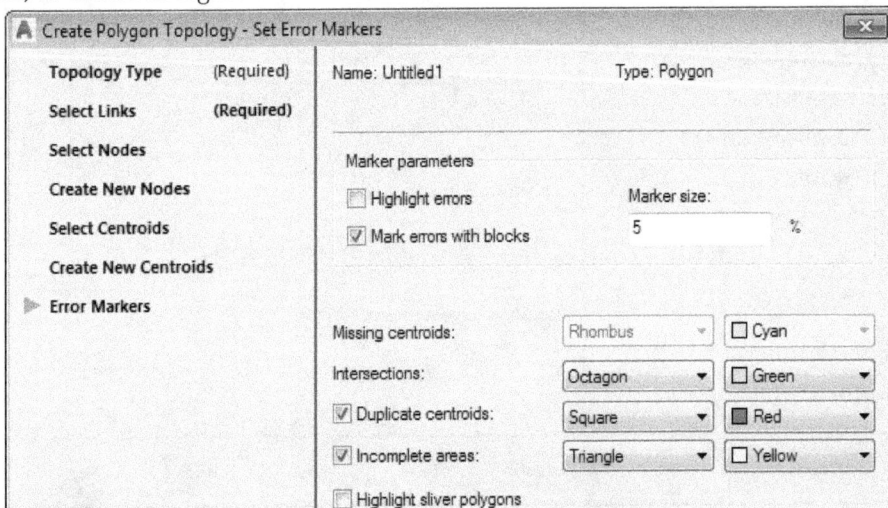

*Figure 8-13 Partial view of the **Set Error Markers** page of the **Create Polygon Topology** wizard*

Set Error Markers Page

In the **Set Error Markers** page of the **Create Polygon Topology** wizard, you can specify the parameters to display errors that may occur while creating a topology. In this page, you can also specify the color and geometric symbol properties for the error markers that will be used to mark the errors in the drawing window.

Tip
Error marker is a drawing block, which is placed at the point location of occurrence of error while creating topologies.

In the **Marker parameters** area of the **Set Error Markers** page, you can specify the display properties of an error marker. The **Highlight errors** check box is clear by default. As a result, the errors if occurred while creating a topology will not be highlighted in the drawing window. If you select this check box, the errors occurred while creating topology will be highlighted after the creation of the topology. The highlighted errors are temporary markers and will disappear when the drawing is regenerated.

The **Mark errors with blocks** check box is selected by default. As a result, the errors occurred while creating the topology will be highlighted with a drawing block. If you clear this check box, the errors will not be highlighted by using a drawing block. In the **Marker size** edit box, which is available on the right of the **Mark errors with blocks** check box, you can specify the size of the marker as a percentage of the screen area. The default percentage value of the marker is set as **5%** of the screen area. If you want to specify some other value as the size of an error marker, enter the required value in the **Marker size** edit box.

You can set the display properties of the errors such as missing centroids, intersections, duplicate centroids, incomplete areas, and highlight sliver polygons that might occur while creating topologies. You can also specify geometric shape with which each of these errors can be represented. To do so, select the **Octagon**, **Triangle**, **Rhombus**, or **Square** option from the drop-down list corresponding to the error type. In addition to setting the geometric shape for displaying an error marker, you can also specify different colors to highlight the errors. To do so, select an option from the color drop-down list corresponding to the error type.

The **Duplicate centroids** and **Incomplete areas** check boxes are selected by default. As a result, the errors occurred in these error types will be highlighted after creating the polygon topologies. If you clear these check boxes, these error types will be hidden after the topology has been created. The **Highlight sliver polygons** check box is clear by default. As a result, the sliver polygons created after creating topologies will be hidden. To highlight the sliver polygons, select this check box.

After specifying all the parameters for creating topology, choose the **Finish** button to create a topology. On doing so, the topology will be created and added to the **Topologies** folder in the **Map Explorer** tab of the **TASK PANE**.

Note
*If the topology created is not displayed by default in the **Topologies** folder, then left-click on the **+** sign corresponding to the folder; the folder will expand and the topology created will be displayed in the **Map Explorer** tab of the **TASK PANE**.*

SLIVER POLYGONS

Sliver polygons are the thin long polygons created erroneously while creating polygon topology or overlaying two polygon topologies. These polygons are usually not visible at normal scale. You have to zoom in to see these polygons.

You can detect sliver polygons created using the **Set Error Markers** page of the **Create Polygon Topology** wizard. To do so, select the **Highlight sliver polygons** check box located at the bottom area of the **Set Error Markers** page in the **Create Polygon Topology** wizard. Then, proceed with creating topologies. You can also detect sliver polygons while overlaying two polygon topologies by analyzing the topology created. To rectify or remove a sliver polygon, select it and then press DELETE; the sliver polygon will be removed.

TOPOLOGY QUERY

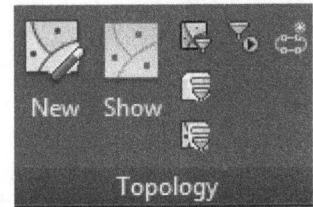

The topology query is used to perform a query to filter objects and their associated data from the current or attached drawing. Unlike standard query, which works with all drawing objects, a topology query works with only one topology. You can query a part or entire topological data as required. The tools used for creating a topology are in the **Topology** panel of the **Create** tab, as shown in Figure 8-14. Some of the tools in the **Topology** panel are discussed next.

Figure 8-14 The Topology tools in the Create tab of the Ribbon

> **Tip**
> *You can also invoke the tools in the Topology panel from the TASK PANE. To do so, right-click on the required topology in the Topologies node in the Map Explorer tab of the TASK PANE; a shortcut menu will be displayed. In this shortcut menu, place the cursor on the Analysis option; a flyout will be displayed. Choose the Topology Query option from the displayed flyout to define a new topology query.*

Defining a Topology Query

Ribbon:	Create > Topology > Define Query
Command:	MAPTOPOQUERY

You can define a topology query to filter the required topology data and create a new topology from the filtered data. To define the topology query, choose the **Define Query** tool from the **Topology** panel; the **Topology Query** dialog box will be displayed, as shown in Figure 8-15. The options in this dialog box are discussed next.

Query Topology Area

The options in the **Query Topology** area are used to specify the source topology data for defining the topology query. If there are any topologies in the current drawing, then these

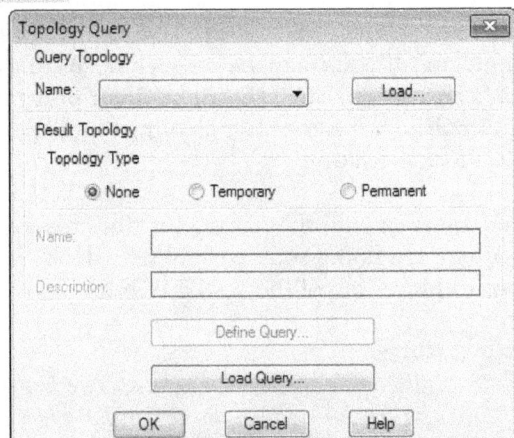

Figure 8-15 The Topology Query dialog box

topologies will be displayed in the **Name** drop-down list. Select the required option from the drop-down list. You can also load the required topology and then perform a query on it. To do so, choose the **Load** button next to the **Name** drop-down list; the **Topology Selection** window will be displayed with a list of topologies that can be loaded, refer to Figure 8-16. Next, select the required topology from the list box and then choose the **OK** button; the **AutoCAD Map Topology Audit** message box will be displayed. Choose the **OK** button in this message box; the selected topology will be added to the **Name** drop-down list in the **Topology Query** dialog box.

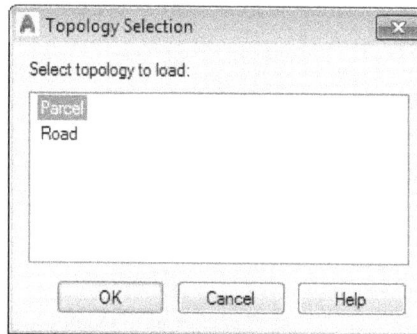

*Figure 8-16 The **Topology Selection** window*

Result Topology Area

The options in the **Result Topology** area are used to specify parameters for the output of the topology query. Specify the output data format of the topology query in the **Topology Type** area. To assign temporary or permanent status to the query output, select the **Temporary** or **Permanent** radio button in the **Topology Type** area; the **Name** and **Description** edit boxes below the **Topology Type** area will be activated. Next, enter the topology label in the **Name** edit box and then enter the description of the current query in the **Description** edit box.

After specifying the output parameters, you can define query to apply the required condition for the filter process. To define the query process, choose the **Define Query** button below the **Result Topology** area; the **Define Query of Topology** dialog box for the loaded topology will be displayed, as shown in Figure 8-17.

In the **Define Query of Topology** dialog box, you can define query parameters using the procedure explained in Chapter 6. Specify all query parameters required for the topology and then choose the **OK** button; the **Define Query of Topology** dialog box will be closed and the **Topology Query** dialog box will be invoked.

Using the **Topology Query** dialog box, you can also load an existing query into the current topological query. To do so, choose the **Load Query** button below the **Define Query** button; the **Load Internal Query** dialog box will be displayed, as shown in Figure 8-18. In this dialog box, the list of available query categories will be displayed in the **Category** drop-down list while the queries available in the selected query category will be displayed in the **Queries** list box of the dialog box. Choose the required query to load and then choose the **OK** button; the query topology with the specified name will be created.

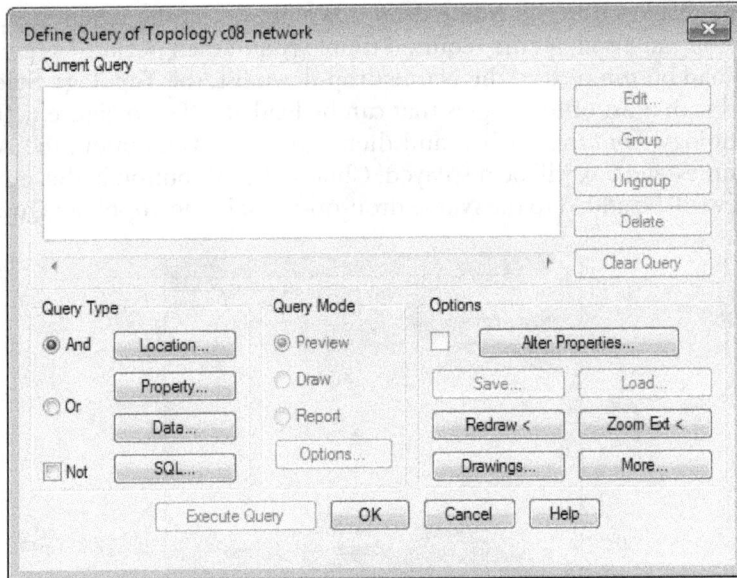

Figure 8-17 *The **Define Query of Topology** dialog box*

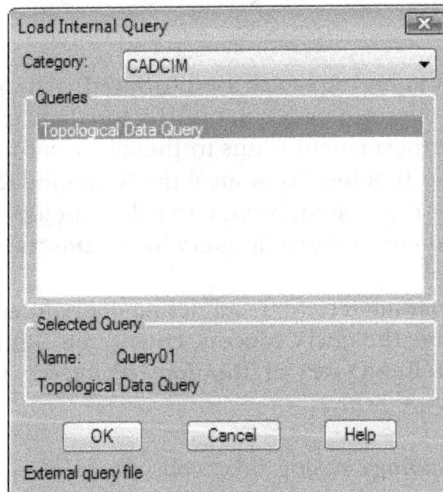

Figure 8-18 *The **Load Internal Query** dialog box with an example query*

Object Thematic Query

Ribbon:	Create > Topology > Object Thematic Query
Command:	MAPTHEMOBJ

The **Object Thematic Query** tool is used to filter the object properties such as color, line format, fill, text, and symbol for creating a thematic map. To invoke this tool, choose the **Object Thematic Query** tool from the **Topology** panel; the **Object Thematic Mapping** dialog box will be displayed, as shown in Figure 8-19. The methods of specifying the parameters in different areas of this dialog box are discussed next.

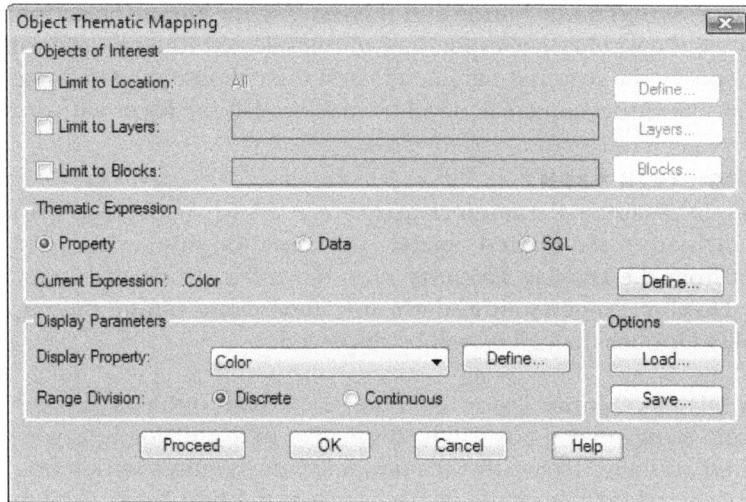

Figure 8-19 *The **Object Thematic Mapping** dialog box*

Objects of Interest Area

The options in the **Objects of Interest** area are used to filter objects in the current dataset. To specify the boundary limitation for the objects to be filtered, select the **Limit to Location** check box; the **Define** button corresponding to the **Limit to Location** check box will be activated. To select a region for locating the objects of interest, choose the **Define** button corresponding to the **Limit to Location** check box; the **Location Condition** dialog box will be displayed. Specify the region containing the objects of interest using the options in this dialog box and then choose the **OK** button; the **Location Condition** dialog box will be closed and the boundary limits will be saved.

> **Tip**
> *You can use more than one filtering option from the **Objects of Interest** area of the **Object Thematic Mapping** dialog box to select the required object from the drawing.*

To filter the objects based on the layer, select the **Limit to Layers** check box; the **Limit to Layers** edit box and the **Layers** button next to this edit box will be activated. To limit the object selection to a layer, either enter the name of the layer in the **Limit to Layers** edit box or select the required layer option by using the **Layers** button next to this edit box.

You can filter objects by limiting them to blocks. To do so, select the **Limit to Blocks** check box in the **Objects of Interest** area; the **Limit to Blocks** edit box and the **Blocks** button next to this edit box will be activated. To select a block type, choose the **Blocks** button next to the **Limit to Blocks** edit box; the **Select** window will be displayed. In this window, select the required block from the list box and then choose the **OK** button; the **Select** window will be closed and the name of the selected block will be displayed in the **Limit to Blocks** edit box.

Thematic Expression Area

The options in the **Thematic Expression** area are used to specify or filter the objects to be used in the thematic mapping. To write expressions for applying conditions to query data, select the **Property**, **Data**, or **SQL** radio button and then choose the **Define** button at the lower right corner of this area; the window corresponding to the selected radio button will be displayed. In this window, specify the required parameters and then choose the **OK** button; this window will be closed and the expression created will be displayed at the lower left corner of this area.

Display Parameters Area

The options in the **Display Parameters** area are used to specify the display properties and the range division of the filtered objects. To define the display properties of an entity, select an option from the **Display Property** drop-down list. To set the limits for the option selected in the **Display Property** drop-down list, choose the **Define** button next to it; the **Thematic Display Options** dialog box will be displayed.

In the **Thematic Display Options** dialog box, choose the **Add** button on the right of the list box; the **Add Thematic Range** dialog box will be displayed. In this dialog box, you can specify the limit of the desired attribute. To specify the limit, select the required option from the list box in the **Add Range** area and then enter the desired value in the **Edit Value** edit box. Next, choose the **OK** button from the **Add Thematic Range** dialog box; this dialog box will be closed and the parameters set will be displayed in the list box in the **Thematic Display Options** dialog box. Next, choose the **OK** button in the **Thematic Display Options** dialog box; the dialog box will be closed and the specified parameters will be applied to the current thematic query. To specify a discrete or continuous range division for the filtered objects, select the **Discrete** or **Continuous** radio button from the **Display Parameters** area of the **Object Thematic Mapping** dialog box.

After specifying all parameters and conditions for the objects to be displayed in the thematic map, choose the **Proceed** button at the bottom of the **Object Thematic Mapping** dialog box; the thematic map will be created.

Topology Thematic Query

Ribbon:	Create > Topology > Topology Thematic Query
Command:	MAPTHEMTOPO

The **Topology Thematic Query** tool is used to filter object to create thematic map based on topology. To filter object properties, choose the **Topology Thematic Query** tool from the **Topology** panel of the **Create** tab of the Ribbon; the **Topology Thematic Mapping** dialog box will be displayed.

The query procedure in the topology thematic query is the same as that of the object thematic query. You will find difference only in selecting the dataset for creating the thematic query. In the topology thematic mapping process, you need to load the required topology for the query process. Then, you need to follow the same procedure to specify the parameters and conditions of the topology, as discussed in the previous section.

TUTORIALS
General instructions for downloading tutorial files:
Before starting the tutorial, you need to download the tutorial data to your computer. To do so, follow the steps given below:

1. Log on to *www.cadcim.com* and browse to *Textbooks > Civil/GIS > Map 3D > Exploring AutoCAD Map 3D 2017*. Next, select *c08_m3d_2017_tut.zip* file from the **Tutorial Files** drop-down. Next, choose the corresponding **Download** button to download the data file.

2. Extract contents of the zip file to the following location:

    C:\m3d_2017

Notice that the *c08_m3d_2017_tut* folder is created within the *m3d_2017* folder.

Tutorial 1   Performing Drawing Cleanup

In this tutorial, you will perform the drawing cleanup process on drawing objects in different layers. **(Expected time: 45 min)**

The following steps are required to complete this tutorial:

a. Open the drawing file.
b. Perform the drawing cleanup on the **Buildings**, **Gas_lines**, and **Gas_points** drawing layers.
c. Rectify errors manually by using the **Interactive** method of correction.
d. Save the current drawing.

Opening the Tutorial Drawing File
1. Choose the **Open** button from the Quick Access Toolbar; the **Select File** dialog box is displayed.

2. In the **Select File** dialog box, browse to the following location:

    C:\m3d_2017\c08_m3d_2017_tut\c08_tut01

3. Now, select the **c08_Tut01.dwg** file from the list box; a preview of the selected file is displayed in the **Preview** area.

4. Choose the **Open** button from the **Select File** dialog box; the dialog box is closed and the drawing is displayed in the drawing window, as shown in Figure 8-20.

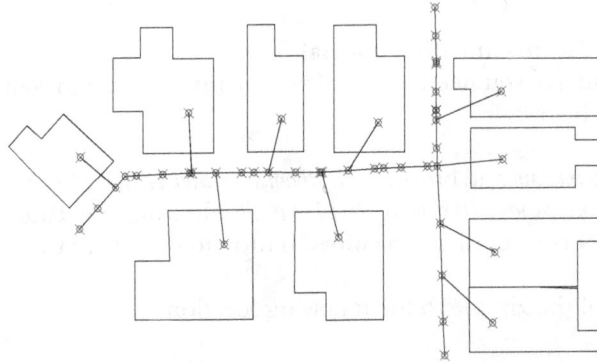

Figure 8-20 *The drawing layers displayed in the drawing window*

Performing Drawing Clean Up

1. Choose the **Drawing Clean Up** tool from the **Map Edit** panel in the **Tools** tab; the **Drawing Cleanup** wizard with the **Select Objects** page is displayed.

2. In the **Select Objects** page, select the **Select all** radio button from the **Objects to include in drawing cleanup** area if not selected by default.

3. In the **Objects to include in drawing cleanup** area of the **Select Objects** page, choose the **Select Layers** button next to the **Layers** edit box; the **Select Layers** dialog box is displayed, as shown in Figure 8-21.

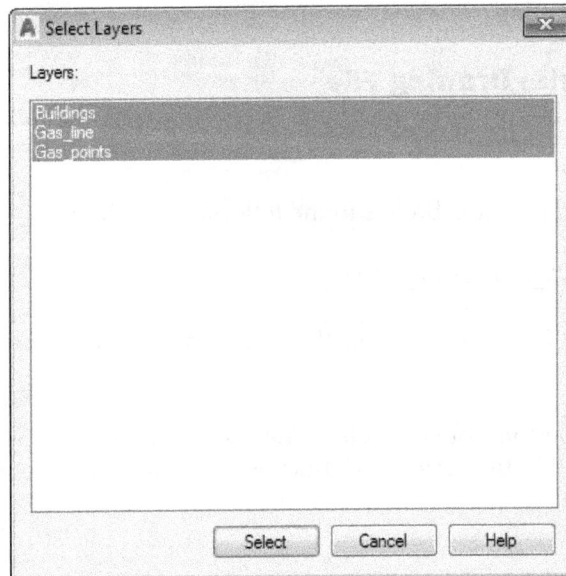

Figure 8-21 *The **Select Layers** window with the list of selected layers*

4. Select all the layers in the **Layers** list box of the **Select Layer** dialog box if they are not selected by default.

5. Choose the **Select** button; the dialog box is closed and the layer names **Buildings, Gas_lines**, and **Gas_points** are displayed in the **Layers** edit box of the **Objects to include in drawing cleanup** area.

6. Choose the **Select manually** button from the **Objects to anchor in drawing cleanup** area of this page; the cursor changes into a selection box in the drawing window.

7. Using this selection box, select the highlighted node objects represented by numbers **1, 2, 3**, and **4**, refer to Figure 8-22.

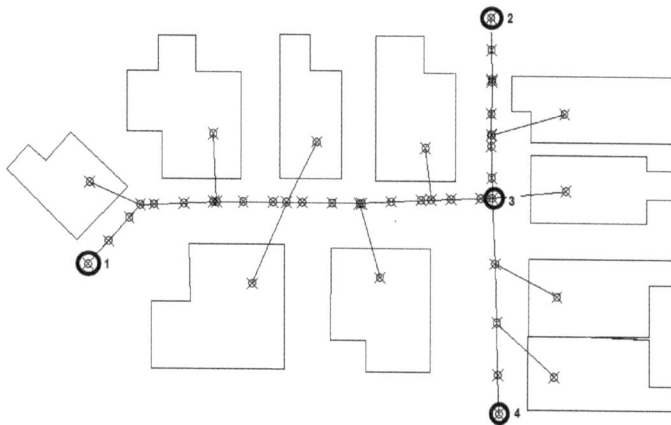

Figure 8-22 The highlighted nodes selected as anchor objects

8. After selecting the highlighted nodes as anchor objects, press ENTER; the statement mentioning the number of objects selected (4 objects) in the drawing is displayed at the bottom of the **Objects to anchor in drawing cleanup** area in the **Select Objects** page.

9. Choose the **Next** button in the **Drawing Cleanup** wizard, the **Cleanup Actions** page of the wizard is displayed.

10. In this page, select the **Delete Duplicates** option from the **Cleanup Actions** list box if not selected by default, and then choose the **Add** button on the right of the list box; the **Delete Duplicates** option is added to the **Selected Actions** list box and the parameters corresponding to this option are displayed in the **Cleanup Parameters** area.

11. In the **Cleanup Parameters** area, enter **0.5** in the **Tolerance** edit box.

12. Repeat the procedure followed in step 10 and to add the **Erase Short Objects, Extend Undershoots**, and **Erase Dangling Objects** cleanup actions to the **Selected Action** list. Also, specify **0.5** in the **Tolerance** edit box for all the cleanup actions mentioned above.

13. Select the **Interactive** radio button from the **Options** area; the **Error Markers** option is activated in the left pane of the **Drawing Cleanup** dialog box.

14. Choose the **Next** button in this page; the **Cleanup Methods** page of the wizard is displayed.

15. In the **Cleanup Methods** area of this page, select the **Modify original objects** radio button if not selected by default.

16. Now, choose the **Next** button in this page; the **Error Markers** page of the **Drawing Cleanup** wizard is displayed.

17. In the **Error Markers** page of the **Drawing Cleanup** wizard, retain the default settings and then choose the **Finish** button; the wizard is closed and the **Drawing Cleanup Errors** dialog box is displayed, as shown in Figure 8-23.

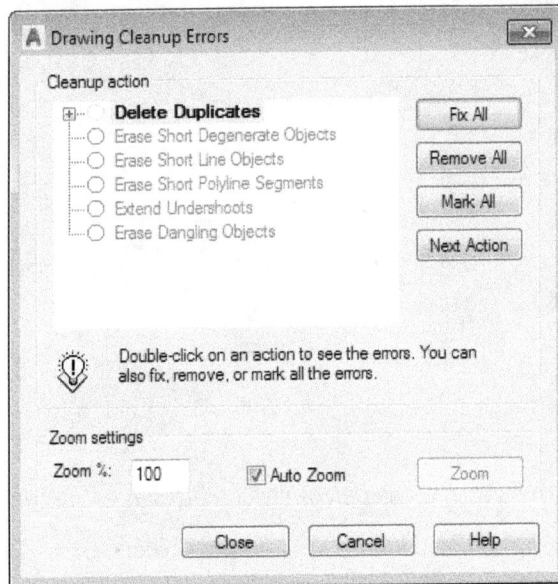

*Figure 8-23 The **Drawing Cleanup Errors** dialog box displaying the errors detected*

Rectifying Errors

1. In the **Cleanup action** area of the **Drawing Cleanup Errors** dialog box, expand the **Delete Duplicate** action by clicking on the [+] node on its left; the **Error 1 of 6** text is displayed in this node.

2. In the **Cleanup action** area of the **Drawing Cleanup Errors** dialog box, select the **Error 1 of 6** statement in the **Delete Duplicates** node by clicking on it; the drawing is zoomed to display the first error in the drawing window. Notice that the error is highlighted by a cyan colored octagon.

Tip
*You can rectify all the errors detected in a cleanup action by selecting the parent node in the **Cleanup action** list box and then choosing the **Fix All** button in the **Drawing Cleanup Errors** dialog box.*

3. In the **Drawing Cleanup Errors** dialog box, choose the **Fix** button displayed on the right of the list box in the **Cleanup action** area; the highlighted error is rectified and the next error gets highlighted.

4. Repeat step 3 to rectify all the errors in the **Delete Duplicates** cleanup action. After you have fixed all the duplicate objects, the **Erase Short Line Objects** option in the **Cleanup action** area is activated, as shown in Figure 8-24.

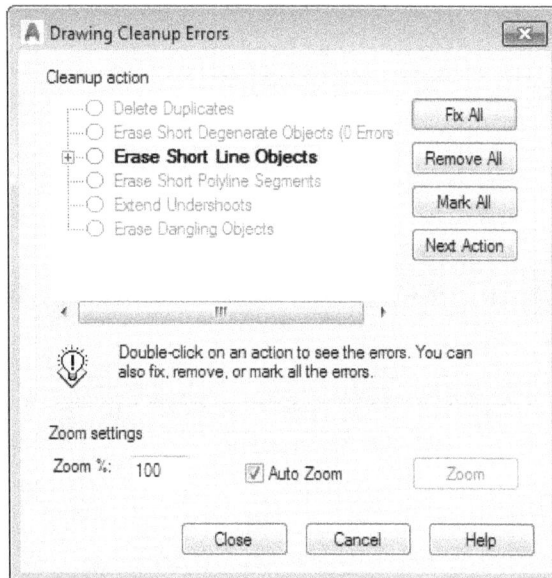

Figure 8-24 *The* ***Drawing Cleanup Errors*** *dialog box with the* ***Erase Short Line Objects*** *cleanup action activated*

5. Next, expand the **Erase Short Line Objects** node in the **Cleanup action** area by clicking on the [+] node; the **Error 1 of 2** statement is displayed in the node. Select the **1 of 2** statement.

6. In the **Zoom settings** area, clear the **Auto Zoom** check box; the **Zoom** button next to it gets activated.

7. Choose the **Zoom** button; the drawing zooms to display the error at the center of the drawing window. Note that this error is highlighted by a red colored octagon, as shown in Figure 8-25.

8. Choose the **Fix** button on the right of the list box in the **Cleanup action** area; the short object error is rectified and the next error is marked in the drawing.

9. In the **Zoom settings** area, choose the **Zoom** button; Map 3D will zoom in the drawing to display the highlighted short object error.

10. In the **Drawing Cleanup Errors** dialog box, choose the **Fix** button on the right of the **Cleanup actions** list box; the error is rectified and the **Erase Short Polyline Segments** cleanup action is activated in the **Cleanup action** area, as shown in Figure 8-26.

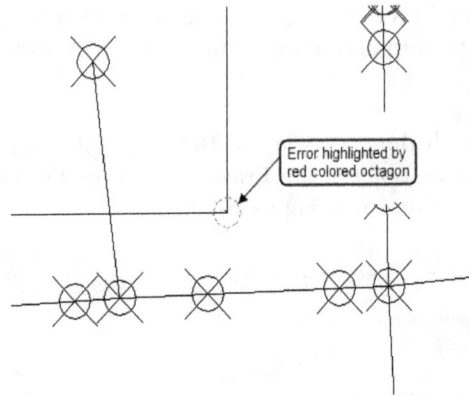

Figure 8-25 The short object error highlighted
by the red colored octagon block

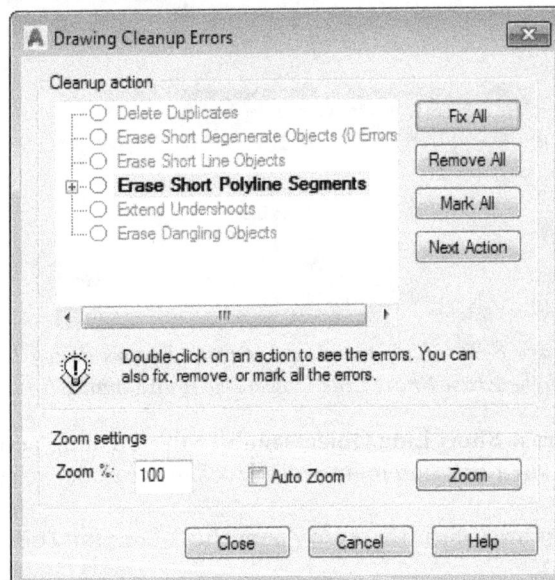

Figure 8-26 The **Erase Short Polyline Segments**
cleanup action activated in the **Cleanup action** area

11. Expand the **Erase Short Polyline Segments** node in the **Cleanup action** list box by clicking
    on the **[+]** node; the **Error 1 of 8** statement is displayed.

12. Select the **Error 1 of 8** statement displayed in the node.

13. In the **Zoom settings** area of the dialog box, choose the **Zoom** button; the error is highlighted
    by a red colored octagon in the drawing window.

14. In the **Cleanup action** area, select the **Erase Short Polyline Segments** node in the list box and
    then choose the **Remove All** button displayed on the right in the list box; all the errors in the
    Erase Short Polyline Segments cleanup action are removed and the **Extend Undershoots**
    cleanup action is activated.

15. In the **Cleanup action** area, choose the **Fix All** button; all the errors in the **Extend Undershoots** cleanup action are rectified and the **Erase Dangling Objects** cleanup action is activated.

16. Choose the **Fix All** button in the **Cleanup action** area; all the errors in the **Erase Dangling Objects** cleanup action are rectified.

17. Next, choose the **Close** button from the **Drawing Cleanup Errors** dialog box; this dialog box is closed and the rectified drawing is displayed in the drawing window, as shown in Figure 8-27.

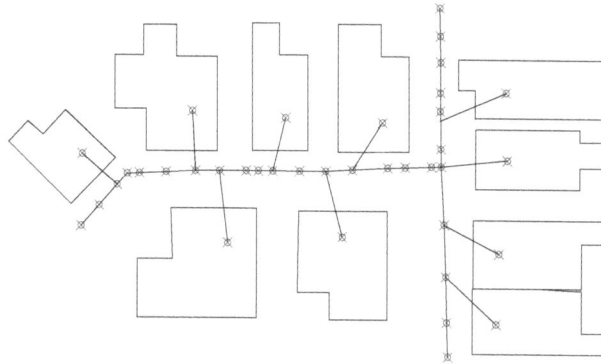

Figure 8-27 *The rectified drawing displayed in the drawing window*

Saving the Rectified Drawing File

1. Choose the **Save As** option from the Application Menu; the **Save Drawing As** dialog box is displayed.

2. In the **Save Drawing As** dialog box, enter the text **c08_Tut01a** in the **File name** edit box.

3. Choose the **Save** button next to the **File name** edit box; the drawing file is saved with the specified name.

Tutorial 2   Creating Topology from the Drawing Objects

In this tutorial, you will create topology data from the drawing objects.   (**Expected time: 30 min**)

The following steps are required to complete this tutorial:

a. Open the drawing file.
b. Create the **c08_network** network topology for the **Gas_lines** and **Gas_points** drawing layers.
c. Display the geometry of the **c08_network** topology created.
d. Save the drawing file.

Opening the Tutorial Drawing

1. Choose the **Open** button from the Quick Access Toolbar; the **Select File** dialog box is displayed.

2. In the **Select File** dialog box, browse to the following location:

    C:\m3d_2017\c08_m3d_2017_tut\c08_tut02

3. In the **Select File** dialog box, choose the **c08_Tut02.dwg** file; a preview of the selected drawing is displayed in the **Preview** area.

4. Choose the **Open** button on the right of the **File name** edit box; the drawing is displayed in the drawing window, as shown in Figure 8-28.

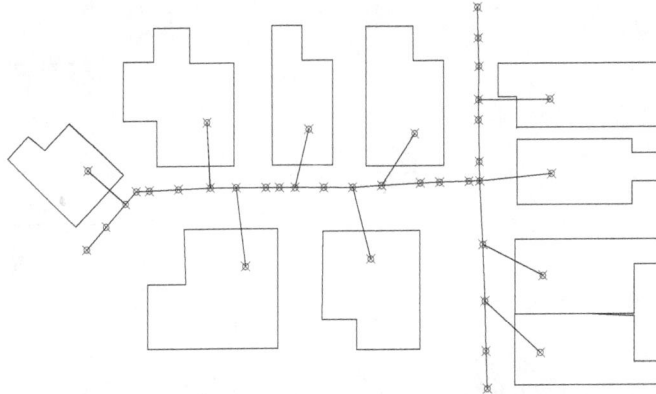

Figure 8-28 Drawing used in creating topology

Creating Topology by Using Drawing Objects

In this section of the tutorial, you will create a network topology by defining the required parameters.

1. Choose the **New** tool from the **Topology** panel in the **Create** tab; the **Create Polygon Topology** wizard with the **Select Topology Type** page is displayed.

2. In this page, select the **Network** radio button in the **Topology type** area. Notice that the left pane of the **Create Polygon Topology** wizard displays the links for the pages required for defining a **Network** topology.

3. Enter the **c08_network** text in the **Topology name** edit box below the **Topology type** area.

4. Choose the **Next** button in this page; the **Select Links** page of the **Create Network Topology** wizard is displayed, as shown in Figure 8-29.

5. In the **Select Links** page, select the **Select all** radio button if not selected by default, and then choose the **Select Layers** button on the right of the **Layers** edit box; the **Select Layers** window is displayed with all the layers selected in the list box.

6. In the **Select Layers** window, select the **Gas_line** layer from the **Layers** list box.

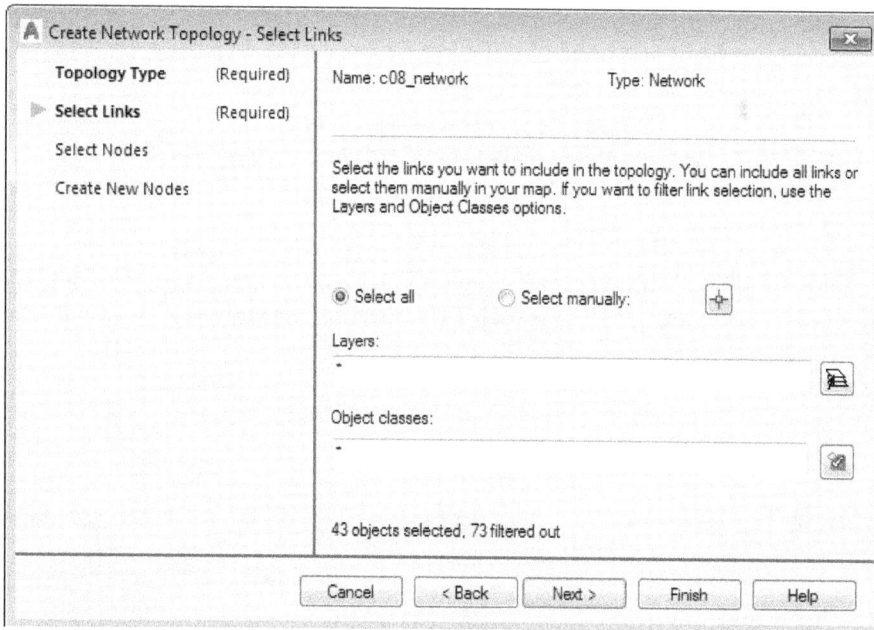

*Figure 8-29 The **Create Network Topology** wizard with the **Select Links** page*

7.   Choose the **Select** button from the **Select Layers** window; the window is closed and the layer name **Gas_line** is displayed in the **Layer** edit box of the **Select Links** page.

8.   Now, choose the **Next** button in this page; the **Select Nodes** page of the **Create Network Topology** wizard is displayed.

9.   In the **Select Nodes** page, select the **Select all** radio button if not selected by default, and then choose the **Select Layers** button on the right of the **Layers** edit box; the **Select Layers** window is displayed.

10. In the **Select Layers** window, click on the **Gas_points** layer in the list box.

11. Choose the **Select** button from the **Select Layers** window; this window is closed and the **Gas_points** layer name is displayed in the **Layers** edit box in the **Select Nodes** page.

12. Now, choose the **Next** button in the **Select Nodes** page; the **Create New Nodes** page of the **Create Network Topology** wizard is displayed.

13. Choose the **Finish** button in this page; the wizard is closed and the network topology is created.

Displaying the Created Topology

In this section of the tutorial, you will display the geometry of the objects in the **c08_network** network topology.

1.   In the **Map Explorer** tab of the **TASK PANE**, click on the [**+**] node corresponding to the **Topologies** node; the **c08_network** topology is displayed in the **Topologies** node.

2. Right-click on the **c08_network** topology; a shortcut menu is displayed. In this shortcut menu, choose the **Show Geometry** option; the network topology created is highlighted in red color in the drawing window.

Saving the Drawing File with the Created Topology

1. Choose the **Save As** option from the Application Menu; the **Save Drawing As** dialog box is displayed.

2. In the **Save Drawing As** dialog box, enter the text **c08_Tut02a.dwg** in the **File name** edit box and select the **AutoCAD 2013 Drawing (*.dwg)** option from the **Files of type** drop-down list, if it is not selected by default.

3. Next, choose the **Save** button next to the **File name** edit box; the drawing file is saved with the specified name.

Tutorial 3   Performing the Topology Query

In this tutorial, you will perform the topology query on a topology.   **(Expected time: 30 min)**

The following steps are required to complete this tutorial:

a. Open the drawing file and load the **c08_network** topology.
b. Perform query on the **c08_network** topology for the region shown in Figure 8-30.
    Query condition: circular buffer zone of radius **7** units from the node **A**, as shown in Figure 8-30.

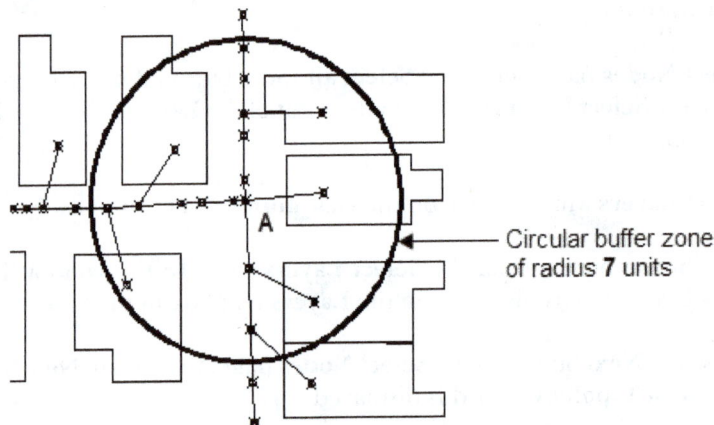

Figure 8-30 The buffer zone for the topological query highlighted by a circle

Opening the Drawing and Loading the Topology in it

1. Choose the **Open** button from the Quick Access Toolbar; the **Select File** dialog box is displayed.

2. In the **Select File** dialog box, browse to the following location:

    C:\m3d_2017\c08_m3d_2017_tut\c08_tut03

3. Next, choose the **c08_Tut03.dwg** file in the **Select File** dialog box; a preview of the selected drawing file is displayed in the **Preview** area of the dialog box.

4. Choose the **Open** button displayed on the right of the **File name** edit box; the drawing is displayed in the drawing window.

5. In the **Map Explorer** tab of the **TASK PANE**, expand the **Topologies** node by clicking on the corresponding [+] symbol; the **c08_network** topology is displayed in the node.

6. Next, right-click on the **c08_network** topology in the **Topologies** folder; a shortcut menu is displayed. In this shortcut menu, place the cursor on the **Administration** option; a flyout is displayed. Choose the **Load Topology** option from the flyout, as shown in Figure 8-31; the **AutoCAD Map Topology Audit** message box is displayed, as shown in Figure 8-32.

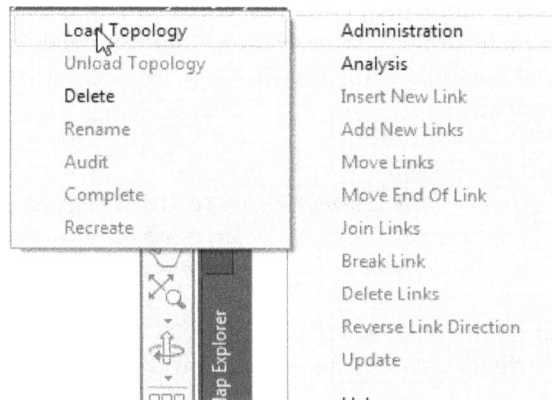

Figure 8-31 *Choosing the **Load Topology** option from the flyout*

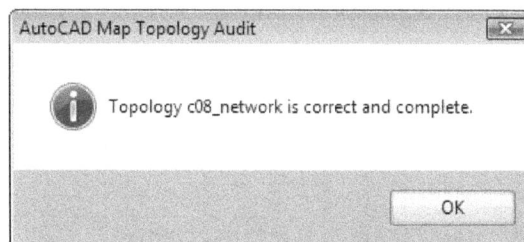

Figure 8-32 *The **AutoCAD Map Topology Audit** message box*

7. Choose the **OK** button in this message box; the topology is loaded and the **c08_network** topology is activated.

Defining and Performing the Topological Query

In this section, you will define and execute a topology query.

1. Choose the **Define Query** tool from the **Topology** panel in the **Create** tab; the **Topology Query** dialog box is displayed.

2. In the **Topology Query** dialog box, ensure that the **c08_network** option is selected in the **Name** drop-down list.

3. In the **Result Topology** area of the **Topology Query** dialog box, select the **Permanent** radio button from the **Topology Type** area.

4. Next, enter the **Topo_query** text in the **Name** edit box.

5. Now, choose the **Define Query** button; the **Define Query of Topology c08_network** dialog box is displayed.

6. In this dialog box, choose the **Location** button in the **Query Type** area; the **Location Condition** dialog box is displayed.

7. Select the **Circle** radio button in the **Boundary Type** area and then choose the **Define** button located at the lower left corner of the **Location Condition** dialog box; you are prompted to specify the center of the circle in the drawing area. Notice, the cursor turns into a crosshair in the drawing window.

Note
*Before proceeding further, make sure that you have activated the **Node** object snap option. To activate the object snap for node, right-click on the **Object Snap** button in the Status Bar; a shortcut menu is displayed. In this shortcut menu, click on the **Node** option; the **Node** object snap is activated.*

8. Place the crosshair on the node **A**, refer to Figure 8-30. When the node is tracked, click on it; the center of the circle is selected and you are prompted to specify the radius of the circle.

9. Specify **7** in the dynamic input edit box, refer to Figure 8-33 and then press ENTER; the **Define Query of Topology c08_network** dialog box is displayed with the **Location: CROSSING CIRCLE** query expression in the list box in the **Current Query** area.

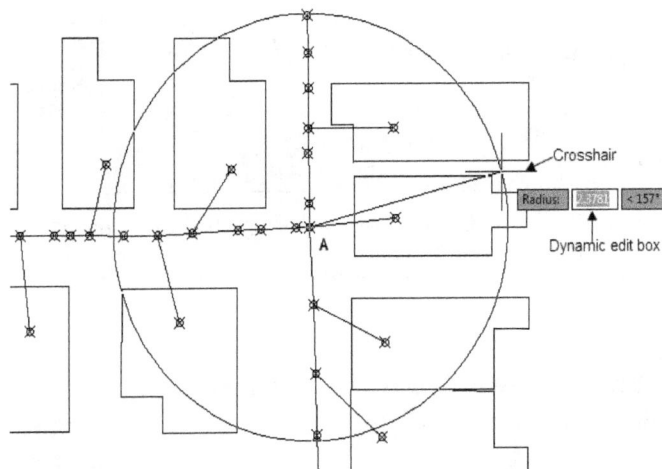

Figure 8-33 Drawing the circular boundary for the topological query

10. Choose the **Execute Query** button at the bottom in the **Define Query of Topology c08_network** dialog box; this dialog box is closed and the result of the current query is displayed in the drawing window, as shown in Figure 8-34.

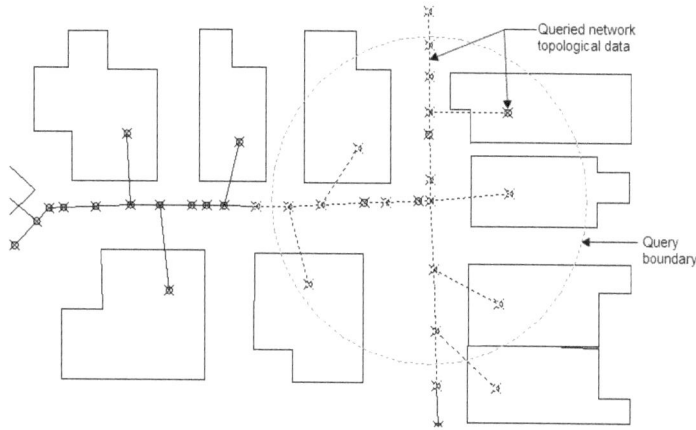

Figure 8-34 *The resulting data after the query is executed*

Saving the Drawing File

1. Choose the **Save As** option from the Application Menu; the **Save Drawing As** dialog box is displayed.

2. In the **Save Drawing As** dialog box, enter the text **c08_Tut03a** in the **File name** edit box.

3. Choose the **Save** button next to the **File name** edit box; the drawing file is saved with the specified name.

Self-Evaluation Test

Answer the following questions and then compare them to those given at the end of this chapter:

1. Which of the following cleanup actions is used to remove all the unwanted small objects?

    (a) **Delete Duplicates**   (b) **Erase Short Objects**
    (c) **Erase Dangling Objects**   (d) **Weed Polylines**

2. Which of the following tools in the **Topology** panel of the **Create** tab is used to create a thematic map that is based on the topology?

    (a) **Object Thematic Query**   (b) **New**
    (c) **Topology Thematic Query**   (d) None of these

3. Which of the following drawing cleanup actions is used to create nodes on objects that cross each other?

    (a) **Break Crossing Objects**   (b) **Dissolve Pseudo nodes**
    (c) **Delete Duplicates**   (d) **Erase Dangling Objects**

4. Which of the following options is not a type of topology?

    (a) **Node**   (b) **Network**
    (c) **Polygon**   (d) None of these

5. You can rectify errors in a drawing manually by selecting the _____ radio button from the **Options** area in the **Select Actions** page of the **Drawing Cleanup** wizard.

6. You can create topologies by using the _____ tool in the **Topology** panel of the **Create** tab.

7. In the **Topology Query** dialog box, you can specify the query condition by choosing the _____ button.

8. The _____ cleanup action is used to identify and delete objects that have same start and end points or have start and end points within the defined tolerance.

9. The drawing cleanup actions selected by the user are displayed in the _____ list box of the **Cleanup Actions** page in the **Drawing Cleanup** wizard.

10. It is not necessary to perform the drawing cleanup process before creating the topology for the drawing objects. (T/F)

11. In the **Select Objects** page of the **Drawing Cleanup** wizard, you can either select drawing objects automatically or manually. (T/F)

12. You can specify the parameters of a cleanup action in the **Cleanup Parameters** area of the **Cleanup Actions** page in the **Drawing Cleanup** wizard. (T/F)

13. You can specify the tolerance value for the **Erase Short Objects** drawing cleanup action. (T/F)

14. You can create the missing centroids while creating a polygon topology. (T/F)

15. In AutoCAD Map 3D, you cannot identify the sliver polygons that are created while creating a polygon topology. (T/F)

Review Questions

Answer the following questions:

1. Which of the following pages is used to specify the display parameters for an error marker?

    (a) **Select Links**   (b) **Select Nodes**
    (c) **Select Centroids**   (d) **Error marker**

2. You can select the blocks in the drawing by using the _____ option in the **Objects of Interest** area of the **Object Thematic Mapping** dialog box.

3. You can load a query by using the _____ button in the **Topology Query** dialog box.

4. You can apply the drawing cleanup to drawing layers or drawing objects as per your requirement. (T/F)

5. You cannot perform the drawing cleanup on a point object. (T/F)

6. You can specify the color and block parameters for an error marker. (T/F)

7. You can detect and correct sliver polygons by using the options in the **Cleanup Methods** page of the **Drawing Cleanup** wizard. (T/F)

8. Sliver polygons are the polygons that are created because of poor digitization. (T/F)

9. In the **Error Marker** page of the **Create Polygon Topology** wizard, you can specify the size of an error marker. (T/F)

10. Network topology is a combined model of nodes and links. (T/F)

EXERCISE

Download *c08_m3d_2017_exe.zip* from *www.cadcim.com* and extract it for the following exercise.

Exercise 1

Extract the **c02_exr01** folder from *c02_m3d_2017_exe.zip*. Select the *c08-m3d-2017-exr01. dwg* drawing file and then perform the drawing cleanup on the drawing objects in the **Road** and **Building** layers. Next, create a network topology using the drawing objects in the **Road** layer and a polygon topology using the drawing objects in the **Building** layer.

(Expected time: 1hr)

Answers to Self-Evaluation Test

1. b, **2.** c, **3.** a, **4.** d, **5.** Interactive, **6.** New, **7.** Define Query, **8.** Delete Duplicates, **9.** Selected Actions, **10.** F, **11.** T, **12.** T, **13.** T, **14.** T, **15.** F

Chapter 9

Data Analysis

Learning Objectives

After completing this chapter, you will be able to:
- *Find information about the object data*
- *Interpret the geospatial feature*
- *Analyze the topological data*
- *Create hillshade effect on a 3D surface*
- *Convert polygon topology to polylines*

INTRODUCTION

In the previous chapters, you learned to create and edit geographical objects (features) in a spatial dataset. You also learned to search and filter data from these datasets by executing the spatial and non-spatial queries.

In this chapter, you will learn about various types of spatial analyses that are used for analyzing feature data. You will also understand the procedure to use the topological data to perform the network, buffer, and overlay analysis.

Moreover, in this chapter, you will learn to collect information about the drawing objects by using various tools that are specific to data analysis. The procedure to convert the data from the polygon topology into polylines is also discussed in this chapter.

OBJECT DATA ANALYSIS

Object data analysis is the process of inspecting, cleaning, measuring, and transforming a dataset and then preparing the output dataset as per the project requirement. You can use various tools in the AutoCAD Map 3D user interface for performing data analysis. The tools used for object data analysis are discussed next.

Line & Arc Information Tool

Ribbon:	Analyze > Inquiry > Line & Arc Information
Command:	MAPCGLIST

The **Line & Arc Information** tool is used to find out the description of the alignment details of the line and arc geometries between two points. To find out the description, choose the **Line & Arc Information** tool from the **Inquiry** panel; the cursor will change into a selection box in the drawing window. By using this selection box, select the points of interest or select the desired polyline in the drawing window. On doing so, the alignment details between two selected points or a polyline will be displayed in the Command prompt. For example, consider a curve drawn between the points **A** and **B**, as shown in Figure 9-1. The procedure to display the alignment details between the two points is discussed next.

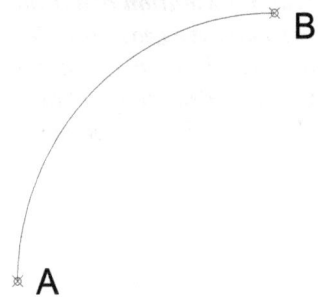

Figure 9-1 A curve drawn between the points A and B

To display the arc information of the curve **AB**, choose the **Line & Arc Information** tool from the **Inquiry** panel; the cursor will change into a selection box in the drawing window. Place this selection box on the curve and then left-click; the alignment information of this curve will be displayed at the Command prompt. Next, press F2; the **AutoCAD Text Window** will be displayed with the arc information, as shown in Figure 9-2.

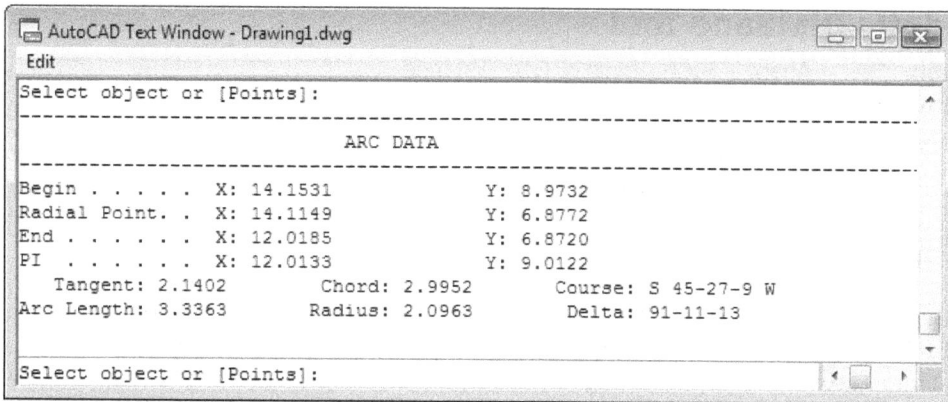

Figure 9-2 The AutoCAD Text Window displaying the arc information of the AB curve

Angle Information Tool

Ribbon:	Analyze > Inquiry > Angle Information
Command:	MAPCGANG

The **Angle Information** tool is used to find out the acute and obtuse angles between two intersecting or non-parallel lines. To find the angles, choose the **Angle Information** tool from the **Inquiry** panel; you will be prompted to select the first line. Select the first line in the drawing area by clicking on the required drawing object; you will be prompted to select the second line. Click on the required line object in the drawing to select it; the values of the acute and obtuse angles between the two selected lines will be displayed at the Command prompt, as given below:

The acute angle is 34-52-42
The obtuse angle is 325-7-18

Note
The angles displayed will be in the Degree-Minute-Second format.

You can also use this tool to find the angle between the three points (start point, vertex, and endpoint) in the current drawing. To do so, choose the **Angle Information** tool from the **Inquiry** panel; the cursor will change into a selection box in the drawing window. Next, type **P** and then press ENTER; the selection box will change into a crosshair in the drawing window and you will be prompted to specify the starting point. Place this crosshair at the start or first point and then click; the first point will be selected and you will be prompted to specify the vertex. Next, place the crosshair at the second point and then click; the vertex will be selected and you will be prompted to select the endpoint. Next, place the crosshair at the endpoint and then click; the endpoint will be selected and the acute and obtuse angles between the three points will be displayed at the Command prompt.

Geodetic Distance Tool

Ribbon: Analyze > Geo Tools > Geo Distance
Command: MAPDIST

The **Geodetic Distance** tool is used to measure the distance between two points in a map. Unlike other distance measuring tools, this tool considers the coordinate system distortion and the curvature of the Earth while calculating the distance between two points. This tool is used for both the geospatial and drawing data.

To measure the actual distance between two points, choose the **Geo Distance** tool from the **Geo Tools** panel; you will be prompted to specify the first point. Note that the cursor will change into a crosshair in the drawing window. Click in the drawing window to specify the start point; you will be prompted to specify the second point. Click in the drawing area to specify the second point; the geodetic distance between the first and second point will be measured and the Distance, Azimuth, Delta X, and Delta Y information will be displayed at the Command prompt. An illustration of the statements displayed at the Command prompt is given below.

Distance = 3.5776 (Meter)
Azimuth = 51 degrees (forward), 231 degrees (reverse)
Delta X = 2.9325, Delta Y = 2.0591

Note
*If the statements given above are not visible in the Command prompt, press F2; the **AutoCAD Text Window** will be displayed. This window displays the statements pertaining to the geodetic distance measurement.*

Coordinate Track Tool

Ribbon: Analyze > Geo Tools > Coordinate Track
Command: MAPTRACKCS

The **Coordinate Track** tool is used to invoke the **TRACK COORDINATES** palette. You can use the options in this palette to track the coordinates of a point/cursor in any coordinate system. Note that the **TRACK COORDINATES** palette can be used to track the coordinates of a point simultaneously in multiple coordinate reference systems.

The option in this palette can also be used to create the Line or MPolygon feature. To create the line or MPolygon, choose the **Coordinate Track** tool from the **Geo Tools** panel; the **TRACK COORDINATES** palette will be displayed. To display the options in the **Coordinate Tracker** toolbar located at the top in the palette, place the cursor on this toolbar; the toolbar will be activated, as shown in Figure 9-3.

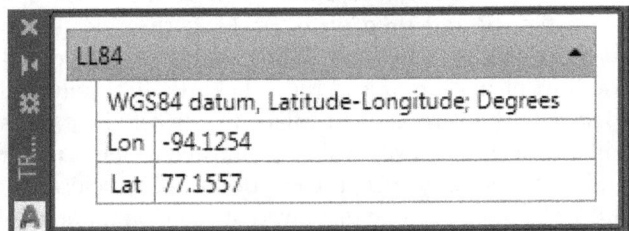

*Figure 9-3 The **TRACK COORDINATES** palette with the **Coordinate Tracker** toolbar activated*

At the top in the **TRACK COORDINATES** palette, you can view the code and description of the assigned coordinate system. To use a predefined coordinate system in AutoCAD, click on the **Coordinate system name**; a pop-up window will be displayed. Figure 9-4 shows the description of various coordinate systems from the pop-up window.

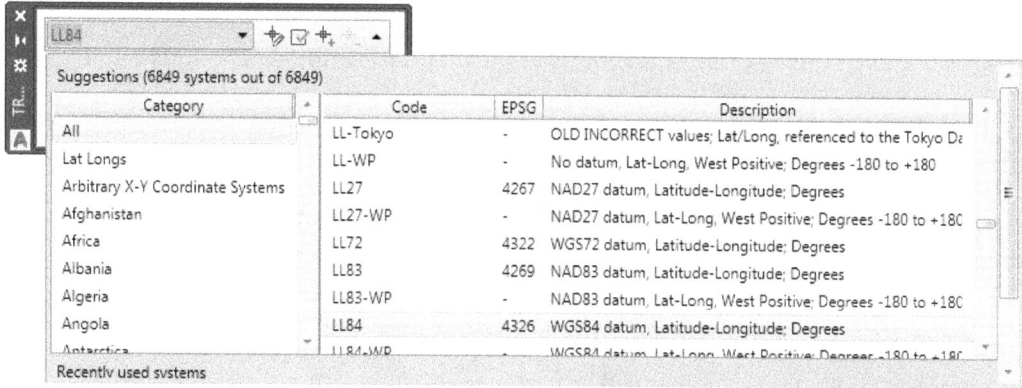

Figure 9-4 *The pop-up window displayed with various coordinate systems in the selected category*

In the pop-up window, select the desired option from the **Category** list box; the coordinate systems in the selected category will be displayed on the right pane of the **Category** list box. Next, place the cursor on the required coordinate system and then double-click on it; the pop-up window will be closed and the code of the selected coordinate system will be displayed in the drop-down list in the **Coordinate Tracker** toolbar of the **TRACK COORDINATES** palette.

You can also specify the options for tracking coordinates. To do so, choose the **Options** button from the **Coordinate Tracker** toolbar of the **TRACK COORDINATES** palette; the **Coordinate Tracker Options** dialog box will be displayed, as shown in Figure 9-5.

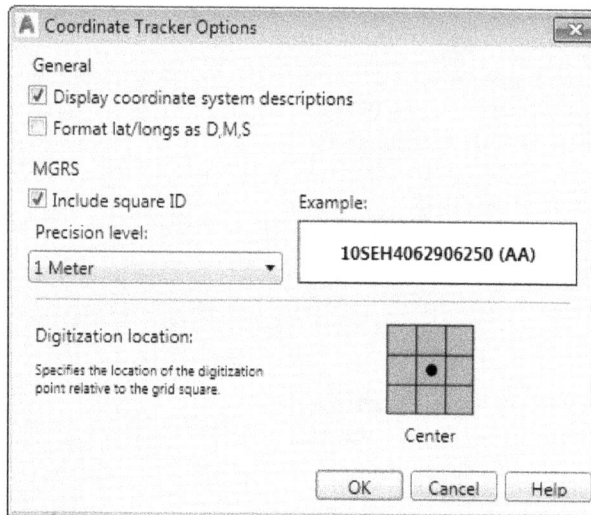

Figure 9-5 *The **Coordinate Tracker Options** dialog box*

In the **General** area of the **Coordinate Tracker Options** dialog box, the **Display coordinate system descriptions** check box is selected by default. As a result, the coordinate system information is displayed below the **Coordinate Tracker** toolbar. If you clear this check box, the description of the coordinate system will not be displayed. In this area, the **Format lat/longs as D,M,S** check box is clear by default. As a result, you can enter latitudes and longitudes in the form of planar coordinates in the **Lat** and **Lon** (or **X** and **Y**) edit boxes below the **Coordinate Tracker** toolbar. If you select this check box, you can enter the coordinates in the Degree, Minute, and Second format. In the **MGRS** area of the **Coordinate Tracker Options** dialog box, you can specify options for the Military Grid Reference System. After specifying options, choose the **OK** button in this dialog box; the modified settings will be applied.

You can add multiple coordinate systems to the **TRACK COORDINATES** palette to track the coordinates in the added coordinate systems. To add another coordinate system to the **TRACK COORDINATES** palette, choose the **Clone this tracker and insert it below** button in the toolbar; a new toolbar with coordinate tracker will be displayed below the existing toolbar. To remove or delete one of these toolbars, choose the **Remove this tracker** button in the corresponding toolbar; this toolbar will be removed.

You can use the **TRACK COORDINATES** palette to track the coordinates in a given coordinate system while creating a feature. To invoke this palette, choose the **Coordinate Track** tool from the **Geo Tools** panel of the **Analyze** tab. On doing so, the **TRACK COORDINATES** palette will be displayed. In this palette, enter the coordinates of the desired point in the **Lon** and **Lat** edit boxes below the **Coordinate Tracker** toolbar and then choose the **Digitize using entered position** button from the toolbar; the entered coordinates of the point will be located in the current drawing.

Continuous Distance Tool

Ribbon:	Analyze > Inquiry > Continuous Distance
Command:	MAPCGCDIST

The **Continuous Distance** tool is used to measure the cumulative distance between a base point and multiple points or the distance between a series of points. Figure 9-6 shows a drawing with the points **A**, **B**, **C**, **D**, **E**, and **F**. The methods of measuring the distance between these points by using the **Base** and **Continuous** options are discussed next.

Base

You can use the **Base** option for measuring the distance between the base point **A** and the point **B** or **C**, refer to Figure 9-6. To do so, choose the **Continuous Distance** tool from the **Inquiry** panel;

Figure 9-6 A drawing with various points drawn in it

the cursor will change into a crosshair in the drawing window. Next, type **B** and then press ENTER; the **Base** option for distance measurement will be invoked and you will be prompted to specify the start point. Place the crosshair on the point **A** and click; this point will be selected

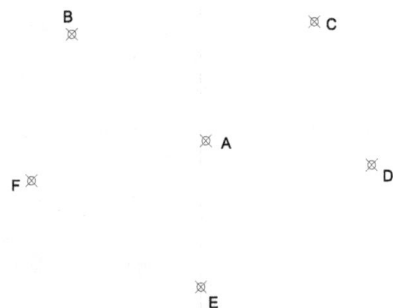

as the base point and you will be prompted to specify the next point. Then, place the crosshair on the point **B** and click; the distance between the points **A** and **B** will be displayed in the Command prompt as given below.

> Command: _mapcgcdist
> Base/Continuous <Base>: b
> Start point:
> Next point:
> Distance = 5.2137

Keep the **Base** option active and place the cursor on the point **C** and click; the distance between the points **A** and **C** will be displayed in the Command prompt. Next, press ENTER; the sum of the measured distances (**AB**+**AC**) will be displayed in the Command prompt. If the Command prompt is not displayed, press F2; the **AutoCAD Text Window** will be displayed with the statements of the measured distances, as shown in Figure 9-7.

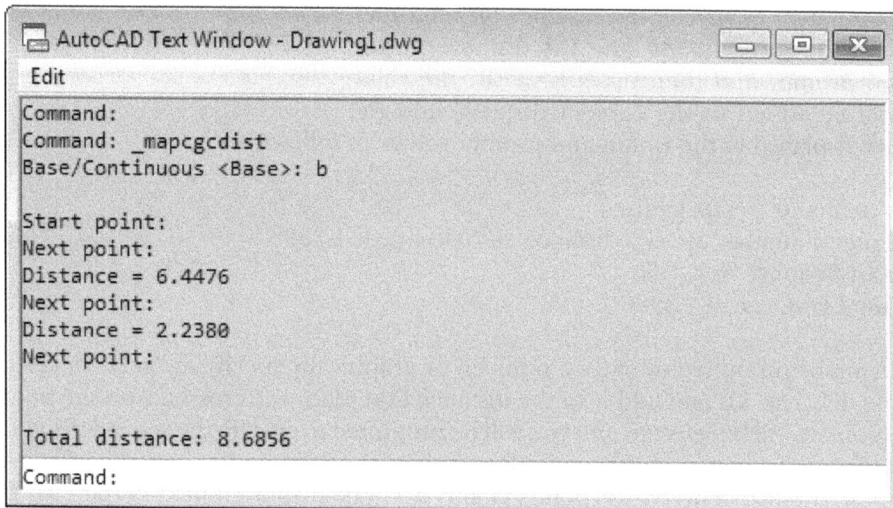

```
AutoCAD Text Window - Drawing1.dwg

Edit

Command:
Command: _mapcgcdist
Base/Continuous <Base>: b

Start point:
Next point:
Distance = 6.4476
Next point:
Distance = 2.2380
Next point:


Total distance: 8.6856
Command:
```

*Figure 9-7 The **AutoCAD Text Window** with the distance measurement statements*

Continuous

You can use the **Continuous** option to measure the distance between multiple points, say **A**, **B**, and **C** in a series. To do so, choose the **Continuous Distance** tool from the **Inquiry** panel of the **Analyze** tab; the cursor will change into a crosshair in the drawing window. Next, type **C** and then press ENTER; the **Continuous** option for distance measurement will be invoked and you will be prompted to select the first point. Place the crosshair on the point **A**, refer to Figure 9-6, and then click; the point **A** will be selected as the origin and you will be prompted to select the next point. Next, drag and place the crosshair on the point **B** and then click; the distance between the points **A** and **B** will be displayed in the Command prompt and you will be prompted to specify the next point. Drag and place the crosshair on the point **C** and then click; the distance between the points **B** and **C** will be displayed in the Command prompt and you will again be prompted to specify the next point. Next, press ENTER; the sum of the distances measured between the points **A**, **B**, and **C** (**AB**+**BC**) in series will be displayed in the Command prompt. If the Command prompt is not displayed, press F2; the **AutoCAD Text Window** will be displayed with the distance measurement statements.

Add Distances Tool

Ribbon:	Analyze > Inquiry > Add Distances
Command:	MAPCGADIST

The **Add Distances** tool is used to add the distances of several disjunct parts by selecting the points in the map or by entering the distance at the Command prompt, or by selecting the numeric text in a drawing. The method of adding the distances between three disjunct objects using various options in the **Add Distances** tool is illustrated next.

Figure 9-8 shows two drawings objects with the distance and text parameters. By using these objects, you can calculate the sum of disjunct distances (**OC=OA+AB+BC**) in this drawing. To calculate distance, choose the **Add Distances** tool from the **Inquiry** panel; the cursor will turn into a crosshair in the drawing window and you will be prompted to enter a number, or specify the distance (graphically), or select a text. Enter **1.3386** (the **OA** distance) in the Command prompt and then press ENTER; the value **1.3386** will be added to the current distance and the statements displayed in the Command prompt will be as follows:

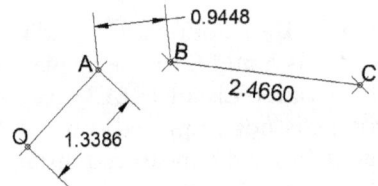

Figure 9-8 The distance, points, and text objects

> Command: _mapcgadist
> Enter a number, specify distance, or Select text: 1.3386
> Last Distance = 1.3386
> Total Distance = 1.3386

You will again be prompted to enter a number, or graphically specify distance, or select a text. To find the distance **AB** and add it to the distance **OA**, place the crosshair on the point **A** and click; the point **A** will be selected and you will be prompted to specify the second distance. Next, drag and place the crosshair on the point **B** and click; the distance between the points **A** and **B**, and the total distance between the points **O** and **B** will be displayed next to the **Last Distance** and **Total Distance** in the Command prompt, respectively (2.2834 = 1.3386 + 0.9448). The statements displayed in the Command prompt are given below:

> Enter a number, specify distance, or Select text:
> Specify second distance:
> Last Distance = 0.9448
> Total Distance = 2.2834

You will be prompted again to enter a number, or specify the distance (graphically), or select a text. Now, you need to add the distance between the points **B** and **C** to the distance **OB** to find **OC**. To specify the distance between the points **B** and **C** by selecting text in the drawing window, type **S** in the Command prompt and then press ENTER; the cursor will change into a selection box. Click on the numeric text **2.4660** above the line **BC**; this selected value will be added to the previous distance and the segment **Total Distance = 4.7494** will be displayed in the Command prompt. If the Command prompt is not displayed, press F2; the **AutoCAD Text Window** will be displayed, as shown in Figure 9-9.

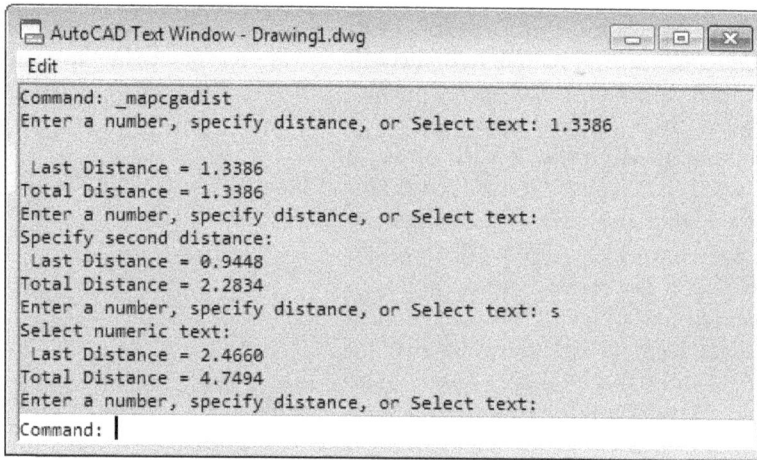

Figure 9-9 AutoCAD Text Window displaying the total distance between the points O and C

Note

While selecting the numeric text in the drawing window, make sure that numbers are written by using the single line text format and not by using the MTEXT option. Otherwise, the numeric text will not be selected.

Measure Tool

Ribbon: Home > Draw > Points drop-down > Measure
Command: MEASURE

The **Measure** tool is used to place the point objects or blocks at desired intervals on an object. While creating drawings, you may need to segment an object at fixed distances without actually dividing it. The procedure of placing points and blocks at desired interval is explained next.

To create segments by using a point object, choose the **Measure** tool from the **Draw** panel; the cursor will change into a selection box in the drawing window and you will be prompted to select the required object to measure. Select the required line or polyline object by clicking on it and then press ENTER; you will be prompted to specify the length of the segment. Specify the segment length and then press ENTER; the points will be placed on the selected line object at equal intervals, as shown in Figure 9-10.

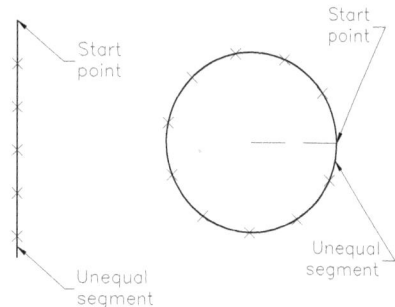

*Figure 9-10 The points placed by using the **Measure** tool*

You can also place blocks at equal intervals on a line object. To do so, choose the **Measure** tool from the **Draw** panel; the cursor will change to a selection box in the drawing window. Place this selection box on the line or polyline object and then click; the line or polyline object will be selected and the **Specify the length of segment or [Block]** statement will be displayed at the

Command prompt. Type **B** and then press ENTER; the statement **Enter name of the block to insert** will be displayed in the Command prompt. Type the name of the block to be inserted and then press ENTER; the statement **Align block with object ? [Yes/No] <Y>** will be displayed in the Command prompt. Type **Y** and then press ENTER to align the block to the selected line object, or type **N** and then press ENTER to insert the blocks without aligning them to the selected line object. On doing so, the statement **Specify length of segment** will be displayed in the Command prompt. Specify the length of the interval between the objects and then press ENTER; the line object with the blocks placed

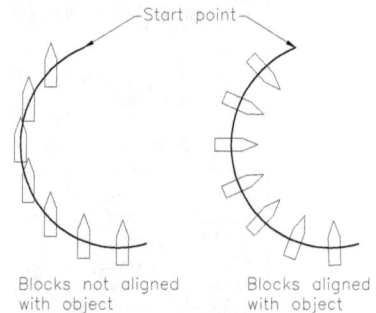

Figure 9-11 Blocks placed by using the Measure tool

on it will be displayed in the drawing window. Figure 9-11 shows two line objects: line with the blocks aligned to it and the other with the blocks not aligned to it.

List Slope Tool

Ribbon:	Analyze > Inquiry > List Slope
Command:	MAPCGSLIST

The **List Slope** tool is used to read information about the height of the first and second points, slope, grade, and the distance of the line in the horizontal plane selected. The procedure to find the slope information between the points of a 3D polyline object is discussed next.

Consider a 3D polyline **AB** with the point **A**(0,0,0) and the point **B**(10,10,10), as shown in Figure 9-12.

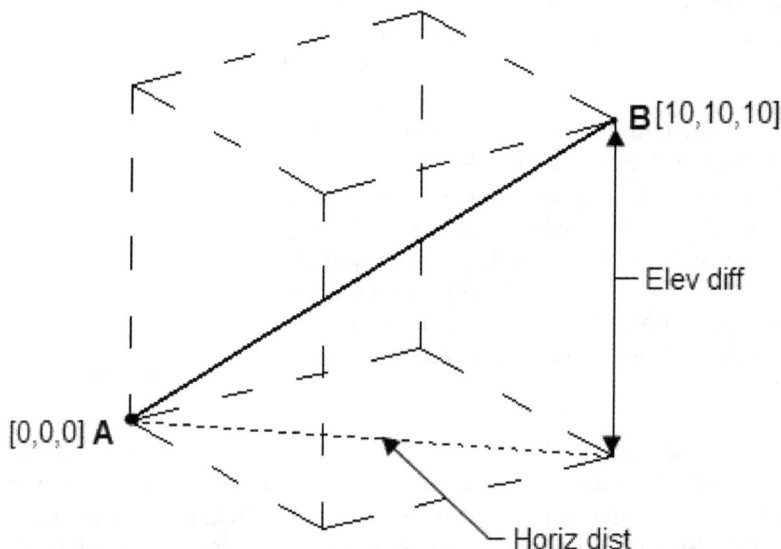

Figure 9-12 The coordinates of the points A and B with the description of dimensions

To read the details related to the slope property, choose the **List Slope** tool from the **Inquiry** panel in the **Analyze** tab; the cursor will change into a selection box in the drawing window and you will be prompted to select the object. To select the object, place this selection box on the polyline **AB** and then click; the information related to the height and distance functions are displayed in the Command prompt. If the Command prompt is not displayed, then press F2; the **AutoCAD Text Window** with slope parameters will be displayed, as shown in Figure 9-13.

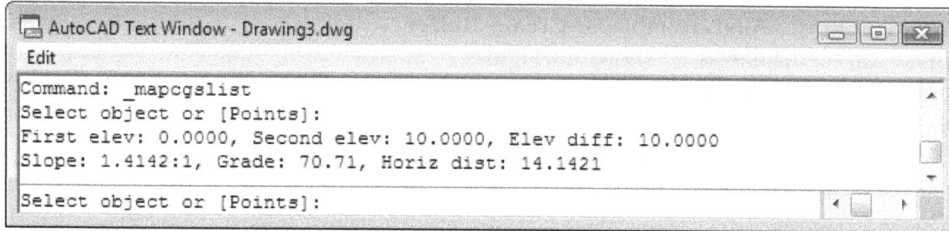

```
AutoCAD Text Window - Drawing3.dwg
Edit
Command: _mapcgslist
Select object or [Points]:
First elev: 0.0000, Second elev: 10.0000, Elev diff: 10.0000
Slope: 1.4142:1, Grade: 70.71, Horiz dist: 14.1421
Select object or [Points]:
```

*Figure 9-13 The **AutoCAD Text Window** with slope parameters*

In the **AutoCAD Text Window**, the **First elev** statement represents the height of the first point (**A**) above the datum plane. Similarly, the **Second elev** statement represents the height of the second point (**B**) above the datum plane. The **Elev diff** statement represents the difference between the first and second points. The **Slope** statement represents the ratio of the horizontal distance between the points **A** and **B** to the difference of heights between the points **A** and **B**. The **Grade** statement represents the percentage of the ratio of the difference of heights between the points **A** and **B** to the horizontal distance between the points **A** and **B**.

> **Tip**
> *The expressions to find the **Slope** and **Grade** entities are:*
>
> $$Slope = \frac{Horizontal\ distance\ between\ two\ points}{Height\ difference\ between\ the\ points} = \frac{Horiz\ dist}{Elev\ diff}$$
>
> $$Grade = \frac{Height\ difference\ between\ two\ points}{Horizontal\ difference\ between\ the\ points}\ X\ 100 = \frac{Elev\ diff}{Horiz\ dist}\ X\ 100$$

FEATURE DATA ANALYSIS

In AutoCAD Map3D, the feature data analysis is used to analyze and interpret the spatial data in a feature dataset. There are two tools used for performing the feature data analysis: **Feature Buffer** and **Feature Overlay**. Note that you cannot use these feature data analysis tools to analyze drawing object data. The application of these analysis tools in analyzing feature data is discussed next.

Feature Buffer Tool

Ribbon:	Analyze > Feature > Feature Buffer
Command:	MAPFDOBUFFERCREATE

The **Feature Buffer** tool is used to create a region of specified width around a feature object. Generally, this tool is used for proximity analysis. Proximity analysis is the process of defining the relation between an object and its neighborhood. The procedure to perform the feature data analysis by using the **Feature Buffer** tool is discussed next.

Consider two points **A** and **B** with a line object between them, as shown in Figure 9-14. You can apply feature buffer to this feature object by using the **Feature Buffer** tool. To do so, choose the **Feature Buffer** tool from the **Feature** panel in the **Analyze** tab; the **Create Buffer** dialog box will be displayed, as shown in Figure 9-15. The options in the **Create Buffer** dialog box are discussed next.

Figure 9-14 A feature line object

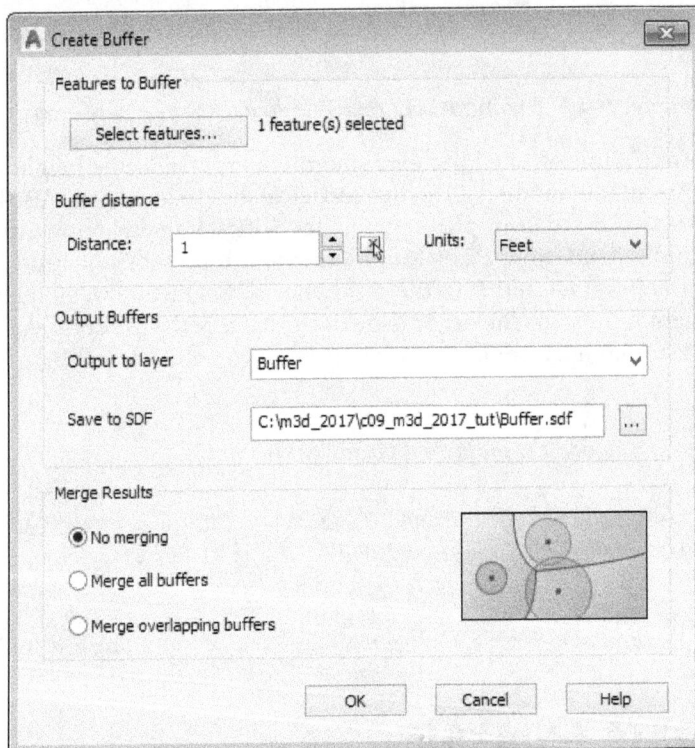

*Figure 9-15 The **Create Buffer** dialog box*

Features to Buffer Area
The option in the **Features to Buffer** area is used to select the feature to buffer. To select the objects to buffer, choose the **Select features** button in this area; the cursor will change into a selection box and you will be prompted to select the feature objects in the drawing. To select the required feature objects, click on the objects in the drawing area. After selecting the required features, press ENTER; the **Create Buffer** dialog box will be displayed. The number of selected objects will be displayed in the **Features to Buffer** area of this dialog box.

Buffer distance Area
The options in the **Buffer distance** area are used to specify the distance between a point and the boundary of the buffer region. To specify the buffer distance between the feature object and

the boundary of the buffer region, enter a value in the **Distance** edit box or set the value by using the spinner on the right of this edit box. You can also specify the buffer distance graphically in the drawing window. To do so, choose the **Pick buffer distance on map** button next to the **Distance** edit box; the **Create Buffer** dialog box will be closed and you will be prompted to specify the first point. Click in the drawing to specify the first point; you will be prompted to specify the second point. Click in the drawing window to specify the second point; the **Create Buffer** dialog box will be displayed again. Note that the **Distance** edit box in the **Create Buffer** dialog box will now display the value of the buffer distance. Next to specify the measuring unit for the buffer distance, select the required option from the **Units** drop-down list.

Output Buffers Area

The options in the **Output Buffers** area are used to specify the layer on which the buffer region will be saved. To create a new layer for the current buffering process, enter a name in the **Output to layer** edit box in this area. You can also save this buffer layer in the *.SDF* format. To do so, choose the Browse button next to the **Save to SDF** edit box; the **Select an SDF File** dialog box will be displayed. In this dialog box, enter a name in the **File name** edit box and then choose the **Save** button; the **Select an SDF File** dialog box will be closed and the file path will be displayed in the **Save to SDF** edit box.

Merge Results Area

The options in the **Merge Results** area are used to specify the criteria for merging the buffer regions. In this area, the **No merging** radio button is selected by default. As a result, the overlapping buffer regions will not merge. If you select the **Merge all buffers** radio button, all buffer regions will merge into a single buffer region. If you select the **Merge overlapping buffers** radio button, only the overlapping buffer regions will merge.

After specifying the required options in the **Create Buffer** dialog box, choose the **OK** button; this dialog box will close and a buffer region will be created. Also, a new feature layer will be created with the name specified in the **Create Buffer** dialog box and the layer will be added to the list box in the **Display Manager** tab of the **TASK PANE**. Figure 9-16 shows a buffer area created for line **AB** with a buffer distance of 25 units.

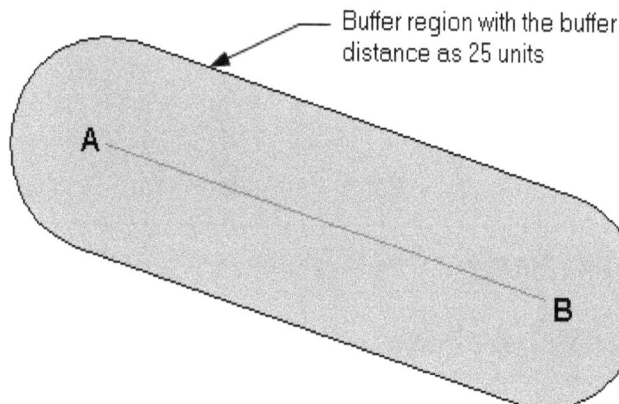

Figure 9-16 The buffer region created for the line feature objects AB

Feature Overlay Tool

Ribbon:	Analyze > Feature > Feature Overlay
Command:	MAPGISOVERLAY

Feature
Overlay

The **Feature Overlay** tool is used to perform analysis by comparing data in two vector layers, which are spatially related to a particular geographical area. In the feature overlay analysis, one of the vector layers is used as a source vector layer and the other vector layer is overlaid on the source vector layer. As a result of the feature overlay analysis, a new feature layer is created. This layer can be saved and used as dataset in your project.

To perform the overlay analysis, choose the **Feature Overlay** tool from the **Feature** panel; the **Sources and Overlay Type** page of the **Overlay Analysis** wizard will be displayed, as shown in Figure 9-17. The options in the **Sources and Overlay Type** and **Set Output and Settings** pages of the **Overlay Analysis** wizard are discussed next.

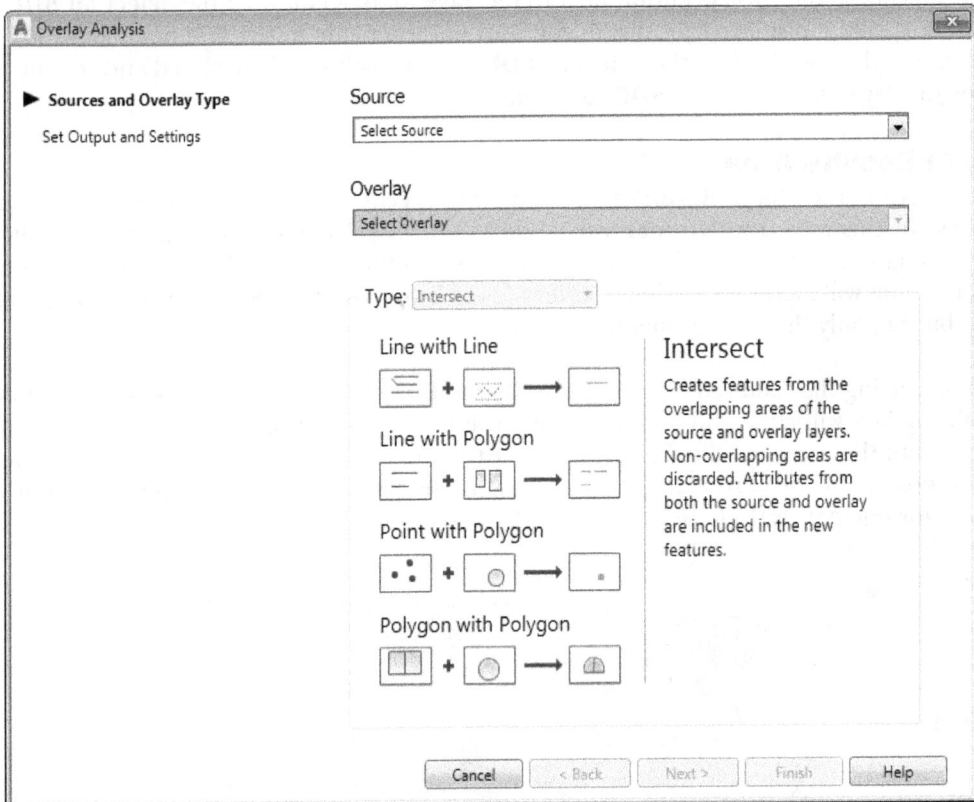

*Figure 9-17 The **Sources and Overlay Type** page of the **Overlay Analysis** wizard*

Sources and Overlay Type Page

In the **Sources and Overlay Type** page of the **Overlay Analysis** wizard, you can select a vector layer to use it as a source. To do so, select a vector layer from the **Source** drop-down list; the selected layer will be displayed in the **Source** drop-down list. To use a vector layer to overlay on the source layer, select an option from the **Overlay** drop-down list; the selected layer will be displayed in the **Overlay** drop-down list. Also, the **Type** drop-down list below the **Overlay** drop-down list will be activated.

Note
*The options in the **Type** drop-down list are displayed based on the source and overlay vector layers selected.*

Based on your analysis requirements, you can use various overlay types such as **Intersect**, **Union**, **Erase**, **Identity**, **Clip**, **Paste**, and **Symmetric Difference** from the **Type** drop-down list. These options are discussed next.

Intersect
The **Intersect** option is an analytical operation that can be used to create a feature from the overlapping areas of source and overlay feature layers. The attributes from both feature layers are included in the new features.

Union
The **Union** option is used to create a feature data by merging the two vector layers used in the process. The resultant layer has attributes from source and overlay feature layers.

Erase
The **Erase** option is used to create a feature data by removing or erasing the features from the source layer that are overlapped by the features in the overlay layer. The new features have attributes data only from the source layer.

Note
*If you use the **Erase** option, the features from only the source vector layer will be retained.*

Identity
The **Identity** option is used to create a new feature data from the common region between the source and overlay vector layers. The new layer inherits attributes from both the layers.

Clip
The **Clip** option is used to create a new feature data from the overlapping areas of the source and overlay feature layers. The feature layer thus created inherits the attributes of the source file only.

Paste
The **Paste** option is used to create a feature layer by pasting the overlay vector layer on the source vector layer. The resulting feature layer will have all the features of the overlay layer. Also, the areas in the source layer that are not covered by the overlay layer become features in the resulting layer.

Symmetric Difference
The **Symmetric Difference** option is used to create a new feature data by eliminating the common region between the source vector layer and the overlay vector layer.

When you select an option from the **Type** drop-down list, an illustration of how the overlay analysis will be performed by using the selected option will be displayed in the **Type** area. Next, choose the **Next** button located at the bottom; the **Set Output and Settings** page of the **Overlay Analysis** wizard will be displayed, as shown in Figure 9-18.

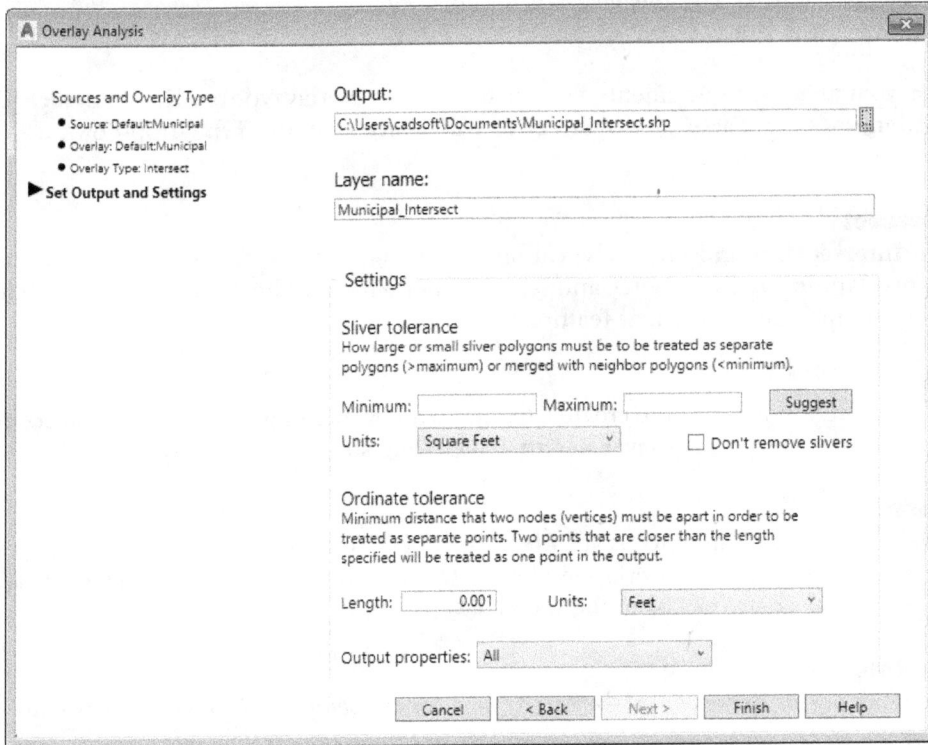

*Figure 9-18 The **Set Output and Settings** page of the **Overlay Analysis** wizard*

Set Output and Settings Page

In the **Set Output and Settings** page of the **Overlay Analysis** wizard, you can specify the options for the data output. In addition, you can specify the parameters of the sliver polygon, ordinate tolerance, and output properties on this page. The options in this page are discussed next.

Output

The **Output** edit box is used to specify the folder path for saving the resultant file of the overlay analysis. To specify the path, choose the Browse button next to the **Output** edit box; the **Save** dialog box will be displayed. In this dialog box, enter a name in the **File name** edit box, and then select the **Autodesk SDF Files (*.sdf)** or **SHP Files (*.shp)** option from the **Save as type** drop-down list. Next, choose the **Save** button from the **Save** dialog box; the dialog box will be closed and a new file with the given name will be created.

Layer name

In the **Layer name** edit box, you can specify the name of the layer to be created.

Settings Area

In the **Settings** area of the **Set Output and Settings** page, you can specify the options for setting the parameters of the sliver polygons, ordinate limit for creating points, and output properties. The options in this area are discussed next.

Sliver tolerance. In this section, you can set the lowest and highest values of area measurements for detecting the sliver polygon by entering the least value in the **Minimum** edit box and the highest value in the **Maximum** edit box.

> **Tip**
> *If you choose the **Suggest** button next to the **Maximum** edit box, the lowest and highest measurement values for detecting the sliver polygons will be calculated automatically by AutoCAD software and the resultant values will be displayed in the **Minimum** and **Maximum** edit boxes.*

You can specify a unit for measuring the area of the sliver polygon by selecting an option from the **Units** drop-down list below the **Minimum** edit box. If you select the **Don't remove slivers** check box on the right of the **Units** drop-down list, the sliver polygons will be retained in the resulting vector layer.

Ordinate tolerance. In this section, you can specify the limits of the ordinate distance for creating points in the resulting vector layer. You can specify the minimum distance between two points so that if the distance between any two points is less than the specified length, then these two points will be treated as one in the output. To specify the length, enter a value in the **Length** edit box and then select an option from the **Units** drop-down list next to it.

Output properties. In this drop-down list, you can specify the desired properties of the source and overlay vector layer to be included in the resulting vector layer. To include all properties in the resulting vector layer, retain the default **All** option from the **Output properties** drop-down list. To include primary keys or feature identifiers such as Feature-ID, select the **Identifiers** option. To include all the properties except identifiers, select the **Non-Identifiers** option from the **Output properties** drop-down list.

After specifying all the parameters in the **Set Output and Settings** page of the **Overlay Analysis** wizard, choose the **Finish** button; the **Overlay** window showing the progress of the process will be displayed, as shown in Figure 9-19. After the processing of the overlay analysis has been completed, the **Overlay** window will close automatically and the resulting vector of the overlay analysis will be displayed in the **Display Manager** tab of the **TASK PANE**. Also, the resulting geometry will be displayed in the drawing window.

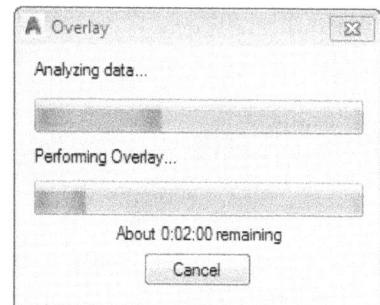

*Figure 9-19 The **Overlay** window*

Surface Hillshade Tool

Ribbon:	Analyze > Feature > Surface Hillshade
Command:	MAPHILLSHADE

The **Surface Hillshade** tool is used to shade the 3D surface. The shading is done by casting the sunlight across the surface at an angle in a direction. Hillshading produces more realistic image and helps to understand various elevation changes on the surface.

To create the hillshade effect, choose the **Surface Hillshade** tool from the **Feature** panel; the **Hillshade Settings** dialog box will be displayed, as shown in Figure 9-20. The options in the **Hillshade Settings** dialog box are discussed next.

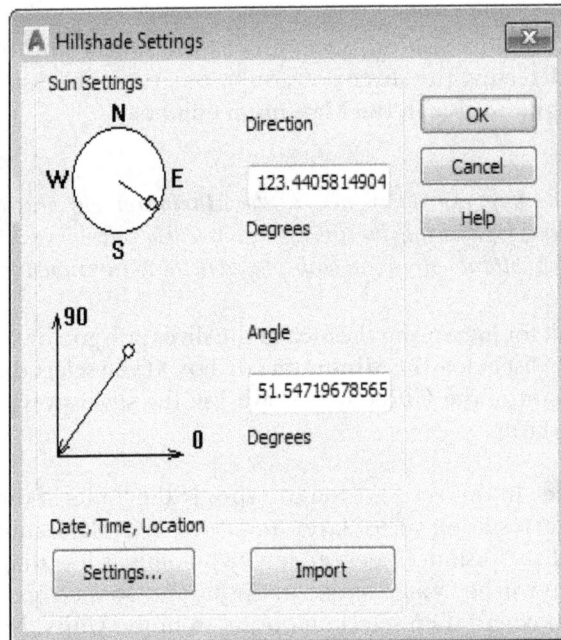

*Figure 9-20 The **Hillshade Settings** dialog box*

Sun Settings Area
In the **Sun Settings** area of the **Hillshade Settings** dialog box, you can specify the direction and angle for the raster-based surface. The options in this area are discussed next.

Direction
In this edit box, you can set the value for the direction from which the light should come. Alternatively, you can drag the yellow disk in the compass, or use **Settings.**

Angle
In this edit box, you can specify a value for the location of the light in the sky, such as near the horizon, directly overhead, or somewhere in between. Alternatively, you can drag the yellow disk to specify an angle, or use **Settings**.

Date, Time, Location Area
In the **Date, Time, Location** area of the **Hillshade Settings** dialog box, you can specify the advanced settings on the geographic location of the Sun.

TOPOLOGICAL DATA ANALYSIS
The topological data analysis is the process of displaying and processing the relationship between the points, lines, and polygons of a geographic region. Also, you can perform the network analysis to find the shortest or desired path between two points.

You need to load the required topology into the current drawing object to perform analysis on it. To load a topology, right-click on it in the **Topologies** node in the **Map Explorer** tab of the **TASK PANE**; a shortcut menu will be displayed. In this shortcut menu, place the cursor on the **Administration** option; a flyout will be displayed. In this flyout, choose the **Load Topology** option, as shown in Figure 9-21; the **AutoCAD Map Topology Audit** message box will be displayed. Choose the **OK** button in this message box; the topology will be loaded into the Modelspace and its name will be activated. The tools in the **Topology** panel of the **Create** tab and the **Drawing Object** panel of the **Analyze** tab, used in the process of topological data analysis, are discussed next.

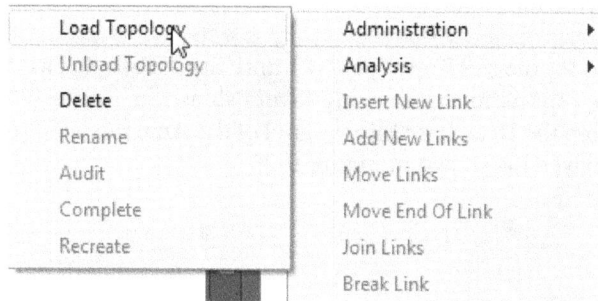

Load Topology	Administration ▶
Unload Topology	Analysis ▶
Delete	Insert New Link
Rename	Add New Links
Audit	Move Links
Complete	Move End Of Link
Recreate	Join Links
	Break Link

*Figure 9-21 Choosing the **Load Topology** option from the **TASK PANE***

Note
*Make sure that the topologies used for the process of analysis are loaded into the Modelspace. If the topologies are not loaded into the current drawing, the statement **No topologies are loaded** will be displayed in the Command prompt.*

Show Topology Tool

| **Ribbon:** | Create > Topology > Show |
| **Command:** | MAPSHOWTOPO |

The **Show** tool is used to display the topologies related to an object in the current or source drawing. To display a topology, choose the **Show** tool from the **Topology** panel; the cursor will change into a selection box in the drawing window. Place this selection box on the required object and then click to select it; the topologies related to the selected object will be displayed at the Command prompt.

Show Geometry Tool

The **Show Geometry** tool is used to display the geometry of the selected topology in the current or source drawing. To display the geometry of a topology, enter **MAPSHOWGEOM** at the Command prompt and then press ENTER; the cursor will change into a crosshair in the drawing window and you will be prompted to enter the topology name. Next, type the name of the required topology and then press ENTER; the geometry of this topology will be displayed in red in the drawing window.

Object Overlay Tool

Ribbon:	Analyze > Drawing Object > Object Overlay
Command:	MAPANOVERLAY

The **Object Overlay** tool is used to compare the relation between two topologies by overlaying the required topology on the source topology. You can use a point, line, or polygon topology as the source topology but the overlay topology must be a polygon topology. Also, you must have two topologies to work with (out of the two topologies, at least one must be a polygon topology). Otherwise, the **AutoCAD Map3D 2017** message box will be displayed with an error message.

For example, create two topologies **Topo_Poly_01** and **Topo_Poly_03**, as shown in Figures 9-22 and 9-23, by using the objects in the drawing layers shown in Figure 9-24. Next, choose the **Object Overlay** tool from the **Drawing Object** panel of the **Analyze** tab; the **Topology Selection** dialog box will be displayed, as shown in Figure 9-25.

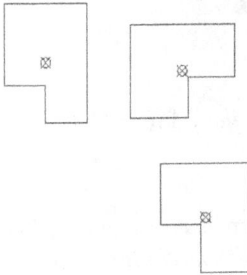

*Figure 9-22 The **Topo_Poly_01** topology*

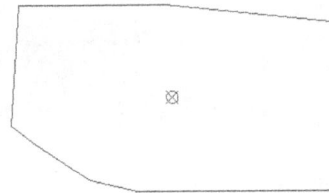

*Figure 9-23 The **Topo_Poly_03** topology*

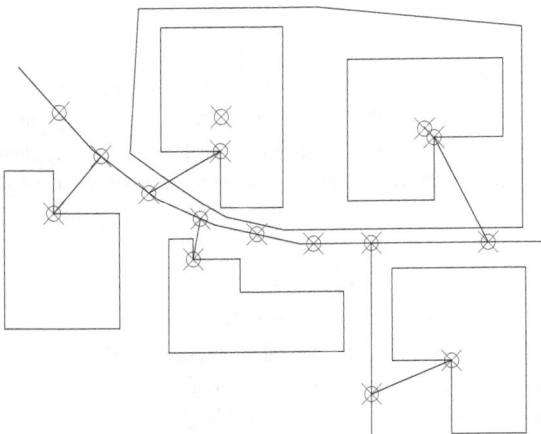

Figure 9-24 The drawing with objects in each layer

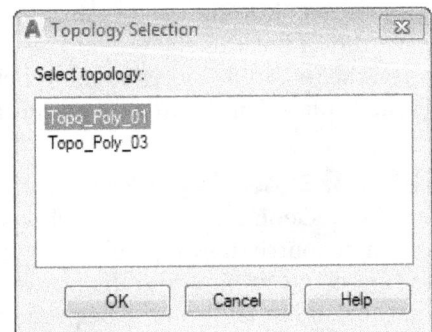

*Figure 9-25 The **Topology Selection** dialog box*

In the **Topology Selection** dialog box, select the required topology from the **Select topology** list box as the source topology. Next, choose the **OK** button; this window will close and the **Topology Overlay Analysis** wizard with the **Analysis Type** page will be displayed, as shown in Figure 9-26.

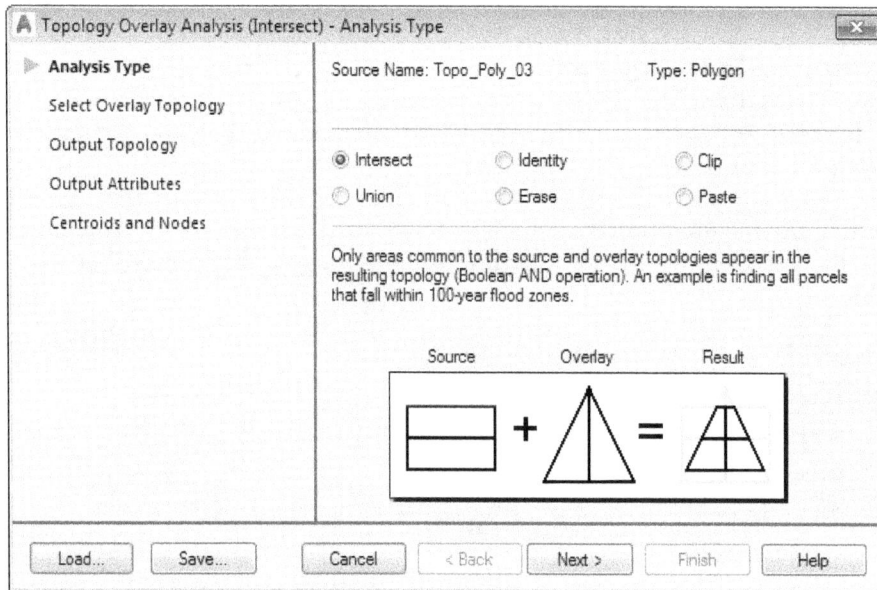

Figure 9-26 The Analysis Type page of the Topology Overlay Analysis wizard

> **Note**
> *The options in the **Topology Overlay Analysis** wizard will depend upon the type of source topology selected in the **Topology Selection** dialog box.*
>
> *In the following section, the polygon topology has been used as the source topology to explain the options in the **Topology Overlay Analysis** wizard.*

Various pages in the **Topology Overlay Analysis** wizard are discussed next.

Analysis Type Page

In the **Analysis Type** page, you can specify the type of overlay operation to be performed on the selected dataset. Also, you can view the description of each analysis operation by selecting the corresponding radio button.

> **Note**
> *The overlay analysis types such as **Intersect**, **Identity**, **Clip**, **Union**, **Erase**, and **Paste** have been discussed in the **Feature Overlay** section of this chapter.*

After selecting the radio button corresponding to the required type for overlay analysis (the **Intersect** radio button is selected by default), choose the **Next** button; **Select Overlay Topology** page will be displayed in the **Topology Overlay Analysis** wizard.

Note

*On selecting the required radio button in the **Analysis Type** page, the name of the selected overlay analysis type is displayed in the heading of the **Topology Overlay Analysis** wizard, (refer to Figure 9-26).*

Select Overlay Topology Page

In the **Select Overlay Topology** page, you can specify a topology that will be used as an overlay topology. To specify the overlay topology, select an option from the **Polygon topology to overlay** drop-down list on this page.

After selecting the required overlay topology, choose the **Next** button; the **Topology Overlay Analysis (Intersect)** wizard with the **Output Topology** page will be displayed, as shown in Figure 9-27.

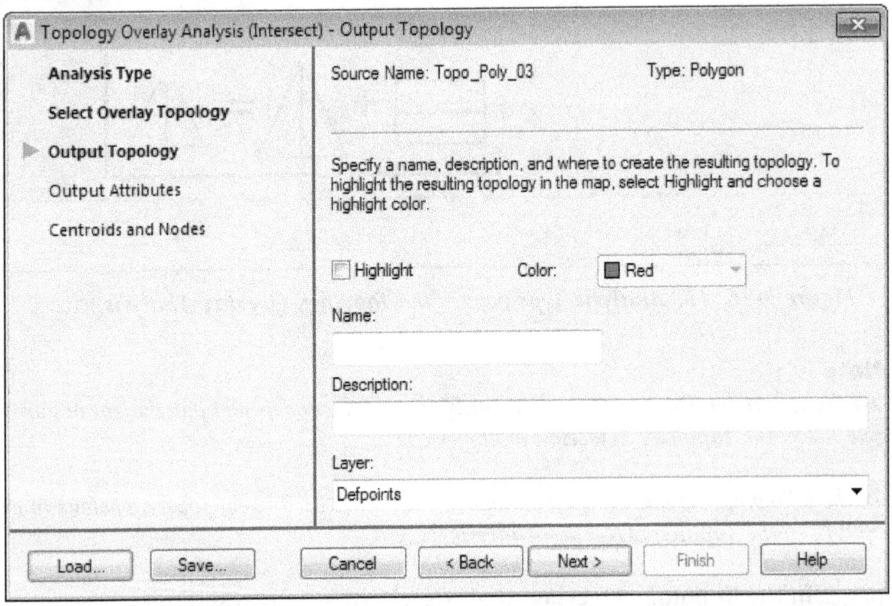

*Figure 9-27 The **Output Topology** page of the **Topology Overlay Analysis (Intersect)** wizard*

Output Topology Page

In the **Output Topology** page, you can specify the display color of the resulting topology, apply a name to it, and specify a layer for the new topology to be created. To highlight the resulting topology with a color in the drawing window, select the **Highlight** check box; the **Color** drop-down list will be activated. Select a color for the resulting topology from this drop-down list. In the **Name** edit box, you can enter a name for the resulting topology. Optionally, you can write the description of the resultant topology in the **Description** text box.

To create the topology in the existing drawing layer, select the required layer name in the **Layer** drop-down list. You can also create a new layer for the topology. To do so, enter the required layer name in the **Layer** edit box.

After specifying the required parameters in the **Output Topology** page, choose the **Next** button; the **Topology Overlay Analysis** wizard with the **Output Attributes** page will be displayed, as shown in Figure 9-28.

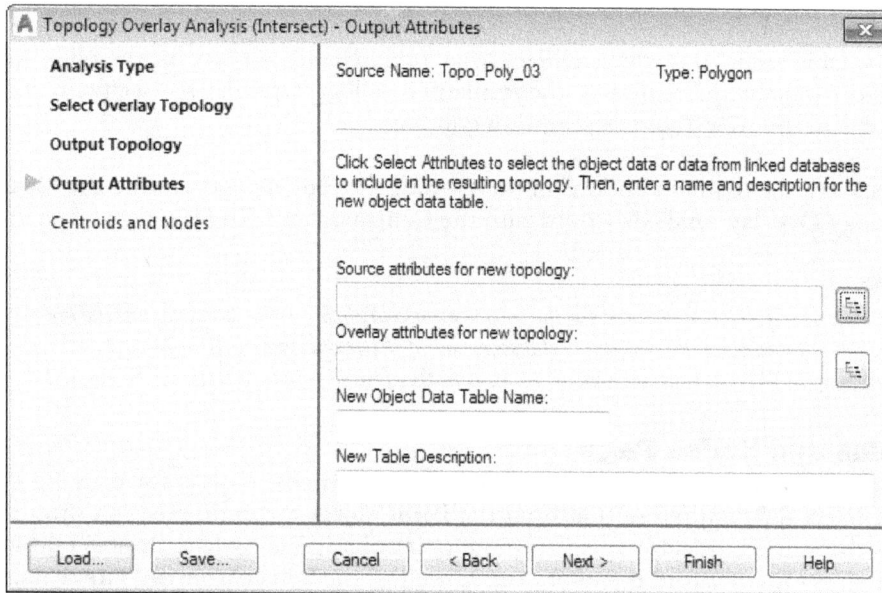

*Figure 9-28 The **Output Attributes** page of the **Topology Overlay Analysis** wizard*

Output Attributes Page

In the **Output Attributes** page, refer to Figure 9-28, you can specify conditions for the attributes to be used for the resulting topology. To do so, choose the **Select Attributes** button next to the **Source attributes for new topology** text box; the **Expression Chooser** dialog box will be displayed.

In the **Expression Chooser** dialog box, you can expand a node by clicking on the [+] symbol corresponding to it in the **Expression** list box. On doing so, a list of options will be displayed in the node. Figure 9-29 shows the **Expression Chooser** dialog box with the **Topologies** node expanded. In the expanded folder, select the check box corresponding to the required option and then choose the **OK** button; the **Expression Chooser** dialog box will close and the expression of the topology selected will be displayed in the **Source attributes for new topology** text box.

Next, to specify the attributes to be copied from the overlay topology, choose the

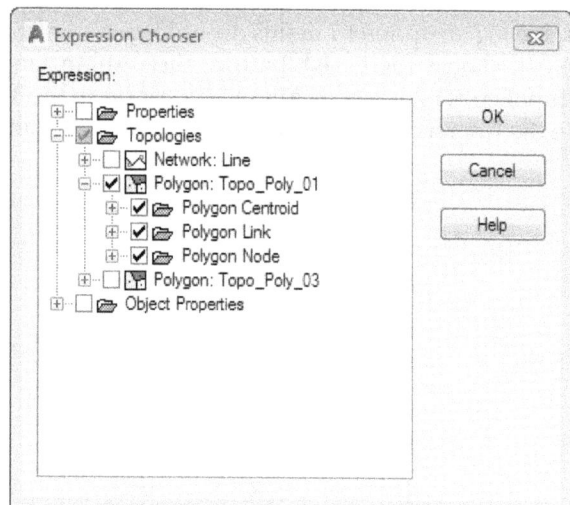

*Figure 9-29 The **Expression Chooser** dialog box with the **Topologies** node expanded*

Select Attributes button displayed next to the **Overlay attributes for new topology** text box; the **Expression Chooser** dialog box will be displayed. Specify the options in the dialog box and choose the **OK** button; the expression for the selected topology will be displayed in the **Overlay attributes for new topology** text box.

Specify the table name in the **New Object Data Table Name** edit box for the object data table that will store the data pertaining to the resultant topology. Optionally, specify the description for this table in the **New Table Description** edit box.

After specifying the required options in the **Output Attributes** page, choose the **Next** button; the **Topology Overlay Analysis** wizard with the **Centroids and Nodes** page will be displayed.

Note
Before specifying an attribute for the source or overlay layer while carrying out the overlay analysis, make sure that you do not select the same set of the property or attribute values for the resulting topology. Otherwise, errors will creep in and the desired topology will not be created.

Centroids and Nodes Page

In the **Centroids and Nodes** page, you can specify the point objects for creating centroids. To do so, select the required option from the **Point object for node creation** drop-down list. Alternatively, to use blocks for specifying the point object, choose the Browse button next to the **Point object for centroid creation** drop-down list; the **Select Drawing File** dialog box will be displayed. Browse and select the drawing file for the required block in this dialog box and then choose the **Open** button; the **Select Drawing File** dialog box will close and the block in the selected DWG file will be used for creating the centroid.

In the **Create new nodes** area of this page, you can also specify whether to create nodes for a new topology. The **Create new nodes for topology** check box in this area is clear by default. As a result, new nodes will not be created for the new topology. To create nodes for a new topology, select this check box. On doing so, the **Point object for node creation** drop-down list with the Browse button next to it will be activated. Specify a point object for the newly created node by selecting an option from this drop-down list or by using the Browse button, as described earlier. Next, choose the **Finish** button; the resulting topology **Topo_result** will be displayed in the **Topologies** folder of the **Map Explorer** tab in the **TASK PANE**, and the geometry corresponding to the **Intersect** overlay analysis will be displayed in the drawing window, as shown in Figure 9-30.

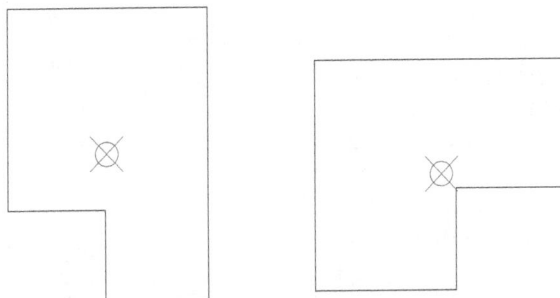

*Figure 9-30 The geometry of the resulting topology **Topo_result***

Dissolve Tool

Ribbon: Analyze > Drawing Object > Dissolve
Command: MAPANDISSOLVE

The **Dissolve** tool is used to combine two polygons or remove nodes between links that name the same attribute or same property value in a specified field. To merge or combine two polygons, choose the **Dissolve** tool from the **Drawing Object** panel; the **Topology Selection** dialog box will be displayed. In this dialog box, select an option from the **Select topology** list box and then choose the **OK** button; the **Topology Dissolve** wizard with the **Set Parameters** page will be displayed. Various pages and options for dissolving two topologies are discussed next.

Set Parameter Page

In the **Set Parameter** page, you can specify the parameters based on which the polygons will be dissolved. To do so, choose the **Select Attributes** button next to the **Dissolve by** edit box; the **Expression Chooser** dialog box will be displayed. In the **Expression Chooser** dialog box, expand the required node and then select an option from the expanded node by clicking on it. Next, choose the **OK** button in the **Expression Chooser** dialog box; the dialog box will be closed and the expression of the parameters set will be displayed in the **Dissolve by** edit box. After setting the parameters, choose the **Next** button; the **New Topology** page will be displayed.

New Topology Page

In the **New Topology** page of the **Topology Dissolve** wizard, you can specify the display color of the resulting topology. You can also assign a name to the resulting topology, and specify a layer for it. If you want the resulting topology with the desired color in the drawing window, select the **Highlight** check box. On doing so, the **Color** drop-down list will be activated. Select a color for the resulting topology from this drop-down list. You can enter a name for the resulting topology in the **Name** edit box and the details in the **Description** text box. You can either create a new layer for the resulting topology by entering the layer name in the **Layer** edit box or select an existing layer by using the down-arrow on the right of the **Layer** edit box. After specifying the parameters for the new topology, choose the **Next** button; the **Object Data** page will be displayed.

Object Data Page

In the **Object Data** page, you can specify a new object data table and an object data field to hold the new topology data of the resulting topology. To do so, choose the **Define** button next to the **Object data field** drop-down list; the **Define Object Data** dialog box will be displayed. Define the object data, as discussed in Chapter 6. After creating or specifying the object data, choose the **Next** button; the **Topology Dissolve** wizard with the **Create New Centroids and Nodes** page will be displayed.

Create New Centroids and Nodes Page

In the **Create New Centroids and Nodes** page, you can specify the options to create new centroids and nodes for the resulting topology. Specify the options in this page, as discussed in the previous section of this chapter. After specifying all parameters in this page, choose the **Finish** button; the **Topology Dissolve** wizard will be closed and a new topology will be created with the given parameters.

Object Buffer Tool

Ribbon:	Analyze > Drawing Object > Object Buffer
Command:	MAPANBUFFER

Object
Buffer

The **Object Buffer** tool is used to create a buffer zone around a topology object. To create a buffer zone around a topology object, choose the **Object Buffer** tool from the **Drawing Object** panel; the **Topology Selection** window with a list of topologies in the current drawing will be displayed. In the **Topology Selection** window, select an option from the **Select topology** list box and then choose the **OK** button; the window will be closed and the **Topology Buffer** wizard with the **Set Buffer Distance** page will be displayed.

In the **Set Buffer Distance** page, you can specify the distance between a topology and the boundary of a buffer. To specify the buffer distance, enter a value in the **Buffer distance** edit box. You can also set the buffer distance by entering an expression in the **Buffer distance** edit box. To do so, choose the **Select Attributes** button next to the **Buffer distance** edit box. On doing so, the **Expression Chooser** dialog box will be displayed.

In this dialog box, expand the required node by clicking on the [+] sign corresponding to it. Next, select an option by selecting the check box corresponding to it. Then, choose the **OK** button; the dialog box will be closed and the expression thus generated will be displayed in the **Buffer distance** edit box of the **Set Buffer Distance** page. Next, choose the **Next** button; the **New Topology** page of the **Topology Buffer** wizard will be displayed.

The parameters in the **New Topology** and **Centroids and Nodes** pages have already been discussed in the previous sections. After setting the parameters in various pages of the **Buffer Topology** wizard, choose the **Finish** button; this wizard will be closed and a topology of the buffer zone will be created around the selected topology. Also, the buffer zone created will be displayed in the drawing window.

Figure 9-31 shows a network topology before applying buffer. Figure 9-32 shows the resultant buffer created around the features in the network topology.

Figure 9-31 A network topology used for buffer analysis

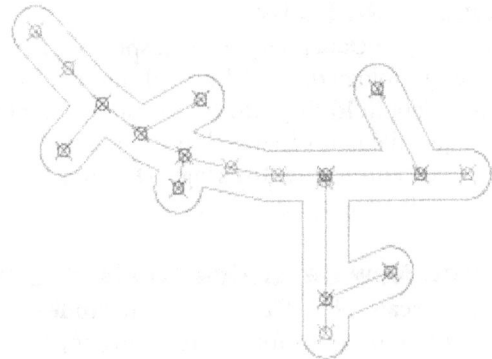

Figure 9-32 A buffer zone created around the features in the network topology

Network Analysis Tool

Ribbon: Analyze > Drawing Object > Network Analysis
Command: MAPANTOPONET

The **Network Analysis** tool is used to perform analyses such as the best route, shortest path, and flood route on a network topology. The applications of the various network analyses are discussed next.

Shortest path
The shortest path network analysis is used to find out the shortest possible route or path between two nodes.

Best route
The best route network analysis is used to find out the optimal or best route between a start point and other intermediate points, and then back to the start point.

Flood trace
The flood trace network analysis is used to find out the selection of nodes and links in the spread area, based on the parameters specified in the selection criteria.

To perform network analysis, choose the **Network Analysis** tool from the **Drawing Object** panel; the **Topology Selection** dialog box will be displayed. In this dialog box, select a network topology from the **Select topology** list box and then choose the **OK** button; this dialog box will close and the **Network Topology Analysis** wizard with the **Select Method** page will be displayed. Various pages and options in the **Network Topology Analysis** wizard are discussed next.

Note

*The options in the **Network Topology Analysis** wizard are discussed for the Shortest path analysis.*

Select Method Page
In the **Select Method** page, you can specify the method for performing the network analysis by selecting the **Shortest path**, **Best route**, or **Flood trace** radio button. On selecting the analysis method, the description and the illustration corresponding to the selected method will be displayed in the page. Next, choose the **Next** button; the **Choose Locations** page of the **Network Topology Analysis** wizard will be displayed, as shown in Figure 9-33.

Choose Locations Page
In the **Choose Locations** page, refer to Figure 9-33, you can use the options to select the start and end points of the path in the loaded topologies. To select the start point, choose the **Select Point** button corresponding to the **Select start point** option; you will be prompted to specify the start point. Click in the drawing at the desired location to specify the start point.

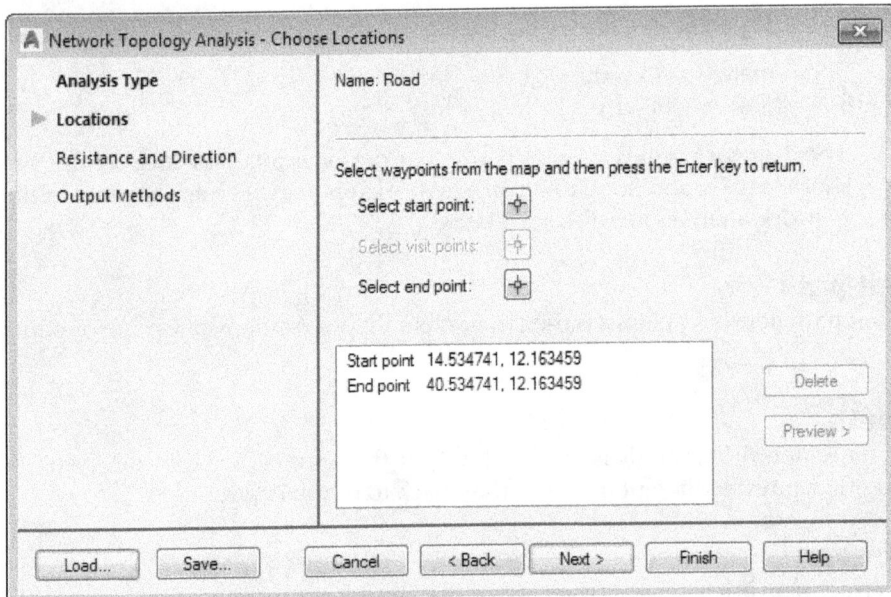

*Figure 9-33 The **Choose Locations** page of the **Network Topology Analysis** wizard*

Tip
*You can use the **Object Snap** option for locating the start point. To do so, enter **OS** at the command prompt; the **Object Snap** tab of the **Drafting Settings** dialog box will be displayed. In this dialog box, select the check box corresponding to the desired object snap option. Next, choose the **OK** button; the **Drafting Settings** dialog box will be closed and you will enter the start point selection mode.*

To confirm the selection of the start point, press ENTER; the **Network Topology Analysis** wizard with the **Choose Location** page will be displayed and the coordinates of the selected point will be displayed in the text box below the **Select end point** option. Next, specify the end point of the path by choosing the **Select Point** button corresponding to the **Select end point** option. On doing so, this wizard will be closed and you will be prompted to specify the end point. Select the end point by following the same procedure used for selecting the start point. On selecting the end point, its coordinates will be displayed below the coordinates of the start point in the list box in the **Choose Locations** page. Next, choose the **Next** button on this page; the **Network Topology Analysis** wizard with the **Resistance and Direction** page will be displayed.

Resistance and Direction Page

In the **Resistance and Direction** page, refer to Figure 9-34, you can specify the direction of links in a network topology. Also, you can specify the resistance of the link and node entities in the topology used for analysis. You can reverse the direction of links by selecting the **Reverse** check box located above the **Link direction** edit box. If you keep the **Link direction** edit box blank, the bidirectional trace will be applied while tracing the links. To specify a direction for tracing the links in the network, you can either enter a constant value in the **Link direction** edit box or apply an expression for including the object data by choosing the **Expression Evaluator** button. On choosing this button, the **Expression Chooser** dialog box will be displayed.

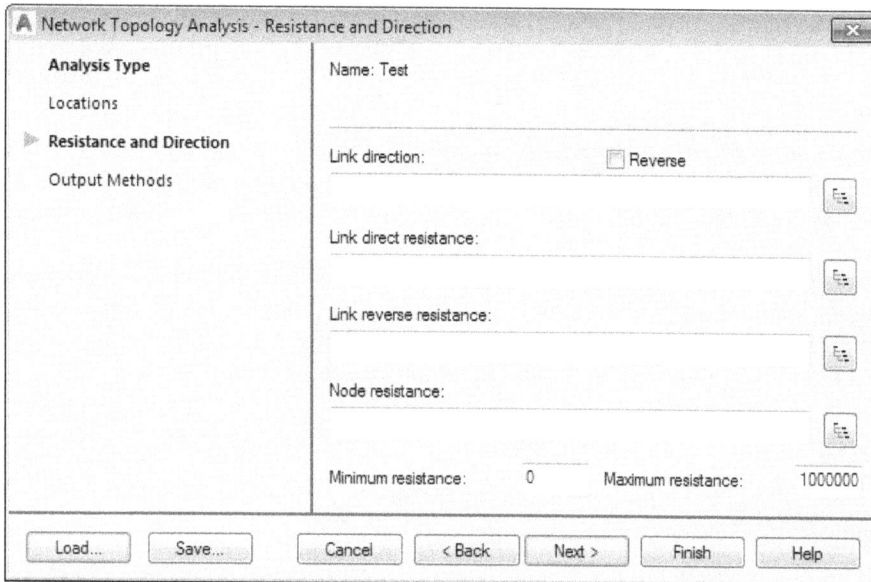

*Figure 9-34 The **Resistance and Direction** page of the **Network Topology Analysis** wizard*

In the **Expression Chooser** dialog box, you can expand a node by clicking on the [+] sign corresponding to the desired option. On doing so, a list of properties or attributes will be displayed. Select an option from the list and choose the **OK** button in the **Expression Chooser** dialog box; this dialog box will be closed and the expression created will be displayed in the **Link direction** edit box of the **Resistance and Direction** page.

You can specify the resistance value for the direct or reverse links in the network topology by using the **Link direct resistance** edit box. If you leave the **Link direct resistance** or **Link reverse resistance** edit boxes blank, then the length of the link in the network topology will be used for the resistance value. To specify the resistance value, you can enter a constant value in the **Link direct resistance** or **Link reverse resistance** edit box. Alternatively, you can apply a desired attribute or property value by using the **Expression Evaluator** button, as discussed in the previous paragraph. To specify the resistance offered by each node in the network topology while tracing through nodes, enter a constant value or select the required attribute or property value by using the **Expression Evaluator** button, as discussed earlier. To specify the lowest resistance value, enter the least allowable resistance value in the **Minimum Resistance** edit box. To specify the maximum value for the resistance, enter the highest allowable resistance value in the **Maximum Resistance** edit box.

After specifying all the parameters for the resistance and the direction of the links and nodes in this page, choose the **Next** button; the **Network Topology Analysis** wizard with the **Output** page will be displayed.

Output Methods Page

In the **Output Methods** page, you can specify the parameters for the output data. The **Highlight** check box is selected by default. As a result, the geometry of the path traced in the network topology will be highlighted in the drawing window.

This path will be highlighted in the color selected from the drop-down list next to the **Highlight** check box. If you clear this check box, then the geometry will not be highlighted in the drawing window.

To save the result of the network analysis as a new topology, select the **Create Topology** check box. On selecting this check box, the **Name** and **Description** edit boxes will be activated. Specify the desired name and description for the new topology in the **Name** and **Description** edit boxes.

After specifying all parameters in this page, choose the **Finish** button; the **Network Topology Analysis** wizard will be closed and the new topology created for the current analysis will be displayed in the **Topologies** folder in the **Map Explorer** tab of the **TASK PANE**. Also, the geometry of the path traced by this topology will be highlighted in the drawing window.

Create Closed Polylines Tool

Ribbon: Create > Topology > Create Closed Polylines
Command: MAPCLPLINE

The **Create Closed Polylines** tool is used to create closed polylines from an existing polygon topology. To create a closed polyline, choose the **Create Closed Polylines** tool from the **Topology** panel of the **Create** tab; the **Create Closed Polylines** dialog box will be displayed, as shown in Figure 9-35.

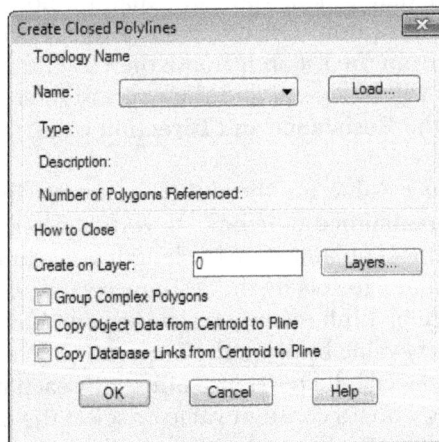

Figure 9-35 The Create Closed Polylines dialog box

Note
If the current drawing has a loaded topology, the name of that topology will be displayed in the **Name** *drop-down list in the* **Topology Name** *area of the* **Create Closed Polylines** *dialog box.*

In this dialog box, you can specify the polygon topology that will be used for creating closed polylines and set parameters for the conversion procedure. The parameters for creating polylines from a polygon topology in the **Topology Name** and **How to Close** areas of this dialog box are discussed next.

Topology Name Area

The options in the **Topology Name** area are used to specify the topology that will be used for creating polylines. To specify a topology, you can select an option from the **Name** drop-down list. On selecting a topology, the information about the **Type**, **Description**, and **Number of Polygons Referenced** parameters will be displayed in this area.

If a topology is present in the current drawing but not loaded, then you can load this topology by using the **Load** button. To do so, choose the **Load** button next to the **Name** drop-down list; the **Topology Selection** dialog box will be displayed with the list of unloaded topologies. Select the name of a topology from the **Select topology to load** list box and then choose the **OK** button; the **AutoCAD Map Topology Audit** message box will be displayed. In this window, choose the **OK** button; the message box will be closed and the name of the loaded topology will be displayed in the **Name** drop-down list in the **Create Closed Polylines** dialog box.

How to Close Area

The options in the **How to Close** area are used to set the parameters for creating closed polylines. In the **Create on Layer** edit box, you can specify the layer on which the polylines are to be created. You can select the name of a layer by choosing the **Layers** button next to the **Create on Layer** edit box. On doing so, the **Select** window will be displayed with a list of layers in the current drawing. Select a layer from the list box in the **Select** window and then choose the **OK** button; the window will be closed and the name of the selected layer will be displayed in the **Create on Layer** edit box. The **Group Complex Polygons** check box is clear by default. As a result, the complex polygons (polygons with islands inside) will not be grouped. To group them, select the **Group Complex Polygons** check box.

To copy all object data attached from the centroid of a polygon to the polyline being created, select the **Copy Object Data from Centroid to Pline** check box. To copy the database links from the centroid of the polygon topology to the polyline being created, select the **Copy Database Links from Centroid to Pline** check box.

After setting all parameters in the **Topology Name** and **How to Close** areas, choose the **OK** button in the **Create Closed Polylines** dialog box; the dialog box will be closed and polylines will be created in the specified drawing layer.

TUTORIALS

General instructions for downloading tutorial Files:

Before starting the tutorials, you need to download the tutorial data to your computer. To do so, follow the steps given below:

1. Log on to the *www.cadcim.com* and browse to *Textbooks > Civil/GIS > Map 3D > Exploring AutoCAD Map 3D 2017*. Next, select *c09_m3d_2017_tut.zip* file from the **Tutorial Files** drop-down list. Next, choose the corresponding **Download** button to download the data file.

2. Extract the contents of the zip file to the following location:

    C:\m3d_2017

Notice that the *c09_m3d_2017_tut* folder is created within the *m3d_2017* folder.

Tutorial 1   Using the Track Coordinate System Tool

In this tutorial, you will insert an image into the current drawing file. Next, you will mark the coordinate points and track them in different coordinate systems by using the **Track Coordinate System** tool.   **(Expected time: 45 min)**

The following steps are required to complete this tutorial:

a. Assign the **NE83 NAD83 Nebraska State Planes, Meter** coordinate system from the **USA, Nebraska** coordinate system category to the current drawing.
b. Insert the *14tpl945240.jpg* image file in the current drawing.
c. Draw the coordinate points using the data given in tabular form below.
d. Track the coordinates of these points in the **LL84** coordinate system.

Point	X Coordinate	Y Coordinate
Point 1	695000.6342	4524381.7988
Point 2	695089.6508	4524422.4577
Point 3	695343.9837	4524386.0341

Assigning a Global Coordinate System
In this part of the tutorial, you will start a new drawing file and then assign it a global coordinate system.

1. Choose the **New** button from the Quick Access Toolbar; a new drawing file is opened in the Workspace.

2. Choose the **Assign** tool from the **Coordinate System** panel in the **Map Setup** tab; the **Coordinate System - Assign** dialog box is displayed.

3. In this dialog box, select the **USA, Nebraska** option from the **Category** drop-down list; a list of coordinate systems in the selected category is displayed in the list box.

4. Next, select the **NE83** code with description **NAD83 Nebraska State Planes, Meter** from the list box.

5. Now, choose the **Assign** button in the **Coordinate System - Assign** dialog box; the dialog box is closed and the coordinate system is assigned to the Workspace.

Inserting the Raster Image in the Current Drawing
In this part of the tutorial, you will insert a raster image in the drawing file using the **Insert** tool.

1. Choose the **Image** tool from the **Image** panel in the **Insert** tab; the **Insert Image** dialog box is displayed.

2. In the **Insert Image** dialog box, browse to the following location:

    C:\m3d_2017\c09_m3d_2017_tut\c09_tut01

3. Now, select the **14tpl945240** image from the list box in this dialog box; the image is displayed in the **Preview** area and properties are displayed on the right in this area.

> **Note**
> *If the preview and properties of the image are not displayed by default, choose the **Information** button at the bottom of the **Insert Image** dialog box to display the information of the image.*

4. Next, choose the **Open** button; the **Image Correlation** dialog box is displayed.

5. In this dialog box, ensure that the **World File** option is selected in the **Correlation Source** drop-down list and then choose the **OK** button; the dialog box is closed and the image is inserted in the current drawing.

> **Note**
> *If the image is not displayed in the drawing window by default, choose the **Extents** tool from the **Navigate** panel of the **View** tab.*

Drawing the Points

In this part of the tutorial, you will digitize the points using the data given in the tutorial. To digitize the points, you will use the **Track Coordinate System** tool.

1. Type **DDPTYPE** in the Command prompt and then press ENTER; the **Point Style** dialog box is displayed.

2. In this dialog box, select a point style ⊠ by clicking on it.

3. Enter **5.0000** in the **Point Size** edit box and then select the **Set Size in Absolute Units** radio button.

4. Next, choose the **OK** button in the **Point Style** dialog box; the dialog box is closed and the point style is saved.

5. Choose the **Coordinate track** tool from the **Geo Tools** panel in the **Analyze** tab; the TRACK COORDINATES palette is displayed.

6. In this palette, place the cursor on the **TRACK COORDINATES** toolbar; the **Coordinate Tracker** toolbar is activated.

7. Next, click on the drop-down list in the **Coordinate Tracker** toolbar; a list of coordinate systems is displayed.

8. In the **Category** list box, select the **USA, Nebraska** category; the coordinate systems in this category are displayed in the list box on the right side.

> **Note**
> *If the coordinate systems in the selected category are not displayed, use the vertical scroll bar at the right of this list box to scroll the list downward.*

9. In this list box, double-click on the **NE83 NAD83 Nebraska State Planes, Meter** option; the **NE83** code is displayed in the drop-down list in the **Coordinate Tracker** toolbar and the **NAD83 Nebraska State Planes, Meter** text is displayed below the drop-down list in the **TRACK COORDINATES** palette.

10. Choose the **Multiple Points** tool from the **Draw** panel in the **Home** tab; the cursor changes into a crosshair in the drawing window.

11. In the **TRACK COORDINATES** palette, delete the existing value in the **X** edit box and then specify **695000.6342**. Similarly, specify **4524381.7988** in the **Y** edit box (X and Y coordinates of **Point 1**).

12. In the activated toolbar of the **Track Coordinate** palettes, choose the **Digitize using entered position** button; a point is created (digitized) in the drawing at the specified coordinates.

13. Digitize **Point 2** and **Point 3** by repeating steps 11 and 12. Figure 9-36 shows the **Point 1**, **Point 2**, and **Point 3** placed on the raster image (create **Point 2** and **Point 3** by manual digitization).

Figure 9-36 Partial image displaying the location of the three points drawn

Tracking Coordinates

1. In the activated toolbar of the **TRACK COORDINATES** palette, choose the **Clone this tracker and insert it below** button; a copy of the existing tracker is displayed below the existing tracker.

2. Next, click on the drop-down list in the **Coordinate Tracker** toolbar of the newly added tracker; a list of categories is displayed, as shown in Figure 9-37.

3. In the **Category** list box of the drop-down list, select the **Lat Longs** category; a list of coordinate systems in this category is displayed on the right.

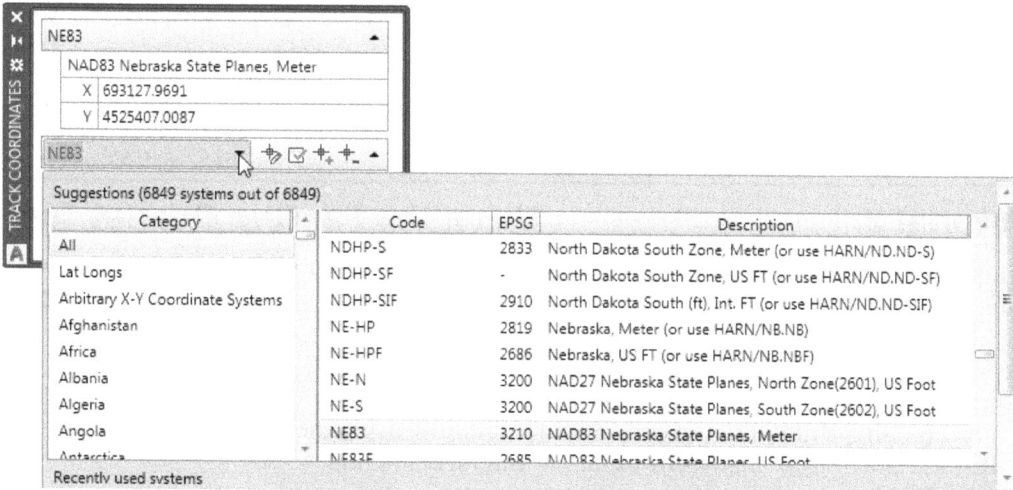

Figure 9-37 The list of categories and the coordinate systems displayed

4. In the coordinate systems list box for the **Lat Longs** category, double-click on the **LL84** in the Code List box with description **WGS84 datum, Latitude-Longitude; Degrees** option; the **LL84** code is displayed in the drop-down list and the description is displayed below this drop-down list, refer to Figure 9-38.

    Note
    *The coordinates shown in the **X** and **Y** edit boxes are in the dynamic mode. So they may vary depending on the location of the cursor.*

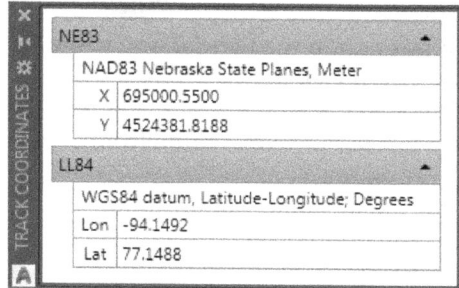

*Figure 9-38 The **TRACK COORDINATES** palette with the **LL84** coordinate tracker*

5. Zoom in the **Point 1** located in the drawing window, refer to Figure 9-36, and then place the cursor on this point; the coordinates of **Point 1** in the **LL84** coordinate system are displayed in the **TRACK COORDINATES** palette.

    Note the coordinates displayed in the **LL84** tracker for the **Point 1**.

6. Repeat step 5 for the **Point 2** and **Point 3** and track the coordinates for these points using the **TRACK COORDINATES** palette. The coordinates tracked in the **LL84** tracker of the **TRACK COORDINATES** palette for all the points are as follows:

Point 1	Long: -94.1492	Lat: 77.1488
Point 2	Long: -94.1464	Lat: 77.1486
Point 3	Long: -94.1389	Lat: 77.1486

Saving the Drawing File

1. Choose the **Save As** option from the Application Menu; the **Save Drawing As** dialog box is displayed.

2. In this dialog box, browse to the desired folder and enter **c09_Tut01a.dwg** in the **File name** edit box.

3. Next, choose the **Save** button in this dialog box; the dialog box is closed and the current drawing file is saved with the given name.

Tutorial 2   Using the Feature Overlay Tool

In this tutorial, you will overlay one feature layer over another layer by using the **Feature Overlay** tool.   **(Expected time: 45 min)**

The following steps are required to complete this tutorial:

a. Load the shape files into the current drawing file by using the **Connect** tool.
b. Perform the overlay analysis by using the **Intersect** method.
c. Save the file.

Loading the Shape Files

In this part of the tutorial, you will start a new drawing file. Next, you will add SHP files using the **DATA CONNECT** wizard.

1. Choose the **New** button from the Quick Access Toolbar; a new drawing file is opened in the Workspace.

2. Choose the **Connect** tool from the **Data** panel in the **Home** tab; the **DATA CONNECT** wizard is displayed.

3. In the **DATA CONNECT** wizard, select the **Add SHP Connection** option from the **Data Connections by provider** list box, if it is not chosen by default; the **OSGeo FDO Provider for SHP** page is displayed in the right pane of the wizard.

4. In this page, choose the second button on the right of the **Source file or folder** edit box; the **Browse For Folder** window is displayed.

5. In this window, browse to the following location:

    C:\m3d_2017\c09_m3d_2017_tut

6. Next, select the **c09_tut02** folder in this window and then choose the **OK** button; the window is closed and the path of the folder is added to the **Source file or folder** edit box in the **OSGeo FDO Provider for SHP** page.

7. Choose the **Connect** button located below the **Source file or folder** edit box; the shape files in the **c09_tut02** folder are displayed in the list box below the **Edit Coordinate Systems** button in the **SHP** page, as shown in Figure 9-39.

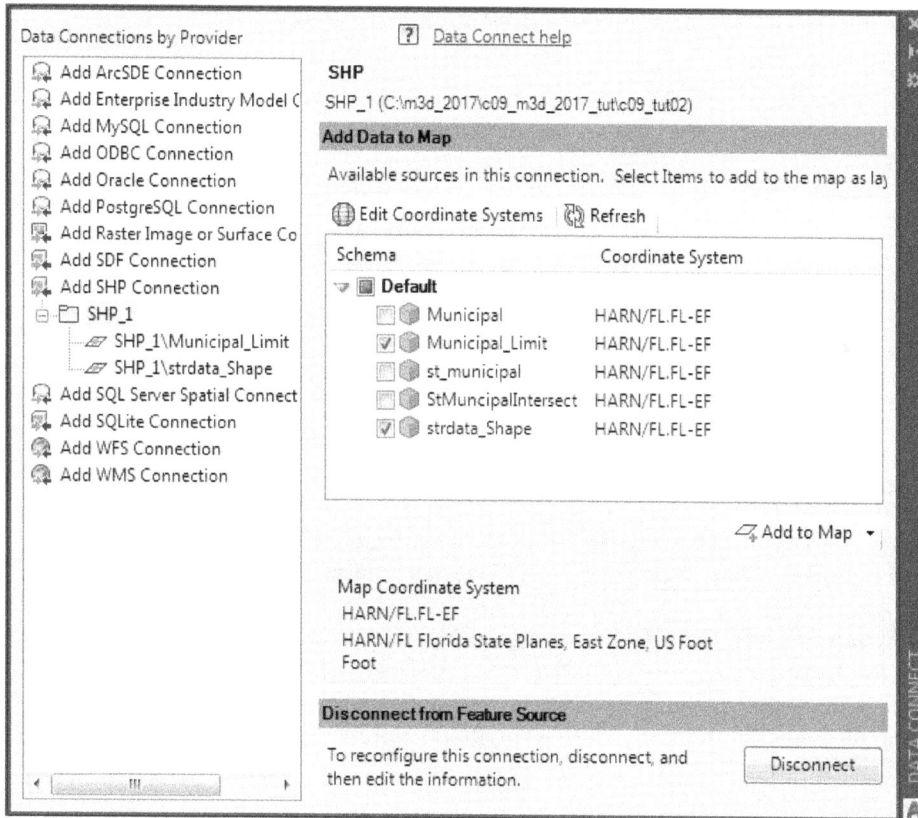

Figure 9-39 The **DATA CONNECT** *wizard displaying the connected shape files in the* *c09_tut02 folder*

8. In the **SHP** page, select the check boxes corresponding to the **Municipal_Limit** and **strdata_Shape** options in the list box; the **Add to Map** button is activated.

9. Choose the **Add to Map** button in the **SHP** page; the **Municipal_Limit** and **strdata_Shape** shape files are added to the current drawing. Also, the geometries in the added shape files are displayed in the drawing window, as shown in Figure 9-40.

Figure 9-40 The geometry of the added shape files

10. Close the **DATA CONNECT** wizard by choosing the **Close** button.

Performing the Overlay Analysis

1. Zoom in the upper right section of the drawing, refer to Figure 9-41.

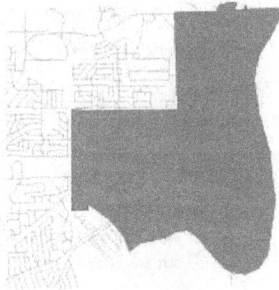

Figure 9-41 *The part of the feature data used for analysis*

2. Choose the **Feature Overlay** tool from the **Feature** panel in the **Analyze** tab; the **Sources and Overlay Type** page of the **Overlay Analysis** wizard is displayed.

3. In the **Sources and Overlay Type** page, select the **strdata_Shape (Lines)** option from the **Layers** hierarchy in the **Source** drop-down list.

4. Next, select the **Municipal_Limit (Polygons)** option from the **Layers** hierarchy in the **Overlay** drop-down list.

5. In the **Type** drop-down list, select the **Intersect** option, if it is not selected by default.

6. Choose the **Next** button in the **Sources and Overlay Type** page; the **Set Output and Settings** page of the **Overlay Analysis** wizard is displayed.

7. In this page, choose the browse button corresponding to the **Output** edit box; the **Save** dialog box is displayed.

8. In this dialog box, browse to the location:

    C:\m3d_2017\c09_m3d_2017_tut\c09_tut02

9. Select **SHP Files (*.shp)** option from the **Save as type** drop-down list.

10. Specify **st_municipal** in the **File name** edit box and choose the **Save** button; the **Save** dialog box is closed and the specified path of the shape file is displayed in the **Output** edit box of the **Overlay Analysis** wizard.

11. Enter **st_municipal** in the **Layer name** edit box.

12. Retain the other settings in the **Set Output and Settings** page and then choose the **Finish** button; the overlay analysis is performed and the final featured data with the **st_municipal** feature layer is added to the **Display Manager** tab of the **TASK PANE**.

13. In the **Display Manager** tab of the **TASK PANE**, clear the check box corresponding to the **SHP_1** option in the list box; the display of all the geometries, except the geometries in the resultant topology of the overlay analysis, is switched off.

14. Zoom in the upper right part of the resultant feature data, refer to Figure 9-41; the resultant feature geometry after the intersection of the **Municipal_Limit** and **strdata_Shape** shape files is displayed, as shown in Figure 9-42.

Figure 9-42 The resultant feature data of the overlay analysis procedure

Saving the Drawing File

1. Choose the **Save As** option from the Application Menu; the **Save Drawing As** dialog box is displayed.

2. In this dialog box, enter **c09_Tut02a.dwg** in the **File name** edit box.

3. Choose the **Save** button in this dialog box; the dialog box is closed and the current drawing file is saved with the given name.

Tutorial 3   Performing Network Analysis - I

In this tutorial, you will perform network analysis to find the shortest route between two points.   **(Expected time: 45 min)**

The following steps are required to complete this tutorial:

a. Open the *c09_Tut03.dwg* file.
b. Perform the network analysis to find the shortest distance between the nodes **A** and **B**.
c. Save the file.

Opening the Tutorial Drawing File

1. Choose the **Open** button in the Quick Access Toolbar; the **Select File** dialog box is displayed.

2. In this dialog box, browse to the following location:

    C:\m3d_2017\c09_m3d_2017_tut\c09_tut03

3. Next, select the **c09_Tut03.dwg** file and then choose the **Open** button; the selected drawing is opened and the geometry is displayed in the drawing window, as shown in Figure 9-43.

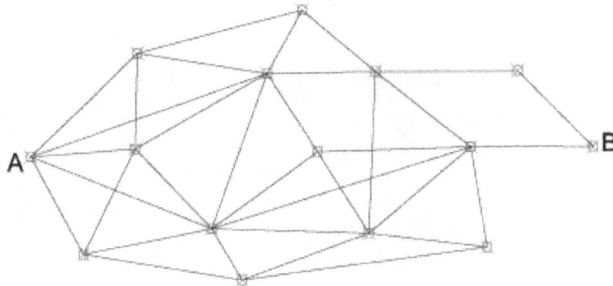

*Figure 9-43 The geometry of the **Network** topology displayed in the drawing window*

4. Expand the **Topologies** folder in the **Map Explorer** tab of the **TASK PANE**. Next, right-click on the **Network** topology in this node; a shortcut menu is displayed. In this menu, place the cursor on the **Administration** option; a cascading menu is displayed.

5. Choose the **Load Topology** option from the flyout; the **AutoCAD Map Topology Audit** message box is displayed. Next, choose the **OK** button in this message box; the topology is loaded in the drawing.

Performing the Network Analysis to Find the Shortest Path

1. Choose the **Network Analysis** tool from the **Drawing Object** panel in the **Analyze** tab; the **Topology Selection** window is displayed with the **Network** topology selected.

    Network Analysis

2. Choose the **OK** button in the **Topology Selection** window; this window is closed and the **Network Topology Analysis** wizard with the **Select Method** page is displayed.

3. In the **Select Method** page, select the **Shortest path** radio button, if it is not selected by default.

4. Now, choose the **Next** button in this page; the **Choose Locations** page of the **Network Topology Analysis** wizard is displayed.

5. In this page, choose the **Select Points** button corresponding to the **Select start point** option; the **Network Topology Analysis** wizard is closed and you are prompted to specify the start point. Notice that the cursor changes into a crosshair in the drawing window.

6. Right-click on the **Object Snap** button in the Status Bar; a shortcut menu is displayed. Now, choose the **Object Snap Settings** option from the displayed shortcut menu; the **Drafting Settings** dialog box is displayed.

7. Next, choose the **Object Snap** tab and select the check box corresponding to the **Node** option in the **Object Snap modes** area and then choose the **OK** button; the dialog box is closed and the **Node** object snap mode is activated prompting you to specify a point.

8. Place the cursor on the node **A**, refer to Figure 9-43, and then click; a cross mark is placed on this node.

9. Press ENTER; the **Network Topology Analysis** wizard with the **Choose Locations** page is displayed and the coordinates of the selected point are displayed in the list box on this page.

10. In the **Choose Locations** page, choose the **Select Points** button corresponding to the **Select end point** option; the cursor changes into a crosshair in the drawing window.

11. Place the crosshair on the **Object Snap** button in the Status Bar and then right-click on it; a shortcut menu is displayed. In this shortcut menu, choose the **Object Snap Settings** option: the **Drafting Settings** dialog box is displayed.

12. Next, choose the **Object Snap** tab and select the check box corresponding to the **Node** option in the **Object Snap modes** area and then choose the **OK** button; the dialog box is closed and the **Node** object snap mode is activated.

13. Place the cursor on the node **B**, refer to Figure 9-43, and then left-click; a cross mark is placed on this node.

14. Press ENTER; the **Network Topology Analysis** wizard with the **Choose Locations** page is displayed and the coordinates of the selected point are displayed in the text box on this page, as shown in Figure 9-44.

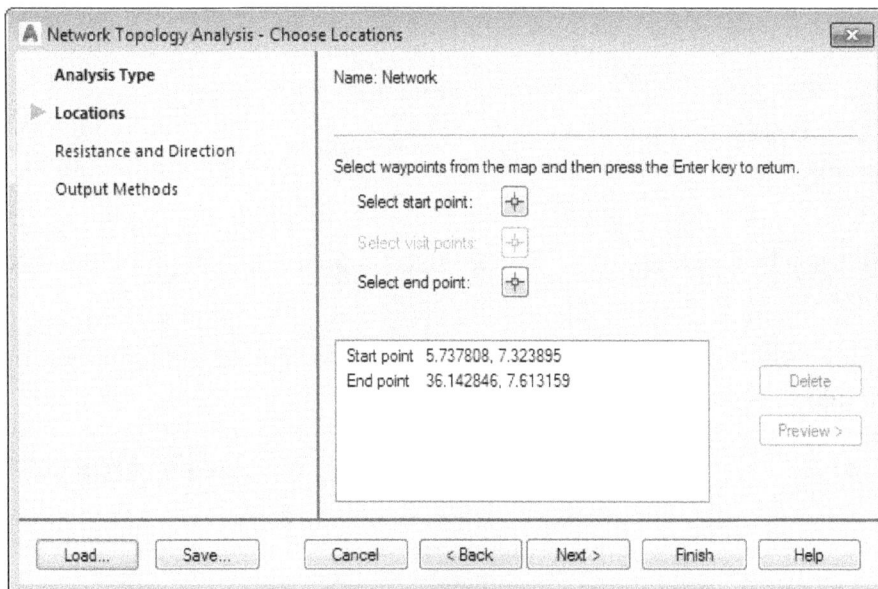

Figure 9-44 The coordinates of the start and end points displayed in the list box

15. Choose the **Next** button in the **Choose Locations** page; the **Resistance and Direction** page of the wizard is displayed.

16. Retain the default settings in the **Resistance and Direction** page and then choose the **Next** button; the **Output** page of the wizard is displayed.

17. In the **Output** page, select the **Create topology** check box; the **Name** and **Description** edit boxes are activated.

18. Enter **shortest_path** in the **Name** edit box and then choose the **Finish** button; the wizard is closed and the **shortest_path** topology is created for the geometry of the shortest path.

19. Right-click on the **shortest_path** topology in the **Topologies** folder in the **Map Explorer** tab of the **TASK PANE**; a shortcut menu is displayed. In this shortcut menu, choose the **Show Geometry** option; the shortest path calculated between the nodes **A** and **B** is highlighted in red color, as shown in Figure 9-45.

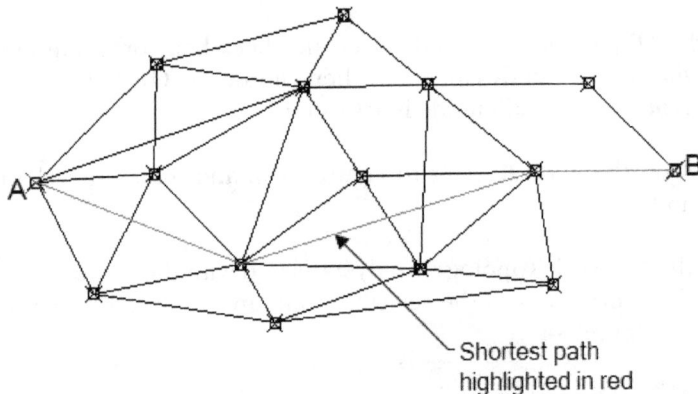

*Figure 9-45 The shortest path between the nodes **A** and **B** highlighted in red*

Saving the Drawing File

1. Choose the **Save As** option from the Application Menu; the **Save Drawing As** dialog box is displayed.

2. In this dialog box, browse to the desired folder and then enter **c09_Tut03a.dwg** in the **File name** edit box.

3. Choose the **Save** button in the **Save Drawing As** dialog box; the dialog box is closed and the current drawing file is saved with the specified name.

Tutorial 4   Performing Network Analysis - II

In this tutorial, you will perform network analysis to find the shortest route between two points. While performing the analysis, you will take into account the effects of direction of the network links and resistance offered by the links and nodes in the network.   **(Expected time: 1hr)**

The following steps are required to complete this tutorial:

a. Open the *c09_Tut04.dwg* file.
b. Perform network analysis to find the shortest distance between the nodes 1 and 5.
c. Apply direction and resistance to the network links and nodes.
d. Perform network analysis to find the shortest route between nodes 1 and 5 with the modified direction and resistance parameters.
e. Save the file.

Opening the Drawing File and Loading the Network Topology

In this section of the tutorial, you will open the drawing file. After opening the drawing file, you will load the network topology in the drawing.

1. Choose the **Open** button in the Quick Access Toolbar; the **Select File** dialog box is displayed.

2. In this dialog box, browse to the following location:

   C:\m3d_2017\c09_m3d_2017_tut\c09_tut04

3. Next, select the **c09_Tut04.dwg** file and then choose the **Open** button; the selected drawing is opened in the drawing window. The drawing contains a road network represented by line geometry and the network topology along with the annotations for links and nodes in the road network. Figure 9-46 shows the network geometry and the annotations in the drawing file.

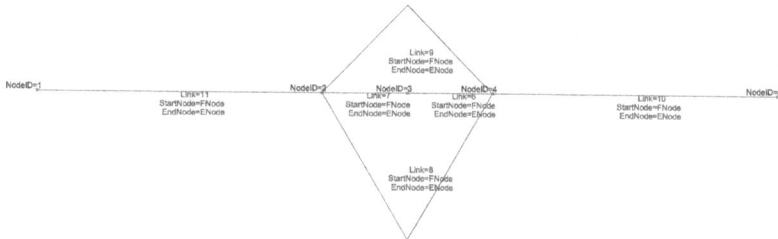

Figure 9-46 *The network geometry and its annotations displayed in the drawing window*

4. Expand the **Topologies** folder in the **Map Explorer** tab of the **TASK PANE**. Next, right-click on the **Road** topology in this node; a shortcut menu is displayed.

5. Next, place the cursor on the **Administration** option of the shortcut menu; a cascading menu is displayed. Choose the **Load Topology** option from the menu, as shown in Figure 9-47; the **AutoCAD Map Topology Audit** message box is displayed with the message that the topology is correct and complete.

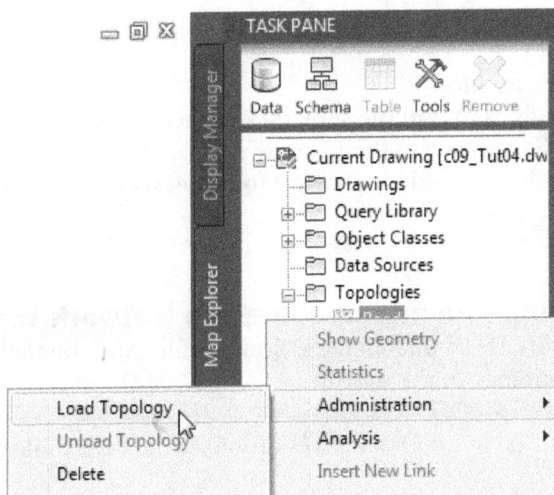

*Figure 9-47 Choosing the **Load Topology** option from the cascading menu*

6. Next, choose the **OK** button in this message box; the message box is closed and the **Road** topology is loaded in the drawing.

Performing the Shortest Path (Network) Analysis

In this part of the tutorial, you will calculate the shortest path between the nodes 1 and 5 of the network topology.

1. Choose the **Network Analysis** tool from the **Drawing Object** panel in the **Analyze** tab; the **Topology Selection** window is displayed.

    Network Analysis

2. Select the **Road** topology in this window, if it is not selected by default and then choose the **OK** button; the **Topology Selection** window is closed and the **Network Topology Analysis** wizard with the **Select Method** page is displayed.

3. In the **Select Method** page of this wizard, ensure that the **Shortest path** radio button is selected.

4. Now, choose the **Next** button in this page; the **Choose Locations** page of the **Network Topology Analysis** wizard is displayed.

5. In this page, choose the **Select Points** button corresponding to the **Select start point** option; you are prompted to specify the start point. Also, notice that the cursor changes into a crosshair in the drawing window.

    Note that the **Node** check box is selected in the **Object Snap** tab of the **Drafting Settings** dialog box.

6. Zoom to the node 1 (NodeID=1) in the drawing window, refer to Figure 9-46; and click on node 1 in the drawing; a red colored cross mark is placed on this node and you are prompted to specify the next point.

7. Press ENTER; the **Network Topology Analysis** wizard is displayed again. Note that the list box in the **Choose Locations** page displays the coordinates of the specified point.

8. Next, choose the **Select Points** button corresponding to the **Select end point** option; you are prompted to specify the end point. Also, notice that the cursor changes into a crosshair in the drawing window.

9. Zoom to the node 5 (NodeID=5) in the drawing window, refer to Figure 9-45, and click on node 5 in the drawing; a red colored cross mark is placed on this node and you are prompted to specify the next point.

10. Press ENTER; the **Network Topology Analysis** wizard is displayed with the coordinates of the specified point in the list box of the **Choose Locations** page, as shown in Figure 9-48.

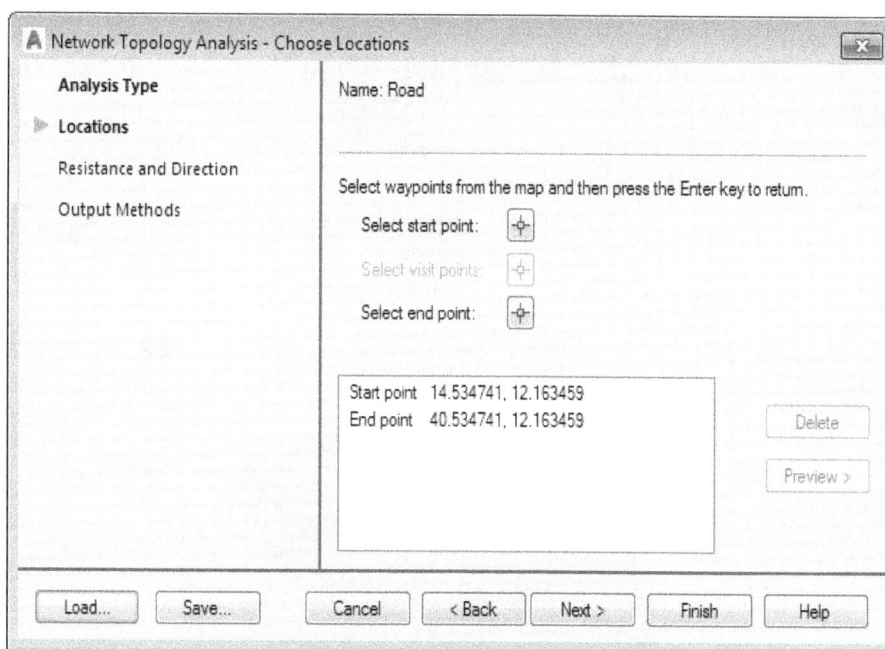

Figure 9-48 *The coordinates of the start and end points displayed in the list box*

11. Choose the **Next** button in the **Choose Locations** page; the **Resistance and Direction** page of the wizard is displayed.

12. Retain the default settings in the **Resistance and Direction** page and then choose the **Next** button; the **Output** page of the wizard is displayed.

13. In the **Output** page, ensure that the **Highlight** check box is selected and then choose the **Finish** button; the **Network Topology Analysis** wizard is closed and the shortest path is highlighted in the drawing window, as shown in Figure 9-49.

14. Right-click on the **Road** topology in the **Topologies** folder in the **Map Explorer** tab of the **TASK PANE**; a shortcut menu is displayed. In this shortcut menu, choose the **Show Geometry**

option; the shortest path calculated between the nodes **A** and **B** is highlighted in red color, as shown in Figure 9-49.

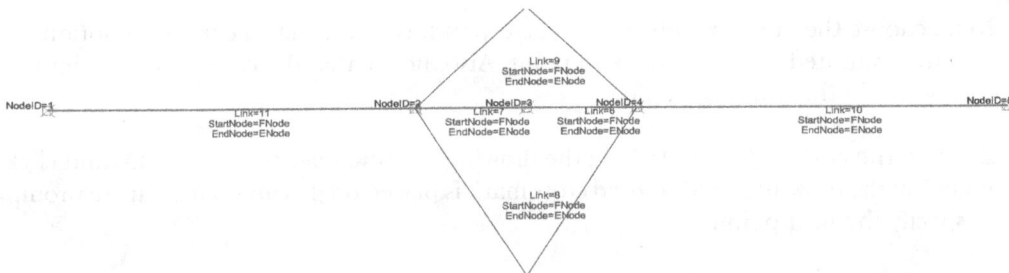

Figure 9-49 The network geometry and its annotation displayed in the drawing window

15. Press **ESC** twice to exit the current selection.

Applying the Node and Link Resistance While Performing the Shortest Path (Network) Analysis

In this part of the tutorial, you will apply the resistance at the node and links of the network and then analyze the network to find the shortest path.

1. Zoom into the drawing and then select the node 3 (NodeID=3), refer to Figure 9-46.

2. Right-click on the node; a shortcut menu will be displayed.

3. Next, choose the **Properties** option from the shortcut menu; the **PROPERTIES** palette showing the properties of the selected node (NodeID=3) is displayed, as shown in Figure 9-50.

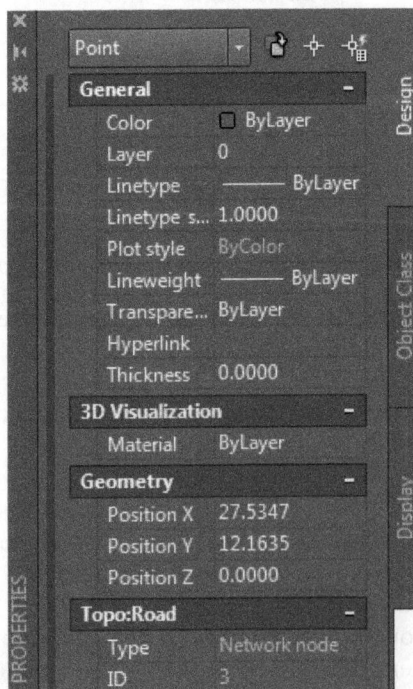

Figure 9-50 The PROPERTIES palette displaying the properties of the selected node

4. Next, in the **Design** tab of the **PROPERTIES** palette, expand the **Topo:Road** head, if it is not expanded by default.

5. Click in the value cell corresponding to **Resistance** in the **Topo:Road** head; the cell becomes editable. Now, specify **10** as the resistance value of the node in this cell and then close the **PROPERTIES** palette.

6. Next, select link 9 (Link=9) between node 2 and 4 and then right-click on the selected link; a shortcut menu is displayed.

7. Choose the **Properties** option from the displayed shortcut menu; the **PROPERTIES** palette showing the properties of the selected link is displayed.

8. Next, in the **Design** tab of the **PROPERTIES** palette, expand the **Topo:Road** head, if it is not expanded by default.

9. In the **Topo:Road** area, specify **15** and **18** as the value for the **Direct resistance** and **Reverse resistance**, respectively. Figure 9-51 shows the **PROPERTIES** palette with the values for direct and reverse resistance for network link with ID = 9.

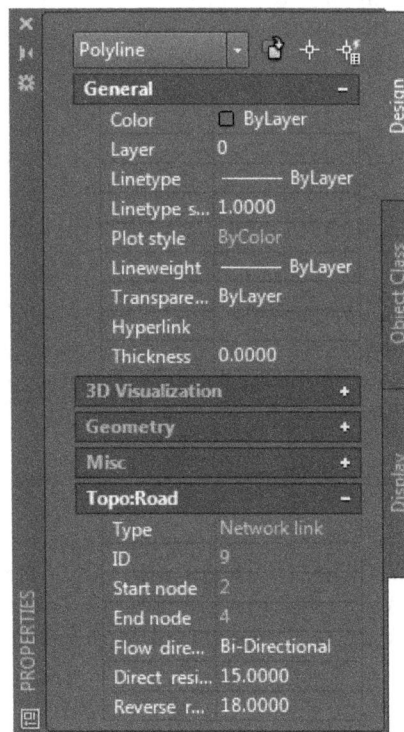

Figure 9-51 *The **PROPERTIES** palette displaying the properties for link with ID=9*

10. Next, repeat the procedure given in steps 6 to 9 and specify the **Direct resistance** and **Reverse resistance** for the network links as follows:

Link ID =6
          Direct resistance = 5
          Reverse resistance = 5

Link ID =7
          Direct resistance = 5
          Reverse resistance = 5

Link ID =8
          Direct resistance = 14
          Reverse resistance = 19

Next, you will find the shortest path in the network to travel between node 1 and node 5 in either direction. While calculating the shortest path, you will take into account the resistance offered by the node and links of the network.

11. Choose the **Network Analysis** tool from the **Drawing Object** panel in the **Analyze** tab; the **Topology Selection** window is displayed.

Network Analysis

12. Select the **Road** topology in this window, if not selected by default and then choose the **OK** button; the **Topology Selection** window is closed and the **Network Topology Analysis** wizard with the **Select Method** page is displayed.

13. In the **Select Method** page of this wizard, ensure that the select the **Shortest path** radio button is selected and then choose the **Next** button; the **Choose Locations** page of the wizard is displayed. Ensure that the list box in this page displays the **Start point** and **End point** coordinates of the previously selected points, refer to Figure 9-48.

> **Tip**
> *If the list box in the **Choose Location** page of the wizard does not display the start and end point coordinates as displayed in Figure 9-48, then specify node 1 as the start point and node 5 as the end point by using the **Select Points** buttons corresponding to the **Select start point** and **Select end point** options in the page.*

14. Now, choose the **Next** button; the **Resistance and Direction** page of the wizard is displayed.

15. In this page of the wizard, choose the **Expression Builder** button corresponding to the **Link direct resistance** edit box; the **Expression Chooser** dialog box is displayed.

16. In the **Expression Chooser** dialog box, choose the **Direct Resistance** option from the **Topologies > Network: Road > Network Link** node, as shown in Figure 9-52.

17. Choose the **OK** button in the **Expression Chooser** dialog box; the dialog box is closed and the expression for direct resistance is displayed in the **Link direct resistance** edit box in the **Resistance and Direction** page of the **Network Topology Analysis** wizard.

18. Choose the **Expression Builder** button corresponding to the **Link reverse resistance** edit box; the **Expression Chooser** dialog box is displayed.

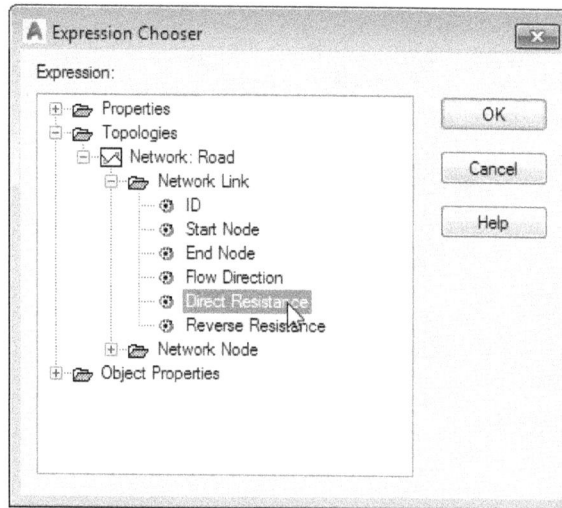

*Figure 9-52 Choosing the **Direct Resistance** option in the **Expression Chooser** dialog box*

19. In this dialog box, choose the **Reverse Resistance** option from the **Topologies > Network: Road > Network Link** node, refer to Figure 9-52.

20. Next, choose the **OK** button in the **Expression Chooser** dialog box; the dialog box is closed and the expression for reverse resistance is displayed in the **Link reverse resistance** edit box in the **Resistance and Direction** page of the **Network Topology Analysis** wizard.

21. Next, choose the **Expression Builder** button corresponding to the **Node resistance** edit box; the **Expression Chooser** dialog box is displayed.

22. In this dialog box, choose the **Resistance** option from the **Topologies > Network: Road > Network Node** node, as shown in Figure 9-53.

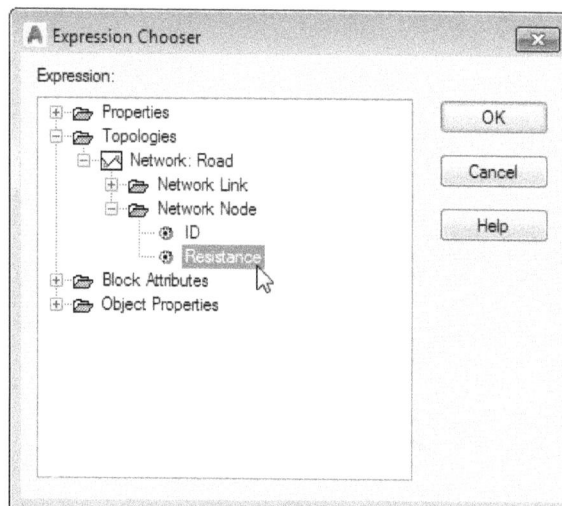

*Figure 9-53 Choosing the **Resistance** option in the **Expression Chooser** dialog box*

23. Next, choose the **OK** button in the **Expression Chooser** dialog box; the dialog box is closed and the resistance expression is displayed in the **Node resistance** edit box of the **Network Topology Analysis** wizard. Figure 9-54 shows the resistance parameters specified for the node and link objects in the network.

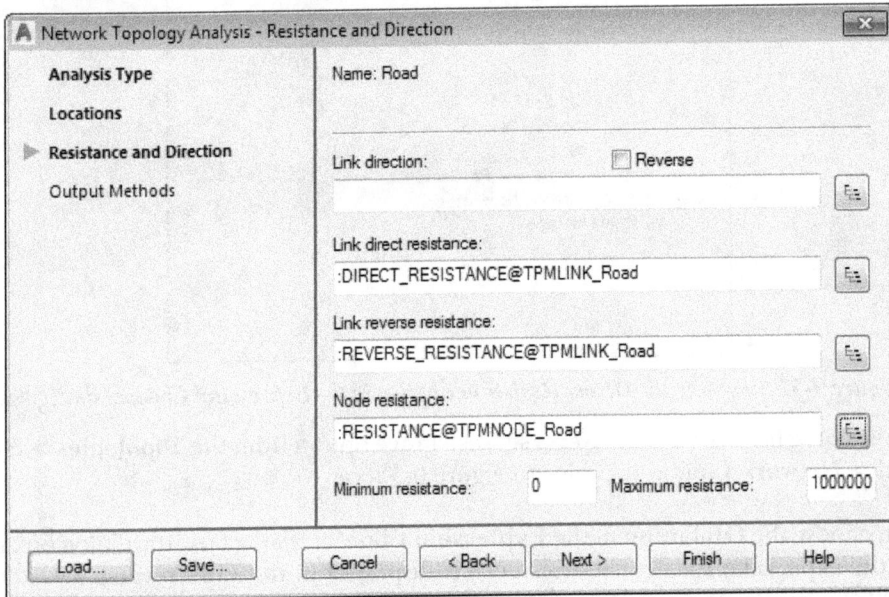

*Figure 9-54 The **Network Topology Analysis** wizard showing the resistance parameters specified for the node and link objects in the network*

24. Now, choose the **Next** button; the **Output** page of the wizard is displayed.

25. In this page, ensure that the **Highlight** check box is selected. Next, choose the **Finish** button in the wizard; the wizard is closed and the shortest path is calculated and highlighted in the drawing window. Figure 9-55 shows the shortest path calculated for traversing from node 1 to node 5.

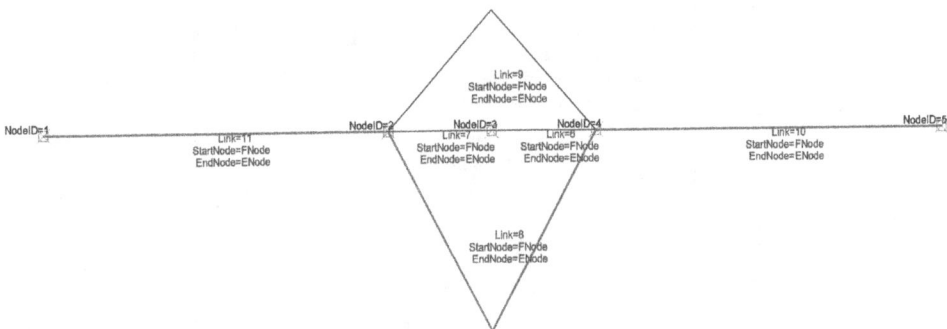

Figure 9-55 The shortest path calculated for traversing from node 1 to node 5

Now, you will find the shortest path between node 1 and node 5 of the network when traversing in the opposite direction (node 5 to node 1).

26. Choose the **Network Analysis** tool from the **Drawing Object** panel in the **Analyze** tab; the **Topology Selection** window is displayed.

Network Analysis

27. Select the **Road** topology in this dialog box, if not selected by default and then choose the **OK** button; the **Topology Selection** window is closed and the **Network Topology Analysis** wizard with the **Select Method** page is displayed.

28. In this page, ensure that the **Shortest path** radio button is selected.

29. Now, choose the **Next** button in this page; the **Choose Locations** page of the **Network Topology Analysis** wizard is displayed. Notice that the list box in this page displays the **Start point** and **End point** coordinates of the points selected previously for the network analysis.

30. Select the coordinates of the **Start Point** and **End Point** displayed in the list box of the **Choose Locations** page; the **Delete** button is activated. Now, choose the **Delete** button; the selected coordinates are deleted.

31. Next, using the **Select Points** button corresponding to the **Select start point** and **Select end point** option in this page, specify node 5 and node 1 as the start point and end point of the path, respectively, and then press ENTER.

32. Now, choose the **Next** button; the **Resistance and Direction** page of the wizard will be displayed.

33. Repeat the procedure given in steps 15 through 23 and specify the expression for **Link direct resistance**, **Link reverse resistance**, and **Node resistance**.

34. Now, choose the **Next** button; the **Output** page of the wizard is displayed.

35. Ensure that the **Highlight** check box is selected in this page and then choose the **Finish** button; the **Network Topology Analysis** wizard is closed and the shortest path to traverse from node 5 to node 1 is displayed in the drawing, as shown in Figure 9-56.

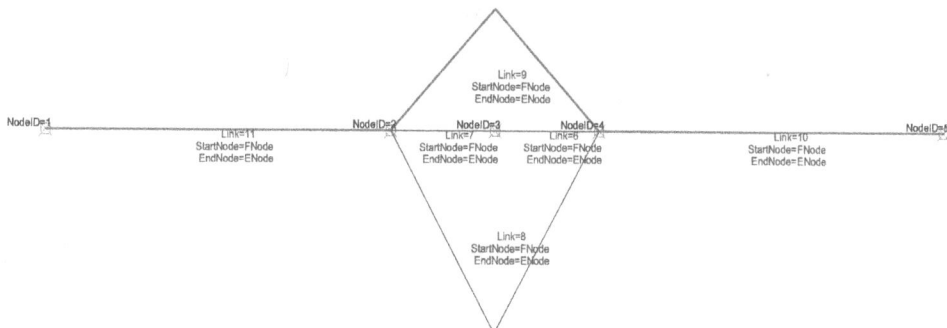

Figure 9-56 The shortest path calculated for traversing from node 5 to node 1

36. Now type **RE** in the Command prompt and press ENTER; the highlight of the shortest path will be removed.

Changing the Direction of the Network Link

In this part of the tutorial, you will use the **PROPERTIES** palette to change the direction of the link in the network topology.

1. Select and right-click on link 8 (Link ID =8) in the network topology; a shortcut menu will be displayed.

2. Choose the **Properties** option from the displayed menu; the **PROPERTIES** palette will be displayed with the properties of the selected list.

3. In the **PROPERTIES** palette, expand the **Topo: Road** area, if not expanded by default. Next, click in the value cell for **Flow direction**; a drop-down list is displayed.

4. Select the **Reverse** option from the displayed drop-down list, as shown in Figure 9-57, and then close the **PROPERTIES** palette.

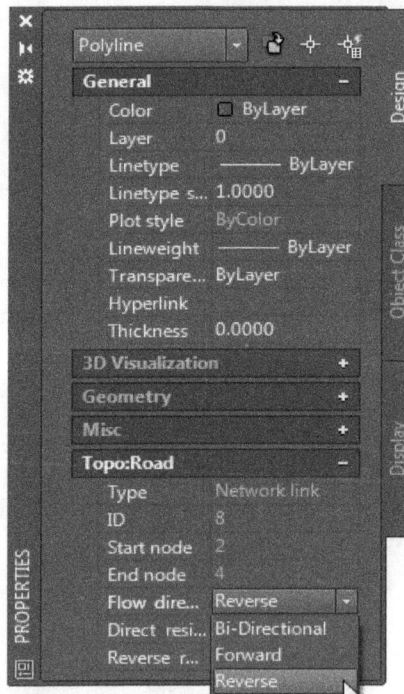

*Figure 9-57 Changing the **Flow direction** of the line*

Performing the Shortest Path Analysis (Using Resistance and Direction of Network Links and Nodes)

In this section of the tutorial, you will find the shortest route for traversing from node 1 to node 5. While calculating the shortest path, you will also consider the effects of node resistance, link resistance (direct and reverse), and link direction.

1. Choose the **Network Analysis** tool from the **Drawing Object** panel in the **Analyze** tab; the **Topology Selection** window is displayed.

Network Analysis

2. Select the **Road** topology in this dialog box, if not selected by default and then choose the **OK** button; the **Topology Selection** window is closed and the **Network Topology Analysis** wizard with the **Select Method** page is displayed.

3. In the **Select Method** page of this wizard, ensure that the **Shortest path** radio button is selected.

4. Now, choose the **Next** button in this page; the **Choose Locations** page of the wizard is displayed. Notice that the list box in this page displays the coordinates of the start and end point of the path selected in previous network analysis.

5. Select the coordinates displayed in the list box of the **Choose Locations** page; the **Delete** button is activated. Choose the **Delete** button; the selected coordinates in the list box are deleted.

6. Next, using the **Select Points** buttons corresponding to the **Select start point** and **Select end point** options in this page, specify node 1 and node 5 as the start point and end point of the path, respectively, refer to Figure 9-48.

7. Now, choose the **Next** button; the **Resistance and Direction** page of the wizard will be displayed.

8. In this page, choose the **Expression Builder** button corresponding to the **Link direction** edit box; the **Expression Chooser** dialog box is displayed.

9. In the **Expression Chooser** dialog box, expand the **Topologies > Network: Road > Network Link** node.

10. Next, choose the **Flow Direction** node in this dialog box and then choose the **OK** button; the **Expression Chooser** dialog box is closed and the expression for the link direction is displayed in the **Link Direction** edit box.

11. Next, specify the expressions for **Link direct resistance**, **Link reverse resistance**, and **Node resistance**. Figure 9-58 shows the direction and resistance parameters for the node and link objects in the network.

12. Now, choose the **Next** button, the **Output** page of the wizard is displayed.

13. Ensure that the **Highlight** check box is selected in this page and then choose the **Finish** button; the **Network Topology Analysis** wizard is closed and the shortest path to traverse from node 1 to node 5 is displayed in the drawing, as shown in Figure 9-59.

Tip
*To specify the resistance parameter for the node, invoke the **Expression Chooser** dialog box by choosing the **Expression Builder** button corresponding to the **Node resistance** edit box. Next, expand **Topologies** > **Network: Road** > **Network Node** and then choose **Resistance** in this node.*

*Similarly, you can use the **Expression Chooser** dialog box to specify the direct resistance and reverse resistance for the links in the network.*

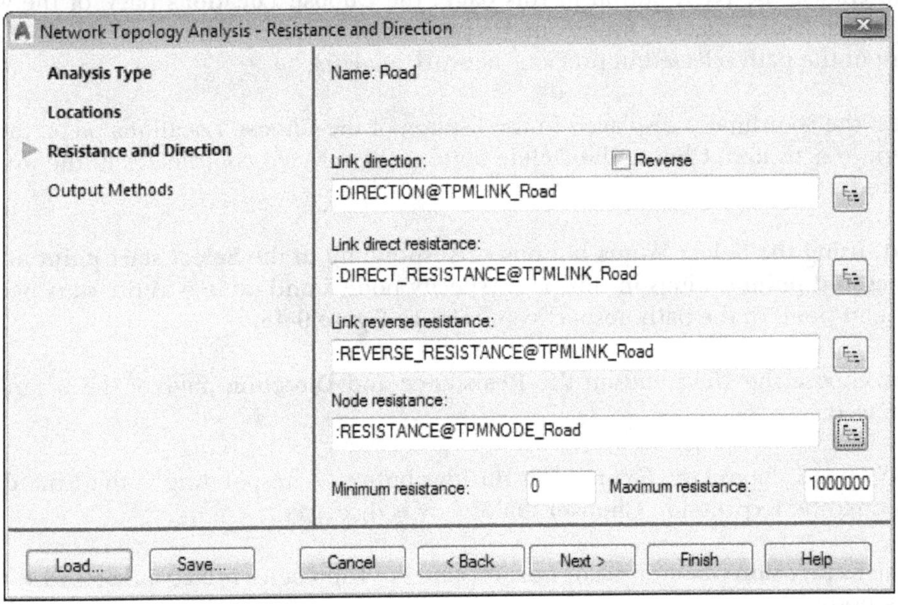

*Figure 9-58 The **Resistance and Direction** page of the **Network Topology Analysis** wizard displaying the resistance and direction parameters*

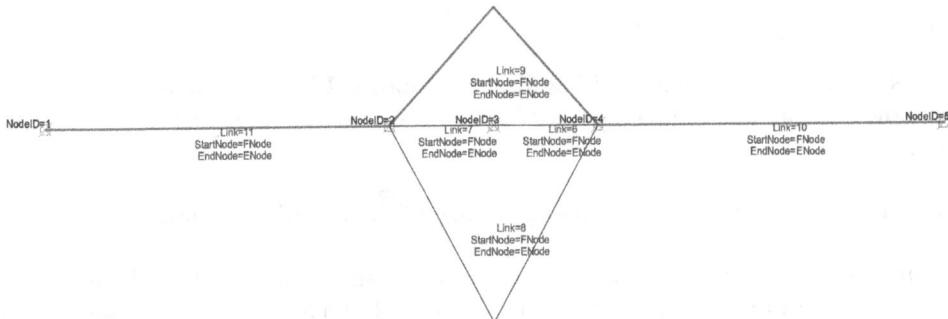

Figure 9-59 The shortest path calculated for traversing from node 1 to node 5

Saving the Drawing File

1. Choose the **Save As** option from the Application Menu; the **Save Drawing As** dialog box is displayed.

2. In this dialog box, browse to the required folder and then enter **c09_Tut04a.dwg** in the **File name** edit box.

3. Choose the **Save** button in the **Save Drawing As** dialog box; this dialog box is closed and the current drawing file is saved with the specified name.

Self-Evaluation Test

Answer the following questions and then compare them to those given at the end of this chapter:

1. Which of the following tools is used to find out the slope and grade parameters?

    (a) **Continuous**   (b) **Measure**
    (c) **List Slope**   (d) **Angle Information**

2. Which of the following tools is used to display the geometry of a topology in a drawing window?

    (a) **Feature Buffer**   (b) **Show Geometry**
    (c) **Show Topology**   (d) **Feature Overlay**

3. Which of the following tools is used to find an optimum route between the start point and given intermediate points?

    (a) **Best route**   (b) **Shortest path**
    (c) **Flood trace**   (d) **Feature Overlay**

4. Which of the following types of overlay analyses is used to create feature only from the overlapping areas of the source and overlay layer?

    (a) **Erase**   (b) **Intersect**
    (c) **Identify**   (d) **Union**

5. Which of the following tools is used to remove boundaries between polygons in a topology based on a specified attribute?

    (a) **Object Overlay**   (b) **Object Buffer**
    (c) **Dissolve**   (d) **Network Analysis**

6. The _____ tool is used to add distance between the disjunct parts of an object.

7. The _____ tool is used to perform overlay analysis of feature class.

8. The Shortest path method in the _____ analysis is used to find the shortest path between the two specified points.

9. The **Create Closed Polylines** tool is used to create _____ from an existing polygon topology.

10. The _____ tool is used to load an existing topology into your current drawing.

11. The **Geo Distance** tool is used to measure the distance between two points, considering the curvature of Earth's surface. (T/F)

12. The **Line & Arc Information** tool is used to find the slope of a line. (T/F)

13. The **Coordinate Track** tool is used to track the cursor in any coordinate system. (T/F)

14. The **Feature Buffer** tool is used to create a buffer around the objects in an existing topology. (T/F)

15. The **Shortest path**, **Best route**, and **Flood trace** are the three network analysis methods that are available in the **Network Topology Analysis** wizard. (T/F)

Review Questions

Answer the following questions:

1. Which of the following tools is used to find the horizontal distance and the height difference between the start and end points of a line?

    (a) **Add Distance**   (b) **Measure**
    (c) **Continuous**   (d) **List Slope**

2. The _____ tool is used to create a region of specified width around a feature layer.

3. The **Line & Arc Information** tool is used to find the geometric information of a _____ or _____ object.

4. The **Angle information** tool is used to find _____ between two lines or points.

5. The _____ palette displays the coordinates of the cursor in multiple coordinate system.

6. The **Add Distance** tool is used to add distance between multiple points in a drawing. (T/F)

7. You can specify the action for the overlapping buffer in the **Create Buffer** dialog box. (T/F)

8. The **Object Overlay** tool is used to perform the overlay analysis on the feature class. (T/F)

9. The **Dissolve** tool is used to merge or combine polygons that have different property or attribute values. (T/F)

10. The flood trace method of network analysis is the same as the shortest path method. (T/F)

EXERCISES
Exercise 1

Download the *c09-m3d-2017-exr01.dwg* from *www.cadcim.com* and perform the **Union** overlay analysis using **TopologyA** and **TopologyB**.   **(Expected time: 45 min)**

Tip: To complete this exercise, you need to load the topologies into your drawing.

Exercise 2

Download the *c09-m3d-2017-exr02.dwg* file from *www.cadcim.com* and then find the shortest path between the nodes **A** and **B**.   **(Expected time: 1 hr)**

Tip: To complete this exercise, you need to load the **Network** topology into your drawing.

Exercise 3

Download the *c09-m3d-2017-exr03.dwg* file from *www.cadcim.com* and perform the network analysis to find the shortest path for traversing between nodes **1** to **5** in either direction. Figure 9-60 shows the geometry of the network to be used for analysis. Use the direction and resistance parameters for the links and nodes as given next.   **(Expected time: 1 hr)**

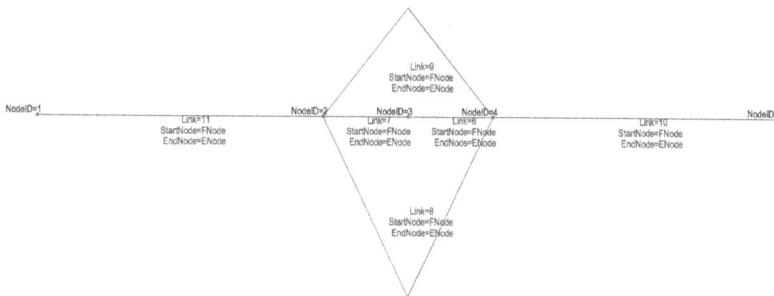

Figure 9-60 The road network

Link ID =6
   Direct resistance = 6
   Reverse resistance = 5
   Flow Direction = Bi-Directional

Link ID =7
   Direct resistance = 5
   Reverse resistance = 6
   Flow Direction = Bi-Directional

Link ID =8
      Direct resistance = 13
      Reverse resistance = 18
      Flow Direction = Reverse

Link ID =9
      Direct resistance = 14
      Reverse resistance = 19
      Flow Direction = Forward

Node ID =3
      Direct resistance = 1

Answers to Self-Evaluation Test
1. c, 2. b, 3. a, 4. b, 5. c, 6. **Add Distance**, 7. **Feature Overlay**, 8. **Network**, 9. polylines, 10. **Load Topology** 11. T, 12. F, 13. T, 14. F, 15. T

Chapter 10

Working with Different Types of Data

Learning Objectives

After completing this chapter, you will be able to:
- *Create index files from a LiDAR point file*
- *Create a 3D surface from an index file*
- *Create a 3D surface from survey point files*
- *Style raster layers*
- *Generate contours from a grid layer*
- *Create and manage Join and Calculation extensions*
- *Manage industry model data or topobase database*

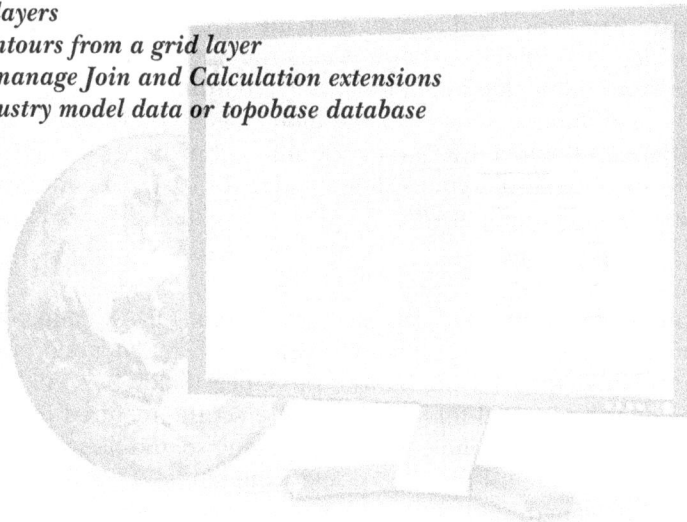

INTRODUCTION

In the previous chapter you learned to analyze vector data. In this chapter, you will learn to work with various tools that will help you to create 3D surfaces, define industry model, and use the online mapping services (Bing maps) to display raster maps in your drawing.

In this chapter, you will learn to use the point cloud (LiDAR) data to create 3D surfaces. You will then use the surface data to generate contours. You will also learn to apply display style and hillshading to produce more realistic image that helps you better understand the elevation changes on the surface.

Moreover, in this chapter, you will learn the usage of the **Join** and **Calculation** extensions for creating a join between two data tables and computing data in a table using the values in another field. This chapter also introduces you to the concept of industry models. The method of creating industry models and using them to create a drawing is also discussed in this chapter.

POINT CLOUD DATA

The point cloud data is a dense dataset that contains a set of points, which are spatially referenced to a geographical location. The point cloud data are usually collected by surveying methods such as the LiDAR survey and multibeam echo sounder survey. Such types of survey are generally conducted to survey the topography of a geographical region.

The LiDAR data is collected by using the LiDAR (Light Detection and Ranging) technology in which laser scanner is used to collect points. This data is used to create a 3D surface and a 3D model. You can load point cloud data into the current model space by using the tools in AutoCAD Map 3D. The different tools used to manage the point cloud data and to create a 3D surface are discussed next.

Point Cloud Manager

Command: MAPPOINTCLOUDMANAGER

Index File

You can use various tools and options from the **POINT CLOUD MANAGER** dialog box to manage the point cloud data. Using the options in this dialog box, you can select the input point cloud data file, assign coordinate system to the data, merge the point cloud data, and generate the index file(*.isd). To invoke this dialog box, type **MAPPOINTCLOUDMANAGER** in the Command prompt; the **POINT CLOUD MANAGER** dialog box will be displayed, as shown in Figure 10-1. The options in this dialog box are discussed next.

Add file

AutoCAD Map 3D can read a point cloud data that has the *.las* (LiDAR data), *.isd* (Point Cloud data), *.xyz* (ASCII data), or any other survey point file format. To add a point cloud file to the point cloud manager, choose the **Add file** button; the **Open** dialog box will be displayed. In the dialog box, browse to the desired folder and then select the required file extension from the drop-down list next to the **File name** edit box. On doing so, the files with the selected file extension will be displayed in the list box. Next, select the required file from the list box and then choose the **Open** button in the **Open** dialog box; this dialog will be closed and the selected file will be added to the list box in the **POINT CLOUD MANAGER** dialog box.

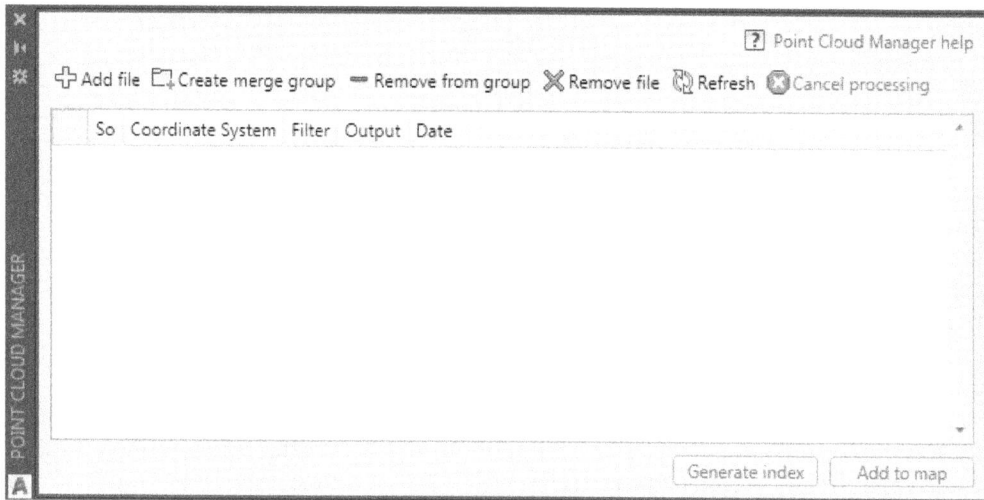

*Figure 10-1 The **POINT CLOUD MANAGER** dialog box*

Create merge group

You can group or merge several point cloud files into one group and create a single index file from the merged group. To create a merge group, choose the **Create merge group** button, refer to Figure 10-1; a new row with the name **Merge group** will be created in the list box. To rename a group, click on the **Source** cell of the group row; an edit box will be displayed. Specify the required name in this edit box.

To add point cloud files into a merge group, select the required merge group; the selected merge group row will be highlighted in blue. Now, add all required point cloud files by using the **Add file** button as discussed in the previous section. You can display the files in a merged group by choosing the [+] node corresponding to required option in the merged group cell.

> **Note**
> *To remove a merge group from the **POINT CLOUD MANAGER** dialog box, select the required merge group; the **Remove group** button will be displayed in the dialog box. Choose this button to delete the group from the dialog box.*

Remove from group

This button is used to segregate a point cloud file from the group and make it a separate point cloud file. To segregate a point cloud file, select the unwanted point cloud file in the merged group and then choose the **Remove from group** button; the selected row will be removed from the merged group.

Remove file

This button is used to remove a point cloud file from the **POINT CLOUD MANAGER** dialog box. To remove a point cloud file, select the file in the dialog box and then choose the **Remove file** button; the selected file will be removed from the dialog box.

Source

The **Source** column in the list box of the **POINT CLOUD MANAGER** dialog box displays the name of the source data file or the name of the merge group that will be used for creating the index file.

Coordinate System

The **Coordinate System** column of the list box displays the coordinate system of the source file or the merge group. If no coordinate system is assigned to the source file of merge group, the **Coordinate System** column will display **<None>**. To assign a coordinate system to a source file or merge group, double-click in the **Coordinate System** cell corresponding to the required source file or the merge group in the list box; a Browse button will be displayed in the cell. Choose the Browse button; the **Coordinate System Library** dialog box will be displayed. Assign the required coordinate system using the options in this dialog box as explained in Chapter 3. After selecting the required coordinate system, the code of the selected coordinate system will be displayed in the **Coordinate System** cell of the selected source file/merge group.

Filter

In the **Filter** column, you can apply a filter to the point cloud data to include only a specific part of the data into the current Workspace. To apply a filter to the point cloud file, double-click in the **Filter** cell corresponding to the required point cloud file in the list box of the **POINT CLOUD MANAGER** dialog box; the **Filter Point Cloud** dialog box will be displayed, as shown in Figure 10-2.

In the **Filter Point Cloud** dialog box, you can specify the filters to be applied to the point cloud file. To specify a filter, select the required options from the **Filter by** drop-down list and then specify the filter parameters in the **Define filter** area of this dialog box.

The data filtering options available in the **Filter Point Cloud** dialog box are discussed next.

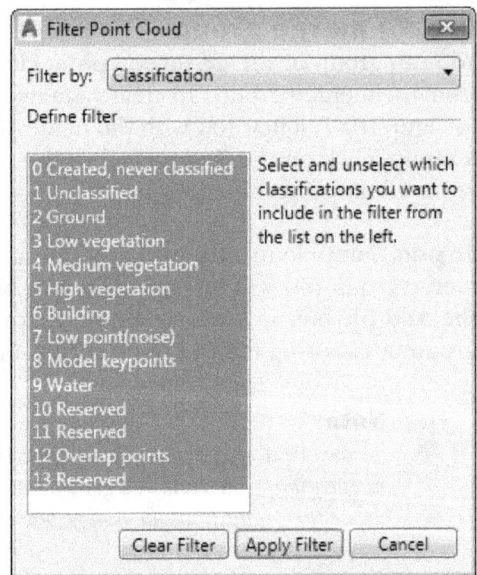

Figure 10-2 The Filter Point Cloud dialog box

Note
*You can apply only one filter to a point cloud file in the **POINT CLOUD MANAGER** dialog box.*

Classification

The **Classification** option is used to classify the current point cloud data based on the predefined filters. The **Classification** option is selected by default in the **Filter by** drop-down list. As a result, various options are displayed in the **Define filter** area, refer to Figure 10-2. To select and apply the desired filter to the current classification, press and hold the CTRL key and then click on the filters that are not required in the **Define filter** area; the selected filters will be highlighted in blue.

Elevation

The **Elevation** option is used to filter the point cloud data based on the specified elevation range. To apply this filter, select the **Elevation** option from the **Filter by** drop-down list. Next, specify the required elevation range in the **Define filter** area. To specify multiple elevation ranges, use a comma as the range separator in the **Define filter** area. For example, to filter data in the elevation range of 0 to 50 and 150 to 200, specify **0-50, 150-200** in the **Define filter** area.

Intensity

The **Intensity** option is used to classify data values into different ranges based on the intensity value calculated by LiDAR technology. To apply this filter, select the **Intensity** option from the **Filter by** drop-down list; the options for setting the parameters of this filter will be displayed in the **Define filter** area. You can specify the range of the **Intensity** option in the minimum value - maximum value pattern, with each range separated by a comma as explained in the previous section.

Spatial

The **Spatial** option is used to filter the point cloud data based on the geographical area. To apply this filter, select the **Spatial** option from the **Filter by** drop-down list; the options for selecting geographical area in this filter will be displayed in the **Define filter** area. Now, you can specify the required boundary in the drawing window by choosing the **Locate on map** button. On choosing this button, a drop-down list will be displayed with the **Circle**, **Polygon**, **Rectangle**, and **Proximity** geometric options. In this drop-down list, select the required geometric option; the cursor will change to a crosshair in the drawing window and you will be prompted to specify the area in the drawing window. Specify the required area in the drawing; the parameters of the geometry drawn in the drawing window will be displayed in the edit box in the **Define filter** area.

Note

*If you want to clear the parameters set for the selected filter, choose the **Clear Filter** button in the **Filter Point Cloud** dialog box; the parameters applied to the selected filter will be removed.*

After specifying the required filter parameters in the **Filter Point Cloud** dialog box, choose the **Apply Filter** button in this dialog box; the dialog box will be closed. The selected data filter will be displayed in the **Filter** cell of the selected source file in the **POINT CLOUD MANAGER** dialog box.

Output

In the **Output** column, you can specify the location and name of the resulting index file. To specify the name of the resulting index file, double-click in the cell in the **Output** column; a Browse button will be displayed in this cell. Choose this button; the **Save As** dialog box will be displayed. In this dialog box, enter a name in the **File name** edit box and then choose the **Save** button; the dialog box will be closed and the path of the saved file will be displayed in the **Output** column.

Date Created

The **Date Created** column in the list box of the **POINT CLOUD MANAGER** dialog box displays the date on which a file is created or modified.

Generate Index

You can create a new index file (*.isd) from an existing point cloud file in the list box. To generate an index file, choose the required source file in the list box of the **POINT CLOUD MANAGER** dialog box; the **Generate index** button will be activated. Next, choose this button; a text will be displayed in the bottom left corner of the **POINT CLOUD MANAGER** dialog box displaying the status of index generation. After completing the index generation, a tick mark will be displayed against the selected source file name in the **POINT CLOUD MANAGER** dialog box.

To create an index file after modifying its settings, select the point cloud file and then choose the **Generate Index** button; the **Existing File** window will be displayed. In this window, you can either choose the **Continue and overwrite the file** option to overwrite an existing file or choose the **Cancel the processing and change the file name** option to create an index file with a different name. On choosing the **Continue and overwrite the file** option in the **Existing File** window, this window will be closed and an index file will be created.

Add to map

The **Add to map** button is used to add an index file to the current drawing or map. To add an index file to the current drawing or map, select the required index file in the **POINT CLOUD MANAGER** dialog box; the **Add to map** button will be activated. Next, choose this button; the name of the selected index file will be displayed in the list box in the **Display Manager** tab of the **TASK PANE** and the geometry of the point cloud index data will be displayed in the drawing window.

> **Note**
> *If the objects of the index file created are not displayed in the drawing window, then right-click on the name of this index file in the **Display Manager** tab of the **TASK PANE**; a shortcut menu will be displayed. In this shortcut menu, choose the **Zoom to Extents** option; the index file will be displayed in the drawing window.*

Creating a 3D Surface from the Point Cloud Data

Command: MAPSURFACEMANAGER

Create from
Point Cloud
In Map 3D, you can create a 3D surface from the point cloud data or LiDAR data and display the surface in the drawing window. You can create a 3D surface by using the **SURFACE MANAGER** dialog box. To invoke this dialog box, type **MAPSURFACEMANAGER** in the Command prompt; the **SURFACE MANAGER** dialog box will be displayed, refer to Figure 10-3.

You can use the tools and columns such as **Add file**, **Create merge group**, **Source**, and so on in this dialog box as discussed in the earlier section. The **Parameters** column in the list box of the **SURFACE MANAGER** dialog box is discussed next.

Parameters Column

The **Parameters** column in the list box of the **SURFACE MANAGER** dialog box is used to specify the parameters of the output surface file. To specify the parameters of the output surface file, double-click on the cell corresponding to the required point cloud file in the **Parameters** column; the **Grid Parameters** dialog box will be displayed, as shown in Figure 10-4. Various parameters for setting the output grid file in different sections of the **Grid Parameters** dialog box are discussed next.

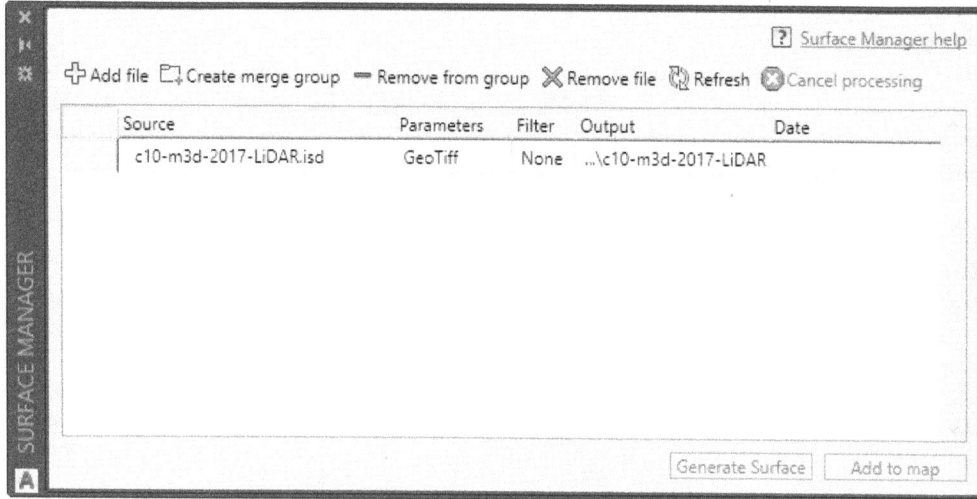

Figure 10-3 *The SURFACE MANAGER dialog box*

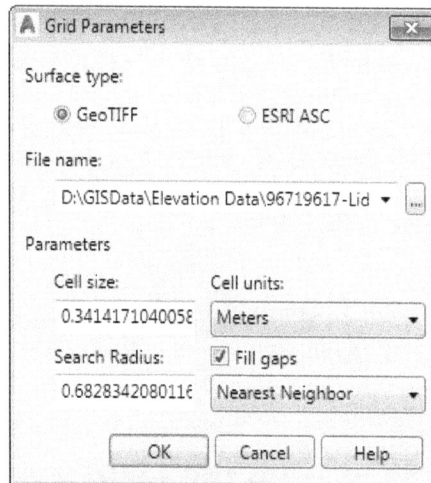

Figure 10-4 *The **Grid Parameters** dialog box*

Surface type

The options in the **Surface type** section are used to specify the file format for an output surface. In this section, the **GeoTIFF** radio button is selected by default. As a result, the output grid file will be created with the *.tif* file extension. To create the ESRI grid file with the *.asc* file extension, select the **ESRI ASC** radio button.

File name

The **File name** edit box is used to specify the name and location of the output grid file. To specify the name and location of the file, choose the Browse button next to the edit box in this section; the **Save As** dialog box will be displayed. In this dialog box, browse to the desired folder location and then enter the name of the output grid file in the **File name** edit box.

Next, choose the **Save** button in the **Save As** dialog box; the dialog box will be closed and the location path of the saved file will be displayed in the edit box in the **File name** edit box.

Parameters

The options in the **Parameters** area are used to specify the raster setting of the output file. To specify the pixel size of the output raster image, enter the pixel size in the **Cell size** edit box. To specify the measuring unit of the pixel size, select an option from the **Cell units** drop-down list. The **Fill gaps** check box below the **Cell units** drop-down list is selected by default. As a result, the pixel data of the blank areas or gaps in the point cloud file will be calculated by using the analysis method selected in the drop-down list below the **Fill gaps** check box. While filling the gaps, the height for each cell is calculated by weighing the elevation value of each point that falls within the cell. All the points within the cell are weighed with respect to the distance from the center of the cell. The points closer to the center are weighed more than the points that are far. Increasing the value in the **Search Radius** edit box will give more weight to the elevation values of the remote points (points away from the center of the cell) and decreasing the value in the **Search Radius** edit box will give less weight.

After specifying all parameters in the **Grid Parameters** dialog box, choose the **OK** button; the dialog box will be closed and the parameters applied for the output grid file will be saved. Then, you can apply filters to the output grid file, as discussed in the previous section. Next, choose the **Generate Surface** button from the **SURFACE MANAGER** dialog box; the status of surface generation will be displayed at the bottom left of the **SURFACE MANAGER** dialog box. After the completion of surface generation process, a tick mark will be displayed in the first cell of the selected row. Next, select the row for which you have generated the surface in the list box of the **SURFACE MANAGER** dialog box; the **Add to Map** button will be activated. Next, choose this button; the surface will be added to the **Display Manager** tab of the **TASK PANE** and the image will be inserted into the drawing window.

Tip
*If the image of the 3D surface generated is not displayed by default, then right-click on the added raster layer in the **Display Manager** tab of the **TASK PANE**; a shortcut menu will be displayed. In this shortcut menu, choose the **Zoom to Extents** option; the added raster image will be displayed in the drawing window.*

Creating a 3D Surface from the Field Survey Data

Ribbon:	Create > 3D Surface > Create from Points
Command:	MAPCREATESURFACE

Create from Points

You can create a 3D surface from the point data collected by field surveying or the point data collected by using the GPS instrument. To create a 3D surface from the point data, choose the **Create from Points** tool from the **3D Surface** panel of the **Create** tab; the **Create Surface** dialog box will be displayed, as shown in Figure 10-5. Various parameters in this dialog box are discussed next.

Source Area

The options in the **Source** area are used to specify the parameters for the source file. These source files can be the data files acquired from the field survey instruments such as total station.

You can use the field survey data in several formats such as *.prn, .csv, .xyz,* and so on based on the method of the point data collection.

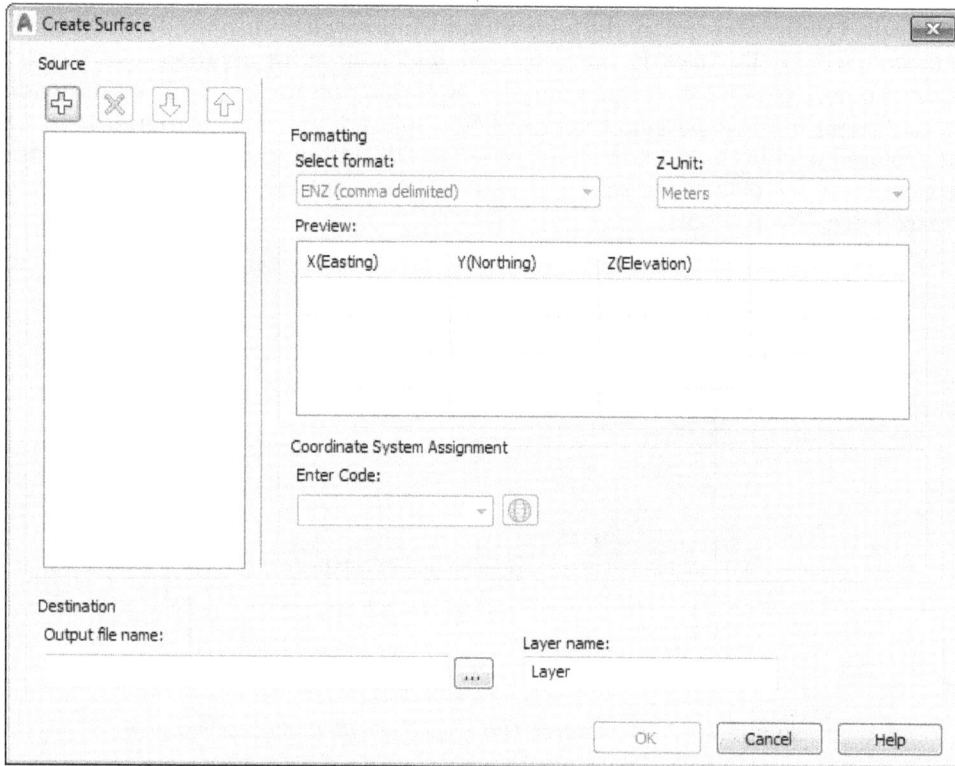

*Figure 10-5 The **Create Surface** dialog box*

As mentioned earlier, you can use survey point datasets to create a 3D surface. To load a point file, choose the **add** button located at the upper left corner in the **Source** area; a flyout will be displayed, as shown in Figure 10-6. In this flyout, you can use any of the displayed options to load the point data by using different methods. The methods of loading a point dataset by using different options in the flyout are discussed next.

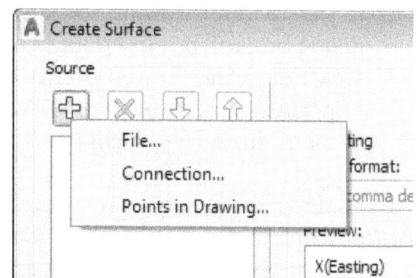

Figure 10-6 The flyout displaying different options to load point data

Using the File Option

You can choose the **File** option in the flyout to load a point file at an external folder location. On choosing the **File** option, the **Select a Point File** dialog box will be displayed. In this dialog box, browse to the folder containing the required point file and then select the file extension of the point data file in the **Files of type** drop-down list. Next, select the point file from the list box and then choose the **Open** button in the **Select a Point File** dialog box; this dialog box will be closed and the selected point file will be loaded into the list box located in the **Source** area. Also, the data in the selected point file will be displayed in the **Preview** display box in the **Formatting** area.

Using the Connection Option

You can use the **Connection** option from the flyout to select the feature data that is already connected to the current drawing as the source data for creating a 3D surface. When you choose the **Connection** option, the **Source Data** window will be displayed and the connected data store will be displayed in the **Select the data source for creating a surface** list box in this window. Expand the required folder by clicking on the [+] node corresponding to it; the dataset in the expanded folder will be displayed, as shown in Figure 10-7. Choose the required data in the list box and choose the **OK** button; the **Source Data** window will be closed and the data in the selected source will be displayed in the **Preview** area of the **Create Source** display box.

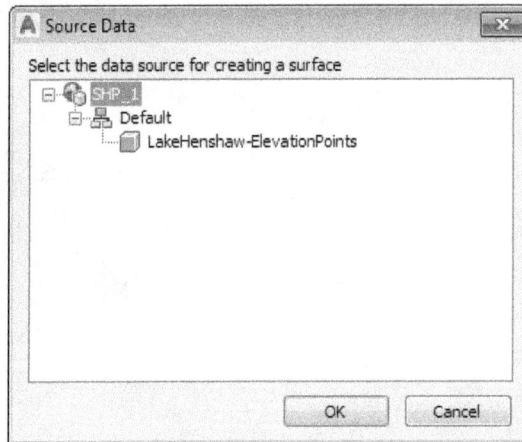

*Figure 10-7 The **Source Data** window with a source connection*

Using the Points in Drawing Option

You can choose the **Points in Drawing** option from the flyout to use the points existing in the current drawing for surface creation. On choosing this option, the **Create Surface** dialog box will be closed and you will be prompted to select the point objects in the drawing area. Select the desired points in the drawing window, and then press ENTER; the number of points selected in the list box will be displayed in the **Source** area and the coordinates of the point selected will be displayed in the **Preview** display box in the **Formatting** area.

After selecting the required source data in the **Create Surface** dialog box, you can specify the options in the **Formatting** and **Coordinate System Assignment** areas. The **Formatting** and **Coordinate System Assignment** areas are discussed next.

Formatting Area

The options in the **Formatting** area are used to specify a format and measuring unit for the survey points loaded.

To apply a format to the current point data file, select an option from the **Select format** drop-down list. Next, select a measuring unit for the height value from the **Z-Unit** drop-down list. On doing so, the format of the selected option will be displayed in the **Preview** display box in this area.

Note
*The **Select format** and **Z-Unit** drop-down lists are activated only when the point data source has been specified by choosing the **File** option from the flyout.*

Coordinate System Assignment Area

The options in the **Coordinate System Assignment** area are used to assign a coordinate system to the selected point data. To assign a coordinate system, enter the code of the coordinate system in the **Enter Code** edit box. Alternatively, choose the **Select Coordinate System** button next to the **Enter Code** edit box; the **Coordinate System Library** dialog box will be displayed. In this dialog box, select the required coordinate system from the list box and then choose the **Select** button; the dialog box will close and the code of the selected coordinate system will be displayed in the **Enter Code** edit box.

Destination Area

The options in the **Destination** area are used to specify the name of an output raster image and the name of the layer created. To specify the name of the output file, choose the Browse button next to the **Output file name** edit box; the **Save as Raster Surface** dialog box will be displayed. In this dialog box, enter the name of the raster image to be created in the **File name** edit box and then choose the **Save** button next to it; this dialog box will be closed and the file name and location will be displayed in the **Output file name** edit box in the **Destination** area. You can also enter the name of the layer in the **Layer name** edit box.

After specifying all parameters in the **Source** and **Destination** areas, choose the **OK** button; a 3D surface will be created by evaluating the point data and the surface created will be displayed in the drawing window. Also, the raster layer created will be displayed in the **Display Manager** tab of the **TASK PANE**.

VIEWING AND ANALYZING 3D SURFACES

You can use the **Style Editor** tool in the **Raster Layer** tab (contextual) to specify the display settings for the surface data created using the point data. To invoke this contextual tab, select the required raster layer in the list box of the **Display Manager** tab by clicking on it; the **Raster Layer** tab (contextual) will be displayed, as shown in Figure 10-8.

*Figure 10-8 The **Raster Layer** contextual tab displayed*

You can use the **Style Editor** tool in the **Raster Layer** tab to create the visual theme for the surface which helps you to analyze the elevation, slope, and aspect. The **Raster Layer** tab (contextual) also contains the **Contour Layer** and **Metadata** tools. These tools are used to generate contours and metadata using the created surface data. These tools are discussed next.

Styling a Raster Layer

Ribbon: Raster Layer (Contextual tab) > Style > Style Editor

You can style a raster layer to enhance its appearance by modifying its display properties such as brightness, contrast, color, and so on. You can perform these modifications by using the options in the **STYLE EDITOR** dialog box. To invoke this dialog box, select the raster layer in the **Display Manager** tab; the **Raster Layer** tab (contextual) will be displayed. Next, choose the **Style Editor** tool from the **Style** panel of the **Raster Layer** tab; the **STYLE EDITOR** dialog box will be displayed. Figure 10-9 shows the **STYLE EDITOR** dialog box displayed for a raster layer with the name given to each of the areas. The methods of applying style to a raster layer in the **Scale Ranges for Layer rasters** and **Raster Style for the Selected Scale Range** areas are discussed next.

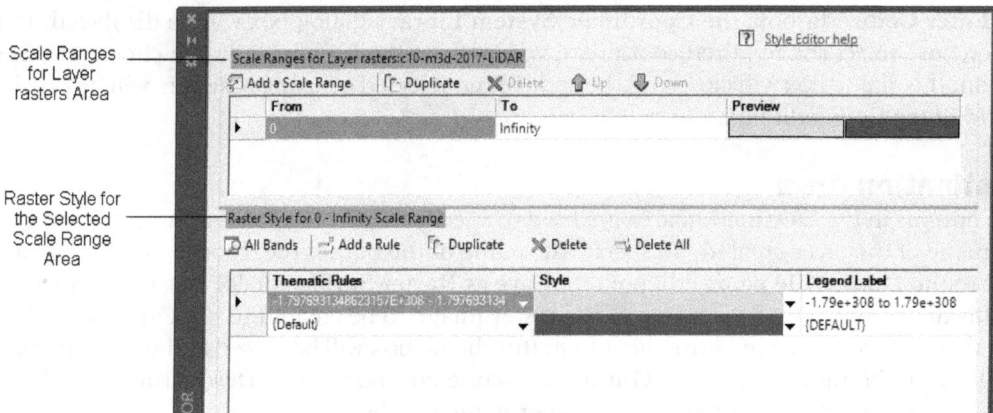

*Figure 10-9 Partial view of the **STYLE EDITOR** dialog box with the name given to each area*

Scale Ranges for Layer rasters Area

The options in the **Scale Ranges for Layer rasters** area are used to customize the scale ranges for the selected raster layer. You can customize the scale ranges by using the **Add a Scale Range**, **Duplicate**, **Delete**, **Up**, and **Down** buttons in this area, as discussed in Chapter 5.

Raster Style for the Selected Scale Range Area

The options in the **Raster Style for the Selected Scale Range** area are used to apply or edit the theme of the selected raster layer. There are many options in the tool panel of this area, which can be used to style the selected raster layer. Figure 10-10 shows the tool panel displayed in the **Raster Style for the Selected Scale Range** area on choosing the **All Bands** button. The tools in this panel are discussed next.

*Figure 10-10 The tool panel displayed in the **Raster Style for the Selected Scale Range** area*

Band Detail

The **Band Detail** tool is used to display the details of the selected band such as the minimum and maximum values of the data range in the list box located below the tool panel. To display the details of the selected band, choose the **Band Detail** tool in the tool panel; the details of the selected band will be displayed in the list box and the **Band Detail** tool will change into the **All Bands** tool from the tool panel. Also, you can add a new rule into the list box by choosing the **Add a Rule** tool in the tool panel. On doing so, a new row will be inserted into the list box. In this inserted row, you can specify the range of the grid data in the cell corresponding to the selected row in the **Thematic Rules** column.

To apply a desired color, click in the cell corresponding to the selected row in the **Style** column; the drop-down list will be displayed in the cell. Select an option from the drop-down list to specify a color. To apply a label to the selected thematic rule, click in the cell corresponding to the selected row in the **Legend Label** column; the selected cell will change into an edit box. Enter the desired name in this edit box, and then choose the **Apply** button located at the lower right corner of the **STYLE EDITOR** dialog box; the changes made will be applied to the current band. To display all bands in the current raster layer, choose the **All Bands** tool from the tool panel; all bands in the current raster layer will be displayed in the list box below the tool panel.

Hillshade Band

The **Hillshade Band** drop-down list is used to apply shading to 3D surface. To apply a hillshade to a surface, select the required band from the **Hillshade Band** drop-down list and then choose the **Apply** button from the **STYLE EDITOR** dialog box; the hillshade will be applied to the selected band. To turn off the hillshade, select the **none** option from the **Hillshade Band** drop-down list and then choose the **Apply** button.

Elevation Band

The **Elevation Band** drop-down list is used to apply height value from a band to the selected band in the list box below the tool panel. To apply the height value, first select the band to which you need to apply height value and then select the band whose height you want to apply to this band from the **Elevation Band** drop-down list. Next, choose the **Apply** button from the **STYLE EDITOR** dialog box; the elevation will be applied to the selected band.

Brightness

The **Brightness** edit box is used to apply brightness to the pixels in a raster image. To do so, enter a numeric value between **0** and **50** in the **Brightness** edit box and then choose the **Apply** button located at the lower right corner of the **STYLE EDITOR** dialog box; the brightness of the raster image will be adjusted in the drawing window based on the value specified.

Contrast

The **Contrast** edit box is used to apply sharpness to the boundary edges in a raster image. To do so, enter a numeric value between **0** and **50** in the **Contrast** edit box and then choose the **Apply** button from the **STYLE EDITOR** dialog box; the specified contrast value will be applied to the selected band in the drawing window.

Transparent

The **Transparent** tool in the tool panel is used to apply transparency to the specific grids in a raster image. To do so, choose the **Transparent** tool in the tool panel; the **Transparency Color** window will be displayed, as shown in Figure 10-11.

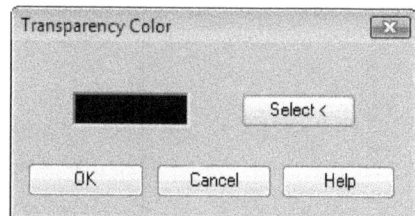

Figure 10-11 The Transparency Color window

In the **Transparency Color** window, choose the **Select<** button; the cursor will change into a crosshair in the drawing window. Place the crosshair on the desired grid in the drawing window and then click; the color of the selected grid will be displayed in the display box in the **Transparency Color** window.

Next, choose the **OK** button in the **Transparency Color** window; this window will be closed and the selected color will be used to set transparency of the pixels in the raster image. Choose the **Apply** button from the **STYLE EDITOR** dialog box; the grids in the raster image that contain the value of the selected color will be set to transparent.

Performing Height, Slope, and Aspect Analyses

To perform the height, slope, or aspect analysis on a 3D surface, ensure that the band details are displayed in the **Raster Style for the Selected Scale Range** area of the **STYLE EDITOR** dialog box. If this area does not display the band details, choose the **All Bands** tool from the tool panel of this area, refer to Figure 10-9; all the bands in the current raster layer will be displayed in the list box. Also, the **Band Details** tool will be displayed in the tool panel of the **Raster Style for the Selected Scale Range** area. Next, click in the **Style** cell corresponding to the required band; a drop-down list will be displayed. Choose the **Theme** option from the displayed drop-down list, as shown in Figure 10-12; the **Theme** dialog box will be displayed, as shown in Figure 10-13.

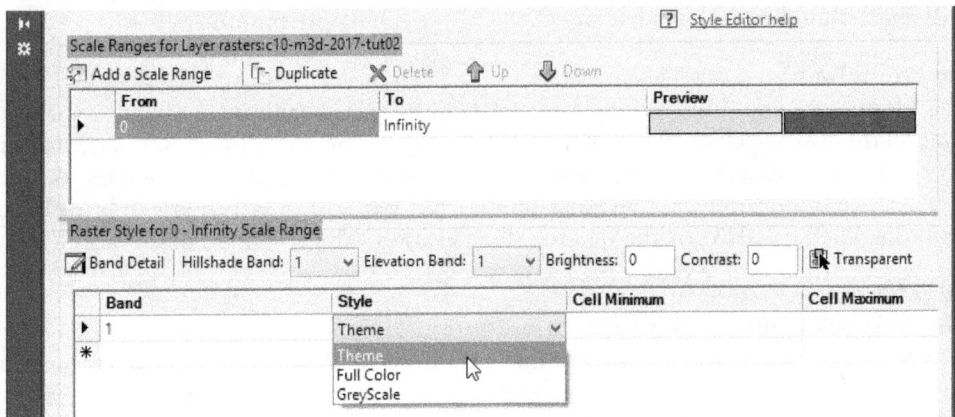

*Figure 10-12 Selecting the **Theme** option from the drop-down list*

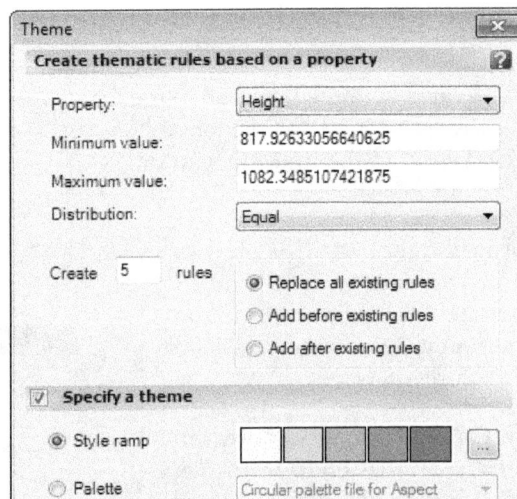

*Figure 10-13 Partial View of the **Theme** dialog box*

To perform the slope analysis, select the **Slope** option from the **Property** drop-down list of the **Theme** dialog box. Next, specify the minimum and maximum values for the slope in the **Minimum value** and **Maximum value** edit boxes, respectively. Choose the distribution method for the analysis by selecting the required option from the **Distribution** drop-down list and then specify the number of classes (rules) that you want to create for the analysis output in the **Create rules** edit box.

Tip
*To perform the height or aspect analysis on the selected surface, choose the **Height** or **Aspect** option from the **Property** drop-down list in the **Theme** dialog box.*

To specify the color theme for displaying the analysis result, ensure that the **Specify a theme** check box is selected in the dialog box. Note that on selecting this check box, the options for specifying the color theme will be activated. To display the analysis result using a color ramp, select the **Style ramp** radio button. Next, choose the browse button next to the **Style ramp** radio button; the **Style Band** dialog box will be displayed, as shown in Figure 10-14. In this dialog box, to specify the range for the color ramp, select the required color from the **From** and **To** drop-down lists and then choose the **OK** button; the dialog box will be closed and the specified color ramp will be displayed in the **Theme** dialog box.

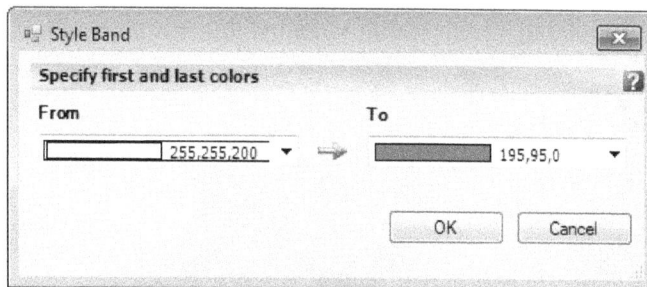

*Figure 10-14 The **Style Band** dialog box*

Note that you can also select a predefined color palette for displaying the analysis results. To do so, select the **Palette** radio button in the **Theme** dialog box; the drop-down list corresponding to this radio button will be activated and the **Create rule** edit box will be deactivated. Next, select the required color palette from this drop-down list.

To create a legend for the result of the analysis, select the **Create legend labels** check box in the **Theme** dialog box; the options for specifying legend parameters will be activated in the dialog box. Next, specify the text that you want to display as the label text for the legend in the **Label text** edit box and then choose the format for the label by choosing the required option from the **Label format** drop-down list.

After specifying the required options in the **Theme** dialog box, choose the **OK** button; the dialog box will be closed and the theme specified will be displayed in the list box of the **STYLE EDITOR** dialog box. Next, choose the **Apply** button in this dialog box; AutoCAD Map 3D performs the analysis and displays the result in the drawing area using the specified color theme. Figures 10-15 through 10-17 show the results of slope, aspect, and height analyses for a study region.

Figure 10-15 The result of slope analysis

Figure 10-16 The result of aspect analysis

Figure 10-17 The result of height analysis

Creating a Contour Layer

Ribbon:	Raster Layer (Contextual tab) > Create > Contour Layer

Contour is a line that connects the points of equal elevation. A contour layer is a feature layer that contains the elevation data. You can create a contour layer from surface layer containing the elevation data (DTM and DEM). To create contours from a surface layer (DTM or DEM), select the required layer from the list box in the **Display Manager** of the **TASK PANE**; the **Raster Layer** tab (contextual) will be displayed. Choose the **Contour Layer** tool from the **Create** panel; the **Generate Contour** dialog box will be displayed, as shown in Figure 10-18.

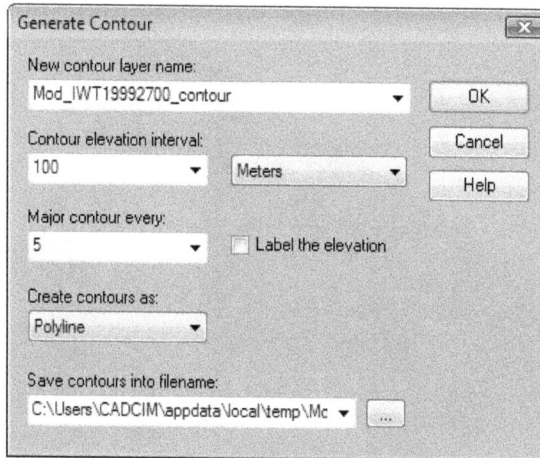

Figure 10-18 The Generate Contour dialog box

In the **Generate Contour** dialog box, you can specify the parameters for creating a contour. You can specify a name for the contour file in the **New contour layer name** edit box. Specify the contour interval by entering desired value in the **Contour elevation interval** edit box. You can also specify the contour interval by selecting an option from the list displayed by clicking on the down-arrow on the right of the **Contour elevation interval** edit box.

You can apply the measuring unit to the contours to be created by selecting an option from the drop-down list next to the **Contour elevation interval** edit box. To define a major contour, you can either enter the number of contours between two major contours in the **Major contour every** edit box. Alternatively, select an option from the drop-down list displayed by clicking on the down-arrow on the right of this edit box. The **Label the elevation** check box is cleared by default. As a result, the major contour will not be labeled in the drawing window. To display the contour value of the major contour, select the **Label the elevation** check box. To create the contours as the polyline or polygon feature data, select the required option from the **Create contours as** drop-down list.

To specify the location for saving the contour file, choose the Browse button next to the **Save contours into filename** edit box; the **Save As** dialog box will be displayed. In the **Save As** dialog box, browse to the desired location and then enter the name of the contour file to be created in the **File name** edit box. Next, choose the **Save** button from the **Save As** dialog box; the dialog box will be closed and the path of the file will be displayed in the **Save contours into filename** edit box in the **Generate Contour** dialog box. Now, choose the **OK** button to close this dialog box. The vector layer of the contour will be created at the specified location and will be added to the list box in the **Display Manager** tab of the **TASK PANE**. The created contours will now be visible in the drawing window. Figure 10-19 displays contours created from a 3D surface.

Figure 10-19 *The contour created from the 3D surface*

Setting the Hillshade Parameters

Ribbon:	Analyze > Feature > Surface Hillshade
Command:	MAPHILLSHADE

You can set the parameters of the hillshade to enhance the appearance of a grid layer as per your requirement. To specify the parameters of the hillshade, choose the **Surface Hillshade** tool from the **Feature** panel; the **Hillshade Settings** dialog box will be displayed, as shown in Figure 10-20. The various options for specifying the settings in the **Hillshade Settings** dialog box are discussed next.

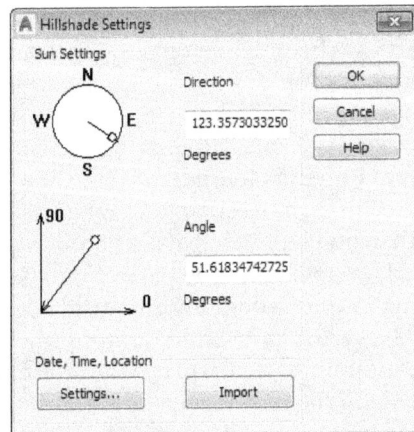

Figure 10-20 *The **Hillshade Settings** dialog box*

Sun Settings Area

The options in the **Sun Settings** area are used to specify the horizontal and vertical positions of the Sun (a source of light) with respect to the surface. To specify the horizontal position of the Sun with reference to a horizontal plane, press and hold the yellow dot in the circle diagram in the upper-left section of this area and then drag the yellow dot to the desired position; the angle in the **Direction** edit box will change accordingly. Alternatively, you can specify the angle of the horizontal position by entering a decimal degree in the **Direction** edit box. To specify the vertical position of the Sun, press the yellow dot in the angle figure displayed in the **Sun Settings** area and then drag it to the desired position; the decimal degree in the **Angle** edit box will change accordingly. You can also specify the vertical position of the Sun by entering a value for the angle in decimal degree format in the **Angle** edit box.

Date, Time, Location Area

The options in the **Date, Time, Location** area are used to specify the advanced settings of the Sun, rendered shadow, and geographic location. To set the parameters of the Sun and shadow elements, choose the **Settings** button in this area; the **Hillshade Settings** dialog box will be closed and the **SUN PROPERTIES** palette will be displayed, as shown in Figure 10-21.

In the **SUN PROPERTIES** palette, you can specify the parameters to set the properties of the sun. For example, to specify the **Height** parameter in the **Horizon** section, click in the cell corresponding to it; an edit box will be displayed in the cell. Specify a new value in the edit box.

After setting the required parameters in the **SUN PROPERTIES** palette, close the palette; the modified settings will be saved.

To apply the settings that you have specified in the **SUN PROPERTIES** palette, invoke the **Hillshade Settings** dialog box as explained earlier. In the **Hillshade Settings** dialog box, choose the **Import** button in the **Date, Time, Location** area; the sun settings specified in the **SUN PROPERTIES** palette will be imported. Next, choose the **OK** button; the dialog box will be closed and the hillshade parameters will be applied to the surface.

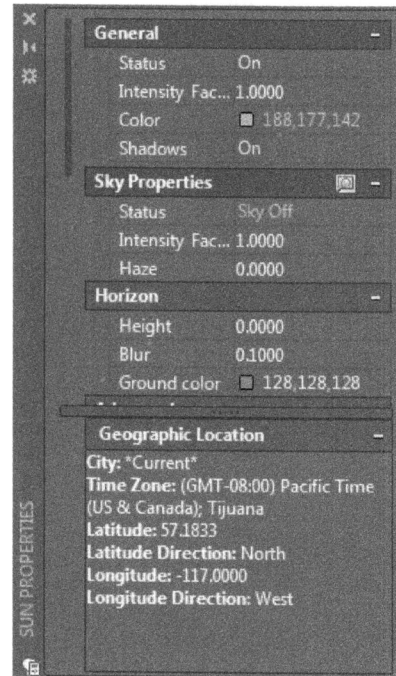

Figure 10-21 The SUN PROPERTIES palette

Vertical Exaggeration

You can apply vertical exaggeration to the surface data. Vertical exaggeration helps you alter the scale of the features in the vertical direction, thereby emphasizing the elevation of the features. To apply exaggeration, choose the required scale from the **Vertical Exaggeration** option in the **Drawing Status** bar displayed below the drawing window.

ATTRIBUTE DATA

In this section, you will learn how to manage dataset in different file formats by using various tools in AutoCAD Map 3D. Also, you will learn how to copy, merge, and calculate attribute data as well as to retrieve the required data. The methods of data management are discussed next.

Creating a Bulk Copy

Ribbon: Create > Feature Data Store > Bulk Copy

The **Bulk Copy** tool is used to copy the property, attributes, and filters of a file created in the geospatial format to another feature dataset. For example, you can copy the feature data in the *.sdf* file format to a feature data in the *.shp* file format. To do so, choose the **Bulk Copy** tool from the **Feature Data Store** panel of the **Create** tab; the **Bulk Copy** dialog box will be displayed, as shown in Figure 10-22. In this dialog box, you can specify the parameters for copying or converting a file from one geospatial format into another. Various options in the **Bulk Copy** dialog box are discussed next.

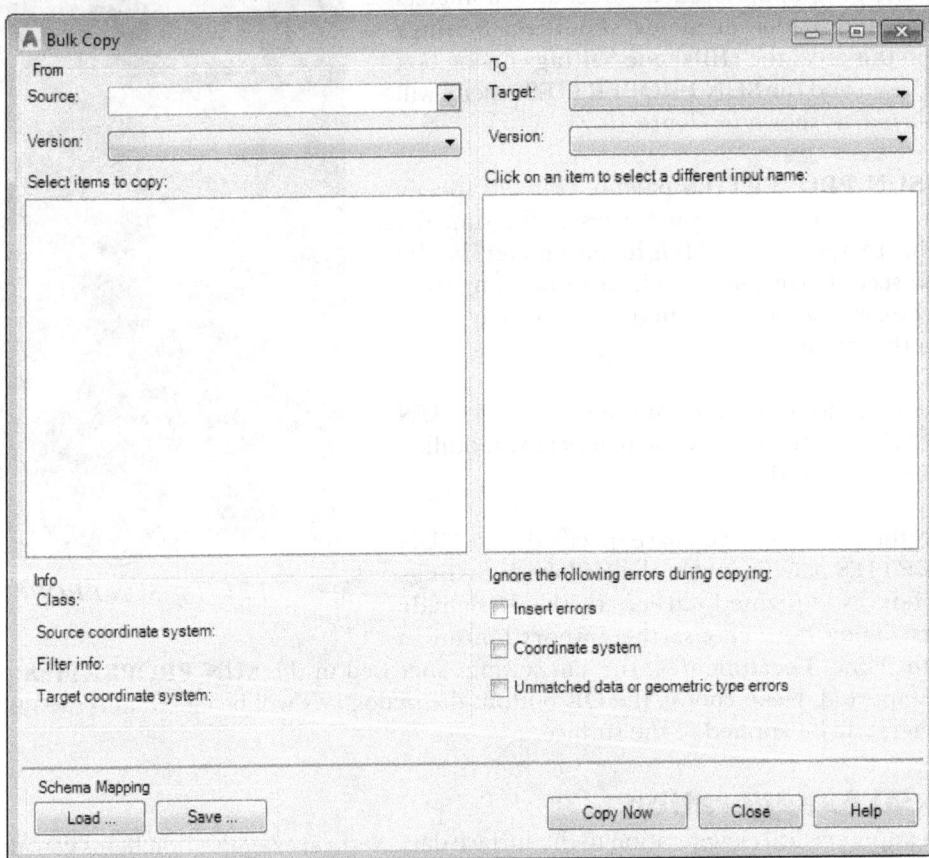

*Figure 10-22 The **Bulk Copy** dialog box*

From Area

The options in the **From** area are used to specify the parameters of the source file from which you will copy the geospatial data. To select an input feature layer, click on the down-arrow on the right of the **Source** drop-down list; a list of the feature layers connected to the current Workspace will be displayed. In this list, select the required feature layer or the FDO connection by clicking on it; the schema, attributes, properties, and geometric properties of the selected feature layer will be displayed in the **Select items to copy** list box in this area.

To include the required entities into the current conversion process, select the check box corresponding to the entities in the **Select items to copy** list box; the label of the selected items will be displayed in the **Click on an item to select a different input name** list box in the **To** area.

Info Area

The **Info** area in the **Bulk Copy** dialog box displays the general information of the selected connection or feature class. The information includes source coordinate system, filters, and target coordinate system of the selected connection or feature class.

To Area

The options in the **To** area are used to specify the parameters and set conditions on the output bulk copy file. To specify an output feature layer, select an option from the **Target** drop-down list. Next, to map the source item to its destination item, click on the arrow for the required option in the **Click on an item to select a different input name** list box; a list will be displayed. Next, select an option from this list; the source item is mapped to its destination item.

The options in the **Ignore the following errors during copying** section are used to specify the errors that should be ignored while creating a bulk copy. To continue with bulk copying by skipping the objects that fail to be inserted, select the **Insert errors** check box. To skip the object geometries or properties that are not supported by the output data file, select the **Unmatched data or geometric type errors** check box. To copy the geometry without performing the coordinate system transformation, select the **Coordinate system** check box.

Schema Mapping Area

The options in the **Schema Mapping** area are used to specify the bulk copy settings from another template file or save the current settings as a template file. You can either load an existing schema by using the **Load** button or save the settings specified for the current bulk copy as a template file by using the **Save** button.

After specifying all parameters for creating the bulk copy, choose the **Copy Now** button from the **Bulk Copy** dialog box; the **Continue Bulk Copy** window will be displayed. In this window, choose the **Continue Bulk Copy** button; the **Bulk Copy Results** window will be displayed with the results for copying. Choose the **OK** button in the **Bulk Copy Results** window; this window will be closed and the bulk copy will be created.

Tip
*You can display the geometry of a newly created bulk copy file by clearing the check box corresponding to other vector layers in the **Display Manager** tab of the **TASK PANE**.*

Managing the Attributes in a Vector Layer

AutoCAD Map 3D allows you to calculate new values in your data table based on existing values. You can bring in data from an external table (secondary table) into the data table of a feature class in the current map (primary table) without affecting the data in the external table. The methods of managing a vector layer by using the **Joins**, **Calculations**, and **Manage Extension** tools are discussed next.

Joins

Ribbon: Vector Layer (Contextual tab) > Create > Joins

The **Joins** tool is used to specify the parameters for joining a table with a feature layer (primary table). To invoke this tool, select a feature layer from the list box in the **Display Manager** tab of the **TASK PANE**; the **Vector Layer** tab (contextual) will be displayed. Choose the **Joins** tool from the **Create** panel of this contextual tab; the **Create a Join** dialog box will be displayed, as shown in Figure 10-23.

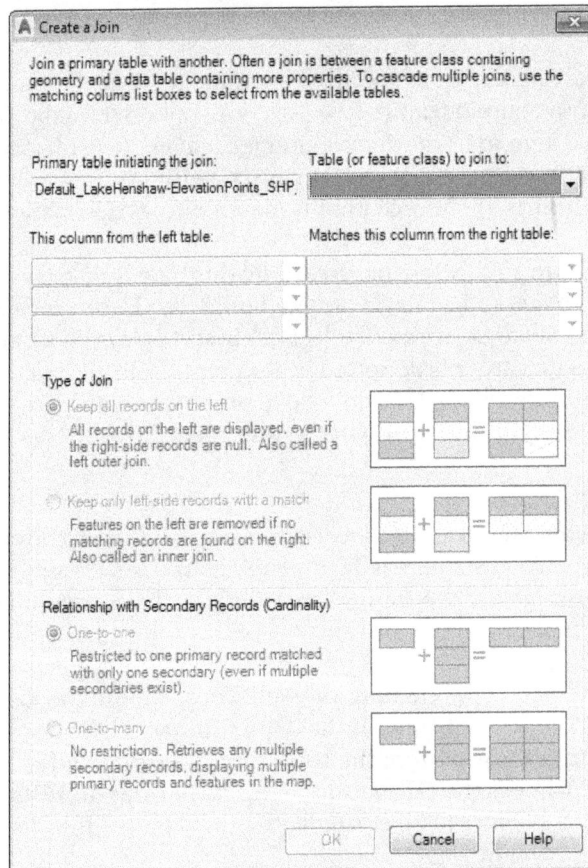

*Figure 10-23 The **Create a Join** dialog box*

In this dialog box, you can specify the parameters for joining attribute data from an external table (secondary table) with the table in the current map (primary table). The **Primary table initiating the join** text box will display the table name to which you will join the secondary table. To select the secondary table, which is to be joined to the primary table, click on the down-arrow on the right of the **Table (or feature class) to join to** drop-down list; a list of connected sources will be displayed in a hierarchy. Next, select the required table or feature class from the displayed hierarchy; the source of the selected database will be displayed in the **Table (or feature class) to join to** drop-down list. Also, the first drop-down list in the **This column from the left table** column will become active.

Note
*Using the options in the **Create a Join** dialog box, you can create a self join (join table to itself). To do so, specify the same table name in the **Table (or feature class) to join to** drop-down list and the **Primary table initiating the join** text box. On doing so, the **Confirm join** message box will be displayed. Choose the **Yes** button in this message box to proceed.*

Next, specify the key column in the primary table by selecting an option from the activated drop-down list in the **This column from the left table** column; the corresponding drop-down list in the **Matches this column from the right table** will become active. Next, select the required option from the activated drop-down list. Also, the **Type of Join** and **Relationship with Secondary Records (Cardinality)** areas will become active. The options in these areas are discussed next.

Type of Join Area

The options in the **Type of Join** area are used to specify how tables will be joined. In this area, the **Keep all records on the left** radio button is selected by default. As a result, all attribute data in the selected column of the primary table will be retained and displayed, even where there is no attribute data (Null value) corresponding to it in the column selected in the secondary table. On selecting the **Keep only left-side records with a match** radio button, the attribute data in the column selected from the primary table will be removed if there is no attribute corresponding to it in the column selected from the secondary table.

Relationship with Secondary Records (Cardinality) Area

The options in the **Relationship with Secondary Records (Cardinality)** area are used to define the functional relationship between attributes from the primary and secondary tables. In this area, the **One-to-one** radio button is selected by default. As a result, only single attribute value in the secondary table matching to the attribute value in the primary table will be displayed. If you select the **One-to-many** radio button, then on joining the data, multiple attribute values in the secondary table matching to the attribute value in the primary table will be displayed.

After specifying the options in the **Type of Join** and **Relationship with Secondary Records (Cardinality)** areas, choose the **OK** button; the join will be created with the specified parameters. The modified attribute data can be viewed in the **DATA TABLE** corresponding to the modified vector layer.

Calculations

Ribbon:	Vector Layer (Contextual tab) > Create > Calculations

Calculations

You can use the **Calculations** tool to calculate new values in a field based on the existing attribute values in the DATA TABLE. You can then use the calculated values to filter data. To calculate the values using the **Calculations** tool, select a vector layer in the **Display Manager** tab of the **TASK PANE**; the **Vector Layer** tab (contextual) will be displayed. Next, choose the **Calculations** tool from the **Create** panel; the **Create a Calculation** dialog box will be displayed, as shown in Figure 10-24.

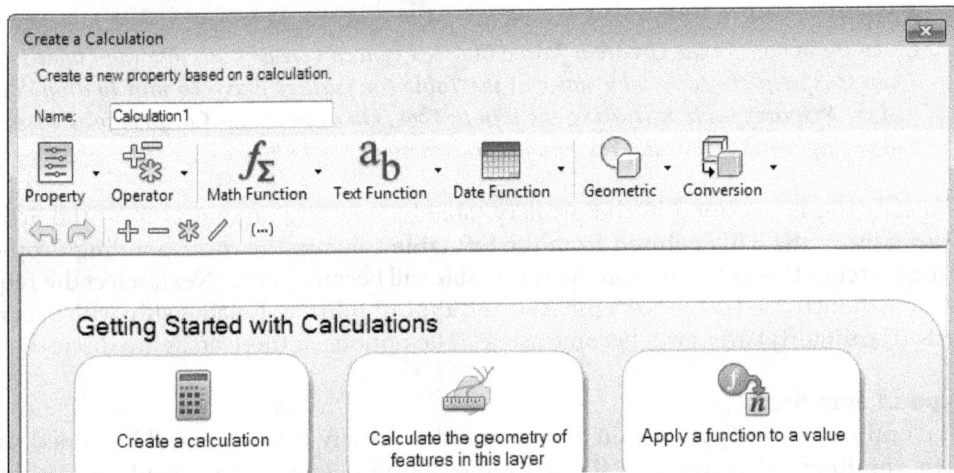

Figure 10-24 Partial view of the **Create a Calculation** *dialog box*

In the **Create a Calculation** dialog box, you can write an expression to calculate values. The calculated values are stored in a new property (attribute field) column in the data table of the feature layer. To specify the name of this attribute field, enter the required name in the **Name** edit box of the **Create a Calculation** dialog box.

Next, write an expression in the **Create a Calculation** dialog box. The method of writing the expression in this dialog box is similar to the method of writing expressions in the **Create Query** window, as discussed in Chapter 5. You can also use the syntax for writing an expression. To do so, choose the **Create a calculation**, **Calculate the geometry of features in this layer**, or **Apply a function to a value** option in the **Getting Started with Calculations** page; the expression syntax corresponding to the selected option will be displayed in the **Create a Calculation** dialog box. Edit the syntax to formulate the required expression as discussed in Chapter 5.

After writing the expressions, validate it by choosing the **Validate** button. On validating the expression, choose the **OK** button in the **Create a Calculation** dialog box; a new property column with the specified name will be added to the data table of the feature layer. To view the updates in the data table of the feature layer, invoke the **DATA TABLE** window by choosing the **Table** tool in the **Display manager** of the **TASK PANE**.

Manage Extension

Ribbon:	Vector Layer (Contextual tab) > Create > Manage Extension

You can use the **Manage Extension** tool to manage the property joins created by using the **Joins** tool. You can also modify, edit, and delete fields calculated in the data table using the **Calculations** tool. In addition to managing the existing extensions, you can also create new extensions in the **Manage Layer Data** dialog box.

To invoke the **Manage Layer Data** dialog box, choose the **Manage Extension** tool from the **Create** panel of the **Vector Layer** (contextual) tab; the **Manage Layer Data** dialog box will be displayed. Figure 10-25 shows the **Manage Layer Data** dialog box with the Joins and Calculations extensions.

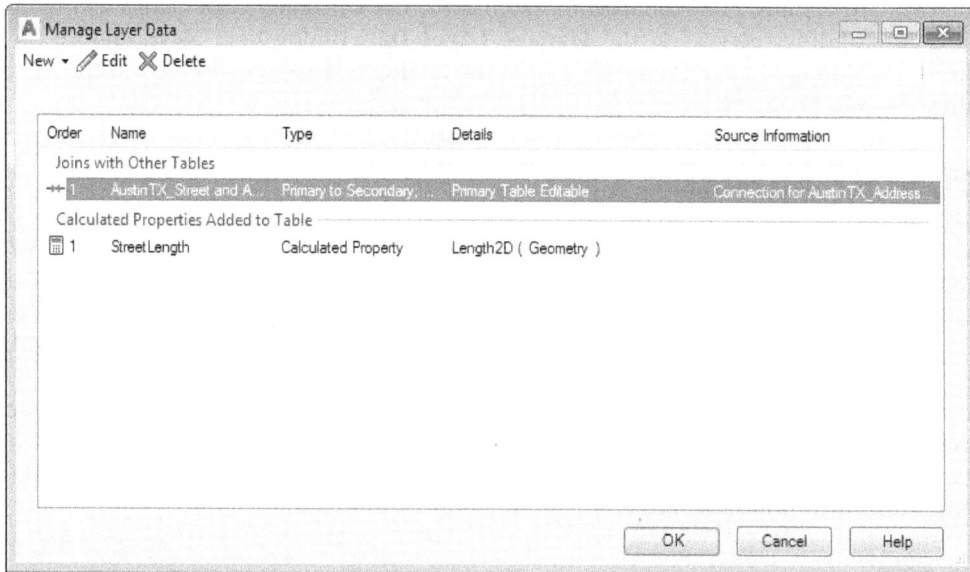

*Figure 10-25 The **Manage Layer Data** dialog box with the Joins and Calculations extensions*

Note

*The **Manage Layer Data** dialog box will not be displayed if you do not have any Join or Calculation extensions associated with the selected vector layer.*

For editing a Joins extension, select the required extension from the **Joins with Other Tables** section in the **Manage Layer Data** dialog box. Next, choose the **Edit** button from the **Manage Layer Data** dialog box; the **Edit a Join** dialog box will be displayed. Edit the selected join extension in the **Edit a Join** dialog box, as discussed in the previous section.

To edit a Calculations extension, select the required extension in the **Calculated Properties Added to Table** section; the selected extension will be highlighted in blue. Next, choose the **Edit** button from the **Manage Layer Data** dialog box; the **Modify a Calculation** dialog box with the expression for the selected extension will be displayed. Edit the expressions as per your requirement and validate the expression, as discussed in Chapter 5. Next, choose the **OK** button in the **Modify a Calculation** dialog box; the modified expression will be saved.

You can also create a new Join extension and a new Calculation extension by using the options in the **Manage Layer Data** dialog box. To create a new Join extension, select the **Join** option from the **New** drop-down list in this dialog box, as shown in Figure 10-26; the **Create a Join** dialog box will be displayed. In this dialog box, specify the parameters for joining the attributes, as discussed in the previous section. Similarly, to create a new Calculation, choose the **Calculation** option from the **New** drop-down list of the **Manage Layer Data** dialog box; the **Create a Calculation** dialog box will be displayed. In this dialog box, specify the condition for creating a new property column, as discussed in the previous section.

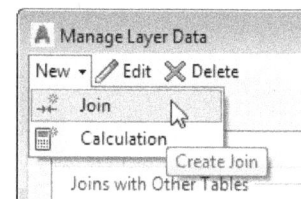

*Figure 10-26 Selecting the **Join** option from the **New** drop-down list*

After creating the Joins and Calculations extensions, the calculated extensions will be added to the corresponding sections in the **Manage Layer Data** dialog box. Next, choose the **OK** button in the **Manage Layer Data** dialog box; this dialog box will be closed and the specified modifications will be saved.

INDUSTRY MODEL

Industry model is a schema that contains predefined feature class, rules, and relationships for a given industry such as water and waste water. In AutoCAD Map 3D, file based industry models can be saved as drawing or template files (DWG, DWT). While using the file based industry model, the features in the model are stored inside the DWG file. You can also create industry model by using a database such as Oracle. These models are called as **Enterprise Industry Model**.

You can create, edit, and connect an industry model using the **Autodesk Infrastructure Administrator** application. To start the **Autodesk Infrastructure Administrator** application, choose **Programs > Autodesk > Autodesk Infrastructure Administrator 2017 > Autodesk Infrastructure Administrator 2017**.

Note
*While installing AutoCAD Map 3D 2017, the check box corresponding to **Autodesk Infrastructure Administrator 2017** is not selected by default. As a result, the Autodesk Infrastructure Administrator will not be installed by default. Select the check box to install this application.*

To create, edit, and maintain the industry model data, you need to use the tools in the **Autodesk Infrastructure Administrator**. The methods of creating, editing, and maintaining an industry model project are discussed next.

Creating the Industry Model Data or Topobase

As mentioned earlier, an industry model contains industry-specific schemas that define the feature class, data attributes, rules, and relationship. To create an industry model, start the **Autodesk Infrastructure Administrator** application interface as described earlier. Next, choose the **New** option from the **File** menu of this interface; the **Create new industry model** window will be displayed with the **Modules** tab chosen in the left pane. The various options in this tab are displayed in the right pane of the **Create new industry model** window, as shown in Figure 10-27.

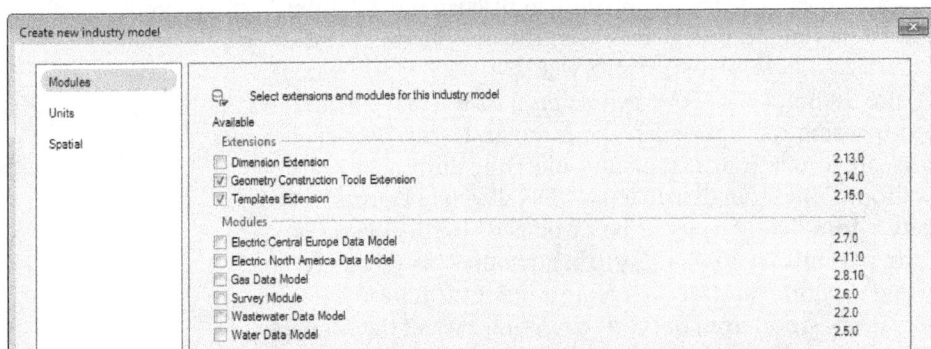

*Figure 10-27 Partial view of the **Create new industry model** window*

To specify the extension, select the check boxes corresponding to the required extensions in the **Extensions** area. Similarly, to create the model based on existing data model template, select the check boxes corresponding to the required module in the **Modules** area. Next, choose the **Units** option in the left pane of the **Create new industry model** window; the **Select the units to be used for the industry model** page will be displayed in the right pane of the window. In this page, specify the required units for the parameters in the industry model. After specifying the units, choose the **Spatial** option in the left pane; the **Spatial settings for the feature in the industry model** page will be displayed. Specify the options in this page to set the spatial settings for the model and then choose the **OK** button; the **Update Modules and DataModels** window will be displayed. Next, choose the **Update** button in this window; AutoCAD Map 3D will execute the update process. Close the **Update Modules and DataModels** window; the created industrial model will be displayed in the **Autodesk Infrastructure Administrator** interface.

Using this interface, you can create new topics, feature class, domains, forms, and so on in your model. After creating the required elements in your model, you need to save the model. To do so, choose the **Save as** option from the **File** menu of the **Autodesk Infrastructure Administrator** interface; the **Save drawing as** dialog box will be displayed. In this dialog box, select the type of file for saving the industry model by selecting an option in the **Save as type** drop-down list. Specify the name for the model in the **File name** edit box. Next, choose the **Save** button; the model will be saved at the given location with the specified name and file type.

Creating Drawings in Industry Model

After creating an industry model template, you can use it to create a drawing with an industry model. To create a drawing file using a file-based industry model template, start the AutoCAD Map 3D application. Next, choose the **Open** tool from the Quick Access Toolbar; the **Select File** dialog box will be displayed. In this dialog box, select the **Drawing Template (*.dwt)** option from the **Files of type** drop-down list. Next, browse to the location of the industry model template that you want to use for creating drawing and then select it. Next, choose the **Open** button; a new drawing file will be opened in the AutoCAD Map 3D drawing window with the selected template file. Figure 10-28 shows the AutoCAD Map 3D 2017 application with the **IM_WATER** industry model template file opened.

*Figure 10-28 The AutoCAD Map 3D 2017 Application showing the **IM_WATER** template file opened*

Note

You can use the predefined industry model template provided by AutoCAD. These templates can be found at the following location:

C:\Users\<Username>\AppData\Local\Autodesk\AutoCAD Map 3D 2017\R21.0\enu\Template\ Industry Templates

You can now create the feature data in the feature layer using the defined feature class and the attribute table in the drawing. The drawing thus created will comply with the industry model defined in the template used.

AutoCAD Map 3D also allows you to import the existing feature data into your industry model. The procedure to convert existing feature data into industry model is discussed next.

Converting Data to Industry Model

To convert an existing feature data into an industry model, open a project or the industry model template. Next, switch to the **Maintenance** workspace and then choose the **Convert to Industry Model** tool from the **Import** panel of the **Insert** tab; the **Convert to Industry model** window will be displayed. In this window, choose the **Add file-based sources to convert** tool to specify the (file-based) data file/s to be converted into the industry data model. Similarly, choose the **Add FDO connection sources to convert** tool or the **Add industry model sources to convert** tool to convert data from an FDO connection or database into the industry model.

The data source for conversion will be displayed in the left pane of the **Convert to Industry model** window. To map the attribute data for conversion, select the data source in the left pane of the window; the selected data source will be displayed in the right pane of the **Convert to Industry Model** window. Now, define the mapping between items in the source data and feature classes and attributes in the industry model. Next, choose the **Convert** button in the **Convert to Industry Model** window; the selected data source will be converted.

WORKING WITH LIVE MAPS (BING MAPS)

Autodesk has introduced the Geolocation feature in its AutoCAD based software, including the AutoCAD Map 3D that allows the user to use the Bing Maps as the default mapping service. Using this feature, users can now integrate the spatial data (raster and road) provided by Bing maps within the project environment. You can use the Bing maps services for displaying the high resolution aerial and satellite imagery of your project area, use them as the data source for creating (digitizing) spatial data and for geocoding locations.

The Bing map services are free services but to use them you are required to Sign In to your Autodesk 360 account. Also, as this is an online mapping service, you will require an internet connection to connect to the Bing map data.

Displaying Bing Map Data into the Project

As discussed earlier, you can use the Bing Map services to display the high resolution satellite image or road data into your project. To do so, first ensure that your project has a coordinate system assigned to it. Next, login to the Autodesk 360 account using your login credentials.

Tip

To login to your Autodesk 360 account, choose the **Sign In** *button in the* **InfoCenter** *bar; a drop-down list will be displayed. Next, select the* **Sign In to Autodesk 360** *option from this list; the* **Autodesk-Sign In** *dialog box will be displayed. Enter your credentials in this dialog box and choose the* **Sign In** *button; Autodesk will validate your credentials and will provide access to your account.*

After you login to your account, enter the **GEOMAP** command in the Command prompt and then press ENTER; you are prompted to specify the map type that you want to add. To add a high resolution satellite/aerial image, enter **Aerial** in the Command prompt. Next, enter **Road** in the Command prompt to display the Bing maps road data into your project. If you require to display the aerial photo along with the road data, enter **Hybrid** in the Command prompt. After specifying the required map type, press ENTER; the selected map will be displayed in the drawing. Figure 10-29 shows the aerial, road, and hybrid type of Bing maps that you can add to your project.

| *(a) Aerial map* | *(b) Road map* | *(c) Hybrid map* |

Figure 10-29 The various types of Bing maps

TUTORIALS

General instructions for downloading tutorial files:

Before starting the tutorials, you need to download the tutorial data to your computer. To do so, follow the steps given below:

1. Log on to *www.cadcim.com* and browse to *Textbooks > Civil/GIS > Map 3D > Exploring AutoCAD Map 3D 2017*. Next, select the *c10_m3d_2017_tut.zip* file from the **Tutorial Files** drop-down list. Next, choose the corresponding **Download** button to download the data file.

2. Extract the contents of the zip file to the following location:

    C:\m3d_2017

Notice that the *c10_m3d_2017_tut* folder is created within the *m3d_2017* folder.

Tutorial 1   Creating a 3D Surface from the LiDAR Data

In this tutorial, you will create a 3D surface from a LiDAR data set and then apply hillshade to the surface.   **(Expected time: 30 min)**

The following steps are required to complete this tutorial:

a. Generate an index file with *.isd* extension for the *c10-m3d-2017-LiDAR.las* file and assign the **UTM84-10N UTM-WGS 1984 datum, Zone 10 North, Meter; Cent. Meridian 123d W** coordinate system to the index file created.
b. Create a 3D surface from the index file.
c. Apply vertical exaggeration to the 3D surface.
d. Save the drawing file.

Generating the Index Point File for the LiDAR Data

In this section of the tutorial, you will open a new drawing file and then open a LiDAR data set using the **Index File** tool and create indexed point cloud data files.

1. Choose the **New** button from the Quick Access Toolbar; a new drawing is opened.

2. Type **MAPPOINTCLOUDMANAGER** in the Command prompt; the **POINT CLOUD MANAGER** dialog box is displayed.

3. In this dialog box, choose the **Add file** button from the tool panel located above the list box; the **Open** dialog box is displayed.

4. In the **Open** dialog box, browse to the following location:

    C:\m3d_2017\c10_m3d_2017_tut\c10_tut01

5. Next, choose the **c10-m3d-2017-LiDAR.las** file in the dialog box and then choose the **Open** button; the dialog box is closed and the selected file is added to the list box in the **POINT CLOUD MANAGER** dialog box, as shown in Figure 10-30.

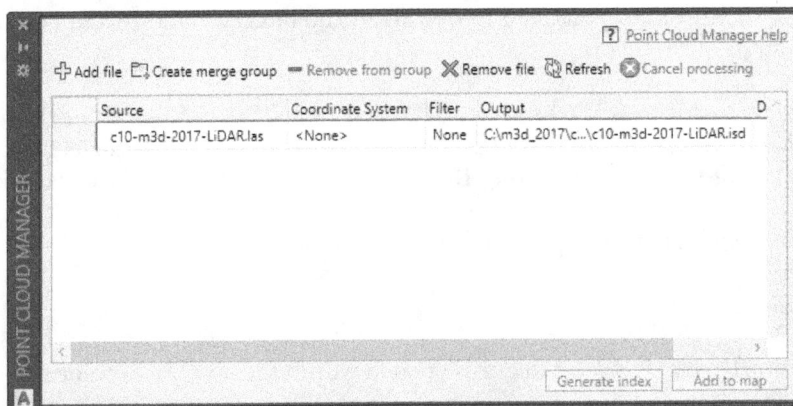

*Figure 10-30 The **POINT CLOUD MANAGER** dialog box with the added LiDAR datafile*

6. In the **POINT CLOUD MANAGER** dialog box, double-click on the cell in the **Coordinate System** column; the cell changes to an edit box and a Browse button is displayed in the cell.

7. Choose the Browse button displayed in the cell; the **Coordinate System Library** dialog box is displayed.

8. In the dialog box, select the **UTM, WGS84 Datum** option from the **Category** drop-down list; the coordinate systems in the selected category are displayed in the list box of the **Coordinate System Library** dialog box.

9. Select the **UTM84-10N** code with the description **UTM-WGS 1984 datum, Zone 10 North, Meter; Cent. Meridian 123d W** from the list box and then choose the **Select** button; the dialog box is closed and the **UTM84-10N** code is displayed in the **Coordinate System** cell of the **POINT CLOUD MANAGER** dialog box.

Tip
*You can select the added point file in the **POINT CLOUD MANAGER** dialog box, if it is not selected. On doing so, the **Generate index** button in this dialog box will be activated.*

10. Choose the **Generate index** button from the **POINT CLOUD MANAGER** dialog box; AutoCAD Map 3D begins the process of generating index file.

Note
*If the **Existing File** message box is displayed on choosing the **Generate index** button, choose the **Continue and overwrite the file** option in this message box; AutoCAD Map 3D will begin the process of generating index file.*

The status of the index file generation will be displayed at the bottom of the **POINT CLOUD MANAGER** dialog box. After the completion of the generation process, a tick mark is displayed in the first column cell of the data file. Also, the **Add to map** button is activated at the lower right corner of this dialog box.

Note
*If the **Add to map** button in the **POINT CLOUD MANAGER** dialog box is not activated on completion of the index generation process, select the row corresponding to the tick mark; the **Add to map button** will be activated.*

12. Choose the **Add to map** button from the **POINT CLOUD MANAGER** dialog box; the *c10-m3d-2017-LiDAR* index layer is added to the list box in the **Display Manager** tab of the **TASK PANE**.

13. Close the **POINT CLOUD MANAGER** dialog box.

14. Select the index layer in the **Display Manager** tab of the **TASK PANE** and then right-click on it; a shortcut menu is displayed. In this shortcut menu, choose the **Zoom to Extents** option; the point data of the selected file is displayed in the drawing window.

Creating a 3D Surface from the Index Point File

In this part of the tutorial, you will create a surface using the index file created. After creating the surface, you will apply a 25x vertical exaggeration.

1. Type **MAPSURFACEMANAGER** in the Command prompt and press ENTER; the **SURFACE MANAGER** dialog box is displayed.

   Note
   *If the list box in the **SURFACE MANAGER** dialog box displays any existing file (row), you can remove it by selecting it in the list box and then choosing the **Remove file** button.*

2. Choose the **Add file** button from this dialog box; the **Open** dialog box is displayed.

3. In the **Open** dialog box, browse to the following location:

   C:\m3d_2017\c10_m3d_2017_tut\c10_tut01

4. Select the file *c10-m3d-2017-LiDAR.isd* from the list box in this dialog box and then choose the **Open** button; the **Open** dialog box is closed and the selected index file is added to the list box in the **SURFACE MANAGER** dialog box.

5. In the **SURFACE MANAGER** dialog box, select the added file by clicking on the first cell corresponding to the added row; the selected row is highlighted in blue. Also, the **Generate Surface** button located at the lower right corner of this dialog box is activated.

6. Choose the **Generate Surface** button from the **SURFACE MANAGER** dialog box; the process of generation of the 3D surface starts.

   Note
   *If the **Existing File** message box is displayed on choosing the **Generate Surface** button, choose the **Continue and overwrite the file** option in this message box; AutoCAD Map 3D will begin the process of generating the surface.*

After the 3D surface is generated, a tick mark is displayed in the first cell of the row corresponding to the added file. Also, the **Add to map** button next to the **Generate Surface** button is activated.

7. Choose the **Add to map** button from the **SURFACE MANAGER** dialog box; the **c10-m3d-2017-LiDAR** raster layer is added to the list box in the **Display Manager** tab of the **TASK PANE**.

8. Close the **SURFACE MANAGER** dialog box.

9. Click on the **c10-m3d-2017-LiDAR** raster layer in the **Display Manager** tab of the **TASK PANE**; the **Raster Layer** tab (contextual) is displayed.

10. Choose the **Zoom to Extents** tool from the **View** panel in the **Raster Layer** tab; the 3D surface generated is displayed in the drawing window.

Applying Vertical Exaggeration to the 3D surface

1. Next, click on the down arrow next to the **Vertical Exaggeration** option in the **Status Bar** at the bottom of the drawing window; a flyout is displayed. From this flyout, choose the **25x** option, as shown in Figure 10-31; the raster image is exaggerated vertically, as shown in Figure 10-32.

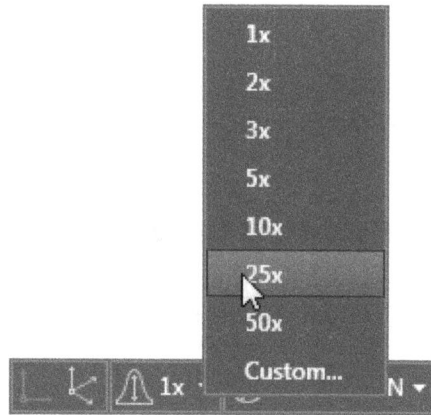

Figure 10-31 *Choosing the **25x** option from the flyout*

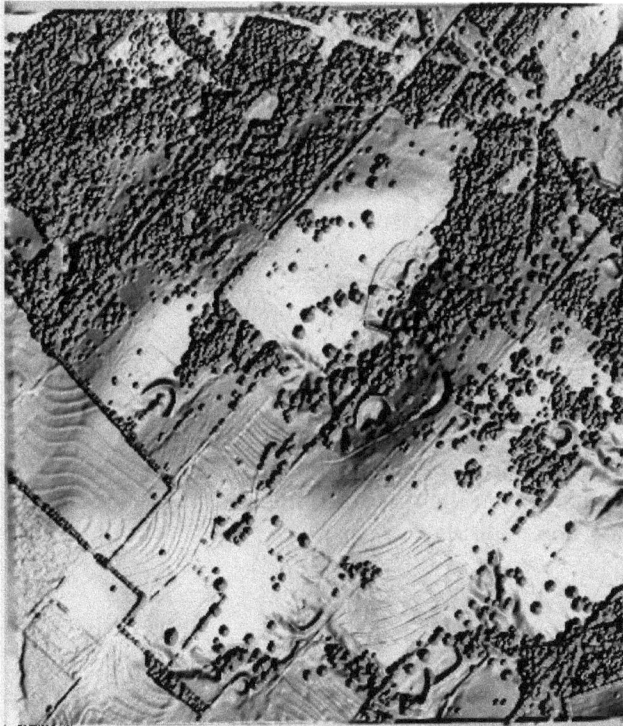

Figure 10-32 *Vertically exaggerated raster image*

Saving the Drawing File

1. Choose the **Save As** option from the Application Menu; the **Save Drawing As** dialog box is displayed.

2. In this dialog box, browse to the following location:

    C:\m3d_2017\c10_m3d_2017_tut\c10_tut01

3. Enter **c10_Tut01a.dwg** in the **File name** edit box and choose the **Save** button from this dialog box; the dialog box is closed and the current drawing file is saved with the name entered in the text box.

Tutorial 2	Creating Contours from the Point Data

In this tutorial, you will create a 3D surface from the survey point data, apply hillshade to the 3D Surface, and generate contours from the raster layer.   **(Expected time: 45 min)**

The following steps are required to complete this tutorial:

a. Create a 3D surface from the **c10-m3d-2017-tut02.XYZ** survey point file by assigning the **USA Washington WA83-N NAD83 Washington State Planes, North Zone, Meter** coordinate system.
b. Apply theme to the raster layer created by using the **-268 - 0**, **0-100**, and **100-1800** ranges and represent these ranges by yellow, brown, and red colors, respectively.
c. Generate contours for the raster layer created with **75m** interval and major contour at every 8th contour. Also, display the value of each major contour.
d. Save the drawing file.

Creating a 3D Surface from the Point File

In this part of the tutorial, you will create a surface using the point data file.

1. Choose the **New** button from the Quick Access Toolbar; a new drawing is opened.

2. Choose the **Create from Points** tool from the **3D Surface** panel in the **Create** tab; the **Create Surface** dialog box is displayed.

3. In the **Create Surface** dialog box, choose the add button; a menu is displayed. Choose the **File** option from the displayed menu, refer to Figure 10-33; the **Select a Point File** dialog box is displayed. Next, select the ***.xyz** option from the **Files of type** drop-down list.

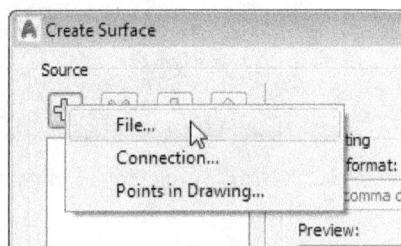

*Figure 10-33 Choosing the **File** option from the flyout*

4. In the **Select a Point File** dialog box, browse to the following location:

   C:\m3d_2017\c10_m3d_2017_tut\c10_tut02

5. In the **Select a Point File** dialog box, select the **c10-m3d-2017-tut02.XYZ** file in the list box and then choose the **Open** button next to the **File name** edit box; the **Select a Point File** dialog box is closed and the source of the selected point file is displayed in the list box in the **Source** area of the **Create Surface** dialog box, as shown in Figure 10-34.

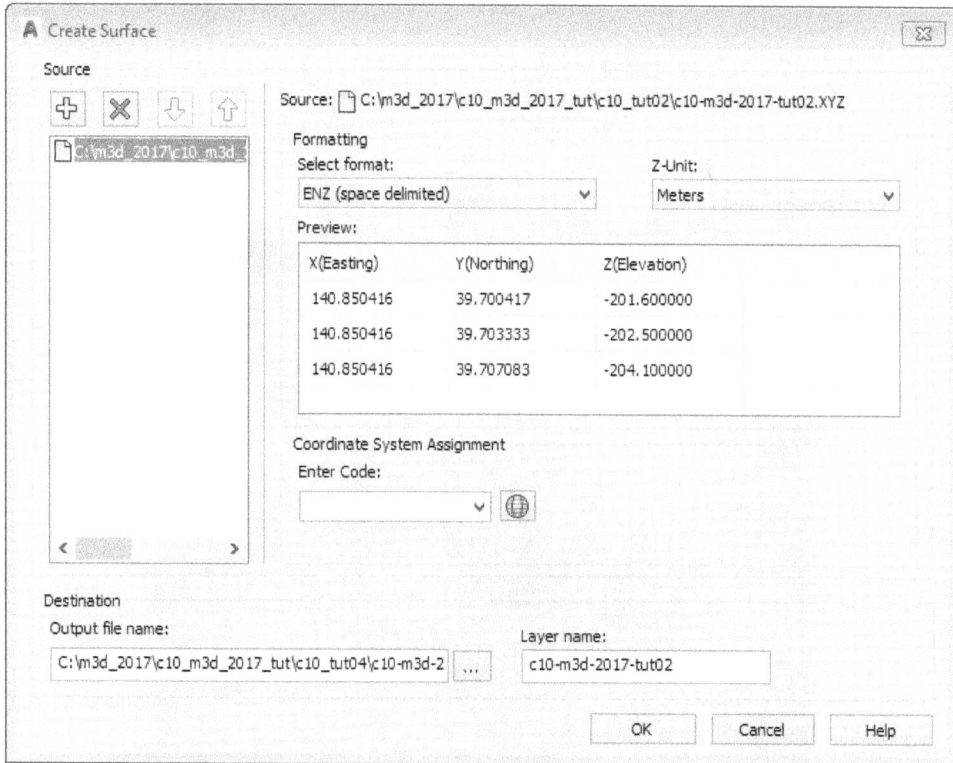

*Figure 10-34 The **Create Surface** dialog box with the loaded point file*

6. In the **Coordinate System Assignment** area of the **Create Surface** dialog box, choose the **Select Coordinate System** button; the **Coordinate System Library** dialog box is displayed.

7. In the **Coordinate System Library** dialog box, select the **USA, Washington** option from the **Category** drop-down list; the coordinate systems in the selected category are displayed in the list box.

8. Select the **WA83-N** code with the description **NAD83 Washington State Planes, North Zone, Meter** in the list box and then choose the **Select** button; the **Coordinate System Library** dialog box is closed and the **WA83-N** code is displayed in the **Enter Code** edit box of the **Create Surface** dialog box.

9. Next, choose the Browse button next to the **Output file name** edit box; the **Save as Raster Surface** dialog box is displayed.

10. In the **Save as Raster Surface** dialog box, browse to the following location:

    C:\m3d_2017\c10_m3d_2017_tut\c10_tut02

11. Retain the default file name in the **File name** edit box and then choose the **Save** button; this dialog box is closed and the path of the file is displayed in the **Output file name** edit box in the **Destination** area of the **Create Surface** dialog box.

12. Next, retain the default settings for all parameters and then choose the **OK** button; the **Create Surface** dialog box is closed and the 3D raster surface created from the point file is displayed in the drawing window, as shown in Figure 10-35.

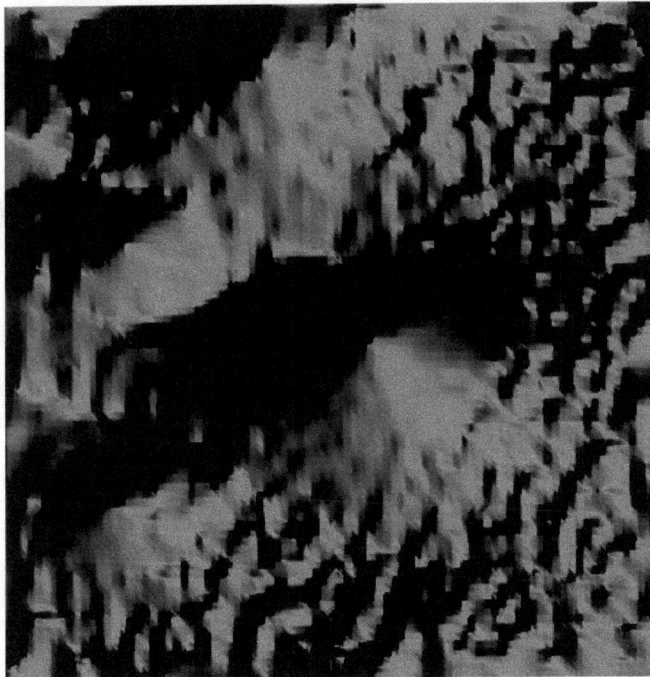

Figure 10-35 The 3D raster surface created from the point file

Applying Theme to the 3D Surface

1. Select the **c10-m3d-2017-tut02** raster layer from the list box in the **Display Manager** tab of the **TASK PANE**; the **Raster Layer** tab (contextual) is displayed.

2. Choose the **Style Editor** tool from the **Style** panel in the **Raster Layer** tab; the **STYLE EDITOR** dialog box is displayed.

3. In this dialog box, choose the **Band Detail** button from the **Raster Style for 0 - Infinity Scale Range** area; the details of the selected band are displayed in this area.

Note

*If the **Band Detail** button is not displayed, choose the **All Bands** button; the **Band Detail** button will be displayed in place of the **All Bands** button.*

4. In the **Raster Style for 0 - Infinity Scale Range** area of the **STYLE EDITOR** dialog box, choose the **Add a Rule** button twice; two new rules are added to the list box in this area, as shown in Figure 10-36.

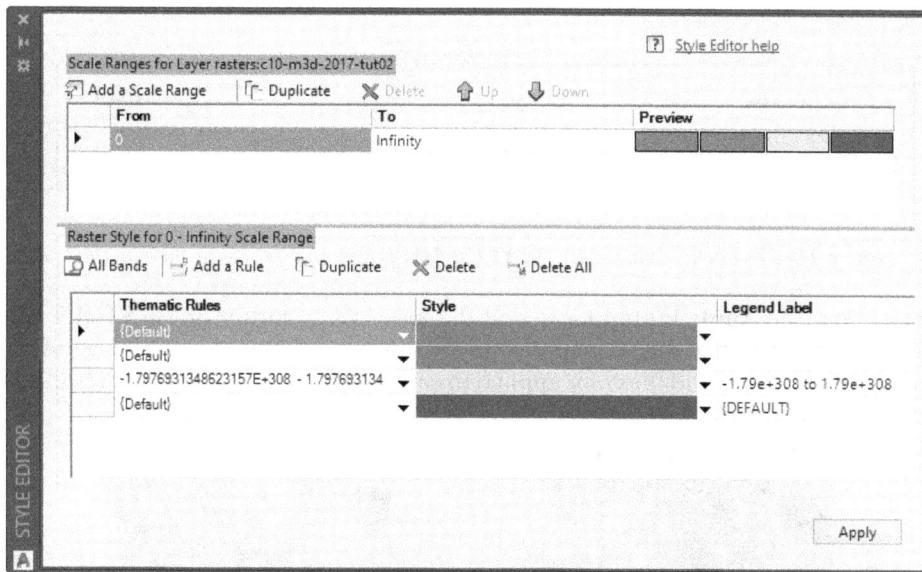

*Figure 10-36 The **STYLE EDITOR** dialog box with the two new rules added*

5. In this area, click in the cell corresponding to the first row in the **Thematic Rules** column of the list box; the cell changes into an edit box. In this edit box, enter the text **-268 - 0** and then click on the cell below this edit box; the added range is saved and the cell below it turns into an edit box. In this edit box, enter the text **0 - 100**. Next, click on the cell below this edit box; the added range is saved and the cell below it changes into an edit box. In this edit box, enter the text **100 - 1800**.

6. In the **Style** column of the **Raster Style for 0 - Infinity Scale Range** area, click on the down-arrow on the right in the cell corresponding to the **-268 - 0** range; a list of colors is displayed. Select the **255, 255, 0** color from the list of colors; the yellow color with the value **255,255,0** is displayed in the selected cell.

7. Again, in the **Style** column, click on the down arrow on the right in the cell corresponding to the **0 - 100** range; a list of colors is displayed. From the list of colors, select the **255, 150, 0** color; the dark orange color with the value **255,150,0** is displayed in the selected cell.

8. Again, in the **Style** column, click on the down arrow on the right in the cell corresponding to the **100 - 1800** range; a list of colors is displayed. From the list of colors, select the **255,0,0** color; the red color with the value **255,0,0** is displayed in the selected cell. Figure 10-37 shows the modified theme applied to the selected raster layer.

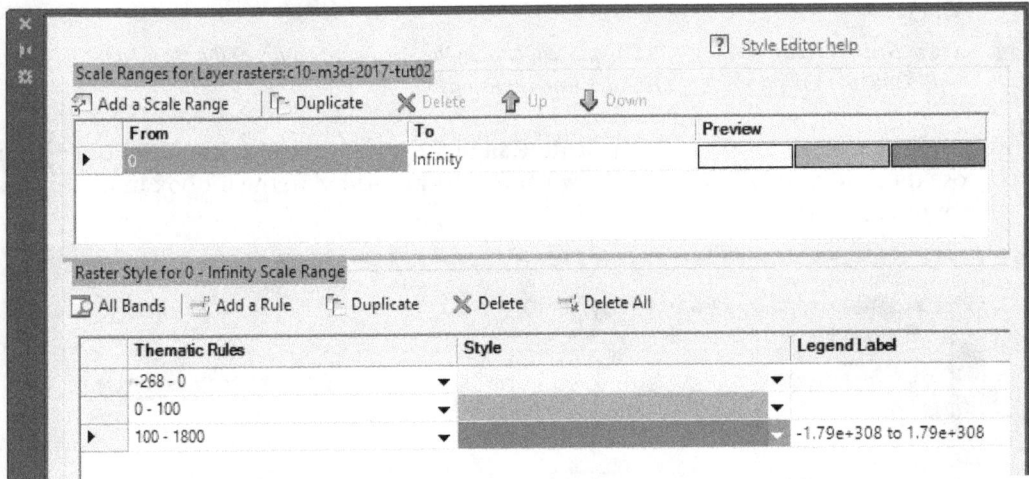

Figure 10-37 Partial view of the STYLE EDITOR dialog box with the modified rules

9. Next, choose the **Apply** button located at the lower right corner of the **STYLE EDITOR** dialog box; the modified theme is applied to the raster layer. Also, the raster layer is classified into three data ranges and the color applied to each layer is displayed in the drawing window, as shown in Figure 10-38.

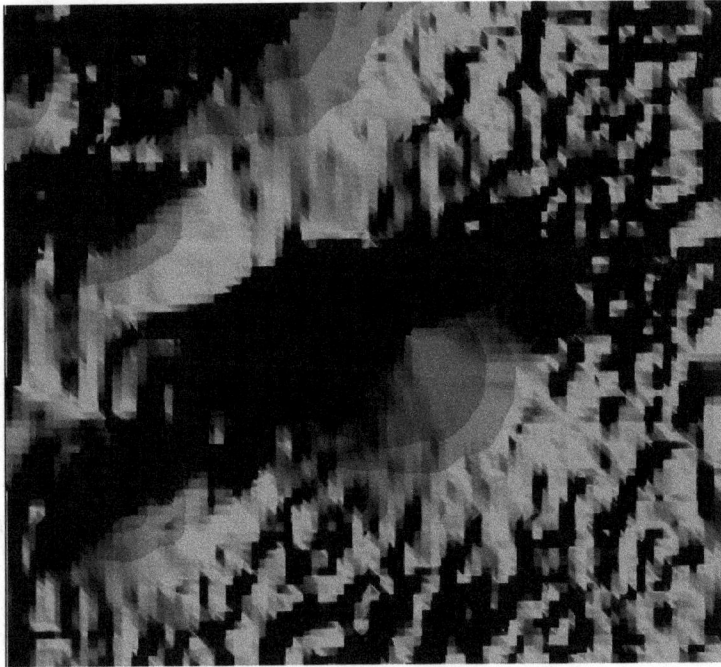

Figure 10-38 The raster layer after applying the modified theme

10. Close the **STYLE EDITOR** dialog box.

Generating Contours from the Raster Layer

In this section of the tutorial, you will create contours using the surface data.

1. Select the **c10-m3d-2017-tut02** raster layer from the list box in the **Display Manager** tab of the **TASK PANE** if it is not selected by default; the **Raster Layer** tab (contextual) is displayed.

2. Choose the **Contour Layer** tool from the **Create** panel in the **Raster Layer** tab; the **Generate Contour** dialog box is displayed.

3. In this dialog box, enter **75** in the **Contour elevation interval** edit box, **8** in the **Major contour every** edit box, and then select the **Label the elevation** check box.

4. Retain the default settings for other options, and then choose the **OK** button in the **Generate Contour** dialog box; a window displays the progress of contour creation.

5. On completion of the process, the **Generate Contour** dialog box is closed and the **c10-m3d-2017-tut02_contour** feature layer representing the generated contours is added to the list box in the **Display Manager** tab of the **TASK PANE**. Also, the feature layer representing the generated contours draped over the raster layer is displayed in the drawing window, as shown in Figure 10-39.

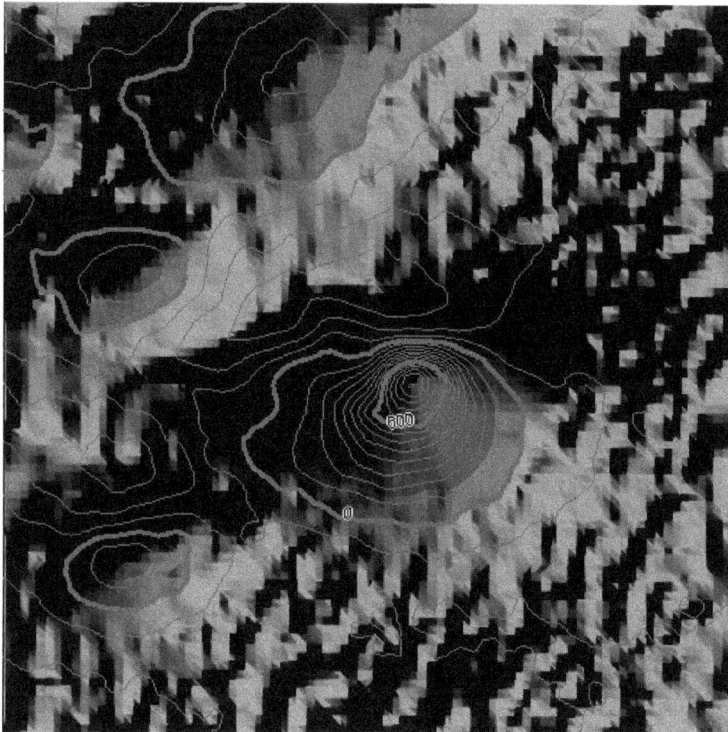

Figure 10-39 Generated contours draped over the raster layer

Saving the Drawing File

1. Choose the **Save As** option from the Application Menu; the **Save Drawing As** dialog box is displayed.

2. In this dialog box, enter **c10_Tut02a.dwg** in the **File name** edit box and choose the **Save** button next to the **File name** edit box; the current drawing file is saved with the name entered in the edit box.

Tutorial 3   Creating Bulk Copy, Join, and Calculation

In this tutorial, you will create a Bulk Copy, a Join, and a Calculation by using a feature layer.
(Expected time: 45 min)

The following steps are required to complete this tutorial:

a. Load the *c10-m3d-2017-tut03.shp* shape file into the current drawing.
b. Create a bulk copy in the *.sdf* file format by copying the **Featid**, **Geometry**, and **SQKM_ADMIN** properties from the *c10-m3d-2017-tut03.shp* file.
c. Create a Join.
d. Calculate the population density of each region in the *c10-m3d-2017-tut03.shp* vector layer.

> **Tip**
> *Population Density:*
>
> $$Population\ Density\ (No.\ of\ persons\ per\ sq.km) = \frac{Population\ of\ region\ (POP\_ADMIN)}{Area\ of\ region\ (SQKM\_ADMIN)}$$

e. Save the drawing file.

Starting a New Drawing and Assigning Coordinate System to It

1. Choose the **New** button from the Quick Access Toolbar; a new drawing is opened.

2. Choose the **Assign** tool from the **Coordinate System** panel in the **Map Setup** tab; the **Coordinate System - Assign** dialog box is displayed.

3. In the **Show** area of the **Coordinate System - Assign** dialog box, select the **Iceland** option from the **Category** drop-down list; the coordinate systems in the selected category are displayed in the list box of this dialog box.

4. Select the **Hjorsey.IcelandGrid** code with the description **Iceland Grid of 1955** from the list box.

5. Choose the **Assign** button in the **Coordinate System - Assign** dialog box; this dialog box is closed and the selected coordinate system is assigned to the current drawing.

Loading the Vector Layer

1. Choose the **Connect** tool from the **Data** panel in the **Home** tab; the **DATA CONNECT** wizard is displayed. Ensure that the **Add SHP Connection** option is selected in the left pane of the wizard.

2. In the **OSGeo FDO Provider for SHP** page, choose the **SHP** button next to the **Source file or folder** edit box; the **Open** dialog box is displayed.

3. In the **Open** dialog box, browse to the following location:

    C:\m3d_2017\c10_m3d_2017_tut\c10_tut03

4. In the **Open** dialog box, select the **c10-m3d-2017-tut03.shp** file in the list box and then choose the **Open** button next to the **File name** edit box; the **Open** dialog box is closed and the path of the selected file is displayed in the **Source file or folder** edit box in the **OSGeo FDO Provider for SHP** page.

5. Choose the **Connect** button located below the **Source file or folder** edit box in the **OSGeo FDO Provider for SHP** page; the **SHP** page with the **c10-m3d-2017-tut03** file added to the list box is displayed.

6. In the **SHP** page, choose the **Edit Coordinate Systems** button at the top of the list box; the **Edit Spatial Contexts** dialog box is displayed.

7. In the **Edit Spatial Contexts** dialog box, select the first row; the **Edit** button gets activated. Next, choose the **Edit** button; the **Coordinate System Library** dialog box is displayed.

8. In the **Coordinate System Library** dialog box, select the **Iceland** category from the **Category** drop-down list; the coordinate systems in the selected category are displayed in the list box of this dialog box.

9. In this list box, select the code **HJORSEY.LL** with description **HJORSEY.LL Automatically generated LL system for WKT use** option and then choose the **Select** button; the **Coordinate System Library** dialog box is closed and the **HJORSEY.LL** code is displayed in the first row of the **Override** cell of the **Edit Spatial Contexts** dialog box.

10. Choose the **OK** button in the **Edit Spatial Contexts** dialog box; this dialog box is closed and the coordinate system is assigned to the **c10-m3d-2017-tut03** vector layer.

11. In the **SHP** page of the **DATA CONNECT** wizard, choose the **Add to Map** button; the **c10-m3d-2017-tut03** vector layer is added to the list box in the **Display Manager** tab of the **TASK PANE** and the corresponding geometry is displayed in the drawing window.

12. Close the **DATA CONNECT** wizard.

Creating the SDF File

1. Choose the **SDF** tool from the **Feature Data Store** panel in the **Create** tab; the **Choose Spatial Database File** dialog box is displayed.

2. In the **Choose Spatial Database File** dialog box, browse to the following location:

    C:\m3d_2017\c10_m3d_2017_tut\c10_tut03

3. In this dialog box, enter **c10-m3d-2017-sdf** in the **File name** edit box and then choose the **Save** button; the **Specify Coordinate System** dialog box with the **Hjorsey.IcelandGrid** code in the **Coordinate System** edit box is displayed.

4. In this dialog box, retain the default settings and then choose the **OK** button; the **Schema Editor** dialog box is displayed.

5. In the **Schema** list box of this dialog box, right-click on the **Schema 1** node; a shortcut menu is displayed.

6. In the shortcut menu, choose the **Delete Schema** option; the **Autodesk Schema Editor** message box is displayed. Choose the **Yes** button in this message box; **Schema1** is deleted from the list box.

7. Choose the **Apply** button from the **Schema Editor** dialog box; the **Submit Changes** message box is displayed.

8. Choose the **Yes** button in the **Submit Changes** message box; the message box is closed.

9. Choose the **OK** button in the **Schema Editor** dialog box; the **Submit Changes** message box is displayed again.

10. Choose the **Yes** button in the **Submit Changes** message box; the **Schema Editor** dialog box is closed and the changes are saved.

Creating the Bulk Copy

1. Choose the **Bulk Copy** tool from the **Feature Data Store** panel in the **Create** tab; the **Bulk Copy** dialog box is displayed.

2. In the **From** area of this dialog box, click on the down-arrow on the right of the **Source** drop-down list; a list of sources is displayed.

3. Select the **c10-m3d-2017-tut03** option from the **Source** drop-down list, as shown in Figure 10-40; the items in the selected source are displayed in the **Select items to copy** list box in the **From** area.

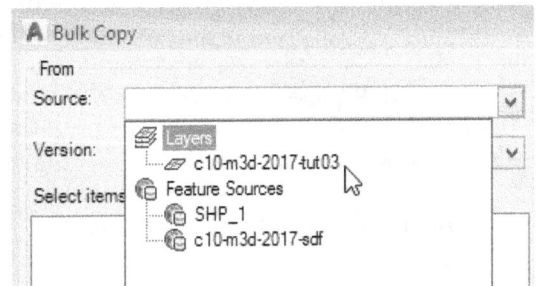

*Figure 10-40 Selecting the **c10-m3d-2017-tut03** option from the **Source** drop-down list*

4. Next, from the **Select items to copy** list box of the **Bulk Copy** dialog box, select the check boxes corresponding to the **FeatId**, **SQKM_ADMIN**, and **Geometry** options in the **c10-m3d-2017-tut03** node; the corresponding options are activated in the list box of the **To** area.

5. In the **To** area of the **Bulk Copy** dialog box, select the **c10-m3d-2017-sdf** option from the **Target** drop-down list; the **c10-m3d-2017-tut03** node is displayed in the **Click on an item to select a different input name** list box, as shown in Figure 10-41.

6. Retain the default settings for all other options, and then choose the **Copy Now** button located at the bottom of the **Bulk Copy** dialog box; the **Continue Bulk Copy** window is displayed.

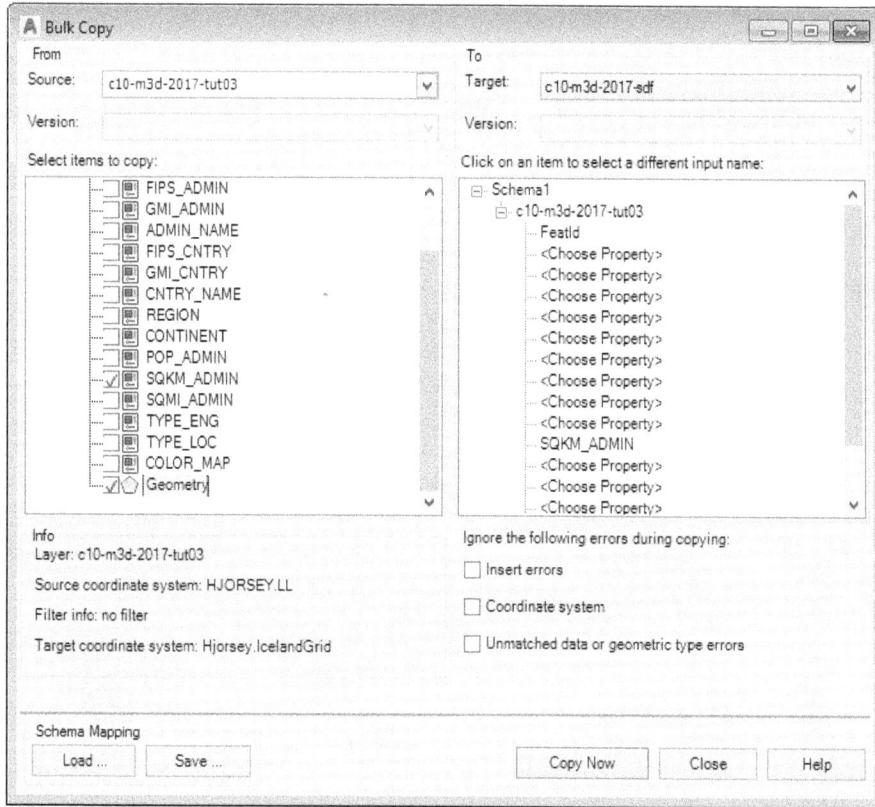

*Figure 10-41 The **Bulk Copy** dialog box displaying the items in the **c10-m3d-2017-tut03** shape file*

7.   Choose the **Continue Bulk Copy** button in the **Continue Bulk Copy** window; the **Bulk Copy Results** window is displayed.

8.   In the **Bulk Copy Results** window, choose the **View Log >>** button; the details of the bulk copy are displayed in the **Log Message** area, as shown in Figure 10-42.

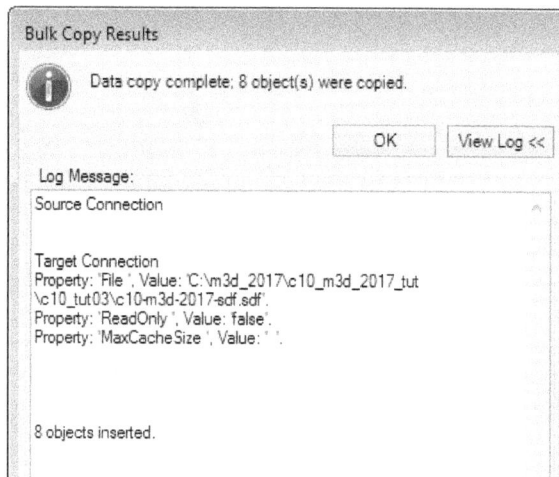

*Figure 10-42 Partial view of the **Bulk Copy Results** window with the details of the bulk copy*

9. Choose the **OK** button in the **Bulk Copy Results** window; this window is closed and the bulk copy is saved.

10. Choose the **Close** button located at the bottom of the **Bulk Copy** dialog box; this dialog box is closed.

Verifying the Bulk Copy

1. Clear all check boxes corresponding to the vector layers in the **Display Manager** tab of the **TASK PANE**; the geometry displayed in the drawing window is turned off.

2. Choose the **Data** button in the **Display Manager** tab; a flyout is displayed. In the flyout, choose the **Connect to Data** option; the **DATA CONNECT** wizard is displayed.

3. In this wizard, choose **c10-m3d-2017-sdf** from **Add SDF Connection** in the **Data Connections by Provider** list box; the **SDF** page is displayed on the right side in this wizard, as shown in Figure 10-43.

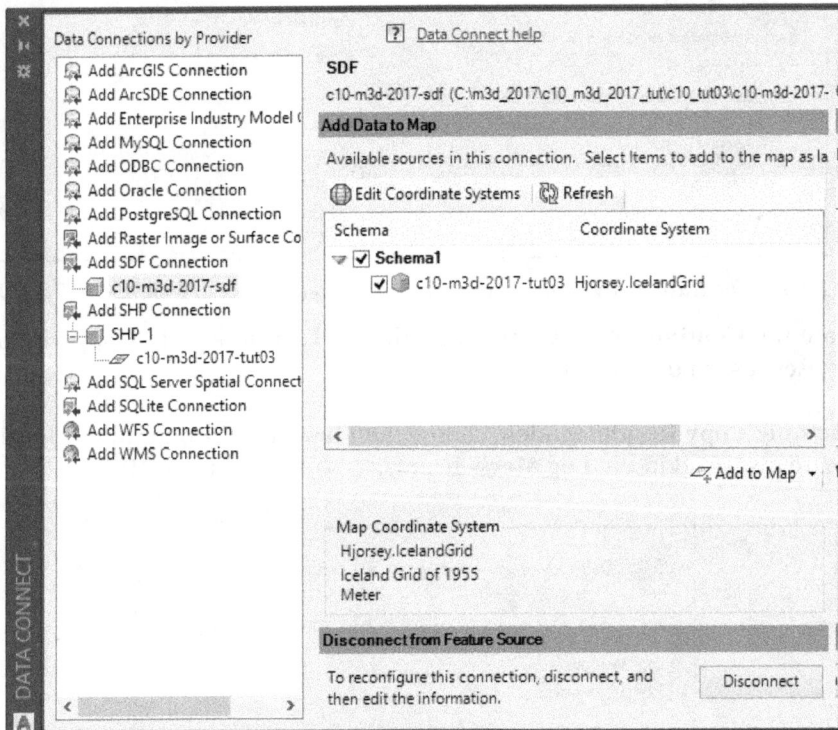

*Figure 10-43 The **DATA CONNECT** wizard with the **SDF** page*

4. Choose the **Add to Map** button from the **SDF** page; the **c10-m3d-2017-tut03** vector layer is added as **c10-m3d-2017-tut03 (1)** in the **Display Manager** tab of the **TASK PANE**. Figure 10-44 shows the geometry for the added vector layer.

5. Close the **DATA CONNECT** wizard.

Figure 10-44 The geometry of the vector layer displayed in the drawing window

6. Next, choose the **c10-m3d-2017-tut03 (1)** layer from the **Display Manager** tab of the **TASK PANE**; the **Vector Layer** tab (contextual) is displayed in the Ribbon.

7. Choose the **Table** tool from the **View** panel in this tab; the **DATA TABLE** windows with the attribute table for the **c10-m3d-2017-tut03 (1)** vector layer is displayed, as shown in Figure 10-45.

Data:	c10-m3d-2017-tut03 (1) ∨		Auto-Zoom	Auto-Scroll

	FeatId	SQKM_ADMIN
	1	21805.119
	2	12975.55
	3	24068.529
	4	9395.404
	5	9668.785
	6	1816.966
	7	22074.961
	8	22074.961

Row   of 8 0   Search to Select   Options ▾

*Figure 10-45 Partial view of the **DATA TABLE** window displaying the attribute data table for the c10-m3d-2017-tut03 (1) vector layer*

8. Close the **DATA TABLE** window.

If you review the geometry in the feature class and its corresponding data table, it is clear that the bulk copy has been created successfully.

Creating a Join

1. In the **Display Manager** tab of the **TASK PANE**, choose the **c10-m3d-2017-tut03** vector layer; the **Vector Layer** tab (contextual) is displayed.

2. Choose the **Joins** tool from the **Create** panel in the **Vector Layer** tab; the **Create a Join** dialog box is displayed.

3. In the **Create a Join** dialog box, click on the down-arrow on the right of the **Table (or feature class) to join to** drop-down list; a list of sources connected to the current drawing is displayed.

4. In this list, select the **c10-m3d-2017-tut03** feature class displayed under the **c10-m3d-2017-sdf** node, as shown in Figure 10-46; the text **c10-m3d-2017-sdf:Schema1: c10-m3d-2017-tut03** is displayed in the **Table (or feature class) to join to** drop-down list in the **Create a Join** dialog box.

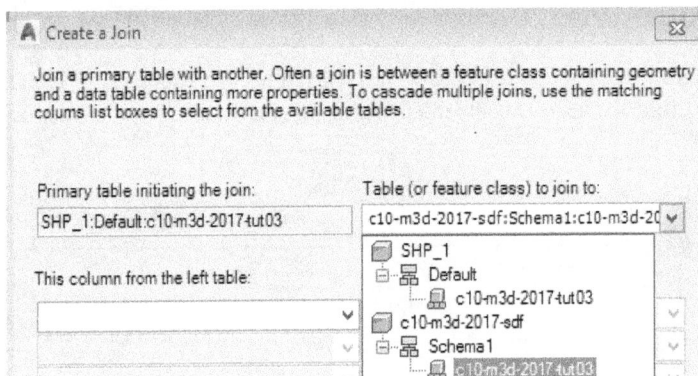

*Figure 10-46 Selecting the **c10-m3d-2017-tut03** feature class from the **Table (or feature class) to join to** drop-down list*

5. In the **Create a Join** dialog box, select the **FeatId** option from the **This column from the left table** drop-down list; the drop-down list next to it is activated.

6. Select the **FeatId** option from the **Matches this column from the right table** drop-down list, if it is not selected by default; the areas below the drop-down list are activated.

7. Retain the default settings in the **Type of Join** and **Relationship with Secondary Records (Cardinality)** areas, and then choose the **OK** button from the **Create a Join** dialog box; a join is created using the attributes in the **FeatId** column of the two data tables.

8. Choose the close button to close the **DATA CONNECT** dialog box.

9. Choose the **Table** tool from the **View** panel in the **Vector Layer** tab; the **DATA TABLE** with the created join is displayed.

    In the **DATA TABLE**, drag the horizontal scroll button to see the last column, as shown in Figure 10-47.

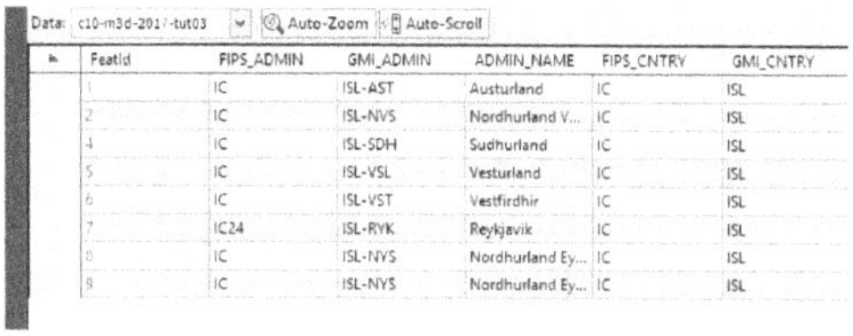

FeatId	FIPS_ADMIN	GMI_ADMIN	ADMIN_NAME	FIPS_CNTRY	GMI_CNTRY
1	IC	ISL-AST	Austurland	IC	ISL
2	IC	ISL-NVS	Nordhurland V...	IC	ISL
4	IC	ISL-SDH	Sudhurland	IC	ISL
5	IC	ISL-VSL	Vesturland	IC	ISL
6	IC	ISL-VST	Vestfirdhir	IC	ISL
7	IC24	ISL-RYK	Reykjavik	IC	ISL
8	IC	ISL-NYS	Nordhurland Ey...	IC	ISL
9	IC	ISL-NYS	Nordhurland Ey...	IC	ISL

Figure 10-47 Partial view of the ***DATA TABLE*** *with the added columns*

Creating the Calculation

1. Select the **c10-m3d-2017-tut03** check box from the **Display Manager** tab of the **TASK PANE** and then choose the **Calculations** tool from the **Create** panel in the **Vector Layer** tab; the **Create a Calculation** dialog box is displayed.

2. In the **Create a Calculation** dialog box, enter the **Pop_density** text in the **Name** edit box located above the tool panel.

3. In the **Getting Started with Calculations** page of the dialog box, choose the **Create a Calculation** option; the **[property] / [property]** expression is displayed in the text box in this dialog box.

Tip
The ***Getting Started with Calculations*** *page will not be displayed if you had selected the check box in the* ***Create a Calculation*** *dialog box in the previous sessions.*

4. Next, select the **[property]** expression and choose the **POP_ADMIN** option from the **Numeric Properties** area in the **Property** drop-down; the **[property]** text in the expression changes to the **POP_ADMIN** expression in the text box.

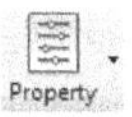

5. Similarly, select the **[property]** expression displayed after the '/' symbol and then choose the **SQKM_ADMIN** option from the **Numeric Properties** section; the **[property]** text in the expression changes to the **SQKM_ADMIN** expression in the text box.

6. Next, choose the **Validate** button; the **The expression is valid** statement is displayed below the text box.

7. Choose the **OK** button in the **Create a Calculation** dialog box; this dialog box is closed and a new property column **Pop_density** is added to the **DATA TABLE**.

8. Choose the **Table** tool from the **View** panel in the **Vector Layer** tab; the **DATA TABLE** corresponding to the *c10-m3d-2017-tut03.shp* vector layer is displayed.

   In the **DATA TABLE**, slide the table toward the left by using the horizontal scroll bar to display the **Pop_density** property, as shown in Figure 10-48.

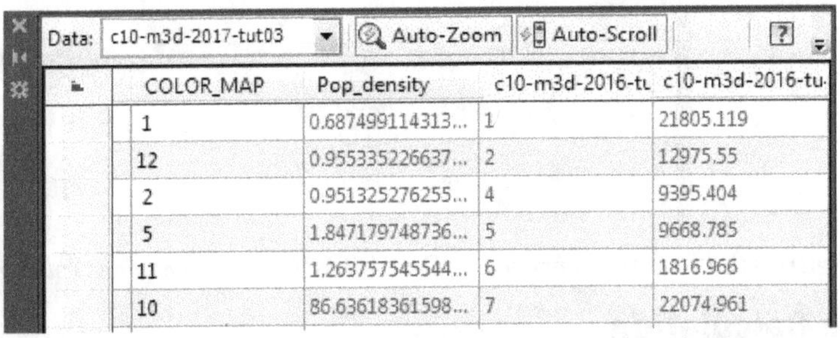

*Figure 10-48 The **DATA TABLE** with the new property calculated*

Saving the Drawing File

1. Choose the **Save As** option from the Application Menu; the **Save Drawing As** dialog box is displayed.

2. In this dialog box, enter **c10_Tut03a.dwg** in the **File name** edit box and choose the **Save** button next to the **File name** edit box; the current drawing file is saved with the name specified in the edit box.

Tutorial 4   Performing Surface Analysis

In this tutorial, you will create a 3D surface from the point shape file and then perform the slope and elevation analysis on the surface.   **(Expected time: 45 min)**

The following steps are required to complete this tutorial:

a. Start a new drawing file and connect to the point shape file.
b. Create a 3D surface from a point shape file.
c. Perform the surface analysis (height) and display the raster image using the predefined color palette.
d. Perform the slope analysis.
e. Save the drawing file.

Starting a New Drawing and Loading the Point Shape File

In this section of the tutorial, you will start with a new drawing and then connect to a point shape file using the **DATA CONNECT** wizard.

1. Choose the **New** button from the Quick Access Toolbar; a new drawing is opened.

2. Choose the **Connect** tool from the **Data** panel in the **Home** tab; the **DATA CONNECT** wizard is displayed. Ensure that the **Add SHP Connection** option is selected in the right pane of the wizard.

3. In the **OSGeo FDO Provider for SHP** page, choose the **SHP** button next to the **Source file or folder** edit box; the **Open** dialog box is displayed.

4. In the **Open** dialog box, browse to the following location:

    C:\m3d_2017\c10_m3d_2017_tut\c10_tut04

5. In the **Open** dialog box, select the **LakeHenshaw-ElevationPoints.shp** file in the list box and then choose the **Open** button; the **Open** dialog box is closed and the path of the selected file is displayed in the **Source file or folder** edit box in the **OSGeo FDO Provider for SHP** page.

6. Choose the **Connect** button located below the **Source file or folder** edit box in the **OSGeo FDO Provider for SHP** page; the **SHP** page is displayed. Note that the list box in this page displays the name of the selected shape file and its coordinate system, as shown in Figure 10-49.

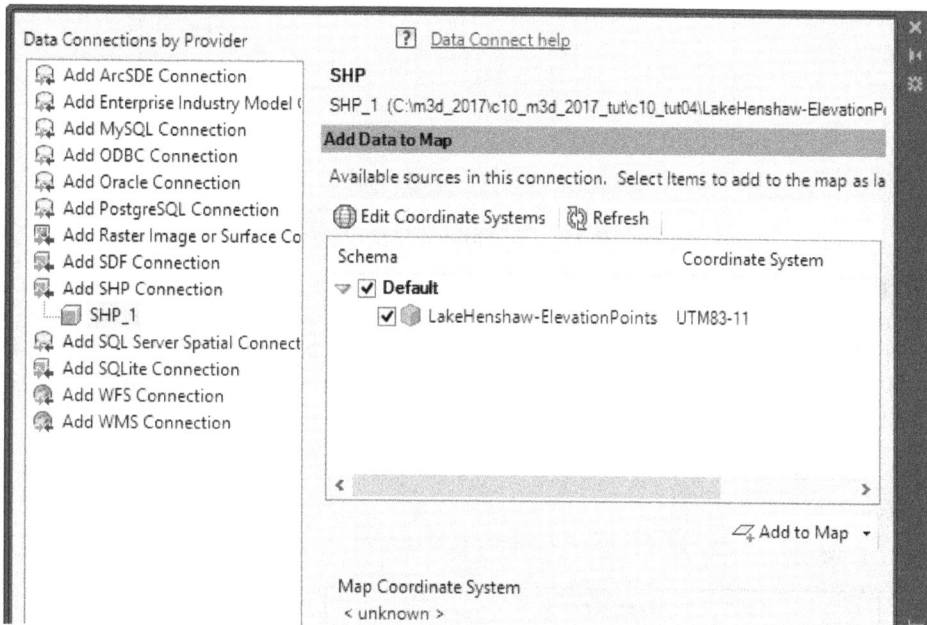

*Figure 10-49 Partial view of the **DATA CONNECT** wizard displaying the selected shape file and its coordinate system*

7. Ensure that the **LakeHenshaw-ElevationPoints** is selected in the list box and then choose the **Add to Map** button; the **LakeHenshaw-ElevationPoints** is added in the **Display Manager** tab of the **TASK PANE**. Also, the point features in the shape file are displayed in the drawing window.

8. Close the **DATA CONNECT** wizard.

Creating a 3D Surface from a Point Shape File

In this part of the tutorial, you will create a 3D surface using the point elevation data from the connected shape file.

1. Choose the **Create from Points** tool from the **3D Surface** panel of the **Create** tab; the **Create Surface** dialog box is displayed.

2. In this dialog box, choose the add button in the **Source** area; a menu is displayed.

3. Choose the **Connection** option from the displayed menu; the **Source Data** dialog box is displayed, as shown in Figure 10-50.

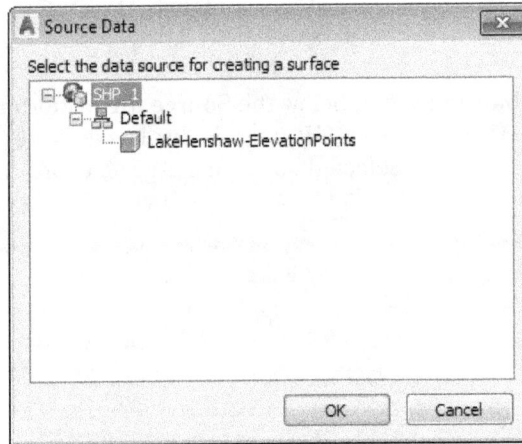

Figure 10-50 *The* *Source Data* *dialog box*

4. Select the **LakeHenshaw-ElevationPoints** option in this dialog box and then choose the **OK** button; the dialog box is closed and the name of the selected file is displayed in the **Source** area of the **Create Surface** dialog box. Also, the **Preview** area of the dialog box displays the xyz data of the points in the shape file. Figure 10-51 shows the **Create Surface** dialog box with the preview of the shape file.

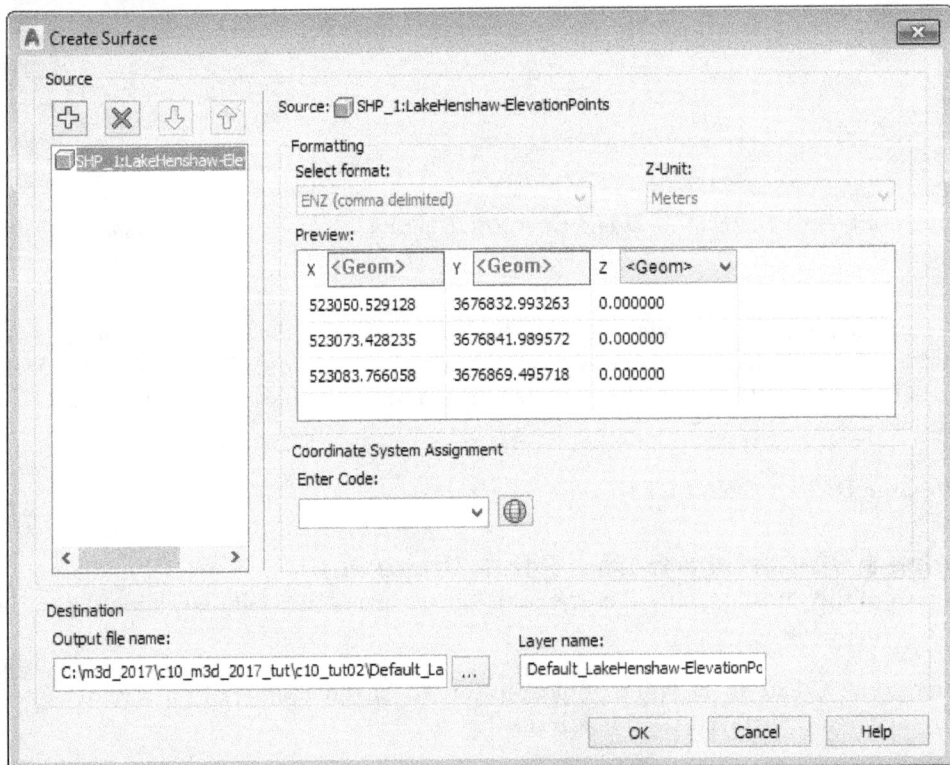

Figure 10-51 *The* *Create Surface* *dialog box showing the preview of the selected point shape file*

Note that the z value should be 0, in the **Create Surface** dialog box for the selected shape file. Now, you need to select the appropriate field in the shape file that contains the elevation values for the point data.

5. Click on the drop-down list in the **Z** column of the **Preview** area; a drop-down list is displayed, as shown in Figure 10-52.

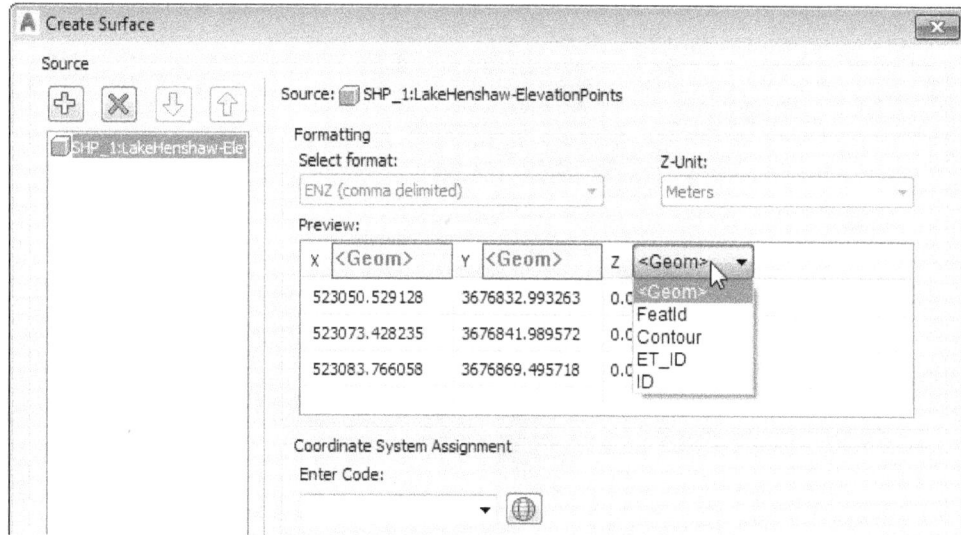

Figure 10-52 *The drop-down list displayed in the* ***Z*** *column of the* ***Create Surface*** *dialog box*

6. Select the **Contour** option from the displayed drop-down list; the cells in the **Z** column display the preview of the elevation values in the selected attribute field of the shape file.

7. Click on the **Enter Code** drop-down list in the **Coordinate System Assignment** area and then select the **UTM83-11 (DataStore)** option; the description of the selected coordinate system is displayed.

8. Choose the Browse button corresponding to the **Output file name** edit box; the **Save as Raster Surface** dialog box is displayed.

9. In this dialog box, ensure that the **TIFF (*.tif)** option is selected in the **Files of types** drop-down list. Next, browse to the following location:

    C:\m3d_2017\c10_m3d_2017_tut\c10_tut04

10. Specify **LakeHenshaw-3Dsurface** in the **File name** edit box and then choose the **Save** button; the **Save as Raster Surface** dialog box is closed and the path of the specified output file is displayed in the **Output file name** edit box of the **Create Surface** dialog box.

11. Ensure that **LakeHenshaw-3Dsurface** is displayed in the **Layer name** edit box and then choose the **OK** button in the **Create Surface** dialog box; the dialog box is closed. AutoCAD Map 3D will process the defined source file and create the **LakeHenshaw-3Dsurface** surface.

After creating the surface, it is added to the **Display Manager** tab of the **TASK PANE**. Also, note that the created surface is displayed in the drawing area of the Map 3D interface, as shown in Figure 10-53.

Figure 10-53 The LakeHenshaw-3Dsurface

12. Turn off the display of the shape file by clearing the check box corresponding to **LakeHenshaw-ElevationPoints** in the **Display Manager** tab of the **TASK PANE**.

Performing the Surface Analysis (Height)

In this part of the tutorial, you will perform the height surface analysis and display the image using a predefined color palette.

1. Select **LakeHenshaw-3Dsurface** in the **Display Manager** tab of the **TASK PANE**; the **Style** tool is activated.

2. Choose the **Style** tool; the **STYLE EDITOR** dialog box is displayed.

3. In the **Raster Style for 0 - Infinity Scale Range** area of this dialog box, ensure that the list box displays the band details in the selected raster. Next, double-click in the **Style** cell corresponding to band **1**; a drop-down list is displayed.

4. Choose the **Theme** option from the displayed drop-down list, as shown in Figure 10-54; the **Theme** dialog box will be displayed.

5. In this dialog box, ensure that the **Height** and **Equal** options are selected in the **Property** and **Distribution** drop-down lists, respectively.

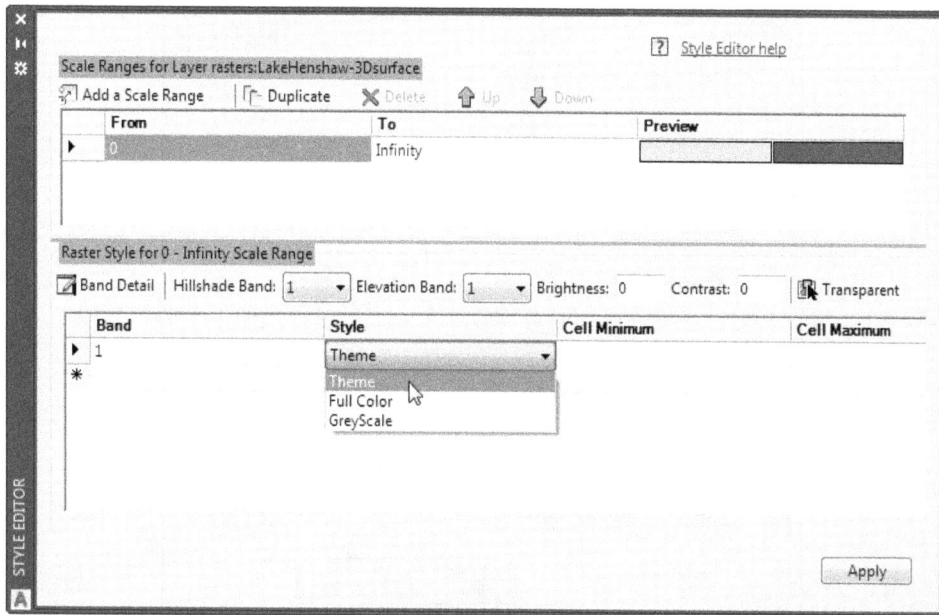

*Figure 10-54 Selecting the **Theme** option from the drop-down list in the **STYLE EDITOR** dialog box*

6. Select the **Specify a theme** and **Create legend labels** check boxes, if not selected by default.

7. Select the **Palette** radio button in the **Specify a theme** area; the corresponding drop-down list is activated.

8. Select the **USGS DEM palette** option from this drop-down list; the preview of the selected palette is displayed above the drop-down list.

9. Enter **Height** in the **Label text** edit box in the **Create legend labels** area.

10. Select the **<Label text> <min> - <max>** option from the **Label format** drop-down list. Figure 10-55 shows the **Theme** dialog box with the specified options.

11. Choose the **OK** button in the **Theme** dialog box; the dialog box is closed and the defined theme is displayed in the **STYLE EDITOR** dialog box.

12. Choose the **Apply** button in this dialog box; the theme (based on the height) is applied to the surface in the drawing. Also note that the **Display Manager** tab now displays the legend corresponding to the current display theme. Figure 10-56 shows the 3D surface with the applied height theme and its legend.

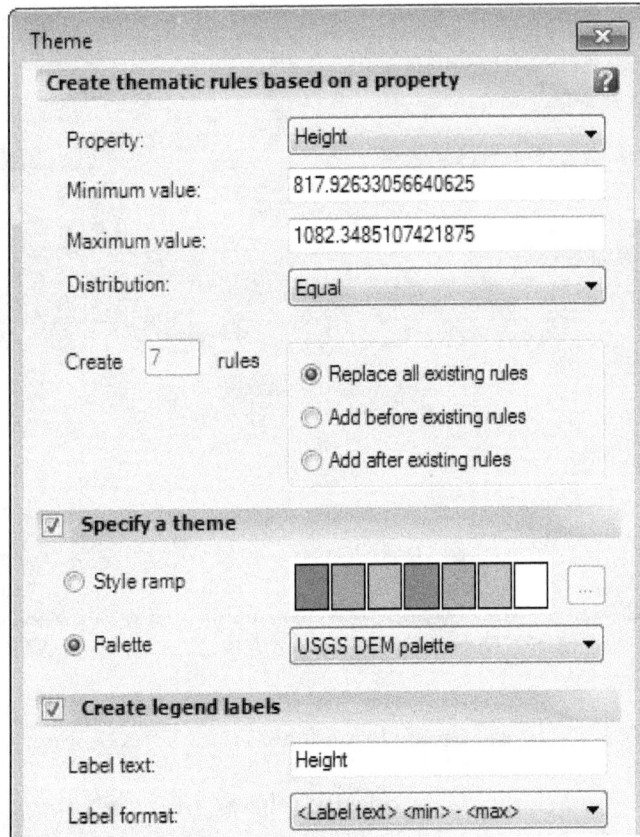

*Figure 10-55 Partial view of the **Theme** dialog box showing the specified parameters*

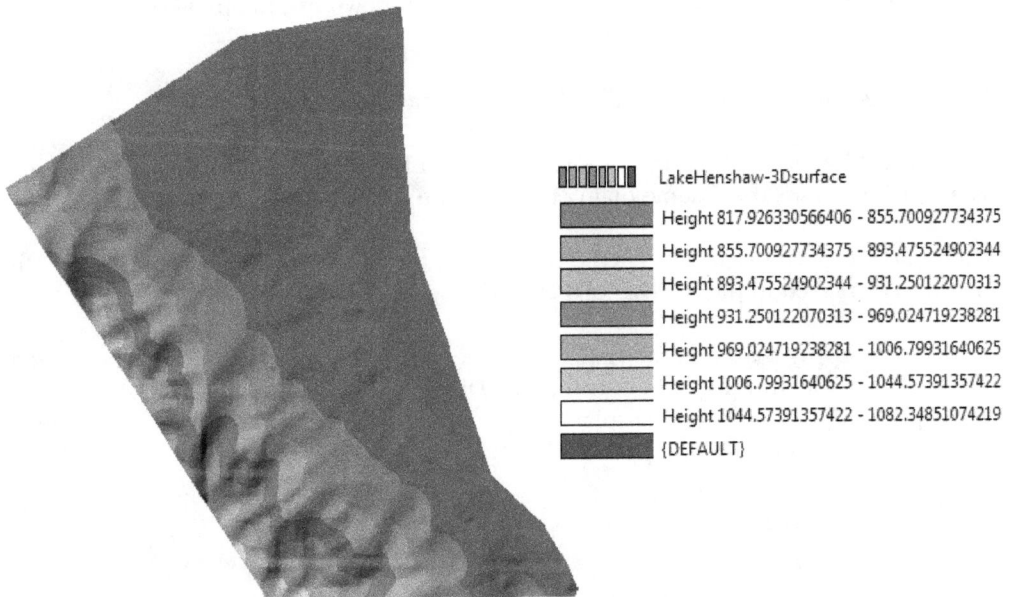

*Figure 10-56 The **3D surface with the applied height theme***

Performing the Slope Analysis

In this section of the tutorial, you will perform a slope analysis on the 3D surface and then apply a predefined color palette to display the result of the analysis.

1. In case you have closed the **STYLE EDITOR** dialog box, select the **LakeHenshaw-3Dsurface** feature layer in the **Display Manager** tab of the **TASK PANE** and then choose the **Style** tool; the **STYLE EDITOR** dialog box is invoked.

2. In the **Raster Style for 0 - Infinity Scale Range** area of this dialog box, double-click in the **Style** cell corresponding to the band **1**; the drop-down list is displayed, refer to Figure 10-54.

3. Choose the **Theme** option from the displayed drop-down list; the **Theme** dialog box is displayed.

4. In the theme dialog box, select the **Slope** option from the **Property** drop-down list.

5. Enter **60** in the **Maximum value** edit box and then choose the **Standard Deviation** option from the **Distribution** drop-down list.

6. Ensure that the **Specify a theme** and **Create legend labels** check boxes are selected. Also, ensure that the **Replace all existing rules** and **Palette** radio buttons are selected in this dialog box.

7. Select the **Slope palette** option from the **Palette** drop-down list in the **Specify a theme** area; preview of the selected color palette is displayed above the drop-down list.

8. Enter **Slope** in the **Label text** edit box and then ensure that the **<Label text> <min> - <max>** option is selected in the **Label format** drop-down list. Figure 10-57 shows the **Theme** dialog box with the parameters specified for the slope analysis of the 3D surface.

9. Choose the **OK** button in the **Theme** dialog box; the dialog box is closed and the defined theme for displaying the slope analysis is displayed in the **STYLE EDITOR** dialog box.

10. Choose the **Apply** button in this dialog box; the defined slope analysis theme is applied to the surface in the drawing. Also note that the **Display Manager** tab now displays the legend corresponding to the slope analysis theme. Figure 10-58 shows the result of the slope analysis.

11. Close the **STYLE EDITOR** dialog box.

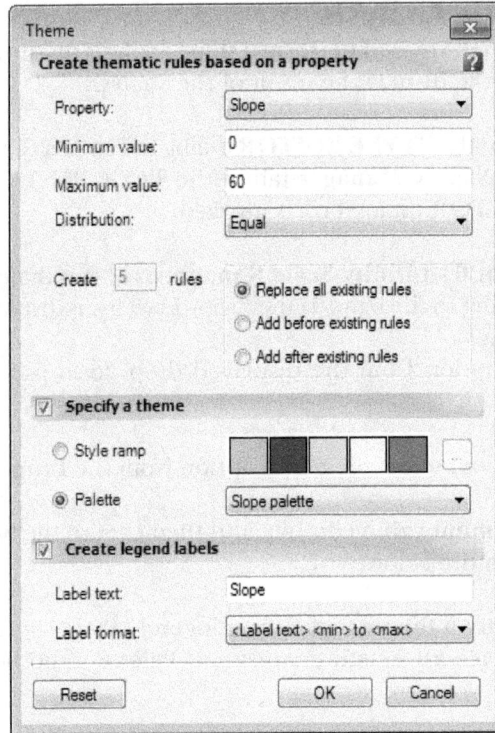

*Figure 10-57 The **Theme** dialog box showing the parameters for slope analysis*

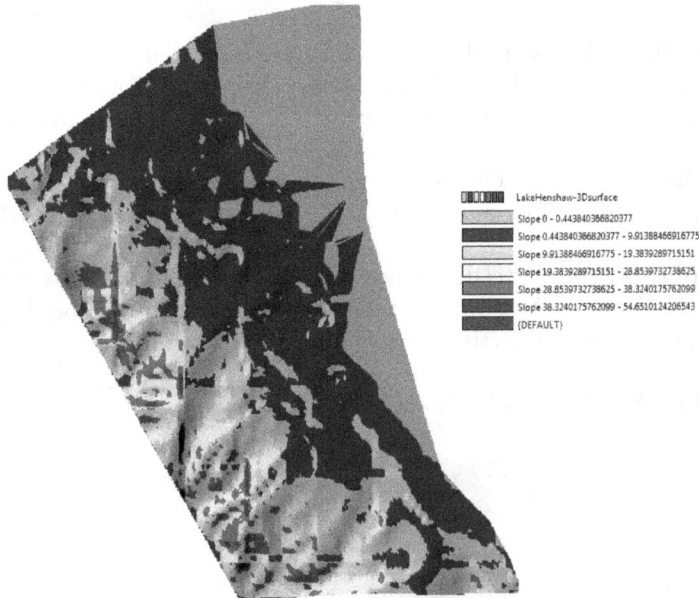

Figure 10-58 The 3D surface displaying the results of slope analysis

Saving the Drawing File

1. Choose the **Save As** option from the Application Menu; the **Save Drawing As** dialog box is displayed.

2. In this dialog box, enter **c10_Tut04a.dwg** in the **File name** edit box and choose the **Save** button next to the **File name** edit box; the current drawing file is saved with the name specified in the edit box.

Self-Evaluation Test

Answer the following questions and then compare them to those given at the end of this chapter:

1. Which of the following dialog boxes is used to create a 3D surface from a point cloud?

   (a) **Point Cloud manager**   (b) **Surface Manager**
   (c) **Create Surface**   (d) None of these

2. Which of the following options is used to copy the settings, template, property, and attributes from one vector layer to another vector layer?

   (a) **Join**   (b) **Bulk Copy**
   (c) **Calculation**   (d) **Attach/Detach Object Data**

3. Which of the following procedures can you perform using the tools and options in **Autodesk Infrastructure Administrator** application?

   (a) Create a new industry model
   (b) Set up a new enterprise or file-based industry model
   (c) Update an industry model
   (d) All of the above

4. You can use the _____ dialog box to create an indexed point cloud data store.

5. You can create a 3D surface for an index file using the _____ tool.

6. In the _____ dialog box, you can use the point files from the GPS instrument.

7. You can manage the Join and Calculation extensions after creating them by using the _____ tool.

8. The _____ tool is used to calculate new values based on the existing values.

9. The point cloud data contains a dense set of point data. (T/F)

10. You can apply an elevation filter condition to filter the point cloud data. (T/F)

11. You cannot create a display theme for a raster data by using the thematic rules that are based on height property of the surface data. (T/F)

12. In AutoCAD Map3D 2017, you can generate contours from a raster image. (T/F)

13. You cannot create a 3D surface using the data from a shape file. (T/F)

14. You can perform a surface analysis to determine the direction of the face of the 3D surface. (T/F)

15. You cannot use a predefined color palette for displaying the result of the slope analysis. (T/F)

Review Questions

Answer the following questions:

1. Which one of the following dialog boxes is used to specify the angle and direction of the Sun while applying style to a raster layer?

    (a) **Hillshade Settings**   (b) **Theme**
    (c) **Style Editor**   (d) None of these

2. You can create a 3D surface from the survey point file by using the _____ tool.

3. In the _____ dialog box, you can create, edit, or delete the Join and Calculation extensions.

4. You can generate contours by using the _____ dialog box.

5. You can create a 3D surface directly from a LiDAR point data file. (T/F)

6. You can create an index file from a single or multiple Point Cloud files. (T/F)

7. You can create the DTM or DEM grid layer from the point cloud data by using the **SURFACE MANAGER** dialog box. (T/F)

8. You can divide an existing scale range of a raster layer, and then style it in a particular scale range. (T/F)

9. The Join extension is used to join property columns from two different vector layers. (T/F)

10. The Calculation extension is used to create the data table for a vector layer. (T/F)

EXERCISES

Exercise 1

Download the *c10-m3d-2017-exr01.xyz* file from *www.cadcim.com*. Create a grid layer using this point data file and apply styling to the grid layer. **(Expected time: 30 min)**

Use the following information to complete this exercise:

a. Create display theme for: Elevation range Color
     -268 - 10   yellow
     -10-10   brown
     10-1800   red

b. Generate contours for the raster layer created with **50m** interval and major contour at every 10th contour. Also, display the value of each major contour.

Exercise 2

Download the *c10-m3d-2017-exr02.dwg* file from *www.cadcim.com* and then calculate the population density per square mile for all the regions. **(Expected time: 30 min)**

Hint: Calculation = (SQMI_ADMIN) / (POP_ADMIN)

Exercise 3

Download the *c10-m3d-2017-exr03.dwg* file from *www.cadcim.com* and then perform the aspect analysis on the 3D surface and then display the result of the analysis using the predefined **Circular palette file for Aspect** color palette. Figure 10-59 displays the result of the aspect analysis of the 3D surface. **(Expected time: 30 min)**

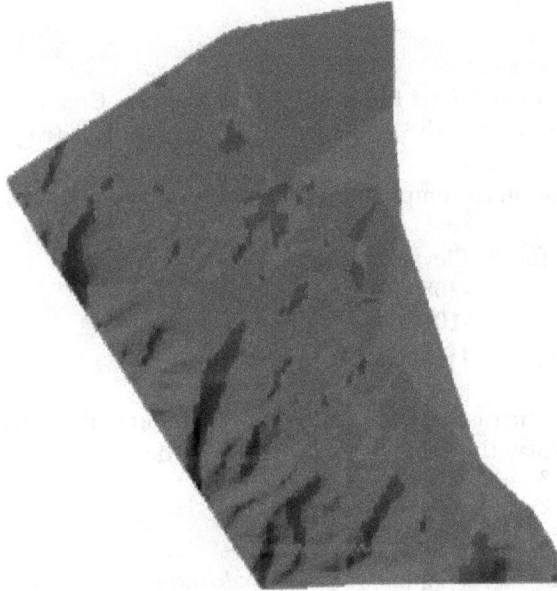

Figure 10-59 *The result of aspect analysis*

Answers to Self-Evaluation Test
1. b, **2.** b, **3.** d, **4. Point Cloud Manager, 5. Create from Point Cloud, 6. Create Surface,**
7. Manage Extension, 8. Calculations, 9. T, **10.** T, **11.** F, **12.** T, **13.** F, **14.** T, **15.** F

Chapter 11

Editing a Map and Creating a Map Book

Learning Objectives

After completing this chapter, you will be able to:
- *Perform the rubber sheeting of drawing objects*
- *Transform the limits of drawing objects*
- *Trim and edit a boundary*
- *Prepare map layout*
- *Modify layout using attribute editor*
- *Create a map book*

INTRODUCTION

In this chapter, you will learn how to perform the rubber sheeting of drawing objects in a map and transform the limits of a drawing object. Also, this chapter discusses how to modify the boundary of a map and erase a portion containing drawing objects from the map. Moreover, you will learn to specify the settings for creating the map elements and the method for creating a map book.

MAP EDITING TOOLS

You can use various map editing tools to edit drawing objects and the boundary of a map. These map editing tools help you to perform rubber sheeting, transform drawing objects, break the boundary of drawing objects, and trim a set of drawing objects. Note that the map editing tools are used to edit only the drawing objects. You cannot edit the geospatial features by using the map editing tools. Some of the map editing tools are discussed next.

Rubber Sheet Tool

Ribbon:	Tools > Map Edit > Rubber Sheet
Command:	ADERSHEET

Rubber sheeting is a procedure that is used to modify the coordinates of a dataset so as to align the data geographically with respect to the known locations. Rubber sheeting process will stretch, shrink, or reorient the dataset to align it with the reference points.

To apply rubber sheet to the drawing objects, choose the **Rubber Sheet** tool from the **Map Edit** panel of the **Tools** tab; the cursor will change into a crosshair in the drawing window and you will be prompted to select the first base point. Place the crosshair on the first base point and then select it by clicking on it; the first base point will be selected and you will be prompted to select the first reference point. Place the crosshair on the first reference point and then click on it; the first reference point will be selected and you will be prompted to select the second base point. Continue selecting the base and reference points as per your requirement.

> **Tip**
> *You can also specify the X and Y coordinates of the base and reference points in the Command prompt to locate them on the map.*

> **Note**
> *It is always recommended that you select a minimum of three base points and three reference points for rubber sheeting. Increasing the number of base points and reference points will increase the accuracy of result for the rubber sheeting process.*

After selecting the desired number of base points and reference points, press ENTER; you will be prompted to specify the method of object selection for rubber sheeting. You can select all objects in the required area. To do so, type **A** and then press ENTER; all the polygons in the base map will be aligned to the reference map. You can also select the drawing objects individually. To do so, type **S** and then press ENTER; the crosshair will change into a selection box and you will be prompted to select the drawing objects. Select the required drawing objects and then press ENTER; the rubber sheeting will be performed on the base map and the base map will be aligned to the boundary of the reference map. Figure 11-1 shows a base map stretched to the limits of a reference map by using the **Rubber Sheet** tool.

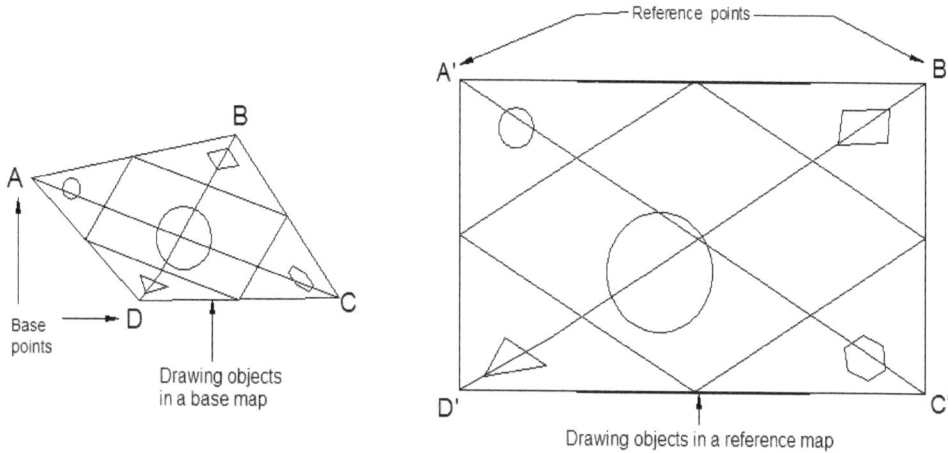

Figure 11-1 *A base map stretched to the limits of a reference map*

Transform Tool

Ribbon:	Tools > Map Edit > Transform
Command:	ADETRANSFORM

The **Transform** tool is used to move, rotate, and scale drawing objects or an entire drawing layer. You can invoke the **Transform** tool by choosing it from the **Map Edit** panel of the **Tools** tab. On doing so, you will be prompted to select the object or the layer that is to be transformed.

To select individual drawing object in the drawing window, enter **S** in the Command prompt and press ENTER; the cursor will change into a selection box in the drawing window and you will be prompted to select the drawing objects. Select the required drawing objects by drawing a selection box or by clicking on the individual objects and press ENTER; you will be prompted to specify the first source point.

To transform all the objects in a layer, enter **L** in the Command prompt; you will be prompted to specify the layer. Enter the layer name in the Command prompt and press ENTER; you will be prompted to specify the first source point.

To specify the source point, place the crosshair at the desired vertex of the drawing object and then click; the first source point will be selected and you will be prompted to select the first destination point. Place the crosshair on the desired location in the current drawing and then click; the first destination point will be selected and you will be prompted to select the second source point. Place the crosshair on the next desired vertex of the drawing object and then click; the second source point will be selected and you will be prompted to select the second destination point. Next, place the crosshair on the second destination point and then click; the selected drawing object will transform to the new extents. Figure 11-2 illustrates the transformation of a drawing object.

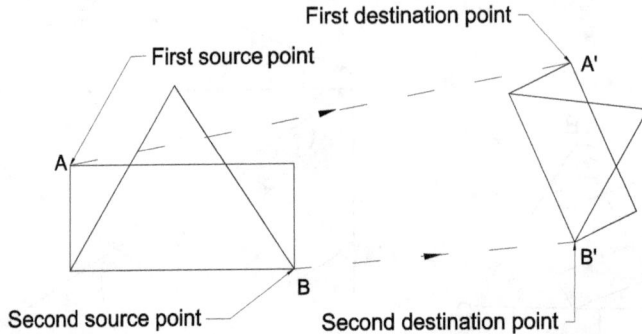

Figure 11-2 Illustration of a drawing object transformation

Note
Once the selected drawing object has been transformed, its basic form will not be displayed.

Boundary Break Tool

Ribbon:	Tools > Map Edit > Boundary Break
Command:	MAPBREAK

You can use the **Boundary Break** tool to break objects along the specified boundary in a map. This tool is also used to cut the drawing objects such as lines, arcs, 2D polylines, and circles that cross the specified boundary in a map.

Note
The BREAK and MAPBREAK commands do not have the same function. The BREAK command is used to split a drawing object whereas the MAPBREAK command is used to break the boundary of an entire map.

You can perform the boundary break operation on a drawing map by using the extents from an active source drawing. To perform this operation, you need to attach the map to the current drawing. To do so, right-click on the **Drawings** folder in the **Map Explorer** tab of the **TASK PANE**; a shortcut menu will be displayed. In this menu, choose the **Attach** option; the **Select drawings to attach** dialog box will be displayed. In this dialog box, browse to the desired folder and select the required drawing file/s. Next, choose the **Add** button and then choose the **OK** button; the dialog box will be closed and the selected drawings will be attached to the current drawing.

After attaching the required drawing files to the current drawing, you need to specify the extents of the drawing. To do so, right-click on the **Drawings** node in the **Map Explorer** tab of the **TASK PANE**; a shortcut menu will be displayed. In this shortcut menu, choose the **Define/Modify Drawing Set** option; the **Define/Modify Drawing Set** dialog box will be displayed. Choose the **Drawing Settings** button located at the lower left corner of the dialog box; the **Drawing Settings** dialog box will be displayed.

In the **Drawing Settings** dialog box, select the required option from the list box in the **Active Drawings** area; the extents of the selected drawing file will be displayed in the list box in the

Save Back Extents area. Next, choose the **Apply** button and then choose the **Close** button in the **Drawing Settings** dialog box; the dialog box will be closed. Next, choose the **OK** button in the **Define/Modify Drawing Set** dialog box; the dialog box will be closed.

To break the boundary of an attached drawing, choose the **Boundary Break** tool from the **Map Edit** panel of the **Tools** tab; the **Break Objects at Boundary** dialog box will be displayed, as shown in Figure 11-3. The options in different areas of this dialog box are discussed next.

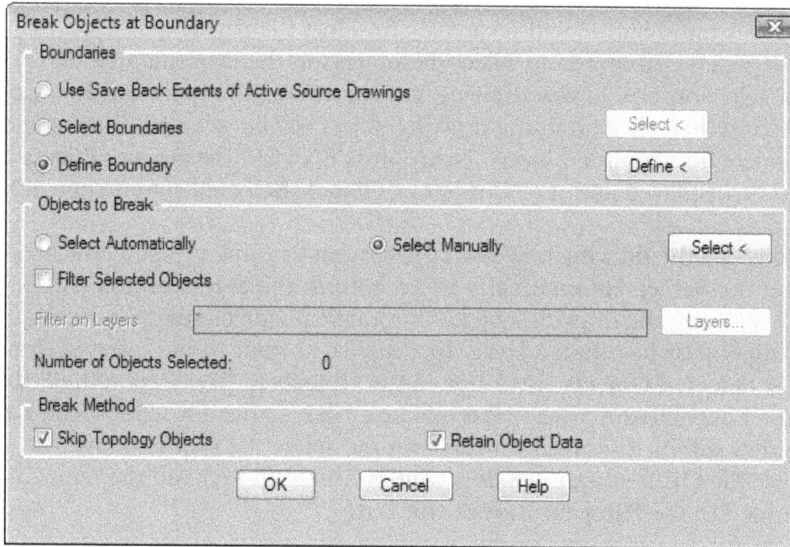

*Figure 11-3 The **Break Objects at Boundary** dialog box*

Boundaries Area

The options in the **Boundaries** area are used to specify the edge for cutting the boundary of a map. To specify the boundary for breaking objects, you can use the extents of the active source drawing, select the existing drawing object from the current drawing, or define a new boundary in the drawing.

To specify the boundary from an active source drawing, select the **Use Save Block Extents of Active Source Drawings** radio button in this area; a rectangle representing the extents of the active source drawing will be displayed in the drawing window.

To specify the boundary by selecting an existing drawing object in the current drawing, select the **Select Boundaries** radio button; the **Select** button corresponding to this radio button will be activated. Choose the **Select** button; the cursor will change into a selection box in the drawing window and you will be prompted to select a drawing object. Click on the required drawing object in the drawing area, the boundary will be created and data boundary will be displayed.

You can also specify a boundary by drawing it in your drawing area. To do so, select the **Define Boundary** radio button in the **Boundaries** area of the **Break Objects at Boundary** dialog box; the **Define** button corresponding to this radio button will be activated. Next, choose the **Define** button; the cursor will change into a crosshair in the drawing window and you will be prompted to specify the first point of the boundary. Place the crosshair on the first point of the boundary and then click; the first point will be specified and you will be prompted to specify the next point of the boundary. Continue specifying the points for defining the boundary until all the required points are located. Next, press ENTER; the boundary will be saved and the **Break Objects at Boundary** dialog box will be displayed.

Objects to Break Area

The options in the **Objects to Break** area are used to specify the conditions for the selection of drawing objects. The **Select Manually** radio button is selected by default. As a result, the **Select** button next to this radio button will be active. To select the drawing objects manually, choose this button; you will be prompted to select the objects in the drawing area and the cursor will change into a selection box in the drawing window. Place this selection box on the desired drawing object and then click on it; the drawing object will be selected. Continue this selection procedure to select the required objects. Next, press ENTER; the **Break Objects at Boundary** dialog box will be displayed and the **Number of Objects Selected** statement will be updated.

You can also select all the drawing objects in the current drawing or in an attached drawing file. To do so, select the **Select Automatically** radio button; the **Select** button will be deactivated. In addition to selecting the drawing objects automatically or manually, you can also filter the selected drawing objects based on a layer. To apply layer filter to the selected drawing objects, select the **Filter Selected Objects** check box in the **Objects to Break** area; the **Filter on Layers** edit box and the **Layers** button next to this edit box will be activated. Choose the **Layers** button; the **Select** window will be displayed. In this window, select the required layer by clicking on it and then choose the **OK** button; the **Select** window will be closed and the name of the selected layer will be added to the **Filter on Layers** edit box.

Break Method Area

The options in the **Break Method** area are used to specify whether or not to include topology and object data in the process of breaking the boundary of a map. The **Skip Topology Objects** check box in this area is selected by default. As a result, the drawing objects with topology data in the selected drawing set will not be included in the process of breaking the boundary of a map. To include the topology data in the current breaking process, clear the **Skip Topology Objects** check box.

The **Retain Object Data** check box in this area is selected by default. As a result, the object data pertaining to a drawing object will be retained after the process of breaking the boundary is completed.

After specifying all parameters in the **Boundaries**, **Objects to Break**, and **Break Method** areas, choose the **OK** button located at the bottom of the **Break Objects at Boundary** dialog box; the dialog box will be closed and the selected drawing objects will split along the specified boundary.

Boundary Trim Tool

Ribbon:	Tools > Map Edit > Boundary Trim
Command:	MAPTRIM

You can use the **Boundary Trim** tool to remove or erase a portion of the drawing objects in a map. To invoke this tool, choose the **Boundary Trim** tool from the **Map Edit** panel;

the **Trim Objects at Boundary** dialog box will be displayed, as shown in Figure 11-4. Various options in different areas of the **Trim Objects at Boundary** dialog box are discussed next.

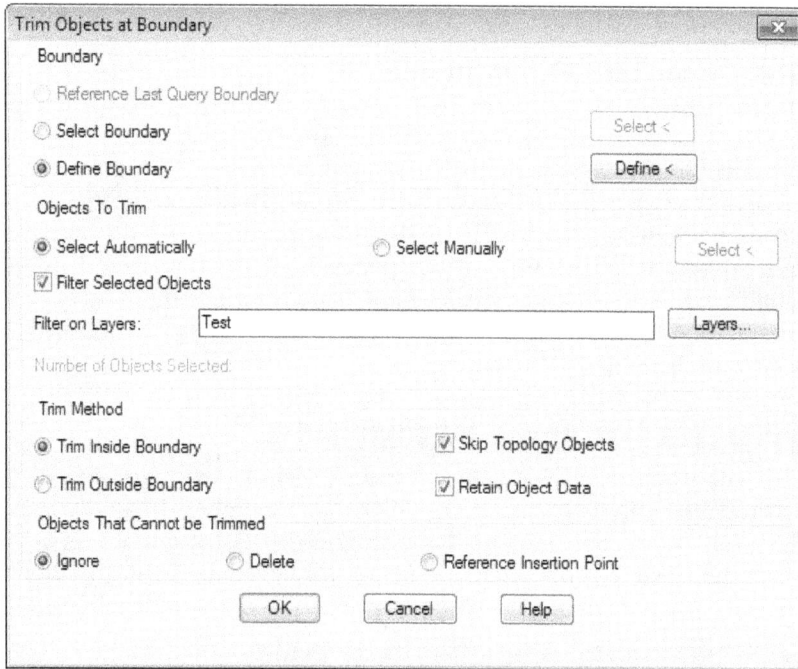

*Figure 11-4 The **Trim Objects at Boundary** dialog box*

Note
*If you have defined a query by using the **Location** condition in the current drawing then the **Reference Last Query Boundary** radio button in the **Boundary** area will be activated.*

Boundary Area

The options in the **Boundary** area are used to specify the boundary of the trimming area. If the **Reference Last Query Boundary** radio button is active then select this radio button to use the boundary specified in the query as the boundary used for trimming. You can also specify the boundary of the trimming area by using the **Select Boundary** and **Define Boundary** options, as discussed in the earlier sections.

Objects to Trim Area

The options in the **Objects to Trim** area are used to specify the objects to be included in the process of trimming. The method of selecting objects for trimming in this area is similar to that of selecting drawing objects, as discussed in the **Objects to Break** area of the **Break Objects at Boundary** dialog box.

Trim Method Area

The options in the **Trim Method** area are used to specify whether to trim the drawing objects inside or outside the boundary. To trim the drawing objects inside the defined boundary area, select the **Trim Inside Boundary** radio button. To trim the drawing objects outside the boundary, select the **Trim Outside Boundary** radio button. You can also specify the options for saving the topology and object data by using the **Skip Topology Objects** and **Retain Object Data** check boxes as discussed in the **Break Method** area of the **Break Objects at Boundary** dialog box.

Objects That Cannot be Trimmed Area

The options in the **Objects That Cannot be Trimmed** area are used to specify the output for the objects that cannot be trimmed. In this area, the **Ignore** radio button is selected by default. As a result, the untrimmed drawing objects in the trimming area will be ignored. You can delete or erase the untrimmed drawing objects by selecting the **Delete** radio button. Also, you can delete the untrimmed objects based on their point of insertion in the drawing by selecting the **Reference insertion point** radio button. After specifying all the parameters in the **Boundary**, **Objects to Trim**, **Trim Method**, and **Objects That Cannot be Trimmed** areas, choose the **OK** button from the **Trim Objects at Boundary** dialog box; this dialog box will be closed and the objects in the current drawing will be erased.

MAP LAYOUT AND MAP ELEMENTS

A map layout is composed of various map elements such as key map, legend, north arrow, scale, and drawing description. To create or edit the map elements in a map layout, you need to switch to the layout space environment. To do so, choose the **Layout** button on the left in the drawing window; the **Layout Tools** tab (contextual) will be displayed in the Ribbon, as shown in Figure 11-5. In the layout space environment, you can create/modify a layout.

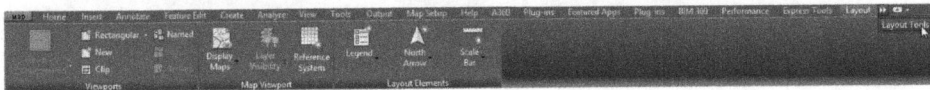

*Figure 11-5 The **Layout Tools** contextual tab*

The layout environment will display the map layout with the map elements. Note that the map elements displayed in the layout will depend on the drawing template that you have selected while creating the drawing. The method of adding map elements to the map layout using the tools in the **Layout Tools** tab (contextual) is discussed next.

Preparing a Map Layout Sheet

AutoCAD Map 3D allows you to create dynamic links between the added map elements. As a result, modifying one map element will result in the change of all the map elements that are associated with it. For example, changing the orientation of the north arrow will rotate the map in the associated view port so that the directional integrity of the map objects is maintained.

You can add various map elements to the map layout using the tools available in the **Layout Tools** tab (contextual). To invoke the **Layout Tools** tab (contextual), switch to the layout environment as described in the previous section. The procedure to add various map elements is discussed next.

Adding Viewports

A viewport is a map element that is used to display the drawing objects in a map. You can have a single viewport in a map layout or add multiple viewports to display the drawing objects. To add a new viewport to your layout, choose the **New** tool from the **Viewports** panel of the **Layout Tools** tab (contextual); the **Viewports** dialog box will be displayed with the **New Viewports** tab chosen.

The **Standard viewports** list box in this tab displays the list of available viewport configurations. Choose the required viewport from this list box; the configuration of the selected viewport will be displayed in the **Preview** area of the **New Viewports** tab. Specify the other required options in this dialog box and then choose the **OK** button; you will be prompted to specify the area in your layout within which you want to fit the selected configuration. Specify the area in the layout; the selected viewport configuration will be displayed in the specified area of the map layout.

You can also create a viewport in a custom defined shape. To do so, choose the **Polygonal** tool from the **Viewports** panel of the **Layout Tools** tab (contextual); you will be prompted to specify the first point. Click in the layout to specify the first point; you will be prompted to specify the second point. Continue specifying the required points for the viewport. After specifying the last point, press ENTER to close the area; the viewport will be created. Now, you can use this viewport to display the map objects.

To modify the content displayed in the viewport and to change the display scale of the content displayed in the viewport, switch to the model space in layout space environment by choosing the **Model or Paper space** button in **Application Status bar**. On switching to model space, the text **MODEL** will be displayed on the **Model or Paper space** button. Next, select the required viewport in the layout and set the required scale for the drawing objects. To resize a viewport, click on the required viewport; blue grip edit markers will be displayed on the selected viewport. Select the required grip edit marker and move it; the viewport will be resized.

Adding Map Legend

The map legend displays various symbols used in the map and the description of the objects they represent. To create a map legend, choose the **Legend** drop-down from the **Layout Elements** panel of the **Layout Tools** tab (contextual). Next, select the required option from the drop-down list; you will be prompted to select a viewport for the legend. Click on the required viewport in the map layout; a floating legend will be displayed and you will be prompted to specify the location of the legend in the map layout. To place the legend in the layout, click at the required location; the legend will be placed at the selected location.

Note
*The map legend added to the layout will display the contents in the **Display Manager** tab of the TASK PANE.*

Adding a North Arrow

You can add a north arrow in your map that will indicate the north direction in your map. To do so, choose the **North Arrow** drop-down in the **Layout Elements** panel of the **Layout Tools** tab (contextual); a list of available north arrows will be displayed. Select the required north arrow from the list; you will be prompted to select a viewport to which this north arrow is to be associated. Click on the required viewport in the map layout; you will be prompted to specify the location of the north arrow in the layout. Click in the layout; the selected north arrow will be placed at the specified location in the map layout.

Adding a Scale Bar

A scale bar shows the scale of the map graphically. To create a scale bar, choose the **Scale Bar** drop-down in the **Layout Elements** panel of the **Layout Tools** tab (contextual); a list of available scale bars will be displayed. Select the required scale bar from the displayed list; you will be prompted to select a viewport to which the scale bar is to be associated. Click on the required viewport in the map layout; the **Scale Bar Properties** dialog box will be displayed, as shown in Figure 11-6.

*Figure 11-6 The **Scale Bar Properties** dialog box*

Specify the value for the division of the scale bar in the **Scale bar division** edit box and select its units for the scale division by selecting an option from the corresponding drop-down list. You can also specify the map scale in the **Scale ratio** edit box. On specifying all the required options in the dialog box, choose the **OK** button; the dialog box will be closed and you will be prompted to specify the location of the scale bar in the map layout. Click in the layout; the selected scale bar will be placed at the specified location in the map layout.

Adding a Reference System

You can add a reference grid to the viewport in your map. To do so, choose the **Reference System** tool from the **Map Viewport** panel of the **Layout Tools** tab (contextual); you will be prompted to select the required viewport. Click on the required viewport in the map layout; the **Create Reference System** dialog box will be displayed, as shown in Figure 11-7.

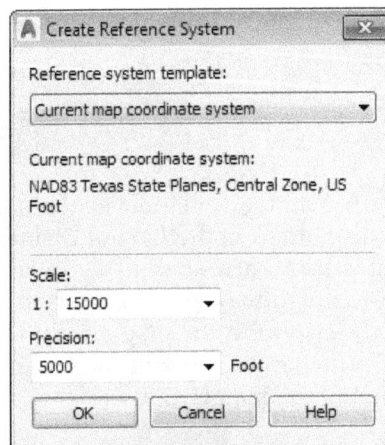

*Figure 11-7 The **Create Reference System** dialog box*

In this dialog box, specify the reference system that you want to create by selecting an option from the **Reference system template** drop-down list. Next, specify the value of scale and precision in the **Scale** and **Precision** edit boxes, respectively and then choose the **OK** button. The reference grid will be created for the selected viewport in the map layout. Figure 11-8 shows an example of a map layout with various map elements.

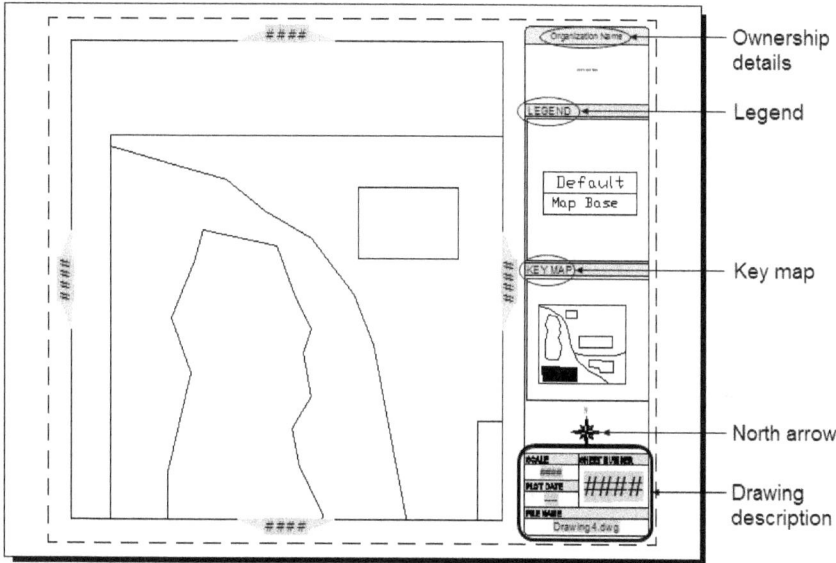

Figure 11-8 *Map layout with map elements*

Enhanced Attribute Editor for Editing the Layout Template

Based on the drawing template used for creating the drawing, a predefined layout may be available for creating the map layout. These layouts are made of blocks and have block attributes associated with them. You can view and edit these block attributes using the **Enhanced Attribute Editor** dialog box.

To do so, place the cursor on the outer frame of the layout; the frame containing map elements will be highlighted. Double-click on the highlighted boundary; the **Enhanced Attribute Editor** dialog box will be displayed, as shown in Figure 11-9. The tabs in this dialog box are discussed next.

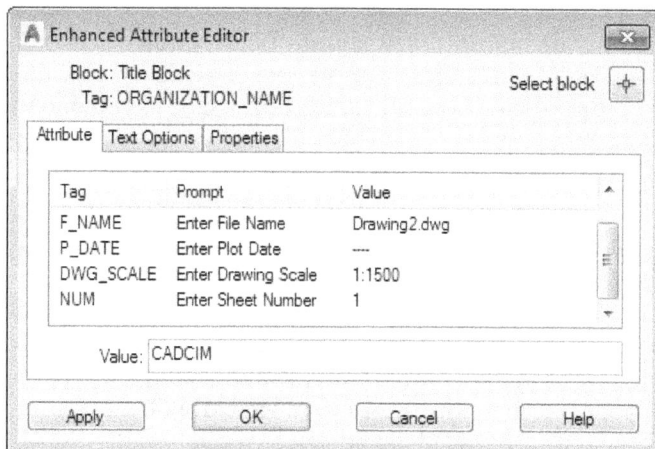

Figure 11-9 *The Enhanced Attribute Editor dialog box*

Attribute Tab

The options in the **Attribute** tab are used to specify the value of each drawing parameter in a map. To specify a value for an option, select the required option from the list box in this tab by clicking on it; the existing value of the selected option will be displayed in the **Value** edit box located below the list box. You can modify this value by specifying the required value in the **Value** edit box. On doing so, the **Apply** button in this dialog box will be activated. Choose the **Apply** button; the modified value will be saved.

Text Options Tab

The options in the **Text Options** tab are used to specify the display properties for the text of the parameter selected in the list box of the **Attribute** tab. To specify the display options for the required parameter, select it in the **Attribute** tab. Next, choose the **Text Option** tab in the **Enhanced Attribute Editor** dialog box; the text options for the selected parameter will be displayed in the **Text Option** tab.

Next, to specify the text style in the **Text Option** tab, select the required option from the **Text Style** drop-down list. To specify the alignment of the text, select an option from the **Justification** drop-down list. The **Backwards** and **Upside down** check boxes are clear by default. As a result, the attribute text is displayed from left to right and it faces upward. If you select the **Backwards** or **Upside down** check box, then the attribute text will be displayed from right to left and will face downward.

You can specify the size of the attribute text by entering the required value in the **Height** edit box and the **Width Factor** edit box. You can place the text at the required angle by specifying the required angle in the **Rotation** edit box. Similarly, to specify the angle for the characters in the text, enter the required value in the **Oblique Angle** edit box. To apply an annotative style to the attribute text, select the **Annotative** check box. After specifying all parameters in the **Text Options** tab, choose the **Apply** button; the modified settings will be saved.

Properties Tab

The options in the **Properties** tab are used to specify the layer, linetype, lineweight, color, and plot style parameters of the attribute text. After specifying the parameters for the attribute text, choose the **Apply** button; the modified settings will be saved.

Once you have set the parameters in the **Attribute**, **Text Options**, and **Properties** tabs, choose the **Apply** button; the modified settings will be saved. Next, choose the **OK** button in the **Enhanced Attribute Editor** dialog box; the dialog box will be closed and the specified settings will be saved for the current map.

MAP BOOK

Map Book is a customized form of a map comprising of a single tile or several tiles. Moreover, map book is a division of a map into several small tiles created by using horizontal and vertical grid lines. Each map tile created by the division can be used as a separate map and can be published separately. To create a map book, choose the **Map Book** tab from the **TASK PANE**; the options in the **Map Book** tab will be displayed. You can use the options in the **Map Book** tab to create a new map book, zoom to the tile or layout view, display the lists of tiles in the selected map book, and edit the settings of the map book selected in the list box. The various methods for creating a map book are discussed next.

Creating a New Map Book

Task Pane: Map Book > New > Map Book
Command: MAPBOOKCREATE

After creating the required drawing file, you can begin with the procedure to create a map book. You can create a new map book by specifying the parameters such as sheet template, tiling scheme, and naming scheme. These parameters are required to create a map book.

Before proceeding with the map book creation, save the current drawing. This will ensure that all the recent updates in the drawings are saved. Next, choose the **New** button in the **Map Book** tab of the **TASK PANE**; a flyout will be displayed. In this flyout, choose the **Map Book** option, as shown in Figure 11-10; the **Create Map Book** dialog box will be displayed, as shown in Figure 11-11. The nodes in this dialog box are discussed next.

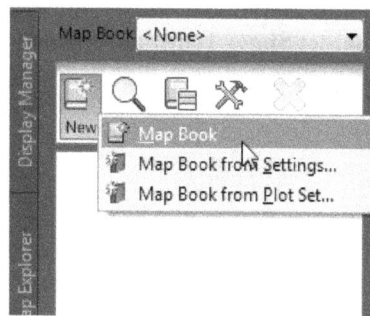

Figure 11-10 *Choosing the* ***Map Book*** *option from the shortcut menu*

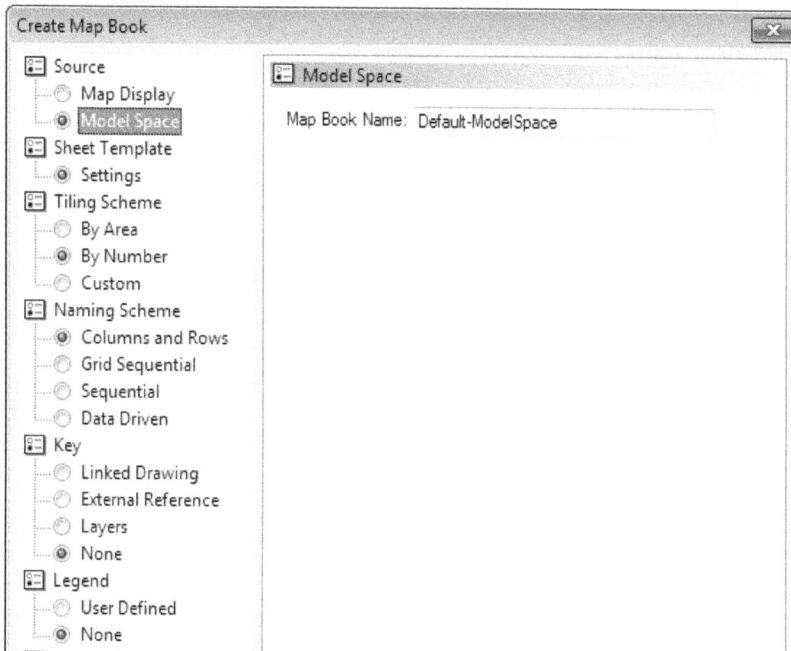

Figure 11-11 *Partial view of the* ***Create Map Book*** *dialog box with the* ***Model Space*** *radio button selected*

Source Node

You can use the options in the **Source** node to specify whether you need to create a map book from the objects in the **Display Manager** tab of the **TASK PANE** or a map book from the current drawing containing drawing objects. The **Model Space** radio button is selected by default. As a result, the **Map Book Name** edit box is displayed in the right pane of the dialog box. You can specify the name of the map book in the **Map Book Name** edit box. If you select the **Map Display** radio button in the **Source** node, the **Map Name** and **Scale** drop-down lists will be displayed along with the **Map Book Name** edit box. Specify the required options in these drop-down lists.

Sheet Template Node

Select the **Settings** radio button in the **Sheet Template** node; the options will be displayed in the right pane. You can use these options to specify a sheet template file and the layout settings for the current map book, as shown in Figure 11-12. To apply the template settings from an existing template file for creating the current map book, choose the Browse button next to the **Choose a Sheet Template** edit box; the **Select Sheet Template Drawing** dialog box will be displayed.

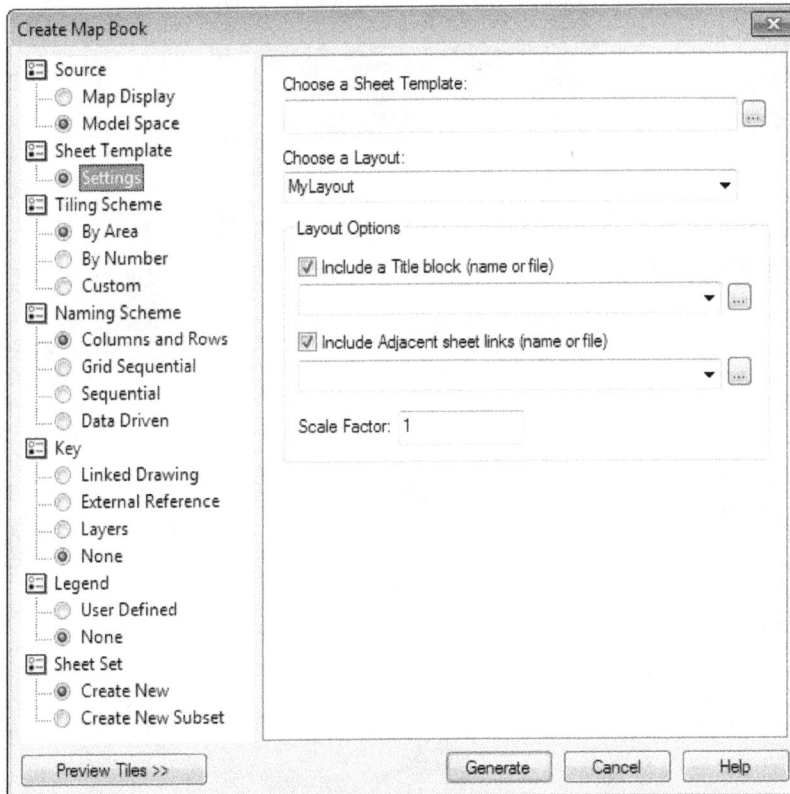

*Figure 11-12 The **Create Map Book** dialog box with the **Settings** radio button selected in the **Sheet Template** node*

In the **Select Sheet Template Drawing** dialog box, browse to the required folder containing the desired template file and then select the template file from the list box; the name of the chosen template file will be displayed in the **File name** edit box. Next, choose the **Open** button in the

Select Sheet Template Drawing dialog box; this dialog box will be closed and the path of the selected template file will be displayed in the **Choose a Sheet Template** edit box. Also, the layouts in the selected template file will be displayed in the **Choose a Layout** edit box.

In the **Layout Options** area, you can modify the settings of the selected layout. The **Include a Title block (name or file)** check box is selected by default. As a result, you can include an existing title block in the sheet template by selecting its name from the **Include a Title block (name or file)** drop-down list. Also, you can use the Browse button next to this drop-down list to choose the desired block for creating the sheet template. The **Include Adjacent sheet links (name or file)** check box is selected by default. As a result, you can include a sheet set linked with an external drawing by selecting it from the **Include Adjacent sheet links (name or file)** drop-down list. Alternatively, you can include the drawing containing the required sheet set by using the Browse button next to this drop-down list. You can also specify the scale for the selected block or an adjacent sheet set by entering the desired scale value in the **Scale Factor** edit box.

Tip
*You can use the **Preview Tiles >>** button to display the preview of a map book in the drawing window. To do so, choose the **Preview Tiles >>** button located at the lower left corner of the **Create Map Book** dialog box; this dialog box will be closed and a preview of the map book will be displayed in the drawing window. To exit the preview mode, press ESC; the **Create Map Book** dialog box will be displayed.*

Tiling Scheme Node
The options in the **Tiling Scheme** node are used to specify the parameters of the tiles to be created in the map book. There are three radio buttons in this node: **By Area**, **By Number**, and **Custom**. These radio buttons are discussed next.

Note
*On selecting the **By Area** or **By Number** radio button, existing drawing layers in the current drawing will be displayed in the **Layer** drop-down list of the page displayed.*

By Area
Select the **By Area** radio button in the **Tiling Scheme** node; the options corresponding to the **By Area** radio button will be displayed in the right pane, as shown in Figure 11-13. You can use these options to specify the rectangular area for creating the tiles. You can specify a layer for creating the map book by selecting an option from the **Layer** drop-down list. To specify the area for the tile, enter the x and y coordinates of the first corner of the rectangular area in the first and second edit boxes corresponding to the **First corner** option. Then, enter the x and y coordinates of the opposite corner of the rectangular area in the first and second edit boxes corresponding to the **Opposite corner** option. Alternatively, you can graphically specify the tile by choosing the **Select area to tile >>** button in this page. On choosing the button, the cursor will change into a crosshair in the drawing window and you will be prompted to specify the first corner point of the rectangular area. Place the crosshair at the desired location and then click; the first corner point of the tile will be selected and you will be prompted to specify the opposite point of the tile. Now, drag the cursor to fix the opposite corner and click to specify the point; the opposite corner of the rectangle will be selected and the **Create Map Book** dialog box will be displayed.

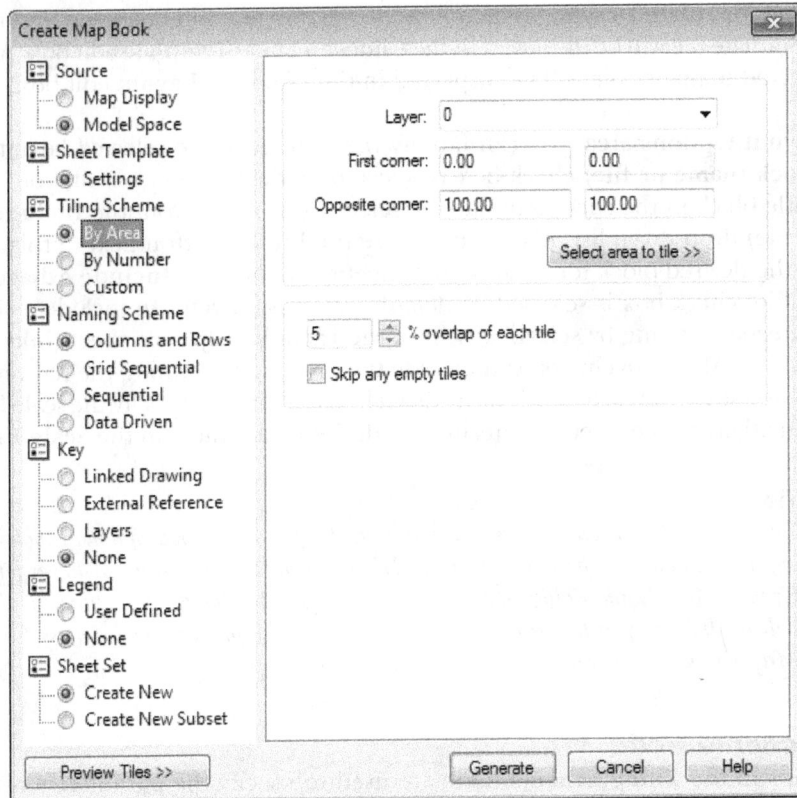

*Figure 11-13 The **By Area** radio button selected in the **Tiling Scheme** node*

In the second area of this page, you can specify the percentage of overlapping area of each tile and ignore the tiles with no data. To specify the overlapping area as a percentage of the map tile, enter the percentage value in the edit box or set the percentage value by using the spinner next to the edit box. The **Skip any empty tiles** check box is cleared by default. Select the **Skip any empty tiles** check box to ignore creating map tiles that do not contain any object data.

By Number
You can select the **By Number** radio button in the **Tiling Scheme** node to specify the number of rows and columns in a map book starting from the upper left corner. The size of the map tile is based on the map scale, the size of the main viewport, and the number of rows and columns specified.

Select a layer from the **Layers** drop-down list to create a map book from this layer. Then, specify the x and y coordinates of the upper left corner of rows and columns by entering the coordinate values in the first and second edit boxes corresponding to the **Upper left** option. Alternatively, you can specify the upper left corner of the tiles by choosing the **Pick Upper Left>>** button. On doing so, the cursor will change into a crosshair in the drawing window and you will be prompted to select a point. Next, place this crosshair at the desired location in the drawing window and then click; the point will be selected and the **Create Map Book** dialog box will be displayed with the coordinates of the selected point displayed in the edit boxes corresponding to the **Upper left** option. Next, specify the number of the rows and columns to be created in the map book by entering the numeric value in the **Columns** and **Rows** edit boxes.

In the lower area of this page, you can specify the percentage value of the overlapping area of each tile. You can also specify whether or not to include the empty tiles in the current map book. The options in the bottom area of this page have been discussed in the earlier sections.

Custom

You can select the **Custom** radio button to specify a custom tile from an existing quadrilateral (square, rectangle, trapezium, parallelogram, and quadrangle) geometry in the current drawing. To do so, choose the **Select Tiles >>** button in the area located at the upper part of this page; the cursor will change into a selection box in the drawing window and you will be prompted to select the drawing objects. Place the selection box on the desired geometry and then click; the geometry will be selected. Next, press ENTER; the object will be selected and the statement **0 selected** will be updated to the statement **X selected** in the upper area of this page. Here, '**X**' represents the number of objects selected. You can specify the options in the area located at the lower section of this page, as discussed earlier.

Naming Scheme Node

You can use the options in the **Naming Scheme** node to specify the naming convention to be used for each of the tiles in the current map book. The methods of specifying the naming convention by using different radio buttons in the **Naming Scheme** node are discussed next.

Columns and Rows

You can select the **Columns and Rows** radio button in the **Naming Scheme** template to specify the naming convention of the tiles in the map book with reference to their row and column values. On selecting this radio button, the **Columns and Rows** page will be displayed, as shown in Figure 11-14. The options in this page are discussed next.

Begin with Drop-down List: The **Begin with** drop-down list is used to specify whether the naming convention will be in the row-column or column-row format. By default, the **Rows** option is selected in this drop-down list. As a result, the naming convention of the map tiles will be in the row-column format. If you select the **Columns** option from this drop-down list, the naming convention of the map tiles will be in the column-row format.

> **Note**
> *When you select the **Column** option from the **Begin with** drop-down list, the name and settings of the **Rows** and **Columns** areas will be interchanged.*

Rows Area: The options in the **Rows** area are used to specify the naming convention for the rows. The **Top to Bottom** option is selected by default in the **Order from** drop-down list. As a result, the numbering is given to the rows from the top row (first row) to the bottom row (last row). If you select the **Bottom to Top** option from the **Order from** drop-down list, then the rows will be numbered from the bottom to top. The **Abc** radio button, next to the **Start with** drop-down list, is selected by default. As a result, the alphabets will be displayed as options in the **Start with** drop-down list. Selecting the **123** radio button will display numbers as options in the **Start with** drop-down list. As a result, you can start the numbering from

1, 10, 100, or 1000 number by selecting an option from the **Start with** drop-down list. You can specify the incremental value of the series in the **Increment by** edit box.

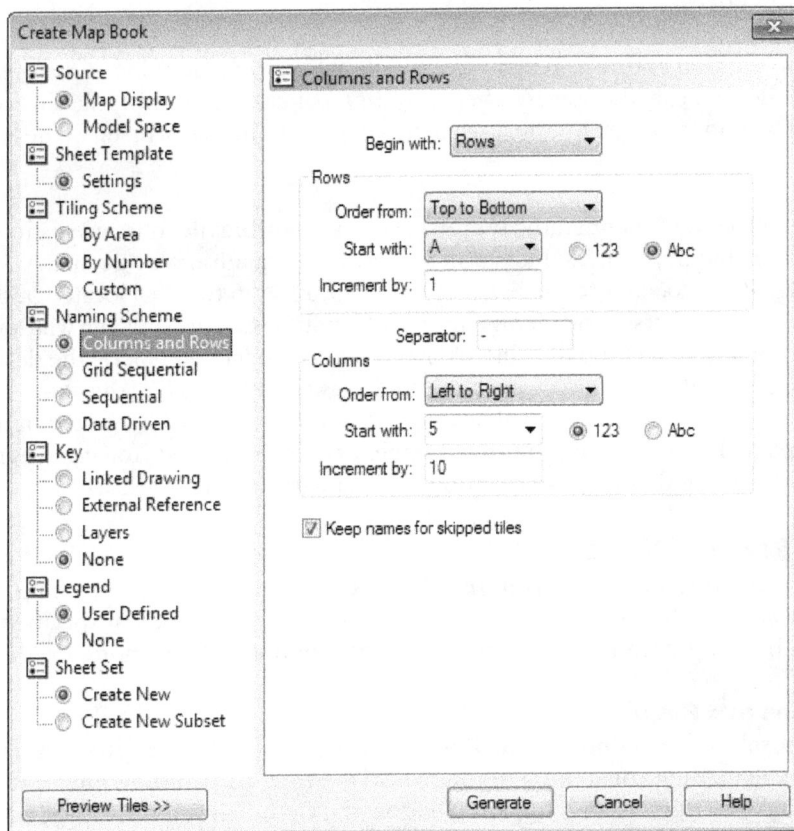

*Figure 11-14 The **Create Map Book** dialog box with the **Columns and Rows** radio button selected in the **Naming Scheme** node*

Separator Edit Box: The **Separator** edit box below the **Rows** area is used to apply a symbol between the row and column values such as -, _, +, and so on.

Columns Area: The options in the **Columns** area are used to specify the naming convention for the columns. The **Left to Right** option is selected by default in the **Order from** drop-down list. As a result, the numbering is given to the columns from the left column (first) to the right column (last). If you select the **Right to Left** option from the **Order from** drop-down list, then the rows will be numbered from the right column to the left column. The **123** radio button, next to the **Start with** drop-down list, is selected by default. As a result, the number **5** will be displayed in the **Start with** drop-down list. You can also select an option from this drop-down list to assign the selected number to the starting number of the column. If you select the **Abc** radio button next to the **123** radio button, the **Start with** drop-down list will display alphabets. Then, you can start naming the columns by selecting an alphabet from the **Start with** drop-down list. You can specify the incremental value of the series in the **Increment by** edit box.

Keep names for skipped tiles Check Box: The **Keep names for skipped tiles** check box is selected by default. As a result, the skipped map tiles are given names. If you clear the **Keep names for skipped tiles** check box, the skipped map tiles will not be given a name.

Grid Sequential

When you select the **Grid Sequential** radio button from the **Naming Scheme** node, corresponding options will be displayed in the page on the right in the dialog box. You can use these options to specify the naming convention for the tiles in the map book by applying numbers to the grids.

In this page, the **Rows** option is selected by default in the **Begin with** drop-down list. As a result, the numbering of the grids will be applied to the rows. If you select the **Columns** option from the **Begin with** drop-down list then the numbering of the grids will be applied to the columns. The **Upper Left to Lower Right** option is selected by default in the **Order from** drop-down list. As a result, the numbering of the grids or the map tiles will start from the upper-left tile in the map book and end at the lower right tile. If you select the **Lower Right to Upper Left** option from the **Order from** drop-down list, then the numbering of the grids or the map tiles will start from the lower right tile and end at the upper left tile in the map book. To start the numbering of the tiles from a specific number, enter the number in the **Start with** edit box below the **Order from** drop-down list. To apply an incremental value to the numbering of tiles in the map book, enter a value in the **Increment by** drop-down list. You can select or clear the **Keep names for skipped tiles** check box as per your requirement.

Sequential

You can select the **Sequential** radio button in the **Naming Scheme** node to display the options for specifying the naming convention for the tiles in series or in sequence. The **Forward** option is selected by default in the drop-down list on the right in the dialog box. As a result, the sequence is named in a forward order. If you select the **Backward** option from this drop-down list, then the map tiles will be named in a backward order by a specific value. To assign a starting value to the sequence, enter desired number in the **Start with** edit box. Next, specify an incremental value in the **Increment by** edit box.

Data Driven

You can select the **Data Driven** radio button in the **Naming Scheme** node to display the options for specifying the naming convention for the map tile based on the data in the current map. To write the expression for naming the map tile, choose the **Expression Chooser** button below the **Expression** edit box; the **Naming Scheme - Data Driven** dialog box will be displayed. In this dialog box, expand the desired node by clicking on the [+] node corresponding to it; the property list will be displayed. In this list, select a property by clicking on it. Next, choose the **OK** button; the **Naming Scheme - Data Driven** dialog box will be closed and the expression created will be displayed in the **Expression** edit box.

Key Node

The options in the **Key** node are used to specify the drawing to be displayed in the key map. The methods of specifying the key map using the options in the **Key** node are discussed next.

Linked Drawing

When you select the **Linked Drawing** radio button, various options will be displayed to specify the linked drawing that can be used for representing the key map.

Note

If you have saved the current map as a drawing file/s (Linked drawing), as discussed in the Key Map section of this chapter, then the list of the saved drawing files will be displayed in the list box in the page displayed.

External Reference

You can select the **External Reference** radio button to display the options used for specifying a drawing in an external location as a key map for the current map book. To specify a drawing in an external location, choose the Browse button next to the **Choose the Map Key Source File** edit box; the **Select Key Reference Drawing** dialog box will be displayed.

In the **Select Key Reference Drawing** dialog box, browse to the folder containing the required drawing file and then choose the drawing option from the list. Next, choose the **Open** button; the **Select Key Reference Drawing** dialog box will be closed and the path of the selected drawing file will be displayed in the **Choose the Map Key Source File** edit box.

Layers

You can select the **Layers** radio button to display the options that are used to specify the drawing layers for representing the key map.

Note

If there are drawing layers in the current drawing other than the default layer, then these layers will be displayed in the All Layers list box.

By default, all the drawing layers are used to represent the key map. To specify a particular drawing layer/s from the current drawing to represent the key map, select the desired drawing layer/s from the **All Layers** list box; the **Add Layers for Map Key** button below this list box will be activated. Next, choose the **Add Layers for Map Key** button; the selected layers will be added to the **Layers for Map Key** list box. You can also add a new layer to the **Layers for Map Key** list box by choosing the **Select Layer** button and delete a layer from the list box by choosing the **Delete** button.

None

To avoid displaying any drawing in the key map, select the **None** radio button in the **Key** node.

Legend Node

The options in the **Legend** node are used to add a desired legend type to the current map book. You can select a radio button in this node to specify the legend type for the current map book. The radio buttons in this node are discussed next.

User Defined

You can select the **User Defined** radio button to invoke the options that are to be used for displaying a particular part of the current map as a legend in the current map book. You can select a specific part of the current map by selecting the rectangular area from the current map. To do so, choose the **Select modelspace bounds >>** button below the **Legend Viewport** area; the **Create Map Book** dialog box will be closed and you will be prompted to select a rectangular area from the drawing window. Next, place the cursor on the first corner

of the rectangular area and then left-click on it; the first corner of the rectangular area will be selected and you will be prompted to select the opposite corner of the rectangular area. Then, place the cursor on the opposite corner of the rectangular area and then left-click on it; the **Create Map Book** dialog box will be displayed with the selected rectangular area appearing in the **Legend Viewport** area. Also, the x and y coordinates of the first and opposite corner of the rectangular area will be displayed in the edit boxes corresponding to the **First corner** and **Opposite corner** options.

You can also locate a rectangular area for using it as a legend in the current map book by specifying the x and y coordinates of the first and opposite corners of the rectangular area. To do so, enter the x and y coordinates of the first and opposite corners of the rectangular area in the **First corner** and **Opposite corner** options, respectively.

None
To avoid displaying a legend in the current map book, select the **None** radio button in the **Legend** node.

Sheet Set Node
The options in the **Sheet Set** node are used to create a sheet set or subset of an existing sheet set from the current map book. You can create a sheet set or subset of an existing sheet set by using the **Create New** and **Create New Subset** radio buttons in this node. These radio buttons are discussed next.

Create New
You can select the **Create New** radio button to display the option for creating a new sheet set from the current map book. To do so, choose the Browse button next to the **New Sheet Set file** edit box in the page displayed; the **Select Sheet Set** dialog box will be displayed. In this dialog box, enter a name for the new sheet set in the **File name** edit box and then choose the **Open** button; the dialog box will be closed and the path of the new sheet set file will be displayed in the **New Sheet Set file** edit box. On generating a new map book, a sheet set will be created for the current map book.

Create New Subset
You can select the **Create New Subset** radio button to display the options for creating a subset in an existing sheet set. To do so, choose the Browse button next to the **Choose Sheet Set file** edit box; the **Select Sheet Set** dialog box will be displayed. In this dialog box, select the desired sheet set option from the list box and then choose the **Open** button; the **Select Sheet Set** dialog box will be closed and the path of the selected sheet set file will be displayed in the **Choose Sheet Set file** edit box. Also, a new subset with the name **New Subset (1)** will be created in the selected sheet set and will be displayed in the **Position new subset** list box.

If there are more than one subsets, you can arrange the position of the subset created by using the **Move Up** or **Move Down** button located below the **Position new subset** list box. To rename a subset, select the subset created and then choose the **Rename** button located below the **Position new subset** list box; the label of the subset created will change into an edit box. Next, enter the desired name in this edit box and then press ENTER; the name entered will be saved.

After specifying all the parameters corresponding to the **Source, Sheet Template, Tiling Scheme, Naming Scheme, Key, Legend,** and **Sheet Set** nodes, choose the **Generate** button located at the bottom of the **Create Map Book** dialog box; a map book will be created with the specified parameters and will be displayed in the **Map Book** tab of the **TASK PANE**. Also, the new sheet sets created with the specified parameters will be displayed at the bottom of the drawing area.

Note
*If you have not specified a sheet template for creating a new map book, the **AutoCAD Map3D 2017** warning message will be displayed.*

You can save the settings of the map book created with the *.mbs* extension. To do so, right-click on the map book created in the **Map Book** tab of the **TASK PANE**; a shortcut menu will be displayed, as shown in Figure 11-15. In this shortcut menu, choose the **Save Settings** option; the **Save Map Book Settings** dialog box will be displayed. In this dialog box, enter a name in the **File name** edit box and then choose the **Save** button; the **Save Map Book Settings** dialog box will be closed and the settings of the map book will be saved with the name entered.

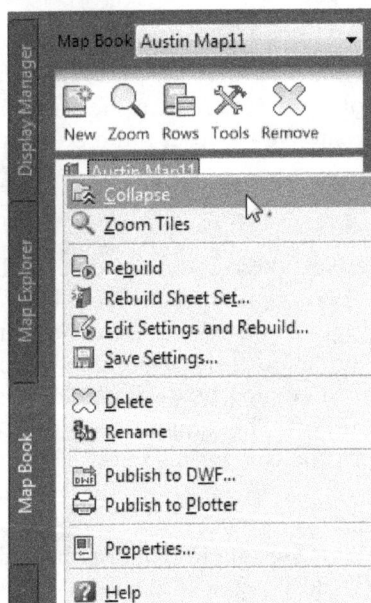

Figure 11-15 The shortcut menu displayed to save settings of the map book

Creating a Map Book from Existing Templates

You can create a map book by importing the settings from an existing map book template. To do so, click on the **New** button in the **Map Book** tab; a shortcut menu will be displayed. In this shortcut menu, choose the **Map Book from Settings** option, as shown in Figure 11-16; the **Select Map Book Settings** dialog box will be displayed.

In the **Select Map Book Settings** dialog box, browse to the folder containing the required map book template file with the *.mbs* extension, and then choose the template file by clicking on it.

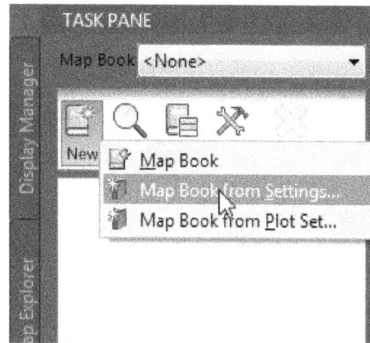

Figure 11-16 *Choosing the* **Map Book from Settings** *option from the shortcut menu*

Next, choose the **Open** button in the **Select Map Book Settings** dialog box; this dialog box will be closed and the **Create Map Book** dialog box will be displayed with the settings from the chosen template file. In addition to the existing settings, you can specify the settings of the current map book in the **Create Map Book** dialog box, as discussed in the earlier section.

Creating a Map Book Using the Plot Settings

You can create a map book by importing the plot settings applied to the current drawing. To do so, choose the **New** tool in the **Map Book** tab of the **TASK PANE**; a shortcut menu will be displayed. In this shortcut menu, choose the **Map Book from Plot Set** option; the plot settings from the current drawing will be applied to the current map book.

TUTORIALS

General instructions for downloading tutorial files:

Before starting the tutorials, you need to download the tutorial data to your computer. To do so, follow the steps given below:

1. Log on to *www.cadcim.com* and then browse to *Textbooks > Civil/GIS > Map 3D > Exploring AutoCAD Map 3D 2017*. Next, select *c11_m3d_2017_tut.zip* file from the **Tutorial Files** drop-down list. Next, choose the corresponding **Download** button to download the data file.

2. Extract the content of the zip file to the following location:

    C:\m3d_2017

Notice that the *c11_m3d_2017_tut* folder is created within the *m3d_2017* folder.

Tutorial 1   Performing the Rubber Sheeting

In this tutorial, you will perform the rubber sheeting of the drawing data by using the **Rubber Sheet** tool.   **(Expected time: 20 min)**

The following steps are required to complete this tutorial:

a.   Open the *c11_Tut01.dwg* file.
b.   Perform the rubber sheeting of the drawing from **A** to **A'**, **B** to **B'**, **C** to **C'**, and **D** to **D'**.
c.   Save the modified drawing.

Opening the Tutorial Drawing

1.   Choose the **Open** button from the Quick Access Toolbar; the **Select File** dialog box is displayed.

2.   In the **Select File** dialog box, browse to the following location:

     C:\m3d_2017\c11_m3d_2017_tut\c11_tut01

3.   Next, select the **c11_Tut01.dwg** file; the preview of the selected drawing file is displayed in the **Preview** area.

4.   Choose the **Open** button on the right of the **File name** edit box; the drawing is displayed in the drawing window, as shown in Figure 11-17.

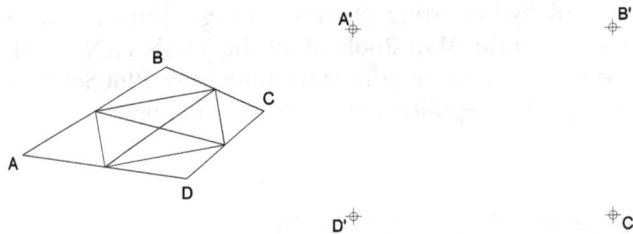

Figure 11-17 The drawing displayed in the drawing window

Performing the Rubber Sheeting of the Tutorial Drawing

1.   Place the cursor on the **Object Snap** button in the Status Bar and then click; a shortcut menu is displayed. In this shortcut menu, choose the **Object Snap Settings** option; the **Drafting Settings** dialog box is displayed with the **Object Snap** tab chosen.

2.   In the **Drafting Settings** dialog box, select the **Object Snap On (F3)** check box.

3.   In the **Object Snap modes** area, select the **Endpoint** and **Node** check boxes and then clear all other check boxes.

4.   Choose the **OK** button in the **Drafting Settings** dialog box; the dialog box is closed.

5.   Choose the **Rubber Sheet** tool from the **Map Edit** panel in the **Tools** tab; the cursor changes into a crosshair and you are prompted to select the first base point.

6.   Move the crosshair to corner **A** and click when the Endpoint object snap is displayed; the first base point is selected and you are prompted to select the first reference point.

7. Drag the crosshair to the node **A'** in the current drawing and click when the **Node** object snap is displayed, the first reference point is selected and you are prompted to select the second base point.

8. Repeat the procedure given in steps 6 and 7 and specify the base points at **B**, **C**, and **D** and their respective reference points at **B'**, **C'**, and **D'** in the drawing window.

9. After specifying all the base and reference points, press ENTER; you are prompted to specify the object selection method for rubber sheeting.

10. Type **S** and then press ENTER; the crosshair changes into a selection box in the drawing window and you are prompted to select the drawing objects for the rubber sheeting.

11. Select all the drawing objects by drawing a selection box, refer to Figure 11-18, and then press ENTER; the rubber sheeting is performed on the selected drawing objects. Figure 11-19 shows the result of rubber sheeting.

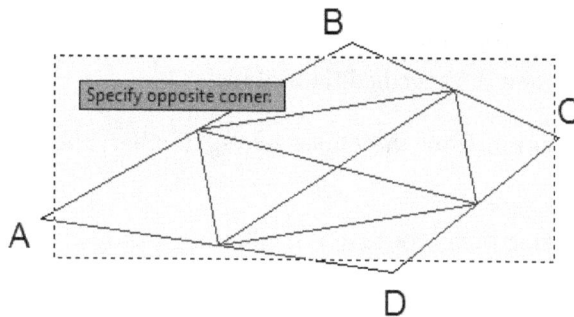

Figure 11-18 Selecting drawing objects

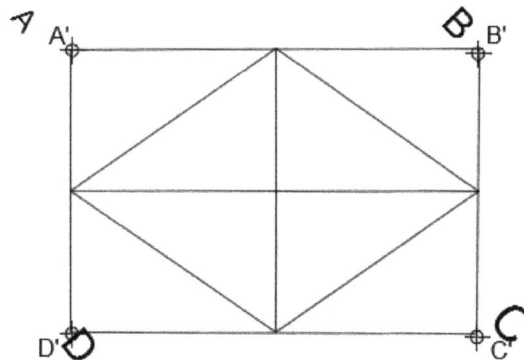

Figure 11-19 The drawing objects after the rubber sheeting has been performed

Saving the Current Drawing

1. Choose the **Save As** option from the Application Menu; the **Save Drawing As** dialog box is displayed.

2. In the **Save Drawing As** dialog box, browse to the desired folder and then enter **c11_Tut01a.dwg** in the **File name** edit box.

3. In the **Save Drawing As** dialog box, choose the **Save** button next to the **File name** edit box; the current drawing file is saved with the specified name.

Tutorial 2   Performing the Boundary Break

In this tutorial, you will break the drawing objects in the drawing map by using the **Boundary Break** tool.   **(Expected time: 1 hr)**

The following steps are required to complete this tutorial:

a. Open the *c11_Tut02.dwg* file.
b. Perform the boundary break on the tutorial drawing by using the boundary displayed in blue.
c. Save the modified drawing.

Opening the Tutorial Drawing
In this section, you will open the specified tutorial file.

1. Choose the **Open** button from the Quick Access Toolbar; the **Select File** dialog box is displayed.

2. In the **Select File** dialog box, browse to the following location:

    C:\m3d_2017\c11_m3d_2017_tut\c11_tut02

3. In the **Select File** dialog box, select the **c11_Tut02.dwg** file from the list box; the preview of the selected drawing file is displayed in the **Preview** area.

4. Next, choose the **Open** button; the drawing is displayed in the drawing window, as shown in Figure 11-20.

Figure 11-20 Geometry of the current drawing with the attached drawing

Performing the Boundary Break on the Current Drawing

In this part of the tutorial, you will break the drawing objects along the edge of the selected object.

1. Choose the **Boundary Break** tool from the **Map Edit** panel in the **Tools** tab; the **Break Objects at Boundary** dialog box is displayed.

2. In the **Boundaries** area of this dialog box, select the **Select Boundaries** radio button; the **Select** button corresponding to this radio button is activated.

3. Choose the **Select** button; the cursor changes into a selection box and you are prompted to select the boundary for breaking the drawing objects.

4. Place the selection box on the rectangular boundary, refer to Figure 11-20, and then click; the rectangle is selected.

5. Press ENTER; the **Break Objects at Boundary** dialog box is displayed again.

6. In the **Objects to Break** area of this dialog box, select the **Select Automatically** radio button.

7. Next, choose the **OK** button from the **Break Objects at Boundary** dialog box; the dialog box is closed and the drawing objects are split along the boundary.

    Note
    The process of breaking objects may take time depending upon your computer speed. Until the process is completed, do not choose any tool in the AutoCAD Map3D 2017 User Interface.

8. Zoom to the drawing object that lies across the rectangular boundary and then place the cursor over the object. On doing so, you will notice that the drawing object has split along the rectangular boundary.

Saving the Drawing

1. Choose the **Save As** option from the Application Menu; the **Save Drawing As** dialog box is displayed.

2. In this dialog box, browse to the desired folder and then enter **c11_Tut02a.dwg** in the **File name** edit box.

3. Choose the **Save** button in this dialog box; the current drawing file is saved with the specified name.

Tutorial 3   Creating a Map Layout

In this tutorial, you will create a map layout by specifying the map elements.

(Expected time: 1 hr 15 min)

The following steps are required to complete this tutorial:

a. Start a new drawing and apply the map template.
b. Add feature data using the **DATA CONNECT** wizard.

   c.   Display required data in the viewport.

   d.   Create a map legend.

   e.   Create a key map.

   f.   Insert a scale bar.

   g.   Insert a reference grid.

   h.   Specify the map information
       1.   Organization name: **CADCIM**
       2.   Scale: **1:1500**.

   i.   Save the drawing.

Starting a New Drawing and Applying the Map Template

1. Choose the **New** button from the Application Menu; the **Select template** dialog box is displayed.

2. In the **Select template** dialog box, browse to the following location:

    C:\m3d_2017\c11_m3d_2017_tut\c11_tut03

3. Ensure that the **Drawing Template (*.dwt)** is selected in the **Files of type** drop-down list and then choose the **CADCIM.dwt** template file from the list box; a preview of the selected template file is displayed in the **Preview** area.

4. Choose the **Open** button on the right of the **File name** edit box; a new drawing file with the **CADCIM** layout is opened, as shown in Figure 11-21.

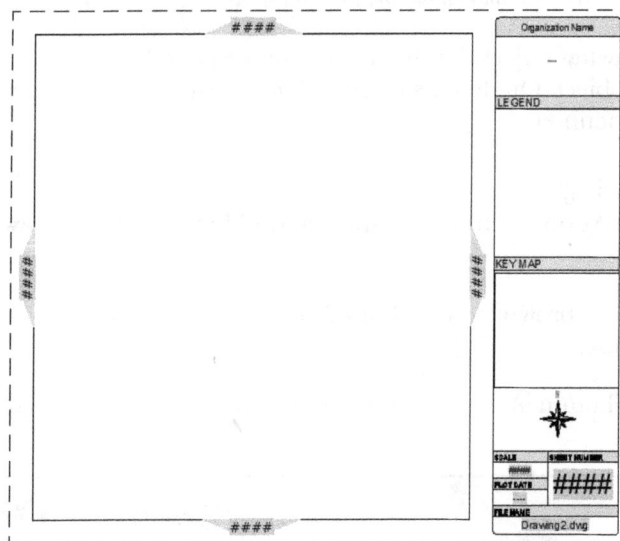

Figure 11-21 *The drawing window with the CADCIM layout*

5. Choose the **Model** tab located at the bottom left of the drawing window to switch to the model environment. Note that the options in the **Display Manager** tab of the **TASK PANE** will now be activated.

Tip
1. In case the Model and CADCIM tabs are not displayed in the drawing window (bottom left) of the current workspace, then you can customize the display properties of the workspace using the options in the Customize User Interface dialog box.

2. You can invoke the Customize User Interface dialog box by specifying the CUI command in the command line.

Adding Data Using the Data Connect Wizard
In this part of the tutorial, you will add a dataset to the drawing by using the **Connect** tool.

1. To add data into the map, choose the **Connect** tool from the **Data** panel in the **Home** tab; the **OSGeo FDO Provider for SHP** page of the **DATA CONNECT** wizard is displayed. Ensure that **Add SHP Connection** option is chosen from the **Data Connections by Provider** list box in the **DATA CONNECT** wizard.

2. In this page, choose the button next to the **SHP** button from the **OSGeo FDO Provider for SHP** page; the **Browse For Folder** dialog box is displayed.

3. Browse and select the *c11_tut03* folder from the following location:

    C:\m3d_2017\c11_m3d_2017_tut

4. Next, choose the **OK** button in the **Browse For Folder** dialog box; the dialog box is closed and the path of the selected folder is displayed in the **Source file or folder** edit box. Notice that the **Connect** button is also activated.

5. Choose the **Connect** button; the **SHP** page is displayed in the right pane of the wizard.

6. Select the check boxes corresponding to the SHP files **AustinTX_Address**, **AustinTX_BuildingFootprints**, and **AustinTX_Street** in the list box of the **SHP** page.

7. Next, choose the **Add to Map** button; the shape files are added to the drawing. Close the **DATA CONNECT** wizard to see the data in the drawing window.

Displaying the Required Data in the Viewport
In this part of the tutorial, you will specify the data that is to be displayed in the map.

1. Choose the **CADCIM** tab displayed at the bottom left of the drawing window; the layout is displayed in the drawing window.

2. Choose the **Model or Paper space** button in the **Status Bar**; the text **MODEL** is displayed on the button indicating that you are in the model space within the layout environment.

3. Next, choose the **Extents** tool from the **Navigate** panel of the **View** tab; all drawing objects are zoomed to fit in the viewport of the map layout.

4. Next, click on the down arrow corresponding to the **Extents** tool; a menu is displayed. Choose the **Window** tool from the menu; you are prompted to specify the first corner point of the zoom window.

5. Specify the area to be zoomed in, refer to Figure 11-22; the specified area is zoomed in the viewport.

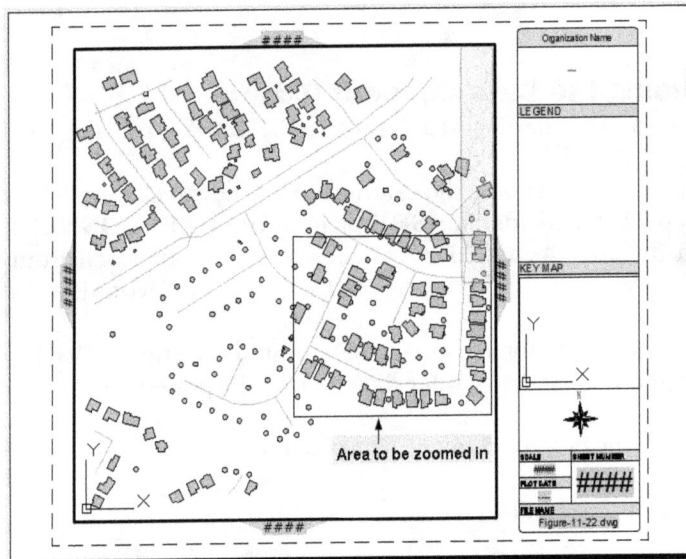

Figure 11-22 The area to be zoomed in the viewport

Creating a Map Legend

In this section of the tutorial, you will assign the required display style to the feature data and then you will create a legend for the map.

1. Choose the **Model** tab displayed at the bottom left of the drawing window; the model environment will be invoked in the drawing window.

2. Select **AustinTX_Street** from the **Display Manager** of the **TASK PANE** and then choose the **Style** button; the **STYLE EDITOR** dialog box is displayed.

3. In this dialog box, choose the Browse button displayed in the **Style** cell of the **Line Style for 0 - Infinity Scale Range** area; the **Style Line** dialog box is displayed.

4. In this dialog box, select the black color from the **Color** drop-down list and then choose the **Apply** button.

5. Choose the **Close** button in the **Style Line** dialog box; the dialog box is closed and the **Style** cell in the **STYLE EDITOR** dialog box displays a black color line.

6. Repeat the procedure given in steps 2 to 5 and change the color of the **AustinTX_ BuildingFootprints** and **AustinTX_Address** feature layers to green and red, respectively.

7. Next, close the **STYLE EDITOR** dialog box and then clear the check box corresponding to **Map Base** in the **Display Manager** of the **TASK PANE**.

8. Choose the **CADCIM** tab displayed at the bottom left of the drawing window; the layout environment is invoked.

9. Next, ensure that the text **PAPER** is displayed on the **Model or Paper Source** button (this indicates that you are in the paper space within the layout environment). If this button displays the text **MODEL**, then click on the button; the button will display the text **PAPER** and the paper space is invoked within the layout environment.

10. Choose the **Legend** tool from the **Layout Elements** panel of the **Layout Tools** tab (contextual); a flyout is displayed.

11. Choose the **Legend** option from the flyout; you are prompted to select a viewport that is to be associated with the legend.

12. Click on the frame of the viewport displaying feature data; you are prompted to specify the location for placing the legend.

13. Click in the **LEGEND** area of the map layout; the legend is placed at the specified location in the **LEGEND** area.

14. Double-click on the frame of the displayed legend; grip edit markers are displayed on the frame of the legend.

15. Click and hold on the square marker and resize the legend so that it fits within the area for the legend.

Creating the Key Map

In this section of the tutorial, you will create a key map that will display the extent of the data in the mapped area.

1. Ensure that the drawing window is in the paper source within the layout area. Next, choose the **Rectangular** tool from the **Viewports** panel of the **Layout Tools** tab (contextual); you are prompted to specify the first corner of the viewport to be created.

2. Next, click on one of the corner of the rectangle in the **KEY MAP** area of the map layout, refer to Figure 11-22; you are prompted to specify the opposite corner.

3. Click on the opposite corner of the rectangle; a new viewport is created and the data is displayed within the viewport.

Inserting the Scale Bar

In this section of the tutorial, you will insert a scale bar that will graphically display the scale of the map in the viewport.

1. Choose the **Scale Bar** tool from the **Layout Elements** panel of the **Layout Tools** tab (contextual); a flyout is displayed.

2.   Choose the **ScaleBar3_Metric** option from the flyout; you are prompted to select a viewport.

3.   Click on the viewport in the map layout; the **Scale Bar Properties** dialog box is displayed.

4.   In this dialog box, enter **100** in the **Scale bar division** edit box and then choose the **OK** button; the **Scale Bar Properties** dialog box is closed and you are prompted to specify the insertion point for the scale bar.

5.   Click below the legend; the scale bar is placed below the legend.

Inserting the Reference Grid

In this section of the tutorial, you will define the parameters of the reference grid and then insert it into the map.

1.   Choose the **Reference System** tool from the **Map Viewport** panel of the **Layout Tools** tab (contextual); you are prompted to select the viewport in the layout.

2.   Click on the viewport in the map layout; the **Create Reference System** dialog box is displayed.

3.   In this dialog box, select the **Current map coordinate system** option from the **Reference system template** drop-down list.

4.   Specify **1500** in the **Scale** edit box and **500** in the **Precision** edit box, respectively.

5.   Next, choose the **OK** button; the **Create Reference System** dialog box is closed and the reference grid is displayed in the viewport, as shown in Figure 11-23.

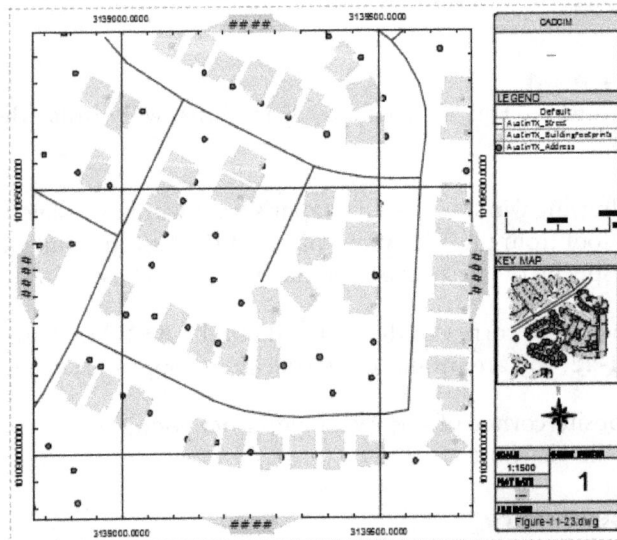

Figure 11-23 The map layout with map legend, key map, scale bar, and reference grid

Specifying the Map Elements

In this part of the tutorial, you will specify the other map elements such as the name of the organization and its logo in the map.

1. Place the cursor on the outer frame containing the map elements, as shown in Figure 11-24, and then double-click on it; the **Enhanced Attribute Editor** dialog box is displayed.

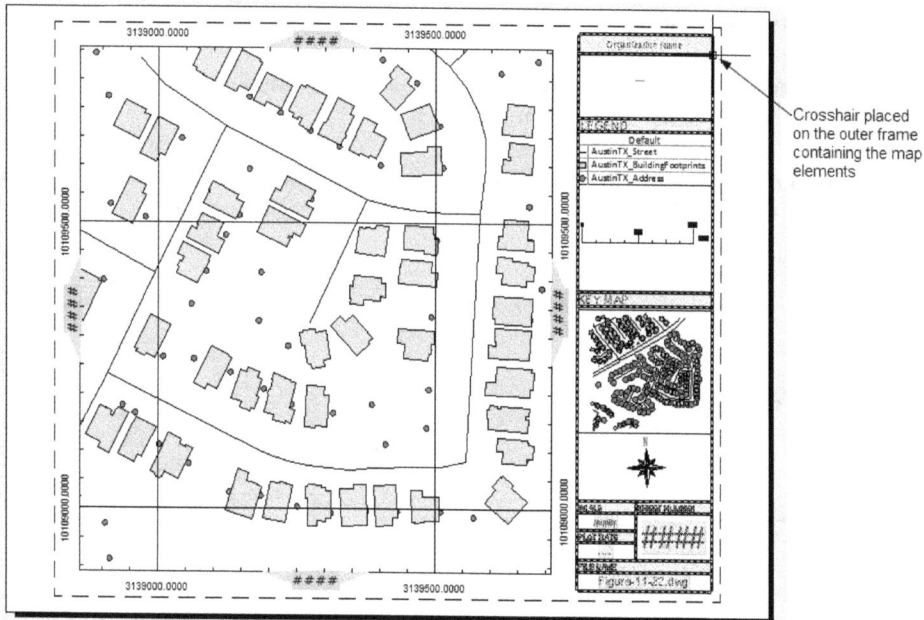

Figure 11-24 Crosshair placed on the outer frame

2. In the **Enhanced Attribute Editor** dialog box, select the **ORGANIZATION_NAME** option in the list box of the **Attribute** tab, and then enter **CADCIM** in the **Value** edit box.

3. Next, select the **DWG_SCALE** option in the list box of the **Attribute** tab; the **####** symbol is displayed in the **Value** edit box. Enter **1:1500** in the **Value** edit box.

4. Select the **NUM** option in the list box of the **Attribute** tab; the **####** symbol is displayed in the **Value** edit box. Enter **1** in the **Value** edit box.

5. Next, choose the **Apply** button in the **Enhanced Attribute Editor** dialog box; the modified values are saved.

6. Choose the **OK** button from this dialog box; the dialog box is closed and the specified values are displayed in the map template.

7. Enter **MAPIINSERT** in the Command prompt and press ENTER; the **Insert Image** dialog box is displayed.

8. In the displayed dialog box, browse to the following location:

    C:\m3d_2017\c11_m3d_2017_tut\c11_tut03

9. Select the **logo-11.JPG** file from the dialog box and then choose the **Open** button; the **Image Correlation** dialog box is displayed.

10. Next, in the **Insertion Values** area, enter **0.25** in the **Scale** edit box. Next, choose the **OK** button; the dialog box is closed.

11. Place the cursor on the logo at the lower left corner of the layout and right-click; a flyout is displayed. Now choose the **Move** option from it.

12. Drag the logo to the **CADCIM** area of the map layout; the logo is placed at the specified location in the **CADCIM** area. Figure 11-25 shows the final map layout.

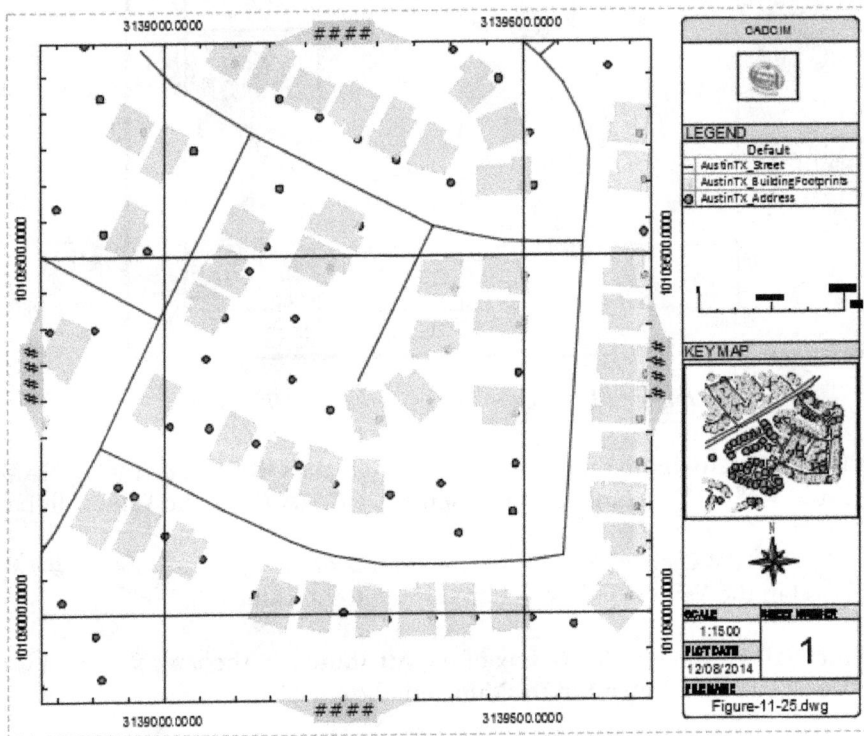

Figure 11-25 The final map layout

Saving the Current Drawing

1. Choose the **Save As** option from the Application Menu; the **Save Drawing As** dialog box is displayed.

2. In this dialog box, browse to the desired folder and then enter **c11_Tut03a.dwg** in the **File name** edit box.

3. Choose the **Save** button in this dialog box; the dialog box is closed and the current drawing file is saved with the name entered in the **File name** edit box.

Tutorial 4   Creating a Map Book

In this tutorial, you will create a map book by using the shape files.   **(Expected time: 20 min)**

The following steps are required to complete this tutorial:

a. Create a new drawing file.
b. Connect the feature data and apply style.
c. Apply display style to the feature data.
d. Update the map legend.
e. Create a map book with the following settings:
    1. Map Book's Name: **Austin**
    2. Template: **CADCIM-MapBook.dwt**
    3. Scale Factor: **100**
    4. Map Tiles: Use the default settings
    5. Key Map: **key.dwg**
    6. View the map book.
f. Save the drawing file.

Creating a New Drawing File

In this section of the tutorial, you will create a new drawing using the **CADCIM-MapBook.dwt** drawing template file.

1. Choose the **New** button from the Application Menu; the **Select template** dialog box is displayed.

2. In the **Open** dialog box, browse to the following location:

    C:\m3d_2017\c11_m3d_2017_tut\c11_tut04

3. Ensure that the **Drawing Template (*.dwt)** is selected in the **Files of type** drop-down list and then select the **CADCIM-MapBook.dwt** file. Next, choose the **Open** button; a new drawing is opened in the paper space of the layout environment.

Connecting the Feature Data

In this section of the tutorial, you will connect the drawing to the required feature data using the **DATA CONNECT** wizard.

1. Type **MS** in the Command prompt and press ENTER; the model space is invoked in the layout environment.

2. Choose the **Connect** tool from the **Data** panel in the **Home** tab; the **DATA CONNECT** wizard with the **OSGeo FDO Provider for SHP** page is displayed.

3. In this page, choose the button next to the **SHP** button; the **Browse for Folder** dialog box is displayed.

4. In this dialog box, browse to the location *C:\m3d_2017\c11_m3d_2017_tut*. Next, select the **c11_tut04** folder and then choose the **OK** button; the **Browse for Folder** dialog box is closed and the path for the selected folder is displayed in the **Source file or folder** edit box of the **OSGeo FDO Provider for SHP** page.

5. Next, choose the **Connect** button in this page; the **SHP** page of the wizard is displayed. Note that the list box in this page displays the names of the feature data files in the connected folder.

6. Select the check boxes corresponding to the three feature layers and then choose the **Add to Map** button from the **SHP** page; the selected feature data is added to the list box in the **Display Manager** tab of the **TASK PANE** and the feature geometry is displayed in the drawing window.

7. Choose the **Close** button in the **DATA CONNECT** wizard; the wizard is closed.

8. Next, choose the **Extents** tool in the **Navigate** panel of the **View** tab; the drawing zooms to the extents, as shown in Figure 11-26.

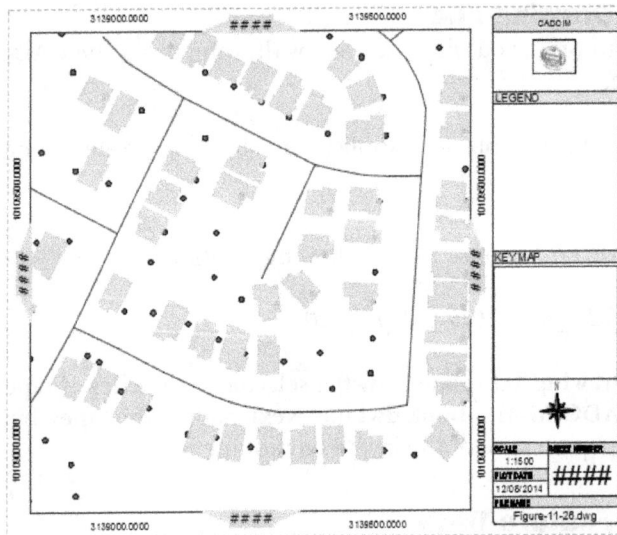

Figure 11-26 *The feature geometry displayed in the drawing window*

Applying Style to the Feature Layer

In this section of the tutorial, you will apply a display style to the feature data in the drawing.

1. Ensure that you are in the model space and then choose the **AustinTX_BuildingFootprints** vector layer in the **Display Manager** tab of the **TASK PANE**.

2. Next, choose the **Style** tool from the **Display Manager** tab of the **TASK PANE**; the **STYLE EDITOR** dialog box is displayed.

3. In the **STYLE EDITOR** dialog box, choose the **New Theme** button in the **Polygon Style for 0 - Infinity Scale Range** area; the **Theme Layer** dialog box is displayed.

4. In the **Theme Layer** dialog box, select the **SHAPE_AREA** option from the **Property** drop-down list.

5. Next, specify **0** and **4000** in the **Minimum value** and **Maximum value** edit boxes, respectively. Ensure that the **Equal** option is selected in the **Distribution** drop-down list.

6. Next, enter **Area** in the **Legend text** edit box in the **Create legend labels** area.

7. Select the **<Label text> <min> - <max>** option from the **Legend format** drop-down list and then choose the **OK** button; a new theme containing several ranges is created and is added to the list box in the **Polygon Style for 0 - Infinity Scale Range** area. Next, delete the default row from the **Thematic Rules** area. Figure 11-27 shows the **STYLE EDITOR** dialog box with the new theme created.

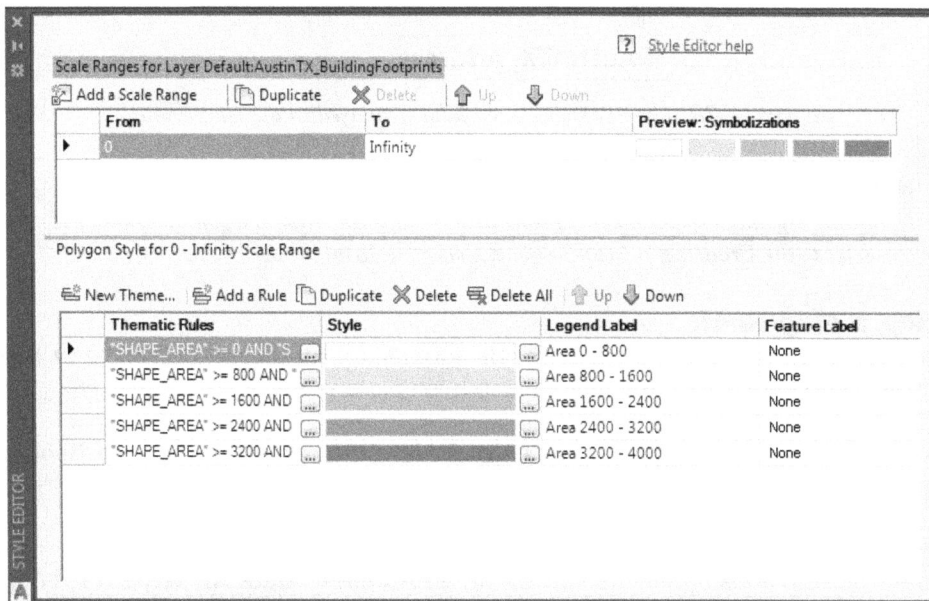

Figure 11-27 *The* **STYLE EDITOR** *dialog box with the new theme created*

8. Choose the **Close** button in the **STYLE EDITOR** dialog box; the dialog box is closed and the feature geometry in the drawing window is classified into five class-intervals.

Updating the Map Legend
In this section of the tutorial, you will update the map legend in the map layout so that it shows the current display style.

1. Clear the check box corresponding to the **Map Base** in the **Display Manager** tab of the **TASK PANE**.

2. Next, choose the **Model** tab displayed at the bottom left of the drawing window; the model space is invoked in the drawing window. Next, choose the **CADCIM** tab displayed at the bottom left corner of the drawing window; the drawing switches back to the model space in the layout environment. Notice that the map legend is updated in the map layout, as shown in Figure 11-28.

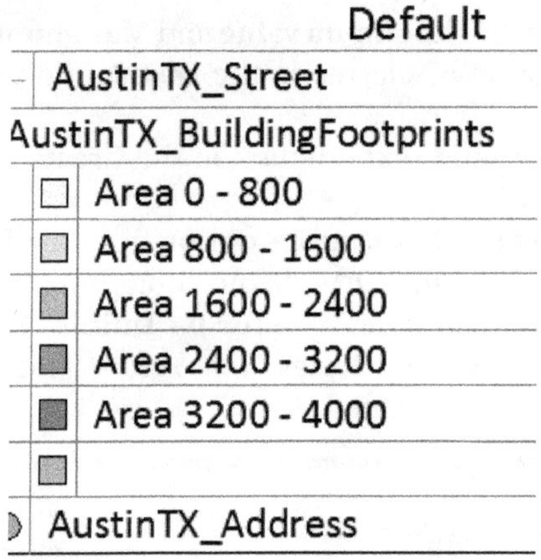

Default

	AustinTX_Street
	AustinTX_BuildingFootprints
☐	Area 0 - 800
☐	Area 800 - 1600
☐	Area 1600 - 2400
☐	Area 2400 - 3200
☐	Area 3200 - 4000
☐	
▷	AustinTX_Address

Figure 11-28 The updated legend table displayed in the map layout

Note
*If the map legend does not update automatically, you can create a legend manually. To do so, you can refer to the **Creating a Map Legend** section of **Tutorial 3**.*

Creating a Map Book

In this section of the tutorial, you will define the parameters for the creation of a map book and then create a map book using the defined map book parameters.

1. Choose the **Map Book** tab from the **TASK PANE**; the options in the **Map Book** tab are displayed.

2. Next, choose the **New** tool from the **Map Book** tab of the **TASK PANE**; a flyout is displayed. Choose the **Map Book** option from the flyout; the **Autodesk AutoCAD Map 3D 2017** message box is displayed, prompting you to save the drawing.

3. In the message box, choose the **Yes** button; the **Save Drawing As** dialog box is displayed.

4. Specify **c11_Tut04a.dwg** in the **File name** edit box and then choose the **Save** button; the drawing file is saved with the specified name and the **Create Map Book** dialog box is displayed.

5. In this dialog box, select the **Map Display** radio button in the **Source** node; the options corresponding to this radio button are displayed in the right pane.

6.   Enter **Austin Map** in the **Map Book Name** edit box.

7.   Select the **Settings** radio button in the **Sheet Template** node; the options corresponding to this radio button are displayed in the right pane.

8.   Choose the Browse button next to the **Choose a Sheet Template** edit box; the **Select Sheet Template Drawing** dialog box is displayed.

9.   In the **Select Sheet Template Drawing** dialog box, browse to the following location:

     C:\m3d_2017\c11_m3d_2017_tut\c11_tut04

10. Next, choose the **CADCIM-MapBook.dwt** file and then choose the **Open** button; the dialog box is closed and the path of the chosen template is displayed in the **Choose a Sheet Template** edit box.

11. In the **Layout Options** area, enter **100** in the **Scale Factor** edit box.

12. Select the **By number** radio button in the **Tiling Scheme** node; the options corresponding to this radio button are displayed on the right side.

13. Specify **3137875** and **10110388** in the first and second edit boxes respectively corresponding to the **Upper left** parameter.

14. Next, select the **Skip any empty tiles** check box.

15. Select the **External Reference** radio button in the **Key** node; the **Choose the Map Key Source File** edit box is displayed on the right side.

16. Choose the Browse button next to the **Choose the Map Key Source File** edit box; the **Select Key Reference Drawing** dialog box is displayed.

17. In the dialog box, browse to the following location:

     C:\m3d_2017\c11_m3d_2017_tut\c11_tut04

18. In the **Select Key Reference Drawing** dialog box, select the **key.dwg** file from the list box and then choose the **Open** button; the dialog box is closed and the path of the selected drawing file is displayed in the **Choose the Map Key Source File** edit box.

19. Choose the **Generate** button from the **Create Map Book** dialog box; the dialog box is closed and the map book is created. The map tiles created are added to the **Map Book** tab of the **TASK PANE**, as shown in Figure 11-29. Also, the map tiles are displayed in a black colored frame in the drawing window.

20. Choose the tab **Austin Map-B-15** displayed at the bottom of the map; the map tile is displayed in the drawing window, as shown in Figure 11-30. Similarly, click on the other tabs to view the map tile in the drawing window.

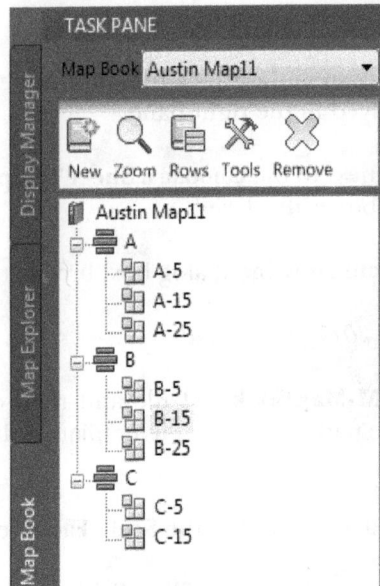

*Figure 11-29 The **Map Book** tab of the
TASK PANE displaying the map tiles*

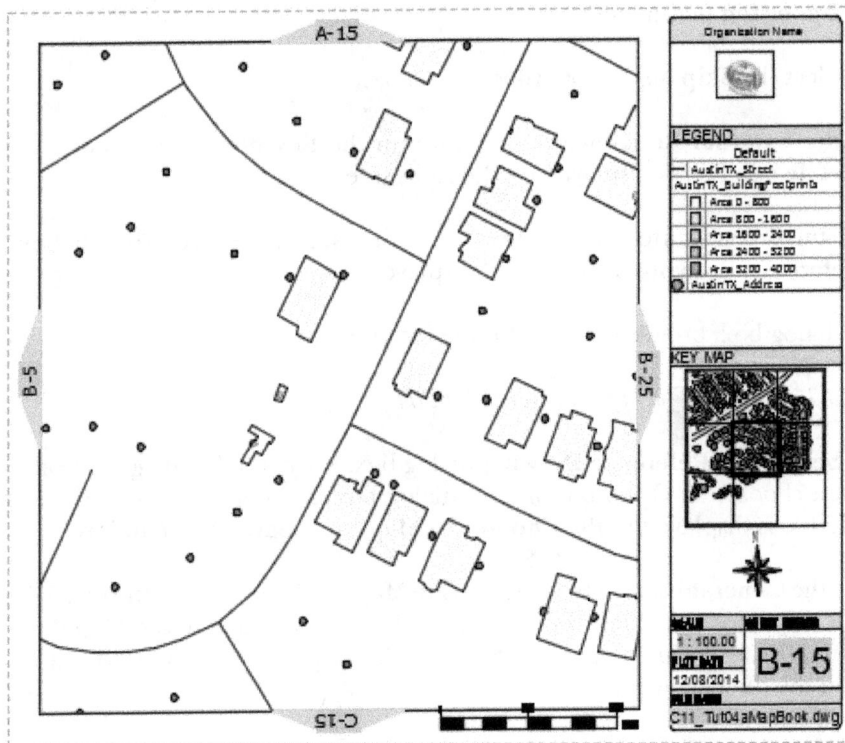

*Figure 11-30 The **Austin Map-B-15** map tile displayed in the drawing window*

Saving the Drawing

1. Choose the **Save As** option from the Application Menu; the **Save Drawing As** dialog box is displayed.

2. In this dialog box, browse to the desired folder and then enter **c11_Tut04aMapBook.dwg** in the **File name** edit box.

3. Next, choose the **Save** button in this dialog box; the current drawing file is saved with the name entered in the edit box.

Self-Evaluation Test

Answer the following questions and then compare them to those given at the end of this chapter:

1. Which of the following tools is used to trim objects inside or outside a given region?

   (a) **Rubber Sheet**   (b) **Transform**
   (c) **Boundary Break**   (d) **Boundary Trim**

2. Which of the following items is not a map element?

   (a) Legend   (b) Key Map
   (c) North Arrow   (d) Map Book

3. The _____ tool is used to break objects along a specified boundary.

4. The **Rubber Sheet** tool is used to split a drawing object into two parts. (T/F)

5. It is always recommended not to use the **Rubber Sheet** tool to modify the drawing objects, where accuracy is a major concern. (T/F)

6. The **Transform** tool requires more than five pairs of source and destination points. (T/F)

7. While breaking the boundary of a map, you can specify the boundary from an attached drawing. (T/F)

8. The **Boundary Trim** tool is used to erase a small portion of the drawing objects in the current map. (T/F)

9. A map book is a customized division of a map into several map tiles. (T/F)

10. You can create a map book by importing settings from an existing map book. (T/F)

Review Questions

Answer the following questions:

1. Which of the following tools is used to erase a portion of a map?

    (a) **Boundary Break**   (b) **Boundary Trim**
    (c) **Transform**   (d) **Rubber Sheet**

2. Which of the following nodes in the **Create Map Book** dialog box is used to create a sheet set or a subset of an existing sheet set for the current map book?

    (a) **Key**   (b) **Sheet Template**
    (c) **Tiling Scheme**   (d) **Sheet Set**

3. You can use the _____ tool to move, rotate, and scale a drawing object or a drawing layer.

4. You can view and edit the attributes for the blocks in the map layout using the _____ dialog box.

5. You must specify the settings of a map template in the _____ node while creating a map book.

6. You can perform rubber sheeting of a drawing object only by using more than one pair of the base and reference points. (T/F)

7. While using the **Boundary Break** tool, you can specify the drawing objects to be used in the process. (T/F)

8. The map elements are used to specify the drawing description of a map and a map book. (T/F)

9. You cannot create a legend and a key map as these map elements are created by the software automatically. (T/F)

10. You can use an external DWG file to be displayed as a key map. (T/F)

EXERCISES

Exercise 1

Download the *c11_exr01* folder from *www.cadcim.com*. Next, open the *c11-m3d-2017-exr01.dwg* file from the downloaded folder and perform the rubber sheeting on this drawing by using a minimum of three pairs of the base and reference points.   **(Expected time: 25 min)**

Exercise 2

Download the *c11_exr02* folder from *www.cadcim.com*. Next, open the *c11-m3d-2017-exr02.dwg* file and trim the drawing objects outside the rectangular boundary, refer to Figure 11-31.

(Expected time: 20 min)

Rectangular boundary used for trimming objects

Figure 11-31 *The boundary used for trimming the objects*

Exercise 3

Download the *c11-m3d-2017-exr03.dwg* file from *www.cadcim.com* and then create a map book using the following parameters: **(Expected time: 1 hr)**

1. Map Book's Name: **EXERCISE MAP**
2. Template: **Map Book Template - 11x17 Elegant**
3. Scale factor: **150**
4. Map Tiles: **3 x 3**
5. Legend: Create after styling

Answers to Self-Evaluation Test

1. d, 2. d, 3. Boundary Break, 4. F, 5. T, 6. F, 7. T, 8. T, 9. T, 10. T

Project *1*

Site Suitability
Study

In this project, you will study the suitability of a site for constructing wind turbines capable of generating 2MW electricity in the Larimer County of the Colorado state (USA).

To work on this project, you need to download the *m3d_2017_Project.zip* file from *www.cadcim.com* and then extract the contents of this file to *C:\m3d_2017* on your computer.

(Expected time: 2 hr 30 min)

Site requirements (site selection parameters)-
> Fire hazard: Lowest
> Wind speed: 14m/s (recommended)
> Site slope: <15 degrees (recommended)
> Setback (Minimum distance between wind turbine and human habitation): 2700 ft

Other selection criteria:
> Wind turbine cannot be constructed on reserved land and forested areas.
> The selected site should be accessible by road.

Wind turbine specifications:

Rated power:	2.1 MW
Cut-in wind speed:	4 m/s
Rated wind speed:	14 m/s
Cut-out wind speed:	25 m/s
50 years gust wind speed:	59.5 m/s
Hub height:	100 m
Rotational speed:	15.1 - 17.7 rpm

Classes of wind power density at 50 m above the ground level (AGL):

Wind Power Class (WPC) (50 m AGL)	Wind Power Density (W/sq.m)	Speed (b) m/s (mph)
1	200	5.6 (12.5)
2	300	6.4 (14.3)
3	400	7.0 (15.7)
4	500	7.5 (16.8)
5	600	8.0 (17.9)
6	800	8.8 (19.7)
7	2000	11.9 (26.6)

Data to be used in the project:

a. Land zone data: **LandZone.shp** (Data field- **Zone**)

Description of the attributes in the **Zone** data field of the **LandZone.shp** file:

Attribute	Description
ACCOMMODATIONS	Human habitation
TOURIST	Tourist areas
FORESTRY	Forest Area
RESTRICTED	Reserved areas

b. Wind data: **Wind.shp** (Data field- **WPC**)

c. Wildfire data: **Wildfire.shp** (Data field- **Hazard**)

d. Road data: **Road.shp**

e. Elevation Data: **NED41W106.tif** (Coordinate Reference System LL84)

The following steps are required to complete this tutorial:

a.   Create a new drawing file and assign coordinate system (**CO83-NF**).
b.   Connect the wind and wildfire feature data (**Wind.shp** and **Wildfire.shp**).
c.   Find areas with suitable wind speed (WPC = 7).
d.   Find areas with lowest wildfire hazard (Hazard = Lowest).
e.   Find areas having suitable wind speed and minimum risk of wildfire:
        (Hint: use Clip Overlay Analysis)
f.   Check the accessibility of the areas that have suitable wind and fire criterion using the existing road data (Limit to visual analysis).
g.   Connect to the **LandZone.shp** file and then check the following criterion for the most accessible site:
        Minimum setback from the areas of habitation (2700ft)
        The areas do not fall in the restricted, forest, or tourist areas.
h.   Connect to the elevation data and perform the Slope Analysis.
i.   Save the drawing.

> **Note**
> *Ignore any warning message if displayed while performing the steps mentioned in the site suitability project.*

Creating a New Drawing File and Assigning a Coordinate System

In this section of the project, you will create a new drawing using the **CADCIM.dwt** drawing template file.

1.   Choose the **New** button from the Application Menu; the **Select template** dialog box is displayed.

2.   In the **Open** dialog box, browse to the following location:

        C:\m3d_2017\m3d_2017_Project

3.   Next, select the **CADCIM.dwt** file and then choose the **Open** button; the drawing window displays the drawing layout in the paper space, as shown in Figure P1-1.

4.   Choose the **Model** tab located at the bottom-left of the drawing window to switch to the model environment. Notice that the options in the **Display Manager** tab of the **TASK PANE** are now activated.

> **Tip**
> *In case the **Model** and **CADCIM** tabs are not displayed in the drawing window (bottom left) of the current workspace, you can customize the display properties of the workspace using the options in the **Customize User Interface** dialog box.*
>
> *You can invoke the **Customize User Interface** dialog box by specifying the **CUI** command in the command line.*

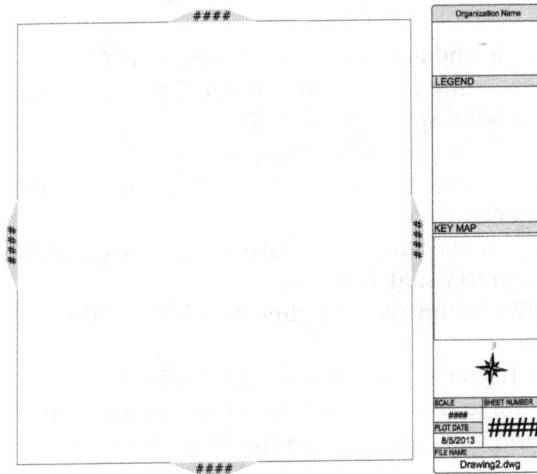

Figure P1-1 The drawing layout displayed in the drawing window

5.   Choose the **Assign** tool from the **Coordinate System** panel in the **Map Setup** tab; the **Coordinate System - Assign** dialog box is displayed.

6.   In the **Show** area of the dialog box, select the **USA, Colorado** option from the **Category** drop-down list; a list of coordinate systems in this category is displayed in the list box of the dialog box.

7.   Next, select the coordinate system with the code **CO83-NF** from the list box, and then choose the **Assign** button; the **Coordinate System - Assign** dialog box is closed and the selected coordinate system is assigned to the drawing.

Connecting the Wind and Wildfire Data

In this section of the project, you will use the **Connect** tool to load the wind and wildfire datasets into the drawing.

1.   Choose the **Connect** tool from the **Data** panel in the **Home** tab; the **DATA CONNECT** wizard is displayed.

2.   In the **DATA CONNECT** wizard, select the **Add SHP Connection** option from the **Data Connections by Provider** list box; the **OSGeo FDO Provider for SHP** page is displayed in the right pane of the wizard.

3.   In the **OSGeo FDO Provider for SHP** page, choose the browse button to the right of the **Source file or folder** edit box; the **Browse For Folder** dialog box is displayed.

4.   In this dialog box, browse to the location: *C:\m3d_2017* and select the folder *m3d_2017_Project*. Next, choose the **OK** button; the **Browse for Folder** dialog box is closed and the path of the selected folder is displayed in the **Source file or folder** edit box in the **OSGeo FDO Provider for SHP** page.

5. Choose the **Connect** button; the shape files in the selected folder are displayed in the **Add Data to Map** list box.

6. Select the check box corresponding to the **WildFire** and **Wind** shape files and then choose the **Add to Map** button; the selected shape files are connected to the drawing. Note that the **Display Manager** tab of the **TASK PANE** displays the names of the added shape files.

7. Close the **DATA CONNECT** wizard.

Filtering the Wind Dataset

In this section of the project, you will apply data filters to the **Wind** shape file to find the areas having Wind Power Class (WPC) 7.

1. Select the **Wind** feature layer in the **Display Manager** tab of the **TASK PANE** if not selected by default; the **Vector Layer** contextual tab is displayed.

2. Choose the **Query to Filter** tool from the **View** panel in the **Vector Layer** tab; the **Create Query** window with the **Getting Started with Filters** page ⊑ Query to Filter is displayed.

3. In this page, enter the following query syntax:

    WPC = 7

4. Choose the **Validate** button at the lower left portion of the **Create Query** dialog box; the statement **The expression is valid** is displayed, confirming that the written expression is a valid query statement.

5. Choose the **OK** button; the regions with WPC 7 are filtered. Figure P1-2 shows the areas in the Larimer County with Wind Power Class (WPC) 7. Note that to view the result of the query, you will require to switch on/off the display of the required feature layers in the **TASK PANE**.

Filtering the WildFire Dataset

In this section of the project, you will apply data filters to the **WildFire** shape file to find the areas that are least prone to the wildfire.

1. Select the **WildFire** feature layer in the **Display Manager** tab of the **TASK PANE**; the **Vector Layer** contextual tab is displayed.

2. Choose the **Query to Filter** tool from the **View** panel in the **Vector Layer** tab; the **Create Query** window with the **Getting Started with Filters** page ⊑ Query to Filter is displayed.

3. In this page, enter the following query syntax:

    HAZARD = 'LOWEST'

Figure P1-2 *Areas in the Larimer County having Wind Power Class (WPC) 7*

4.   Choose the **Validate** button at the lower left portion of the **Create Query** dialog box; the statement **The expression is valid** is displayed, confirming that the written expression is a valid query statement.

5.   Choose the **OK** button; the regions least prone to fire hazard are filtered. Figure P1-3 shows the areas in the Larimer County region that are least prone to wildfire.

Figure P1-3 *Areas in the Larimer County that are least prone to wildfire hazard*

Finding Regions with Least Wildfire Hazard and WPC 7 in the Larimer County

In this section of the project, for constructing a 2 MW wind turbine, you will find the areas in the Larimer County that have a suitable WPC (WPC=7) and are at the minimum risk of wildfire.

1. Choose the **Feature Overlay** tool from the **Feature** panel of the **Analyze** tab; the **Overlay Analysis** wizard with the **Source and Overlay Type** page is displayed.

2. In this page of the wizard, select the **WildFire (Polygons)** option from the **Source** drop-down list and the **Wind (Polygons)** option from the **Overlay** drop-down list.

3. Select the **Clip** option from the **Type** drop-down list. Choose the **Next** button; the **Set Output and Settings** page of the wizard is displayed.

4. In this page, choose the browse button corresponding to the **Output** edit box; the **Save** dialog box is displayed.

5. In this **Save** dialog box, browse to the *C:\m3d_2017\m3d_2017_Project* location.

6. Select the **SHP Files (*.shp)** option from the **Save as type** drop-down list.

7. Enter **WFireLowestWPC7** in the **File name** edit box and then choose the **Save** button; the dialog box is closed and the path of the specified file is displayed in the **Output** edit box of the **Set Output and Settings** page.

8. Make sure that the **Layer name** edit box displays **WFireLowestWPC7** as the output file name, else specify it in the edit box.

9. In the **Settings** area of this page, enter **0** and **500** in the **Minimum** and **Maximum** edit boxes, respectively.

10. Make sure that the **Square Meters** option is selected in the **Units** drop-down list.

11. Retain the other default parameters for the options in the **Set Output and Settings** page and then choose the **Finish** button; the **Overlay Analysis** wizard is closed and the progress of the overlay analysis is displayed in the **Overlay** message box.

After completing the analysis, the **WFireLowestWPC7** shape file is added to the **Display Manager** of the **TASK PANE**. This shape file contains those areas in the Larimer County that satisfy the required wind and wildfire conditions for constructing the 2 MW wind turbine.

Figure P1-4 shows the result of overlay analysis. The areas displayed in the figure have WPC of 7 and pose the lowest risk of wildfire.

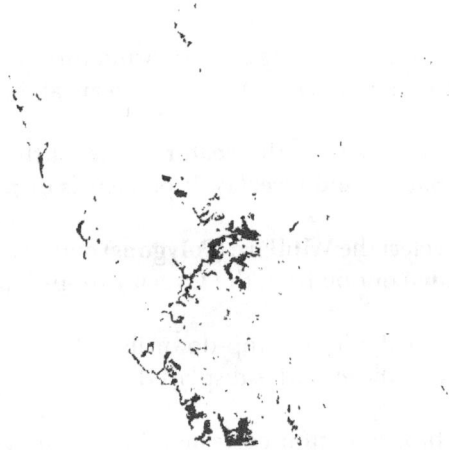

Figure P1-4 *The areas in the Larimer County that have WPC 7 and are least prone to wildfire hazard*

Determining the Accessibility of Areas Satisfying the Wind and Wildfire Criteria

In this section of the project, you will first connect the road data to the drawing. Next, you will visually check the distance from the nearest road to the areas that are most suitable for wind turbine construction.

1. Choose the **Connect** tool from the **Data** panel in the **Home** tab; the **DATA CONNECT** wizard is displayed.

2. In the **DATA CONNECT** wizard, select the **Add SHP Connection** option from the **Data Connections by Provider** list box; the **OSGeo FDO Provider for SHP** page is displayed in the right pane of the wizard.

3. In the **OSGeo FDO Provider for SHP** page, choose the browse button next to the **Source file or folder** edit box; the **Open** dialog box is displayed. In this dialog box, browse to the location: *C:\m3d_2017\m3d_2017_Project*.

4. Next, select the **Road.shp** file from the list box and then choose the **Open** button; the **Open** dialog box is closed and the path of the shape file is added to the **Source file or folder** edit box in the **OSGeo FDO Provider for SHP** page.

5. Choose the **Connect** button; the selected shape file is displayed in the **Add Data to Map** list box.

6. Choose the **Add to Map** button; the shape file is connected to the drawing. Also, the name of the added shape file is displayed in the **Display Manager** tab of the **TASK PANE**.

7. Close the **DATA CONNECT** wizard.

8. Clear all the feature layer check boxes except **Road** and **WFireLowestWPC7** in the **Display Manager** tab of the **TASK PANE**; the drawing window displays the features in the **Road** and **WFireLowestWPC7** feature layers, as shown in Figure P1-5. In this figure, the site with suitable wind and wildfire conditions for constructing the wind turbines are indicated by rectangle **A** and **B**.

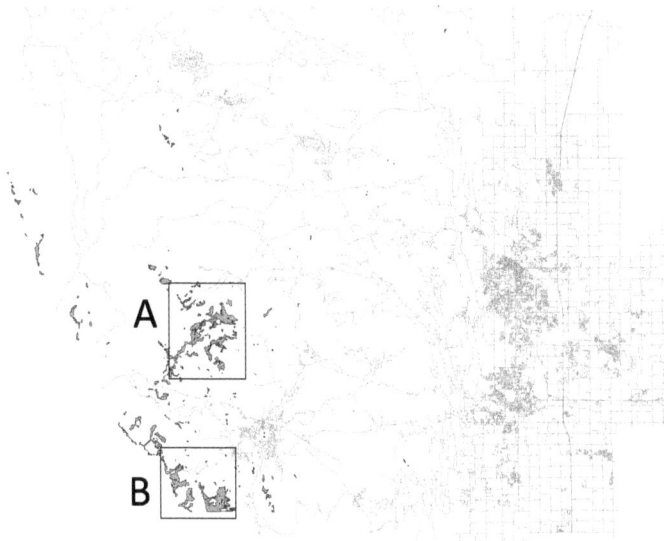

Figure P1-5 *The existing road network and the areas with suitable WPC and wildfire hazard requirements for wind turbine construction in the Larimer County*

9. Zoom to the area at sight **A**, refer to Figure P1-5. Figure P1-6 shows the road data and the areas suitable for wind turbine construction at site **A**. Note that there is no direct access to the site as there is no road passing through.

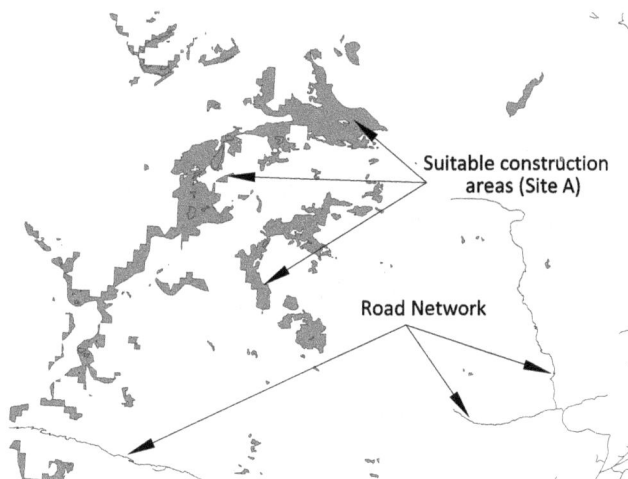

Figure P1-6 *The existing roads and the areas suitable for wind turbine construction at site A*

10. Zoom to the area at sight **B**, refer to Figure P1-5. Figure P1-7 shows the road data and the areas suitable for wind turbine construction at site **B**.

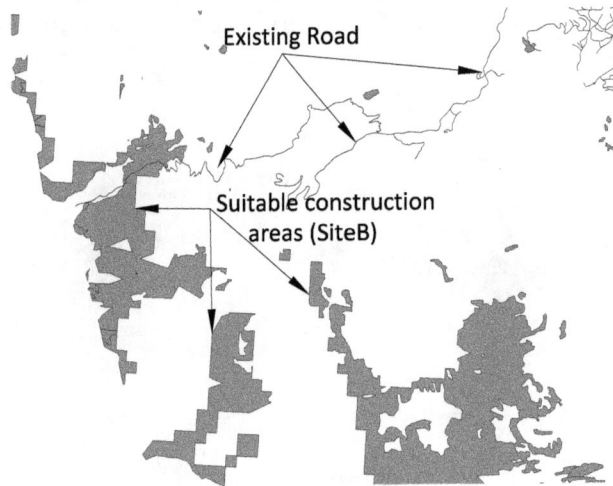

Figure P1-7 *The existing roads and the areas suitable for wind turbine construction at site **B***

Notice that a road passes through the site **B**. As a result, this site is more accessible than site **A**.

Connecting the LandZone Feature Data

In this section of the project, you will add the land zone feature data to the drawing using the **Connect** tool.

1. Choose the **Connect** tool from the **Data** panel in the **Home** tab; the **DATA CONNECT** wizard is displayed.

2. In the **DATA CONNECT** wizard, select the **Add SHP Connection** option from the **Data Connections by Provider** list box; the **OSGeo FDO Provider for SHP** page is displayed in the right pane of the wizard.

3. In the **OSGeo FDO Provider for SHP** page, choose the **SHP** button next to the **Source file or folder** edit box; the **Open** dialog box is displayed. In this dialog box, browse to the location: *C:\m3d_2017\m3d_2017_Project*.

4. Next, select the **LandZone.shp** file from the list box and then choose the **Open** button; the **Open** dialog box is closed and the path of the shape file is added to the **Source file or folder** edit box in the **OSGeo FDO Provider for SHP** page.

5. Choose the **Connect** button; the selected shape file is displayed in the **Add Data to Map** list box.

6. Choose the **Add to Map** button; the shape file is connected to the drawing. Also, the name of the added feature class is displayed in the **Display Manager** tab of the **TASK PANE**.

7. Close the **DATA CONNECT** wizard.

Checking the Minimum Setback Criteria at Site B

In this section of the project, you will filter the inhabited area from the **LandZone** shape file and then apply the buffer of **2700** ft to check the minimum setback condition between suitable construction areas at Site **B** and the areas having human habitation.

1. Select the **LandZone** feature layer in the **Display Manager** tab of the **TASK PANE**; the **Vector Layer** contextual tab is displayed.

2. Choose the **Query to Filter** tool from the **View** panel in the **Vector Layer** tab; the **Create Query** window with the **Getting Started with Filters** page is displayed.

3. In this page, enter the following query syntax

    Zone   =   'ACCOMMODATIONS'

4. Choose the **Validate** button at the lower left portion of the **Create Query** dialog box; the statement **The expression is valid** is displayed, confirming that the written expression is a valid query statement.

5. Choose the **OK** button; the polygon features with the attribute **Accommodations** in the **LandZone** shape file are filtered. Figure P1-8 shows the areas of habitation at site **B**.

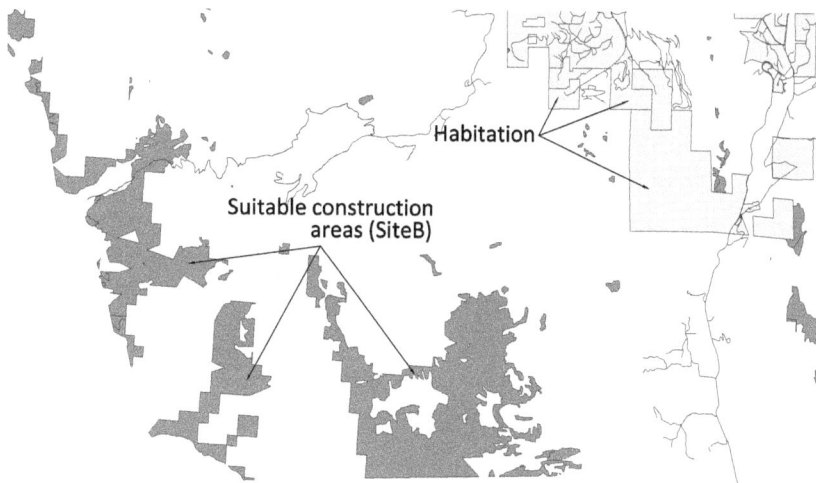

Figure P1-8 *The areas of habitation at site B*

Next, you will create a buffer that will represent a setback (2700 ft) around the areas of human habitation at site **B**, refer to Figure P1-8.

6. Choose the **Feature Buffer** tool from the **Feature** panel of the **Analyze** tab; the **Create Buffer** dialog box is displayed.

7. In this dialog box, choose the **Select features** button; the dialog box is closed and you are prompted to select the required features in the drawing window.

8. Select all the inhabited polygons from the **LandZone** shape file (closest to the most suitable areas for construction, refer to Figure P1-8) at site **B** and then press ENTER; the **Create Buffer** dialog box is displayed.

9. Next, make sure that the **Feet** option is selected in the **Units** drop-down list and then enter **2700** in the **Distance** edit box.

10. Enter **Setback2700** in the **Output to layer** edit box.

11. Next, specify the output path for the buffer file as *C:\m3d_2017\m3d_2017_Project* in the **Save to SDF** edit box.

12. Next, select the **Merge all buffers** radio button in the **Merge Results** area. Figure P1-9 shows the **Create Buffer** dialog box with specified parameters.

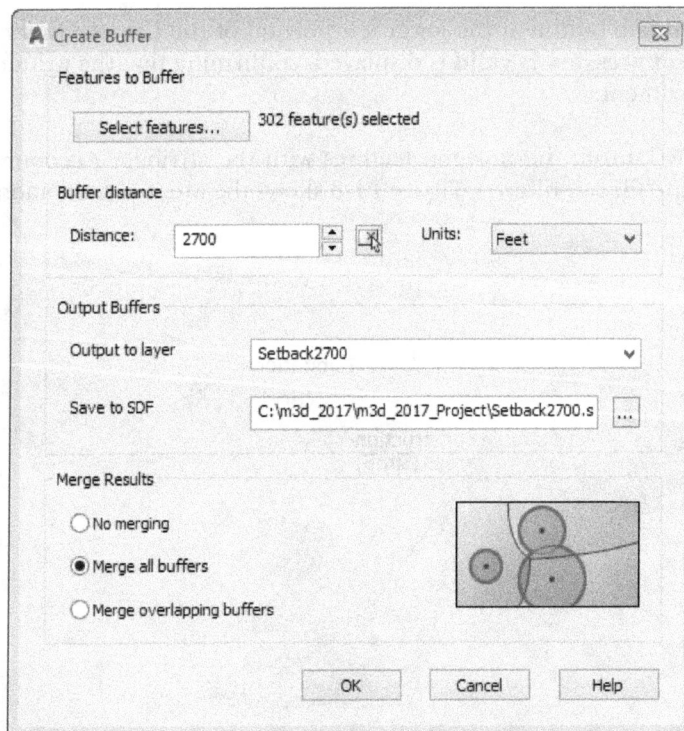

Figure P1-9 *The **Create Buffer** dialog box with the specified buffer parameters*

13. Next, choose the **OK** button. AutoCAD Map 3D processes the specified parameters and creates a buffer around the selected polygons. Figure P1-10 shows the result of the Buffer Analysis. Note that the areas suitable for wind turbine construction are well beyond the minimum required setback.

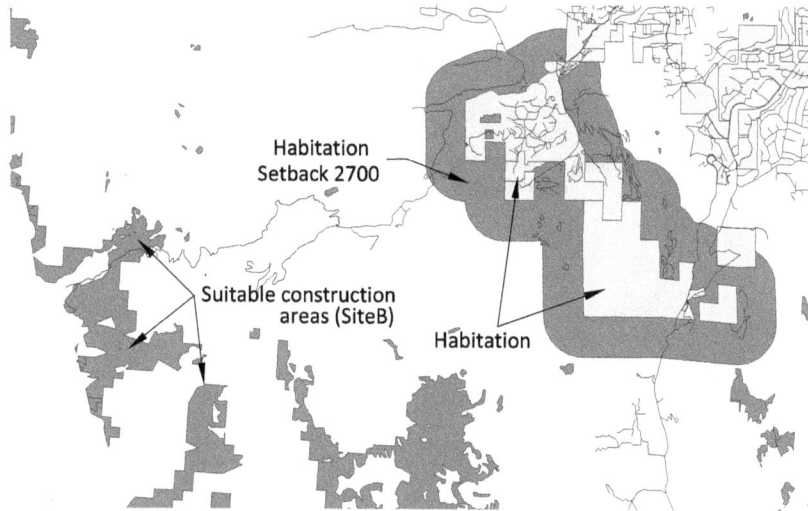

*Figure P1-10 Setback of 2700 ft at site **B***

Checking if the Suitable Construction Areas at Site B are Outside Reserved/Forest/Tourist Area

In this section of the project, you will verify that the suitable areas for construction are outside the tourist/forest/reserved areas.

1. Select the **LandZone** feature layer in the **Display Manager** tab of the **TASK PANE**; the **Vector Layer** contextual tab is displayed.

2. Choose the **Query to Filter** tool from the **View** panel in the **Vector Layer** tab; the **Modify Query** window is displayed with the previously specified query.

3. Clear the previous query and then specify the following query syntax:

    Zone = 'TOURIST' OR Zone = 'FORESTRY' OR Zone = 'RESTRICTED'

4. Choose the **Validate** button at the lower-left portion of the **Create Query** dialog box; the statement **The expression is valid** is displayed, confirming that the written expression is a valid query statement.

5. Choose the **OK** button; the polygon features for the specified query is displayed in the drawing window. Figure P1-11 shows the reserved/ tourist/ forest area at site **B**. Notice that the site **B** is outside these areas and therefore is a suitable construction site.

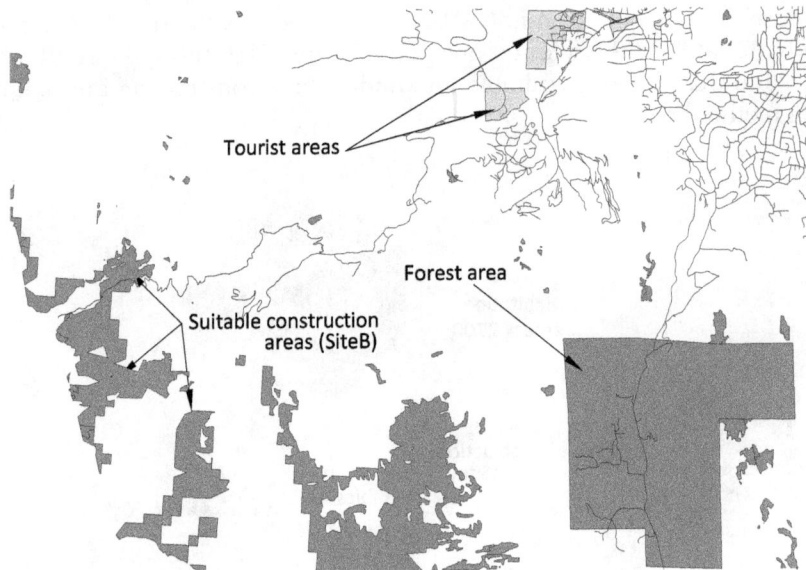

Figure P1-11 *The tourist and forest areas at site B*

Checking the Slope at Site B

In this section of the project, you will check the slope of the surface at site **B**. To do so, you will first connect the elevation data to the drawing and then perform the Slope Analysis.

1. Choose the **Connect** tool from the **Data** panel in the **Home** tab; the **DATA CONNECT** wizard is displayed.

2. In the **DATA CONNECT** wizard, select the **Add Raster Image or Surface Connection** option from the **Data Connections by Provider** list box; the **Autodesk FDO Provider for Raster** page is displayed in the right pane of the wizard.

3. In this page, choose the button next to the **Source file or folder** edit box; the **Open** dialog box is displayed. In this dialog box, browse to the location: *C:\m3d_2017* *m3d_2017_Project*.

4. Next, select the **NED41W106.tif** file from the list box and then choose the **Open** button; the **Open** dialog box is closed and the path of the shape file is added into the **Source file or folder** edit box in the **Autodesk FDO Provider for Raster** page.

5. Choose the **Connect** button; the selected raster file is displayed in the **Add Data to Map** list box.

6. Choose the **Add to Map** button in the **DATA CONNECT** wizard; the wizard is closed and the data is connected to the drawing. Notice that the **Display Manager** tab of the **TASK PANE** displays the name of the added elevation data.

7. Close the **DATA CONNECT** wizard.

Next, you will perform the Slope Analysis to find out the slope at site **B**.

8. Click on the **NED41W106** raster layer in the **Display Manager** tab of the **TASK PANE**; the **Raster Layer** (contextual) tab is displayed.

9. Choose the **Style Editor** tool from the **Style** panel of the **Raster Layer** tab; the **STYLE EDITOR** dialog box is displayed.

10. Click on the down-arrow in the **Style** column of the **Raster Style for 0 - Infinity Scale Range** list box; a drop-down list is displayed.

11. Select the **Theme** option from the displayed drop-down list; the **Theme** dialog box is displayed.

12. In this dialog box, select the **Slope** option from the **Property** drop-down list.

13. Enter **0** and **25** in the **Minimum value** and **Maximum value** edit boxes, respectively.

14. Enter **5** in the **Create rules** edit box.

15. Make sure that the **Create legend labels** check box is selected and then choose the **OK** button; the dialog box is closed and the **STYLE EDITOR** dialog box displays the specified theme.

16. Choose the **Apply** button in the **STYLE EDITOR** dialog box; the display theme is applied to the raster data.

17. Close the **STYLE EDITOR** dialog box. Figure P1-12 displays the Slope Analysis at site **B**. This analysis shows that a very small percentage of area at site **B** has a slope less than 15 degrees.

Figure P1-12 *The result of Slope Analysis at site **B***

Saving the Drawing File

1. Choose the **Save As** option in the Application Menu; the **Save Drawing As** dialog box is displayed.

2. In the **Save Drawing As** dialog box, enter **StudentProject** in the **File name** edit box and select the **AutoCAD 2013 Drawing (*.dwg)** option in the **Files of type** drop-down list, if not selected by default. In the **Save Drawing As** dialog box, choose the **Save** button corresponding to the **File name** edit box; the current drawing file is saved with the given name.

Index

This page is intentionally left blank

Other Publications by CADCIM Technologies

The following is the list of some of the publications by CADCIM Technologies. Please visit *www.cadcim.com* for the complete listing.

AutoCAD Map 3D Textbooks
- Exploring AutoCAD Map 3D 2016, 6th Edition
- Exploring AutoCAD Map 3D 2015

AutoCAD Civil 3D Textbooks
- Exploring AutoCAD Civil 3D 2017, 7th Edition
- Exploring AutoCAD Civil 3D 2016, 6th Edition

Autodesk Revit Architecture Textbooks
- Exploring Autodesk Revit 2017 for Architecture, 13th Edition
- Autodesk Revit Architecture 2016 for Architects and Designers, 12th Edition

Autodesk Revit Structure Textbooks
- Exploring Autodesk Revit 2017 for Structure, 7th Edition
- Exploring Autodesk Revit Structure 2016, 6th Edition

Autodesk Revit Navisworks Textbooks
- Exploring Autodesk Navisworks 2016, 3rd Edition
- Exploring Autodesk Navisworks 2015

Autodesk Revit MEP Textbooks
- Exploring Autodesk Revit MEP 2016, 3rd Edition
- Exploring Autodesk Revit MEP 2015
- Exploring Autodesk Revit MEP 2014

Exploring Oracle Primavera Textbook
Exploring Oracle Primavera P6 v7.0

Exploring AutoCAD Raster Design Textbook
Exploring AutoCAD Raster Design 2016

AutoCAD Textbooks
- AutoCAD 2017: A Problem-Solving Approach, Basic and Intermediate, 23rd Edition
- AutoCAD 2017: A Problem-Solving Approach, 3D and Advanced, 23rd Edition
- AutoCAD 2016: A Problem-Solving Approach, Basic and Intermediate, 22nd Edition
- AutoCAD 2016: A Problem-Solving Approach, 3D and Advanced, 22nd Edition
- AutoCAD 2015: A Problem-Solving Approach, Basic and Intermediate, 21st Edition
- AutoCAD 2015: A Problem-Solving Approach, 3D and Advanced, 21st Edition

Autodesk Inventor Textbooks
- Autodesk Inventor 2017 for Designers, 17th Edition
- Autodesk Inventor 2016 for Designers, 16th Edition
- Autodesk Inventor 2015 for Designers, 15th Edition

AutoCAD MEP Textbooks
- AutoCAD MEP 2016 for Designers, 3rd Edition
- AutoCAD MEP 2015 for Designers

NX Textbooks
- NX 10.0 for Designers, 9th Edition
- NX 9.0 for Designers, 8th Edition

SolidWorks Textbooks
- SOLIDWORKS 2016 for Designers, 14th Edition
- SOLIDWORKS 2015 for Designers, 13th Edition
- SolidWorks 2014: A Tutorial Approach
- Learning SolidWorks 2011: A Project Based Approach

Creo Parametric and Pro/ENGINEER Textbooks
- PTC Creo Parametric 3.0 for Designers, 3rd Edition
- Pro/Engineer Wildfire 5.0 for Designers
- Pro/ENGINEER Wildfire 4.0 for Designers

ANSYS Textbooks
- ANSYS Workbench 14.0: A Tutorial Approach
- ANSYS 11.0 for Designers

Creo Direct Textbook
- Creo Direct 2.0 and Beyond for Designers

Autodesk Alias Textbooks
- Learning Autodesk Alias Design 2016, 5th Edition
- Learning Autodesk Alias Design 2015, 4th Edition

AutoCAD Electrical Textbooks
- AutoCAD Electrical 2017 for Electrical Control Designers, 8th Edition
- AutoCAD Electrical 2016 for Electrical Control Designers, 7th Edition
- AutoCAD Electrical 2015 for Electrical Control Designers, 6th Edition

3ds Max Tutorial Design Textbooks
- Autodesk 3ds Max 2017 for Beginners: A Tutorial Approach, 17th Edition
- Autodesk 3ds Max 2016 for Beginners: A Tutorial Approach, 16th Edition
- Autodesk 3ds Max Design 2015: A Tutorial Approach, 15th Edition

3ds Max Textbooks
- Autodesk 3ds Max 2017: A Comprehensive Guide, 17th Edition
- Autodesk 3ds Max 2016: A Comprehensive Guide, 16th Edition
- Autodesk 3ds Max 2016 for Beginners: A Tutorial Approach, 16th Edition
- Autodesk 3ds Max 2015: A Comprehensive Guide, 15th Edition

Autodesk Maya Textbooks
- Autodesk Maya 2016: A Comprehensive Guide, 8th Edition
- Autodesk Maya 2015: A Comprehensive Guide, 7th Edition
- Character Animation: A Tutorial Approach

Fusion Textbooks
- Blackmagic Design Fusion 7 Studio: A Tutorial Approach
- The eyeon Fusion 6.3: A Tutorial Approach

Computer Programming Textbooks
- Introduction to C++ programming
- Learning Oracle 11g
- Learning ASP.NET AJAX
- Learning Java Programming
- Learning Visual Basic.NET 2008
- Introduction to C++ Programming Concepts
- Learning C++ Programming Concepts
- Learning VB.NET Programming Concepts

AutoCAD Textbooks Authored by Prof. Sham Tickoo and Published by Autodesk Press
- AutoCAD: A Problem-Solving Approach: 2013 and Beyond
- AutoCAD 2012: A Problem-Solving Approach
- AutoCAD 2011: A Problem-Solving Approach
- AutoCAD 2010: A Problem-Solving Approach
- Customizing AutoCAD 2010
- AutoCAD 2009: A Problem-Solving Approach

Textbooks Authored by CADCIM Technologies and Published by Other Publishers

3D Studio MAX and VIZ Textbooks
- Learning 3DS Max: A Tutorial Approach, Release 4
  Goodheart-Wilcox Publishers (USA)
- Learning 3D Studio VIZ: A Tutorial Approach
  Goodheart-Wilcox Publishers (USA)

CADCIM Technologies Textbooks Translated in Other Languages

SolidWorks Textbooks
- SolidWorks 2008 for Designers (Serbian Edition)
  Mikro Knjiga Publishing Company, Serbia
- SolidWorks 2006 for Designers (Russian Edition)
  Piter Publishing Press, Russia

NX Textbooks
- NX 6 for Designers (Korean Edition)
  Onsolutions, South Korea
- NX 5 for Designers (Korean Edition)
  Onsolutions, South Korea

Pro/ENGINEER Textbooks
- Pro/ENGINEER Wildfire 4.0 for Designers (Korean Edition)
  HongReung Science Publishing Company, South Korea
- Pro/ENGINEER Wildfire 3.0 for Designers (Korean Edition)
  HongReung Science Publishing Company, South Korea

Autodesk 3ds Max Textbook
- 3ds Max 2008: A Comprehensive Guide (Serbian Edition)
  Mikro Knjiga Publishing Company, Serbia

AutoCAD Textbooks
- AutoCAD 2006 (Russian Edition)
  Piter Publishing Press, Russia
- AutoCAD 2005 (Russian Edition)
  Piter Publishing Press, Russia
- AutoCAD 2000 Fondamenti (Italian Edition)

Coming Soon from CADCIM Technologies
- Exploring RISA 3D
- Exploring ETABS
- SOLIDWORKS Simulation 2016 for Designers
- Mold Wizard using NX 10.0

Online Training Program Offered by CADCIM Technologies
CADCIM Technologies provides effective and affordable virtual online training on various software packages such as CAD/CAM/CAE, Animation, Civil, GIS, and computer programming languages. The training will be delivered 'live' via Internet at any time, any place, and at any pace to individuals, students of colleges, universities, and training centers. For more information, please visit the following link:
http://www.cadcim.com

www.ingramcontent.com/pod-product-compliance
Lightning Source LLC
Chambersburg PA
CBHW060952210326
41598CB00031B/4798